Algorithms for Intelligent Systems

This book series publishes research on the analysis and development of algorithms for intelligent systems with their applications to various real world problems. It covers research related to autonomous agents, multi-agent systems, behavioral modeling, reinforcement learning, game theory, mechanism design, machine learning, meta-heuristic search, optimization, planning and scheduling, artificial neural networks, evolutionary computation, swarm intelligence and other algorithms for intelligent systems.

The book series includes recent advancements, modification and applications of the artificial neural networks, evolutionary computation, swarm intelligence, artificial immune systems, fuzzy system, autonomous and multi agent systems, machine learning and other intelligent systems related areas. The material will be beneficial for the graduate students, post-graduate students as well as the researchers who want a broader view of advances in algorithms for intelligent systems. The contents will also be useful to the researchers from other fields who have no knowledge of the power of intelligent systems, e.g. the researchers in the field of bioinformatics, biochemists, mechanical and chemical engineers, economists, musicians and medical practitioners.

The series publishes monographs, edited volumes, advanced textbooks and selected proceedings.

Indexed by zbMATH.

All books published in the series are submitted for consideration in Web of Science.

Amar Ramdane-Cherif · T. P. Singh · Ravi Tomar ·
Tanupriya Choudhury · Jung-Sup Um
Editors

Machine Intelligence and Data Science Applications

Proceedings of MIDAS 2022

Springer

Editors
Amar Ramdane-Cherif
University of Versailles—Paris-Saclay
Vélizy, France

Ravi Tomar
Senior Architect
Persistent Systems
Pune, India

Jung-Sup Um
Kyungpook National University
Daegu, Korea (Republic of)

T. P. Singh
School of Computer Science Engineering and Technology
Bennett University
Greater Noida, India

Informatics Cluster,
School of Computer Sciences
University of Petroleum and Energy Studies (UPES)
Dehradun, Uttarakhand, India

Tanupriya Choudhury
CSE Department, Symbiosis Institute of Technology
Symbiosis International University
Pune, Maharashtra, India

Informatics Cluster, School of Computer Science
University of Petroleum and Energy Studies (UPES)
Dehradun, Uttarakhand, India

CSE Department
Daffodil International University
Daffodil Smart City, Bangladesh

Director Research (Honorary)
The AI University
Cutbank Montana, US

ISSN 2524-7565 ISSN 2524-7573 (electronic)
Algorithms for Intelligent Systems
ISBN 978-981-99-1619-1 ISBN 978-981-99-1620-7 (eBook)
https://doi.org/10.1007/978-981-99-1620-7

This Springer imprint is published by the registered company Springer Nature Singapore Pte Ltd.
The registered company address is: 152 Beach Road, #21-01/04 Gateway East, Singapore 189721, Singapore

Preface

MIDAS-2022 is an international conference that aims to bring together researchers, academia, and practitioners worldwide to discuss recent advances in Machine and Artificial Intelligence and Data Sciences. This conference provides a platform for professionals, scientists, engineers, educators, and students from all over the world to share their research contributions and ideas on this field of study. Université de Versailles Saint-Quentin-en Yvelines will jointly organize MIDAS-2022 in Paris, France, along with the University of Petroleum & Energy Studies in Dehradun, India. It is also supported by Nikhil Bharat Siksha Parishad (NBSP) India, South Asian Chamber of Scientific Research & Development, International Association of Professional and Fellow Engineers USA, IET UK DLN, etc., making it a truly global event where experts can exchange knowledge about cutting-edge technologies related to Machine Learning/AI/Data Science.

The 3rd edition of MIDAS-2022 promises an exciting program consisting of keynote speeches delivered by renowned personalities who are at the forefront working on various aspects related to AI/ML/DS, technical sessions with papers presented by leading academicians, industry panels discussing current trends impacting technology landscape followed up with panel discussions between participants sharing their experiences which would provide insights into challenges faced while adopting new technologies. In addition, the event will also have workshops covering topics like introduction to machine learning techniques used for solving real-life problems, ethical considerations when using data science algorithms, etc. All these activities are designed keeping one goal in mind—helping people learn more about how they can use the latest advancements made within AI space to become better equipped when trying to solve complex problems encountered during the day-to-day work schedule or even while developing products based on them.

Overall, MIDAS-2022 looks forward to providing its attendee's opportunity to gain valuable insight into current developments taking place within the domain of artificial intelligence and data sciences through a series of lectures delivered by eminent speakers combined with interactive discussions among members present there, thus giving everyone a chance to explore possibilities offered us due to advancement being made in this area right now. The conference on Machine Intelligence and Data

Sciences is an excellent opportunity for researchers to come together and discuss the latest advances in their field. The fact that it is being organized online means that more people worldwide can participate without geographical restrictions. These will lead to a wide range of perspectives on current research topics, which could be beneficial for everyone involved.

The theme of this conference was carefully chosen as it reflects the current trend in academic research—building sophisticated systems using data generation and processing capabilities. With its potential applications ranging from artificial intelligence (AI) to big data analytics, this topic has become increasingly popular among academics over recent years due to its vast scope of exploration possibilities. It appears that the choice of the theme has been well received by those who submitted papers or proposals for presentation at this event; many submissions are related directly or indirectly to machine intelligence and data sciences—showing an active interest in exploring further into these fields within academia today. These indicate just how important such topics have become when considering modern-day developments across multiple disciplines, including computer science, engineering, mathematics, etc., proving why such conferences are essential platforms for discussing new ideas among experts.

The International Conference on Machine Intelligence and Data Science Applications was of great success, along with researchers across 22 countries in the world submitting their research contributions as articles. These provided an excellent platform for exchanging ideas and further strengthening collaborations among those interested in research. In addition, the conference resulted in a compilation of chapters from various presentations, which serve as a memoir to mark this eventful occasion.

The theme of this international conference was Machine Intelligence and Applications, which consisted of four tracks: natural language processing, machine learning, computer vision, and robotics/autonomous systems. Each track focused on topics such as deep learning algorithms for image recognition or independent navigation techniques using reinforcement learning methods, etc., providing attendees with both theoretical knowledge and practical applications that can be used to develop intelligent machines capable of performing complex tasks autonomously without human intervention. Moreover, these sessions also highlighted potential challenges faced by current AI technologies while offering solutions to overcome them through innovative approaches like transfer-learning or meta-learning strategies, etc., thus allowing participants to gain valuable insights into how best they could use existing resources available at hand more effectively when developing new products or services based on AI technology. Overall, it is clear that the International Conference on Machine Intelligence & Applications has been highly beneficial for all involved parties—researchers were able to exchange information about their latest findings while gaining insight into cutting-edge developments within each field discussed during its sessions. Furthermore, it has enabled delegates to become aware of potential opportunities and understand any associated risks so they can make better informed decisions when deploying artificial intelligence technologies within their projects going forward.

The recent explosion of machine learning research has created a need for an up-to-date book on the subject. This book, Machine Intelligence: Theory and Applications, aims to fill this gap by providing comprehensive coverage of all aspects of machine intelligence. The book is split into four tracks: Track 1 addresses the algorithmic part of Machine Intelligence; Track 2 includes the framework and optimization of various algorithms; Track 3 consists of all papers related to broad applications in multiple fields; and finally track 4 consists of interdisciplinary applications.

Track 1 covers all the topics relevant to the supervised learning techniques such as neural networks, decision trees, and Support Vector Machines (SVMs), unsupervised learning methods like clustering algorithms or anomaly detection models, as well as reinforcement learning approaches which allow agents to interact with their environment using trial-and-error strategies. The second track focuses on frameworks that help optimize these algorithms, such as Bayesian optimization or genetic programming, which can be used for hyperparameter tuning tasks to maximize model performance while avoiding overfitting issues.

Finally, tracks 3 and 4 cover practical application areas where machine intelligence can play a crucial role, including natural language processing (NLP) tasks such as image recognition systems or autonomous vehicles powered by deep reinforcement learning technology. In addition, the fourth track also explores how ML technologies are being applied across different disciplines ranging from healthcare diagnosis systems based on biomedical data analysis to financial forecasting tools capable of predicting stock price movements with high accuracy rates. The conference outcome will serve the researcher's purpose when it comes down to enlightening the readers with relevant knowledge regarding both the theoretical foundations behind Machine Learning theories and the real-world scenarios where they could be implemented in a practical situation.

Vélizy, France — Amar Ramdane-Cherif
Greater Noida, India — T. P. Singh
Pune, India — Ravi Tomar
Pune, India — Tanupriya Choudhury
Daegu, Korea (Republic of) — Jung-Sup Um

Contents

About the Editors

Prof. Amar Ramdane-Cherif received his Ph.D. degree from Pierre and Marie Curie University in Paris in 1998. In 2007, he obtained his HDR degree from Versailles University. From 2000 to 2007, he was an associate professor at the University of Versailles and worked in PRISM Laboratory. Since 2008, he is a full professor at the University of Versailles—Paris-Saclay, working in the LISV laboratory. His research interests include software ambient intelligence—semantic knowledge representation, modeling of the ambient environment, multimodal interaction between person, machine, and environment, fusion and fission of events, ambient assistance; and software architecture—software quality, quality evaluation methods, functional and non-functional measurement of real-time, and reactive and software-embedded systems. As of today, he has edited 2 books on computational intelligence and communications. He authored 10 book chapters, 50 international journals, and about 150 international conference papers. He also has supervised 20 doctoral Ph.D. theses and reviewed 30 Ph.D. theses. He managed several projects and has been doing several national and intentional collaborations. He is currently a member of the Council Board of the Graduate School of Computer Science of Paris-Saclay University.

T. P. Singh is currently working as full Professor at School of Computer Science Engineering & Technology, Bennett University, Greater Noida, India. Prior to this he was associated with University of Petroleum & Energy Studies, Dehradun, Uttarakhand, India as Professor and Head of Department at School of Computer Science and Sharda University, Greater Noida, India. Dr. Singh carries double Masters Degrees, once in Computer Applications (1996) while another in Computer Science & Engineering (2009) with very high grades. Further he went on pursuing his Doctorate in Computer Science from prestigious Jamia Millia Islamia University, New Delhi. He carries 27 years of rich experience with him. He started his career with Tata Group as programmer and then joined academics in year 2000 and served Sharda University, Greater Noida, NCR, India. His research interests include machine intelligence, pattern recognition and development of hybrid intelligent systems. He has been instrumental in forming the curriculum of B-Tech program of the Indira Gandhi National Open University (IGNOU), established by an Act of Indian Parliament in

1985. One research scholar has completed her PhD and currently 06 others are working towards their Doctoral degree under him. Out of these 06, two scholars are being jointly supervised with French and UK universities. He has been a widely travelled academician and participated at various platforms across the countries including UK, France, UAE and Singapore. He has also guided 21 Masters Theses. There are more than 50 publications to his credit in various National and International journals. He has been the editor of 10 books on various allied topics of Computer Science and more than 20 book chapters. Dr. Singh is a senior member of IEEE and member of various other professional bodies including IEI, ACM, EAI, ISTE, IAENG, etc. and also on the editorial/reviewer panel of different journals published by reputed publication houses like Elsevier, Springer, InderScience, etc. He is fellow of IETA-India since 2019. He is also in the board of studies of different Indian and abroad Universities. He has been associated with many conferences throughout India and abroad as Keynote Speaker, Chair, Organizing Committee member, TPC member, Session Chair, etc.

Ravi Tomar is working as Senior Architect at Persistent Systems, India. Dr. Tomar is an experienced academician with a history in the higher education industry for a decade. He is skilled in Computer Networking, Stream processing, Python, Oracle Database, C++, Core Java, J2EE, RPA, and CorDApp. His research interests include Wireless Sensor Networks, Image Processing, Data Mining and Warehousing, Computer Networks, Big data technologies, and VANET. He has authored 100+ papers in different research areas, filled four Indian patents, edited 5 books, and has authored 4 books. He has delivered Training to corporates nationally and internationally on Confluent Apache Kafka, Stream Processing, RPA, CordaApp, J2EE, and IoT to clients like KeyBank, Accenture, Union Bank of Philippines, Ernst and Young, and Deloitte. Dr. Tomar is officially recognized as an Instructor for Confluent and CordApp. He has conducted various International conferences in India, France, and Nepal. He has been awarded a young researcher in Computer Science and Engineering by RedInno, India in 2018, Academic Excellence and Research Excellence Award by UPES in 2021, and Young Scientist Award by UCOST, Dehradun.

Dr. Tanupriya Choudhury completed his undergraduate studies in Computer Science and Engineering at the West Bengal University of Technology in Kolkata (2004-2008), India, followed by a Master's Degree in the same field from Dr. M.G.R University in Chennai, India (2008-2010). In 2016, he successfully obtained his PhD degree from Jagannath University Jaipur. With a total of 14 years of experience in both teaching and research, Dr. Choudhury holds the position of Professor at CSE Department, Symbiosis Institute of Technology, Symbiosis International University, Pune, Maharashtra, 412115, India and also he is holding Visiting Professor at Daffodil International University Bangladesh and Director Research (Honorary) at AI University, Montana US.Prior to this role, he served Graphic Era Hill University Dehradun (Research Professor), UPES Dehradun (Professor), Amity University Noida (Assistant Professor), and other prestigious academic institutions (Dronacharya College of Engineering Gurgaon, Lingaya's University Faridabad, Babu Banarsi Das Institute of Technology Ghaziabad, Syscon Solutions Pvt. Ltd. Kolkata etc.). Recently recognized for his outstanding contributions to education with the Global Outreach Education Award for Excellence in Best Young Researcher Award at GOECA 2018. His areas of expertise encompass Human Computing, Soft Computing, Cloud Computing, and Data Mining, among others. Notably accomplished within his field thus far is filing 25 patents and securing copyrights for 16 software programs from MHRD (Ministry of Human Resource Development). He has actively participated as an attendee or speaker at numerous National and International conferences across India and abroad.With over hundred plus quality research papers (Scopus) authored to date on record; Dr. Choudhury has also been invited as a guest lecturer or keynote speaker at esteemed institutions such as Jamia Millia Islamia University India, Maharaja Agersen College (Delhi University), Duy Tan University Vietnam, etc. He has also contributed significantly to various National/International conferences throughout India and abroad serving roles like TPC chair/member and session chairperson.As an active professional within the technical community; Dr.Choudhury holds lifetime membership with IETA (International Engineering & Technology Association) along with being affiliated with IEEE (Institute of Electrical and Electronics Engineers), IET(UK) (Institution of Engineering & Technology UK), and other reputable technical societies. Additionally, he is associated with corporate entities and serves as a Technical Adviser for Deetya Soft Pvt. Ltd., Noida, IVRGURU, and Mydigital360. He is also serving as Editor in reputed Journals. He currently serves as the Honorary Secretary in IETA (Indian Engineering Teacher's Association-India), alongside his role as the Senior Advisor Position in INDO-UK Confederation of Science, Technology and Research Ltd., London, UK and International Association of Professional and Fellow Engineers-Delaware-USA.

Prof. Jung-Sup Um is currently working as an editor-in-chief of Spatial Information Research (Springer Nature) since 2015 and shares his interests in the drone by maintaining strong links with international editorial boards, authors, and reviewer communities. He was awarded thirteen times Outstanding Research Award for significant and lasting research contributions. He is the author of a recent book on Drones as Cyber-Physical Systems, Concepts and Applications for the Fourth Industrial Revolution, co-author of seven books on Spatial Information, and author of more than 100 journal publications (including two patents registered). He is trying to promote the importance of drones as a core infrastructure of the 4th Industrial Revolution era while offering as an online lecture the contents of this book (course title: future industry and job selection). His current research work is on UAV remote sensing as a cyber-physical system, the use of satellite data (GOSAT, OCO-2) in UN-REDD MRV requirements, and synergistic application research that combines with various disciplines such as law, economics, political science, and environmental science. As a photographer and remote sensing practitioner, he began to intensively occupy himself with drones by the revolutionary prospect of being able to take aerial photographs at any time or any place. Over time, with a great deal of enthusiasm for this technology, he published a book about drones (2019, Drones as Cyber-Physical Systems,

Concepts and Applications for the Fourth Industrial Revolution, Springer Nature). He was a principal investigator in the various research projects that led to promoting geographic information to the general public. He was the program co-chair of the Organizing Committee of FOSS4G Seoul from 2013 to 2015. He was an associate editor and member of the editorial board of Spatial Information Research, from 1995 to 2015. His Ph.D. and master's thesis focus on the validity evaluation and application of modern remote sensing techniques for a zero-carbon city. Furthermore, he has been in a university teaching position in Remote Sensing and GIS since 1999. He had worked as a Government Official in the Ministry of Environment in Korea, with a focus on the application of remote sensing and GIS. He was responsible for the initial introduction of remote sensing and image processing for the Ministry of Environment in Korea. He had developed extensive research and laboratory capabilities in the Ministry. His experience includes projects with related central governments in Korea, as well as private companies and research institutions. He has been frequently invited as a consultant for zero-carbon Geomatics planning, to local and central governments.

Blockchain Supported Sustainable Supply Chain in Industry 4.0

Rahul Gupta

1 Introduction

Supply chain practices in Industry 4.0 are equipped with various emerging and novel technologies [41]. These technologies support organizations with enhanced business processes, financial results, and higher customer satisfaction. All this is possible to be achieved through seamless coordination and augmenting the value chain across the supply chain, which is inclusive of the back end from suppliers, manufacturers, financiers, and regulators, and at the front end by logistics support providers and retailers. Due to the complexity of multi-echelon, multination supply chains, disruptions and uncertainty have amplified with increased costs and decreased revenues. Further, these disruptions get exacerbated by the availability of non-trusted and untimely data. Agility, Resiliency, and Transparency are a result of the smooth functioning of supply chains.

Rayome (2019) [32] suggested that numerous emerging computing technologies such as IoT, AI, 5G networks, Robotics, 3D printing, Biometric recognitions, Drones, and AR/VR are process enablers for Industry 4.0. Blockchain and wireless communication devices are emerging recently, are well-positioned, and are accepted for pioneering successful business models. Integration of Blockchain technology in supply chain operations offers transparency, trust, and consensus among stakeholders and offers flexibility in operations. Tomar et al. (2020) [36] show how Blockchain is effective in building security and trust in its application. Enterprises across geographies are experiencing an acceleration in the speed of transactions, which can be managed by using reusable technologies, software, algorithms, and governance models.

Nowinski and Kozma (2017) [29] commented that, in changing business scenarios, the onus of trust has shifted from individuals to smart contracts, analytics,

R. Gupta (✉)
Amity Business School, Amity University Uttar Pradesh, Noida, India
e-mail: rgupta10@amity.edu

A. Ramdane-Cherif et al. (eds.), *Machine Intelligence and Data Science Applications*, Algorithms for Intelligent Systems, https://doi.org/10.1007/978-981-99-1620-7_1

and sharing applications bypassing the need for a central entity. Scholars and business organizations have acknowledged the contribution of technological advancements. However, how to address sustainability is an issue that still needs comprehensive discussion and deliberations.

Technology implementation has detrimental consequences, depending upon transportation to agribusiness and manufacturing organizations. Some treatments are imposed on the natural ecosystem. The understanding of complexities among society, technology, and business needs to be identified and addressed, to attain sustainability challenges. The study focuses on technological contribution in meeting challenges sustainably. The degree of the adverse environmental impact of these technologies may be mitigated by new sustainable opportunities. Stock and Seliger (2016) [33] found a lack of research to understand the impact of technologies in Industry 4.0 on resolving sustainability issues and industrial symbiosis. The study focuses on the opportunities offered by blockchain technology and Industry 4.0 in managing sustainable relationships.

2 Background

The study reviews the literature on the impact of Blockchain technology on the sustainability of Industry 4.0 and supply chain and discusses a brief review of Blockchain, Industry 4.0, and sustainable development. To study the impact of blockchain on the sustainability of the Industry 4.0, supply chain, a brief of these concepts is studied here.

2.1 Blockchain

A distributed ledger shares data on the peer-to-peer network. The nodes, being members, validate and communicate through a predefined protocol. These ledgers can be centralized, offering explicit rights to an individual user, or decentralized offering identical rights. Every node in the distributed ledger has a copy of the ledger. A decentralized or centralized ledger of the blockchain can be designed as per the rights of the users. In the case of decentralized blockchain, decision-making is shared by various users, while in the case of centralized blockchain, a single entity has the responsibility to make decisions.

A cryptocurrency called Bitcoin gets popularized due to peer-to-peer networks. Cryptography helps in creating a blockchain in each block data file, transactions have a cryptographic hash that is linked with the previous block. Casado et al. (2018) [6] said that a block is verified with the percentage of network nodes, gets added to form a blockchain, and is called a public ledger transaction.

Blockchain supports digital administrative control, sharing, and storing of data by converting it into code transparency and mitigating revision or deletion risk. The potential of blockchain supports all digital payments, agreements, and transactions. Upon validation, these records are shared among machines, persons, algorithms, and firms. Lansiti and Lakhani (2017) [21] said lawyers, bankers, and brokers need this less often. Intermediaries act as middlemen for verification and accurately handling data in various industries. Individuals and organizations are trusting computer codes for transaction verification over humans.

Bitcoins are in full control of individuals, the Bitcoin balance is invisible unlike a normal bank balance, which cannot be viewed/manipulated digitally. Athey et al. (2016) [2] said with proper passcodes one may authorize entry on a ledger and transaction to another individual's address. Sharma et al. (2017) [24], and Crosby et al. (2016) [3], said blockchain transaction benefits include low risk, transparency, instantaneous, security, private, and no exchange cost. Barton (2018) [4] discussed the following capabilities depending on the platform to be used for transactions.

2.2 *Sustainable Development*

Korhonen et al. (2018) [50] talked about the lack of any consensus on the circular economy, as it has its origination from both environmental and industrial economics. It is considered as overcoming restrictions of consumption models for optimum resource utilization and linear production processes. The circular economy is considered a mediator to coordinate environmental, social, and economic aspects of sustainability. Chinese industries for a greener production strategy prefer a circular economy where resources are consumed efficiently. Ghisellini et al. (2016) [14] said that countries like the USA, Japan, and European countries consider this as a strategy to alleviate waste management. Pearce and Turner (1990) [30] said that the law of thermodynamics is the foundation of economic system circularity. Energy and matter degradation is described to preserve resources sustainably. A circular economy environment is supported by resources, a sink for waste, emissions, and a life support system. De Bruyn et al. (2018) [11] and Weidema (2015) [46] said that environmental emissions result in economic losses due to the absence of no direct prices for environmental goods like water and air, lately, life cycle assessments try to monetize environmental prices. Ghisellini et al. (2016) [14] said that to mitigate resource consumption producers, consumer responsibilities and stricter environmental policies are employed.

The circular economy is supported by value drivers.

1. Increase the usage cycle of assets.
2. Shared recourses and assets result in enhanced productivity.
3. Remanufacturing, reusing, recycling, and extended usages for cascading asset looping.
4. Natural resources are preserved and regenerated by avoiding leakages of nutrients and recurring to the original ecology.

Prendeville et al. (2017) [31] suggested a framework for implementing these value drivers called ReSOLVE- regenerate, share, optimize, loop, virtualize, and exchange. He said that it is not beyond criticism, due to the uncertainty of its belongingness be it to a micro level or macro level. Few environmental principles may not add direct benefits to environmental concerns, like infinite recycling may lose efficiency and at times consume higher energy, or sharing economy initiatives may not offer benefits as promoted. The circular economy is accepted in a few organizations, and these models may offer more authority to firms rather than society and communities. With the emergence of enhanced technologies, sustainability outcomes improve in organizations by practising a circular economy.

2.3 *Industry 4.0*

Vogel et al. (2016) [40] said in Industry 4.0, production and manufacturing connect through cyber-physical objects using big data. Lom et al. (2018) [49] said that IoT and CPS (cyber-physical system) initiatives make the smart city part of Industry 4.0, these terms are used interchangeably in research. Choudhury et al. (2021) [32] highlighted the application and helpfulness of Blockchain in the IoT ecosystem. Kagermann (2015) [18] suggested that various characteristics support Industry 4.0 by equipping manufacturing with enhanced data acquisition technologies with service innovation and value generation. Sille et al. (2022) [25] review on the security concerns in IoT devices and how Blockchain tackles the same. Zhou et al. (2015) [45] suggested planning objectives, efficient management, security and safety, standardization, reliability, work design, and regulatory framework, for improving resource efficiency. Guo et al. (2021) [19] examine the use of Blockchain and IoT in the agriculture and food sector. Sharma et al. (2020) [12] suggested the use of Blockchain in IoMT and also proposed a novel architecture using Blockchain and IoMT, which has improved performance as compared to traditional approaches.

A four-step strategy transforms the earlier information-age industry to advanced stage IoT, and big data-supported Industry 4.0. The following four steps support this transformation.

1. A CPS network.
2. Developing intelligent production and smart factories.
3. Horizontal, vertical, and end-to-end integration.
4. Achieve planning objectives as discussed above.

Last et al. (2014) [22] said that mass customization, design flexibility, short lead times, resource efficiency, and decentralized and flexible production are a few achievements of Industry 4.0. Research suggests that the industrial revolution offered opportunities for industrial 4.0. The success of blockchain technology is gauzed by its ability to enhance information integration across supply chains. Few other support systems are:

Shared Ledger: A locally distributed data structure shared among participants.

Permissioning: Data transparency and privacy are secured in authenticated transactions.

Smart Contracts: Databases are embedded with business terms and transaction records.

Consensus: Transacted data are endorsed by users ensuring traceability and mutability.

Yli-Huumo et al. (2016) [43] said recent blockchain research describes crypto applications and Bitcoin. Athey et al. (2016) [2] enumerate various other applications of this technology in industries ranging from real estate, energy, and health care. Kane (2017) [44] said blockchain applications are limited but with various key features, it qualifies to be a general-purpose technology. Catalini and Gans (2016) [7] said technologies such as electricity, steam engines, etc. are seeing higher productivity and innovation through the internet along with various industries. these benefits are the results of blockchain; its fruition is a question mark.

Regulatory challenges, scalability, and security are numerous limitations in the mass-level implementation and acceptance of blockchain adoption. Coders and software engineers define rules in a smart contract to govern the system, regarding network structure, applications, and architecture.

Defining the scope and content of smart contracts by engineers faces numerous difficulties while its implementation with regulations and compliances. Computer coding in the smart contract of all possible and complex business scenarios is a challenge, and in case of doubts, traditional legislators and courts are to be relied upon. Von (2016) [43] said blockchain platforms face various legal challenges like verification of the information ledger, connected to the physical world (physical verification of the title holder of the land), apart from adaptation to inflexibility in uncertain scenarios.

Private blockchain for industrial applications faces various critics, being different from public blockchain, and has the status of a secreted blockchain platform. Narayanan (2015) [28] said it is a shared database. These issues disqualify blockchain as innovative technology. Barber et al. (2012) [5] discussed various scalability issues like linear transactions and malfunctioning communication. Conoscenti et al. (2016) [10] talked about the increased number of transactions that face the struggle of consensus protocols upon scalability. Numerous scaling slants are practised in computer science.

3 Sustainable Supply Chain

Supply chain sustainability can be achieved by enhancing the product life cycle by taking support in three broad areas data collection, analysis, and enhanced decision-making. Practicing Industry 4.0, the data collection is supported by devices loaded with sensors, and software and performs real-time analysis using decision support systems across multiple levels. With these advanced devices, an enormous amount of data is generated for endless transactions and recordings. Desired results cannot be achieved in the absence of data or from unprocessed data. Smart factories equipped with software tools, and techniques quantify ecological impacts on ubiquitous production processes and resources. Saberi et al. (2018) [34] suggested that stainable supply chain supports through the following factors.

1. Tracking mitigates rework and recall.
2. Product footprints are tracked to accurately measure carbon footprint and the amount to be charged in the process.
3. Incentivization for recycling is propagated among participants.
4. System fidelity is improved by reducing fraud in emission trading schemes.

Blockchain is capable of (a) green supply chain practices, (b) improved product lifecycle, (c) improved system and operational efficiency, and (d) improved monitoring and reporting sustainability.

3.1 Green Supply Chain Practices

The green supply chain includes reuse, recycling, repair, refurbishing, waste reduction, buying energy-efficient items, sharing, and maintenance. Eco-behavior can support by purchase, usage, and EOD behavior. Coin offerings support financing sustainability. Blockchain is using ICO (Initial Coin Offerings) and token sales. Li and Maan (2013) [23] said funds are raised using ICO as a mechanism to support initial investments. Catalini and Gans (2018) [8] said the business offers token as a promise to perform payments while availing services on the blockchain platform. These platforms may design and promote tokens locally for green behavior among consumers in a product lifecycle. Network users are incentivized with crypto tokens for their eco-behavior. Andoni et al. (2019) [1] said to verify eco-behavior, inspectors, certified vendors, and IoT sensors are being used.

Catalini and Gans (2019) [9] said the role of a token is predefined as either serving as a medium of exchange or an investment mechanism. The value of a token is in a proposal to its supply, size of the platform, and service or product. While adopting policy token economies shall be considered. While designing the rewards and incentives, the behavior and quantification award should include in the smart contract. Designing a platform requires the following steps to be taken.

a. Defining objective.
b. Behavior to be rewarded.
c. Terms to be included.
d. Amount of incentive.
e. Business model and sustainability consequences.

3.2 Improved Product Lifecycle

Blockchain capabilities being decentralized are not limited to checking the prerequisite of central and mediator authority. Immutable and decentralized nature and identity protection offer an opportunity to verify, record, and use the information on the public ledger with concealed identity. Producers can do lifetime monitoring and data collection for improved design, usage, sale, and reverse logistical activities. Lifetime accessibility was not available with traditional supply chain practices. Blockchain stakeholders help in verification, collection, and use during its life cycle. Sustainability principles and corporate integrity of the business model get a boost from sharing. A traceable and transparent product lifecycle sees low waste generation, curved emission, and purposeful engagement of users, stakeholders, and authorities.

Mattila et al. (2016) [26] said the life cycle of capital and durable goods can be managed by managing and sharing product-centric information. Sectors like healthcare, research in the medical field, and controlling counterfeit drugs get immense benefits. Sun et al. (2016) [39] said sharing devices help in managing smart cities. Hackius and Petersen (2017) [17] said blockchain helps in processing paperwork in global shipping, tracking any outbreak through the food chain, checking counterfeit supplies, and tracking shipments through IoT and Sensor equipped devices. Vehicle identity helps in traffic control, checking carbon emissions, improved maintenance, and data sharing of self-driven vehicles.

3.3 The Improved System and Operational Efficiency

Catalini and Gans (2016) [7] identified two cost elements affected by blockchain networking cost and verification cost. Verification cost is important among buyers and sellers as participants must verify the transaction attributes and credentials, product features, with other clauses. As the intermediaries are reduced hence costs are controlled, as power is distributed among network participants. Compared to traditional systems blockchain is cost-effective, fast, and secure. Tushar et al. (2020) [41] have quoted an example of low energy shortages and reduced peak demand in a reliable manner among peer-to-peer electricity networks.

3.4 Improved Monitoring and Reporting Sustainability

Blockchain supports businesses to monitor and manage, their actions effectively, which helps them enhance their sustainability reporting. As reported by PWC in 2018 firms' performances and awareness among shareholders get enhanced with improved capabilities. Over time firms observe increased efficiency and reduced operational expenses. To mitigate the popularity of Bitcoin other cryptocurrencies came into popularity, blockchain 2.0 gained popularity in various other sectors also for solving business problems through new algorithms with lower computational power.

Technological companies are considering blockchain as a service for more than 90 cases in business applications, thereby reducing operational expenses. Batra et al. (2019) [13] said with blockchain facilitates the smooth transaction of records by removing intermediaries. Esmaeilian et al. (2018) [51] said green behavior comprises waste reduction, buying refurbished items, buying local, buying energy-efficient equipment, sharing, and maintenance. Greening et al. (2000) [15] illustrated the "rebound effect" as improved demand for energy services as improvements in efficiency consumption.

4 Use Cases of Blockchain

4.1 Lessening Bullwhip Effect

Rabe et al. (2006) [48] said when each entity wishes to maximize its profitability locally and ignore supply chain profitability, the Bullwhip effect is created in the supply chain and demand fluctuates as you move up in the supply chain from customer to retailer to wholesaler to manufacturer. This can be checked by price stabilization and channel alignment via information sharing with the help of block31chain. Overordering is reduced with smart contracts, with transparent and traceable transactions.

Several unique characteristics of blockchain share data in place of a centralized database. Multi-echelon supply chain members can communicate and share data among authorized members. A set of rules authorizes the level of authority for each member. Agreements for creditable transactions and information are made without a third party. Tamper-proof ledger authentications are done with digital signatures and cryptographic hash. Various users access encrypted information from the ledger, using smart contracts and predefined rules, where users record, access, evaluate, and retrieve data.

4.2 Producer Responsibility

Producer responsibility is extended for product stewardship to guide and own responsibility of their product to the end of life with original equipment manufacturers. They must bear the cost of reverse logistics for sustainable usage. Government charges a fee from OEMs as per their market share. Paper for reuse, recycling, and remanufacturing are submitted to the government, to claim payment for recycling. Blockchain through digital platform plugins several loops and missing information between final processor and OEMs, mitigating EPR limitations and incentivizing reuse, recycling, and remanufacture.

4.3 Shipment and Transportation Tracking

Goods are tracked globally using concepts of blockchain. Sadouskaya (2017) [35] said end-to-end shipment tracking among various parties is possible by developing distributed permission platform accessible through multiple supply chain partners for information exchange and documentation. Sadouskaya (2017) [35] and Smart Log (2017) [47] said information management gives information that is anonymized after filtration and added to a public ledger.

4.4 Traceability of Food Safety and Recall

Blockchain is extensively used in food safety, recall, and traceability in its supply chain. Along with safety, document workflow is reduced leading to automated invoice and billing. Various categories like coffee, beef, etc. are tracked digitally throughout their growing and passing to their supply chain.

4.5 Blockchain and Its Adverse Effects

With its immense potential to offer neumarous benefit to society blockchain, faces certain environmental and social concerns. Hawken et al. (2013) [16] said a fundamental rethinking is required to best use technology focusing on getting the right thing rather than doing something in the right way. In case value-driven models are not linked with blockchain, the results may be disastrous for the environment because of faster material extraction and wrong product offering at low cost to the market.

Barber et al. (2012), [5], said current blockchain is computational-intensive, it faces various scalability challenges like communication breaches among parties, and a significant energy loss is observed while verifying transactions. Mora et al. (2018) [27] compare the Bitcoin adoption rate with other technologies and comment that global warming exceeds by 2 degrees centigrade due to Bitcoin. Users of Bitcoin are divided into clients and miners. Miners create the block and the verification results in mining coins, whereas users do not get involved in mining codes. Ethereum offers another consensus algorithm as Proof-of-Stake (PoS) where the miner validates the next block or delegates Proof-of-Stake (DPoS) with limited block creators, Konstantopoulos (2018) [38].

Krause and Tolaymat (2018) [20] compared the unit dollars required to mine cryptocurrency and the unit dollars required for mining minerals like gold, copper, or aluminum and concluded crypto mining consumes higher energy. Stoll et al. (2018) [37] discussed information on mining websites and geographical regions of miners for studying energy consumption and greenhouse gas emission metrics.

Once technology gets matured, environmental sustainability issues can be managed. Barber et al. (2012) [5] suggested that the scalability issue of Bitcoin can be managed by dividing users into separate groups, as clients (they do not mine, only need transactions payable into their account) and minors (verify transactions and create blocks). The complex nature of Bitcoin, users are decentralized, and immutable human errors are among other limitations of Bitcoin.

Immutable blockchain requires information verification, as once recorded data cannot be reversed, and any human error needs to be checked. Yadav and Singh, (2020) [45], said insufficient regulations and trust among users is another hindrance in its implementation, their accessibility, quality, and safe documentation are considered a success story of blockchain.

5 Discussion

Being in its early stage, its full potential is yet to be achieved after overcoming its limitations. Scalability should be addressed by limiting the number of users and size. Computational efficiency helps in achieving higher security and privacy. Blockchain's human and social dimensions need better understanding. Research adequately does not support various principles and concepts in the blockchain. Sustainability objectives can be studied as follows.

1. Data-driven decision-making.
2. Lifecycle assessment and new performance indicator.
3. Develop new cases.
4. Better product design.
5. Better human–machine interaction.

6 Conclusion

Sustainable supply chain issues are discussed as IoT-enabled energy management, green logistics, and models supporting the sharing economy. Blockchain capabilities are reviewed as an emerging technology with distributed ledgers and enablers for sustainability implementation. Circular economy concepts are studied as green behavior with specialized tokens, visible product lifecycle, improved system efficiency at decreased operational cost, high performance, and sustainable capabilities. These capabilities enhance circular economies through the improved product lifecycle, optimum resource utilization, and less carbon emission. Transparency is maintained while developing product, using raw material, and doing trade, with support from regulatory and economic forces. Research gaps are identified for Industry 4.0 to support sustainability objectives. Further research is required in studying new performance indicators, new data processing, and reasoning techniques, new design strategies, studying cases for best practices for successful sustainability implementation, and analyzing further workforce requirements.

References

1. Andoni M, Robu V, Flynn D, Abram S, Geach D, Jenkins D, McCallum P, Peacock A (2019) Blockchain technology in the energy sector: a systematic review of challenges and opportunities. Renew Sustain Energy Rev 100:143–174
2. Athey S, Parashkevov I, Sarukkai V, Xia J (2016) Bitcoin pricing, adoption, and usages: theory and evidence
3. Crosby M, Pattanayak P, Verma S, Kalyanaraman V (2016). Blockchain technology: beyond Bitcoin. Appl Innov 2:6–10
4. Barton D (2018) The future of finance: how FinTech, AI & blockchain will shape our future. IBM Watson
5. Barber S, Boyen X, Shi E, Uzun E (2012) Bitter to better—how to make Bitcoin a better currency. In: International conference on financial cryptography and data security. Springer, pp 399–414
6. Casado-Vara R, Prieto J, De la Prieta F, Corchado JM (2018) How Blockchain improves the supply chain: case study alimentary supply chain. Procedia Comput Sci 134:393–398
7. Catalini C, Gans JS (2016). Some simple economics of the blockchain. National bureau of economic research
8. Catalini C, Gans JS (2018) Initial coin offerings and the value of crypto tokens. National bureau of economic research
9. Catalini C, Gans J (2019) Some simple economics of the blockchain. Rotman Sch. Manag. Working paper No. 2874598 MIT Sloan Res. Pap. No. 5191-16
10. Conoscenti M, Vetro A, De Martin JC (2016) Blockchain for the internet of things: a systematic literature review. In: 2016 IEEE/ACS 13th international conference of computer systems and applications (AICCSA). IEEE, pp 1–6
11. De Bruyn S, Ahdour S, Bijleveld M, de Graaff L, Schep E, Schroten A, Vergeer R (2018) Environmental prices handbook 2017: methods and numbers for valuation of environmental impacts. CE Delft, Delft
12. Sharma A, Sarishma TR, Chilamkurti N, Kim B-G (2020) Blockchain based smart contracts for internet of medical things in e-healthcare. Electronics 9:1609. https://doi.org/10.3390/electronics9101609
13. Gaurav B, Olson R, Pathak S, Santhanam N, Soundararajan H (2019) Blockchain 2.0: what's in store for the two ends—semiconductors (suppliers) and industrials (consumers)?
14. Ghisellini P, Cialani C, Ulgiati S (2016) A review on circular economy: the expected transition to a balanced interplay of environmental and economic systems. J Clean Rod 114:11–32
15. Greening LA, Greene DL, Difiglio C (2000) Energy efficiency and consumption—the rebound effect—a survey. Energy Policy 28(6–7):389–401
16. Hawken P, Lovins AB, Lovins LH (2013) Natural capitalism: the next industrial revolution. Routledge
17. Hackius N, Petersen M (2017) Blockchain in logistics and supply chain: trick or treat? In: Proceedings of the hamburg international conference of logistics (HICL). E-publication, pp 3–18
18. Kagermann H (2015) Change through digitization—value creation in the age of industry 4.0. management of permanent change. Springer, pp 23–45
19. Guo J, Cengiz K, Tomar R (2021) An IOT and blockchain approach for food traceability system in agriculture. Scalable Comput Pract Exp 22:127–137

20. Krause MJ, Tolaymat T (2018) Quantification of energy and carbon costs for mining cryptocurrencies. Nat Sustain 1(11):711
21. Lansiti M, Lakhani K (2017) The truth about Blockchain. Harv Bus Rev
22. Lasi H, Fettke P, Kemper H-G, Feld T, Hoffmann M (2014) Industry 4.0. Bus Inf Syst Eng 6(4):239–242
23. Li J, Mann W (2018) Initial coin offering and platform building
24. Sharma PK, Moon SY, Park JH (2017) Block-VN: a distributed Blockchain based vehicular network architecture in smart city. J Inf Process Syst 13(1):84
25. Sille R et al (2022) Review study on blockchain frameworks for security issues in IoT devices. In: 2022 international symposium on multidisciplinary studies and innovative technologies (ISMSIT), Ankara, Turkey, 2022, pp 876–881. https://doi.org/10.1109/ISMSIT56059.2022.9932744
26. Mattila J, Seppala T, Holmström J (2016) Product-centric information management: a case study of a shared platform with blockchain technology
27. Mora C, Rollins RL, Taladay K, Kantar MB, Chock MK, Shimada M, Franklin EC (2018) Bitcoin emissions alone could push global warming above 2 C. Nat Clim Chang 8(11):931
28. Narayanan A (2015) 'Private blockchain' is just a confusing name for a shared database. https://freedom-to-tinker.com/2015/09/18/private-blockchainis-just-a-confusing-name-for-a-shared-database/
29. Nowinski W, Kozma M (2017) How can Blockchain technology disrupt the existing business models? Entrep Bus Econ Rev 5(3):173–188
30. Pearce DW, Turner RK (1990) Economics of natural resources and the environment. JHU Press
31. Prendeville S, Cherim E, Bocken N (2017) Circular cities: mapping six cities in transition. Environ Innov Soc Transitions
32. Choudhury T, Khanna A, Toe TT, Khurana M, Nhu NG (2021) Blockchain applications in IoT ecosystem. Springer, Cham, Switzerland
33. Rayome AD (2019) Top 10 emerging technologies of 2019. TechRepublic
34. Stock T, Seliger G (2016) Opportunities of sustainable manufacturing in Industry 4.0. Procedia Cirp 40:536–541
35. Saberi S, Kouhizadeh M, Sarkis J, Shen L (2018) Blockchain technology and its relationships to sustainable supply chain management. Int J Prod Res 1–19
36. Sadouskaya K (2017) Adoption of blockchain technology in supply chain and logistics
37. Tomar R (2020) Maintaining trust in VANETs using blockchain. ACM SIGAda Ada Lett 40:91–96
38. Stoll C, Klaaßen L, Gallersdörfer U (2018) The carbon footprint of Bitcoin
39. Konstantopoulos G (2018) Understanding blockchain fundamentals, part 3: delegated proof of stake. https://medium.com/loom-network/understanding-blockchain-fundamentals-part-3-delegated-proof-of-stakeb385a6b92ef
40. Sun J, Yan J, Zhang KZK (2016) Blockchain-based sharing services: what blockchain technology can contribute to smart cities. Financ Innov 2(1):26
41. Tushar W, Saha TK, Yuen C, Smith D, Poor HV (2020) Peer-to-peer trading in electricity networks: an overview. IEEE Trans Smart Grid
42. Vogel-Heuser B, Hess D (2016) Guest editorial industry 4.0–prerequisites and visions. IEEE Trans Autom Sci Eng 13(2):411–413
43. Von Haller Gronbaek M (2016) Blockchain 2.0, smart contracts and challenges. Comput Law SCL Mag 1–5
44. Negi D, Sah A, Rawat S, Choudhury T, Khanna A (2021) Block chain platforms and smart contracts. In: Choudhury T, Khanna A, Toe TT, Khurana M, Gia Nhu N (eds) Blockchain applications in IoT ecosystem. EAI/Springer innovations in communication and computing. Springer, Cham. https://doi.org/10.1007/978-3-030-65691-1_5
45. Yadav S, Singh SP (2020) Blockchain critical success factors for sustainable supply chain. Resour Conserv Recycl 152:104505
46. Yli-Huumo J, Ko D, Choi S, Park S, Smolander K (2016) Where is current research on Blockchain technology? —a systematic review. PLoS ONE 11(10):e0163477

47. Kane E (2017) Is blockchain a general-purpose technology?
48. Zhou K, Liu T, Zhou L (2015) Industry 4.0: towards future industrial opportunities and challenges. In: 2015 12th international conference on fuzzy systems and knowledge discovery (FSKD). IEEE, pp 2147–2152
49. Weidema BP (2015) Comparing three life cycle impact assessment methods from an endpoint perspective. J Ind Ecol 19(1):20–26
50. Smart Log (2017) Smart log - proof of concept project for IoT blockchain solution in logistics industry. https://www.kinno.fi/en/smartlog
51. Rabe M, Jäkel, F-W, Weinaug, H (2006) Supply chair demonstrator based on federated models and HLA application. SimVi 329–338

COVID-19 Cases in India: A Study on the Effect of Weather Factors and Building the Forecast Model

V. L. Helen Josephine, G. Meshach, and Poorvi Prakash

1 Introduction

In terms of revenue and employment, India's healthcare sector has grown to become one of the country's most important industries. The healthcare sector includes hospitals, medical devices, clinical trials, outsourcing, telemedicine, medical tourism, health insurance, and medical equipment. Because of expanded coverage, services, and increasing spending by both public and private entities, the Indian healthcare system is quickly expanding. India's competitive advantage is its large pool of well-trained medical workers. India is cost competitive when compared to its Asian and Western peers. In India, surgery costs around a tenth of what it does in the United States or Western Europe. The healthcare market might triple in size by 2022, reaching Rs. 8.6 trillion (US$ 133.44 billion). In Budget 2021, India's public expenditure on healthcare as a percentage of GDP was 1.2%. In FY21, gross direct premium income for health insurance companies increased 13.3% year on year to Rs. 58, 572.46 crore (US$ 7.9 billion). The health industry accounts for 29.5% of the country's total gross written premiums. The Indian healthcare market is tremendously diverse, with opportunities for providers, payers, and medical technology in every category. As the rivalry intensifies, businesses are looking for the latest dynamics and trends that may benefit their firm. From Rs. 4 trillion (US$ 61.79 billion) in FY17 to Rs. 8.6 trillion (US$ 132.84 billion) in FY22, India's hospital business is predicted to increase at a CAGR of 16–17%. India's government plans to increase public health spending to 2.5% of GDP by 2025.

V. L. H. Josephine (✉) · G. Meshach · P. Prakash
Business Analytics, School of Business and Management, Christ University, Bengaluru, India
e-mail: helen.josephine@christuniversity.in; helenjose.cbe@gmail.com

P. Prakash
e-mail: poorvi.prakash@mba.christuniversity.in

A. Ramdane-Cherif et al. (eds.), *Machine Intelligence and Data Science Applications*, Algorithms for Intelligent Systems, https://doi.org/10.1007/978-981-99-1620-7_2

2 Literature Review

To understand more about the paper and to know what methods and algorithms can be used, various research papers were downloaded from various websites and used as a reference. These papers help to understand the significance of the project and to know about the various methods that can be used and various algorithms that can be implemented. The scientific evidence presented in this review article demonstrates that climatic variables such as temperature and humidity may influence the spread of the 2019 COVID virus. Warm and humid conditions, it appears, may signal decreased transmission of COVID-19. Considering the topic, which appears to impose both positive and negative data on the likely effect of environmental conditions, particularly humidity and temperature, on the stability and transmissibility of the 2019-COVID, nearly all available information suggested a possible positive effect [1]. By utilizing a semantics-based probabilistic framework to analyze the causal relationship and effects of climate variability on the COVID-19 outbreak, the framework becomes more effective in capturing the unique characteristics of the pandemic. More importantly, the effect analysis framework includes an auxiliary module for quantifying semantic relatedness on a regional basis, allowing for the realistic accounting of multiple climate types within a single spatial region [2].

In Colombian cities like Barranquilla, Bogota, Cali, Cartagena, Leticia, and Medellin, mobility variables had the greatest impact on virus transmission, particularly in the post-quarantine period. In different biogeographical locations, the weather variables have varying effects on virus transmission. One of the characteristics linked to the greatest contamination rate in Leticia was wind speed. In Bogota, where cold/dry conditions prevailed, the largest number of new daily cases were observed [3]. In LAC region, humidity, wind speed, and rainfall indicated a significant association with daily cases, total cases, and death. The results indicate that the spreading of COVID-19 positive and negative was associated with wealth disparity and poverty levels in the cities studied [4].

In the Brazilian state of Rio de Janeiro, the spread of SARS-COVID was found to have a substantial negative connection with solar exposure. There was a negative link between temperature and wind speed. As a result, excessive solar radiation can be identified as the principal climatic factor suppressing COVID-19 spread in the examined tropical condition. High temperatures and high wind speeds are also important factors to consider [5]. In five Brazilian cities (Sao Paulo, Rio de Janeiro, Brasilia, Manaus, and Fortaleza), based on data from April 13, in contrast to reports from the coldest nations or periods under freezing temperatures, the results revealed that greater mean temperatures and average relative humidity enhanced COVID-19 transmission [6].

COVID-19 data and daily average temperature and relative humidity have a significant positive and negative connection, respectively, in the regions of Indonesia. The correlation coefficients, on the other hand, are low. This relationship was repeatedly confirmed by regression analysis [7]. Several climatic variables as well as ambient air pollution were linked to the transmission of SARS-CoV-2 in 3739 locations around the world. It is discovered that there is a moderately negative relationship between estimated reproduction number and temperatures above 25 °C (a decrease of 37% [95% CI 1–54] per additional degree), a U-shaped relationship with outdoor ultraviolet exposure, and weaker positive associations with air pressure, wind speed, precipitation, diurnal temperature, sulphur dioxide, and ozone [8]. The impact of physicochemical variables on virus vitality and spread should be considered in experimental and clinical studies. Another avenue to explore is the effect of diet and its interaction with the immune system on SARS-CoV-2 infection and mortality rate. Over the last year, numerous studies have been conducted on both the virus and pathophysiology of COVID-19, as well as the cellular mechanisms underlying the spread of the virus. The result of the study shows that airborne transmission of SARS-CoV-2 is especially crucial under dry conditions and high temperature, and it needs to be considered more. Cold weather and high humidity can increase the likelihood of surface transmission as an important factor in spreading the COVID-19 outbreak. Hydrophobic surfaces can enable the virus to remain in droplet form for more extended periods, potentially facilitating its spread. In these circumstances, social distancing measures may be beneficial [9]. To assess the potential link between weather patterns and COVID-19 fatality rates in the United States. The primary factors include minimum and maximum daily temperature, ozone concentration, PM2.5 concentrations, precipitation, and UV radiation index. The result observed in the research is that the fatality rate increased when the temperature is minimum and the ozone level is negatively correlated with the deaths when the level is below 38 ppb [10]. According to previous research, high temperatures and humidity limit the transmission of the severe acute respiratory syndrome coronavirus 2 (SARSCoV-2).

As a result, the purpose of this study work is to examine the effects of temperature, precipitation, and ultraviolet (UV) light on community transmission of SARS-CoV-2 using case data from the United States. Negative binomial regression modeling was employed to determine whether daily maximum temperature, precipitation, and UV index correlate with COVID-19 fatality rates. The number of infected cases decreases as the temperature rises to 52°F, and it is lower at warmer temperatures than at cooler temperatures. Temperature and transmission, on the other hand, have a minor relationship, and transmission is likely to remain high at warmer temperatures [11]. This study aimed to build a model to forecast the number of COVID-19 cases in the upcoming 14 days in three different provinces in Indonesia and to analyze the relationship between COVID-19 cases and meteorological variables. The method used to build a model to forecast is Autoregressive Moving Average (ARIMA) model. The Spearman correlation analysis analyzes the relationship between B.1.1.7 cases and meteorological variables such as sunshine, humidity, rainfall, maximum and minimum temperature, and wind speed. The study's findings revealed that the fitted ARIMA models predicted an increase in daily instances in three provinces. In comparison to actual cases, total cases in the three provinces will increase by 36% (West Java), 13.5% (South Sumatra), and 30% (East Kalimantan) until the end of 14 days. The significant contributors to B.1.1.7 cases were temperature, rainfall, and sunshine [12].

3 Methodology

Research methodology refers to the design of a study technique as well as the procedures for gathering data for a research study. It outlines the procedures for acquiring, measuring, and analyzing data in order to achieve the objectives of a research project. The research approach is critical in a research endeavor since it outlines the study strategy. The operational definitions of the concepts used in the research, as well as the parameters that have been included in the study, are defined explicitly in this chapter. The goal of this paper is to provide an overview of the study in terms of the research area, study components, and population. The organization and design of the questionnaire, as well as data collection and analysis procedures, are also discussed in the paper.

3.1 Problem Statement and Data Understanding

Since the number of COVID-19 cases is increasing every day and the cause of the breakout or spread of the virus is unknown, the goal of this study is to see if environmental factors such as weather and air pollution play a role in the growth of COVID-19 cases in India. The major three metropolitan cities, Mumbai, Chennai, and Bangalore, have been picked for this initiative. This study aims to determine whether weather parameters including temperature, rainfall, dew, humidity, wind, and precipitation play a role in the spread or rise of COVID-19 cases. The weather data and the COVID-19 virus confirmed cases data for each city are collected separately and then merged.

3.2 Methodology

The methodology used for this paper is the CRISP-DM methodology. The Cross Industry Standard Process for Data Mining (CRISP-DM) methodology consists of six phases, which explain the life cycle of data analytics.

The six phases of CRISP-DM are:

- Business understanding
- Data understanding
- Data preparation
- Data modeling/ analysis Data evaluation
- Data visualization

– Business understanding: The first phase of the cycle is business understanding, which entails comprehending the project's problem statement and goal. The goal of this experiment is to see if environmental factors like weather and air pollution play a role in the rise of COVID-19 cases in India. This study aims to determine whether weather parameters including temperature, rainfall, dew, humidity, wind, and precipitation play a role in the spread or rise of COVID-19 cases.

- Data understanding: Understanding the data is the second phase of the cycle. Knowing which dataset to collect and comprehending the data are both part of this phase. For this study, two datasets were compiled and combined. Weather is the first dataset, and COVID-19 cases are the second. The information was gathered between April 27, 2020 and September 9, 2021.
- Data preparation: Data preparation, which includes data transformation, aggregation, and other data pre-processing, is the most important and time-consuming stage in this process. Cleaning data and filling in missing numbers, as well as data processing, are all part of this process.
- Data modeling/Analysis: The fourth step is data modeling and analysis, which entails analyzing the paper's domain in order to select the best and most relevant modeling method for the data. The correlation of each variable is discovered using the Spearman rank correlation method after the EDA for the dataset is completed. This aids in the discovery of major factors influencing daily COVID-19 cases in India. After that, using the ARIMA model, the significant variables are used to develop a predictive model to forecast the daily cases.
- Data Evaluation: This procedure entails assessing all of the models that have been prepared and selecting the best model to deploy. The ARIMA model was the sole one employed in this study to create a predictive model for daily COVID-19 cases forecasting.
- Data visualization: Data visualization is the final step in this cycle. The data is represented in this phase to represent each data in a dashboard, and the results are then analyzed to see if the model is performing properly or not.

4 Experiment and Result Analysis

Data analysis is the process of putting order, structure, and importance into the information gathered. Data analysis aims to identify patterns or regularities by observing, examining, organizing, modifying, and modeling the gathered data. It supports the decision-making process by assisting in the production of important insights, and the formation of conclusions. This method of organizing and summarizing data is also used to get answers to questions that will be used to determine whether or not the hypothesis is correct. Exploratory data analysis is a crucial component of data analysis, as it involves comprehending and uncovering the connections between variables within the dataset.

Data interpretation is the process of giving meaning to the processed and analyzed data. It enables us to reach informed and meaningful judgments, deduce the significance of variable relationships, and interpret data patterns.

	A	B	C	D	E	F	G	H
1	Date	State	District	Confirmed	Recovered	Deceased	Other	Tested
2	#########	Andaman a	Unknown	33	11	0	0	
3	#########	Andhra Pra	Anantapur	53	14	4	0	
4	#########	Andhra Pra	Chittoor	73	13	0	0	
5	#########	Andhra Pra	East Goda	39	12	0	0	
6	#########	Andhra Pra	Guntur	214	29	8	0	
7	#########	Andhra Pra	Krishna	177	29	8	0	
8	#########	Andhra Pra	Kurnool	279	31	9	0	
9	#########	Andhra Pra	Prakasam	56	23	0	0	
10	#########	Andhra Pra	S.P.S. Nell	72	23	2	0	
11	#########	Andhra Pra	Srikakulam	3	0	0	0	
12	#########	Andhra Pra	Visakhapa	22	19	0	0	
13	#########	Andhra Pra	West Goda	51	10	0	0	
14	#########	Andhra Pra	Y.S.R. Kada	58	28	0	0	
15	#########	Arunachal	Lohit	1	1	0	0	
16	#########	Assam	Unknown	36	27	1	0	
17	#########	Bihar	Arwal	4	0	0	0	
18	#########	Bihar	Aurangaba	2	0	0	0	
19	#########	Bihar	Banka	2	0	0	0	
20	#########	Bihar	Begusarai	9	1	0	0	
21	#########	Bihar	Bhagalpur	5	1	0	0	
22	#########	Bihar	Bhojpur	2	1	0	0	
23	#########	Bihar	Buxar	25	1	0	0	
24	#########	Bihar	East Cham	5	0	0	0	
25	#########	Bihar	Gaya	6	5	0	0	
26	#########	Bihar	Gopalganj	12	3	0	0	
27	#########	Bihar	Jehanabad	1	0	0	0	
28	#########	Bihar	Kaimur	14	0	0	0	
29	#########	Bihar	Lakhisarai	1	1	0	0	

Fig. 1 Weather factors dataset

4.1 Data Preprocessing

Data preparation is the process of transforming raw data into a format that can be understood. Make sure the data is of good quality before utilizing machine learning or data mining techniques. Two datasets were obtained and combined in this work using Excel. The first is the district-by-district COVID-19 cases dataset, which provides district-by-district data on confirmed, recovered, and dead COVID-19 cases. Weather data is the second dataset, which contains information on weather parameters. As a pre-processing step, these two datasets are blended into one (Figs. 1 and 2).

As part of data cleaning, missing or null value check has been done on the dataset. It was discovered that there were some instances of missing values in the dataset. So drop the column that contains the null value and check for missing values again.

4.2 Exploratory Data Analysis

With the use of statistical measures and graphical representations, exploratory data analysis relates to the crucial process of doing first investigations on data to uncover patterns, spot irregularities, test hypotheses, and check assumptions (Fig. 3).

This visualization shows that there is no null or missing values in the dataset (Figs. 4 and 5).

	A	B	C	D	E	F	G	H	I	J	K
1	Date	Confirmed	cases_per_day	T2MDEW	T2M_MAX	T2M_MIN	T2M	QV2M	RH2M	PRECTOTC	PS
2	26-04-2020	5407		18.88	39.24	24.29	31.06	13.92	53.69	0.01	99.77
3	27-04-2020	5776	369	20.85	38.71	25.6	31.34	15.56	57.94	0.04	99.91
4	28-04-2020	6169	393	20.16	42.44	26.13	32.91	15.14	53.44	0.11	99.76
5	29-04-2020	6644	475	19.93	42.33	26.52	33.07	15.14	53.56	0.23	99.69
6	30-04-2020	7061	417	20.72	41.15	26.49	32.11	15.62	56.81	0.34	99.83
7	01-05-2020	7812	751	21.9	38.98	26.62	31.5	16.66	60.94	1	99.85
8	02-05-2020	8359	547	22.13	39.56	26.33	31.76	16.91	61	1.63	99.69
9	03-05-2020	8800	441	21.05	40.13	26.48	32.23	15.99	57.56	0.49	99.68
10	04-05-2020	9310	510	19.88	39.65	25.37	31.63	14.95	56.5	0.06	99.79
11	05-05-2020	9945	635	22.02	40.69	25.86	32.4	16.91	60.44	0.26	99.7
12	06-05-2020	10714	769	20.6	41.25	26.15	33.02	15.75	55.62	0.07	99.6
13	07-05-2020	11394	680	19.05	41.48	24.95	32.52	14.28	52.69	0	99.63
14	08-05-2020	12142	748	19.15	40.77	25.27	31.99	14.28	53.5	0	99.69
15	09-05-2020	12864	722	19.38	41.58	25.74	32.31	14.53	53.38	0.01	99.7
16	10-05-2020	13739	875	20.03	41.2	25.68	32.27	15.01	54.69	0.06	99.72
17	11-05-2020	14521	782	19.22	41.12	26.4	32.51	14.16	49.62	0.75	99.72
18	12-05-2020	14947	426	18.73	40.29	25.36	32.02	13.98	52.25	0.08	99.69
19	13-05-2020	15747	800	21.53	39.35	25.97	31.89	16.3	58.5	0.06	99.6
20	14-05-2020	16738	991	21.87	41.44	27.35	33.32	16.66	55.38	0.49	99.52
21	15-05-2020	17671	933	20.83	38.38	28.33	32.91	15.56	51.12	0.87	99.52
22	16-05-2020	18555	884	21.2	37.36	28.24	32.33	15.93	54.12	0.32	99.52
23	17-05-2020	20150	1595	21.4	38.14	27.91	32.18	16.24	56.31	0.29	99.49
24	18-05-2020	21335	1185	21.34	38.08	27.66	32.13	16.24	56.94	0.12	99.4
25	19-05-2020	22746	1411	21.48	38.1	27.15	31.71	16.3	58	0.04	99.38
26	20-05-2020	24118	1372	21.57	39.11	26.43	31.8	16.36	58.44	0.01	99.36
27	21-05-2020	25500	1382	22.1	39.51	27.3	32.21	16.91	59	0	99.48
28	22-05-2020	27251	1751	21.28	39.46	27.59	32.42	16.17	56.44	0	99.58
29	23-05-2020	28817	1566	20.61	41.12	25.81	32.5	15.75	57.19	0.01	99.55

Fig. 2 Merged dataset

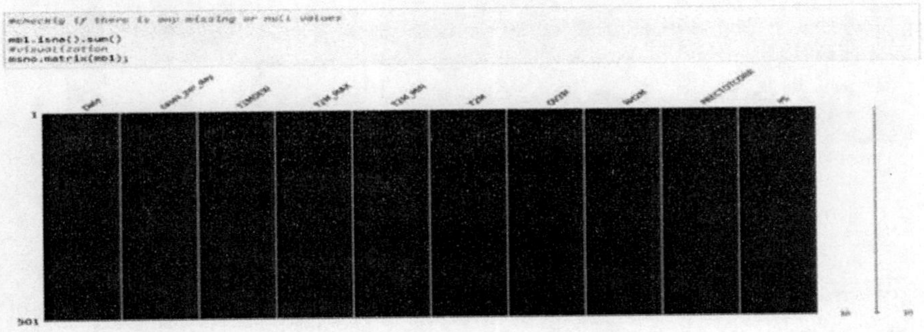

Fig. 3 Visualizing missing or null values

4.3 *Model Building*

4.3.1 Mumbai

The first step in model building is to find the important variables that have some impact on the dependent variable. In this paper, the dependent variable is cases per day. Therefore, a correlation of all the factors with the target variable should be found.

```
# Plotting Correlation Matrix using Heatmap

plt.figure(figsize = (16,14))
sns.heatmap(mb1.corr(), annot =True);
```

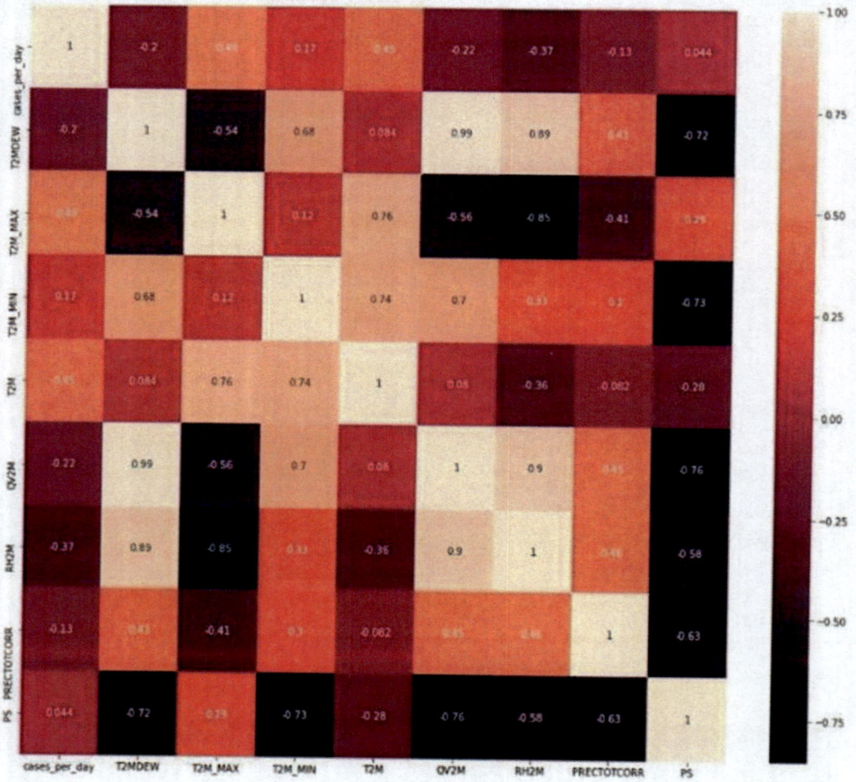

Fig. 4 Correlation between variables in heatmap

The above figure clearly states that only the temperature (T2M) has a significant impact on the cases per day variable. So now, leaving other variables behind, only the temperature variable is needed to build the model. Given that the P-value associated with temperature is less than 0.05, it can be concluded that there is a positive correlation between temperature and the number of daily cases. The above graph shows the forecasted value of corona cases from September 2021 to February 2022 (Figs. 6 and 7).

4.3.2 Chennai

The first step in model building is to find the important variables that actually have some impact on the dependent variable. In this paper, the dependent variable is cases per day. Therefore, a correlation of all the factors with the target variable should be

```
import matplotlib.style as style

#Creating the Line chart
plt.plot(mb1['Date'], mb1['cases_per_day'])
#Adding the aesthetics
plt.xlabel('Date')
plt.ylabel('Confirmed Cases per day')

#Show the plot
plt.show()
```

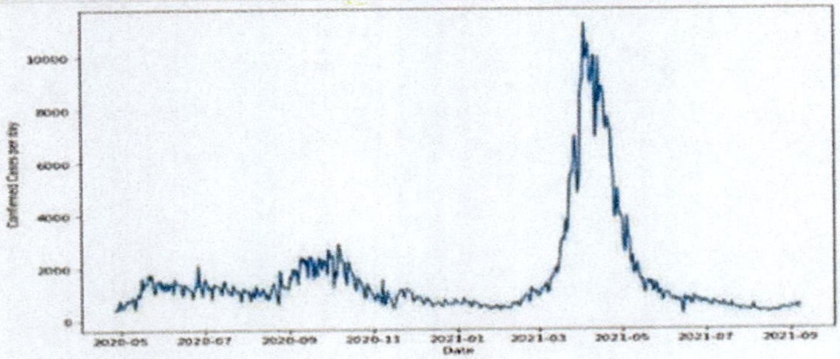

Fig. 5 Line graph (Data Vs cases per day)–coding snippet

```
mb1.corr(method="spearman")
```

	cases_per_day	T2MDEW	T2M_MAX	T2M_MIN	T2M	QV2M	RH2M	PRECTOTCORR	PS
cases_per_day	1.000000	0.002048	0.328064	0.281577	0.510969	0.007404	-0.165322	-0.016271	-0.110775
T2MDEW	0.002048	1.000000	-0.585207	0.818577	0.114745	0.999473	0.861965	0.856312	-0.824766
T2M_MAX	0.328064	-0.585207	1.000000	0.068091	0.612855	-0.582181	-0.877785	-0.637207	0.427621
T2M_MIN	0.281577	0.818577	0.068091	1.000000	0.728567	0.823957	0.258318	0.572070	-0.702826
T2M	0.510969	0.114745	0.612855	0.728567	1.000000	0.123811	-0.300037	0.086269	-0.312683
QV2M	0.007404	0.999473	-0.582181	0.823957	0.123811	1.000000	0.860341	0.851535	-0.835763
RH2M	-0.165322	0.861965	-0.877785	0.258318	-0.300037	0.860341	1.000000	0.813305	-0.679923
PRECTOTCORR	-0.016271	0.856312	-0.637207	0.572070	0.086269	0.851535	0.813305	1.000000	-0.836508
PS	-0.110775	-0.824766	0.427621	-0.702826	-0.312683	-0.835763	-0.679923	-0.836508	1.000000

```
mb2 = mb1[['cases_per_day','T2M']].dropna()
```

```
mb2.describe()
```

	cases_per_day	T2M
count	500.000000	500.000000
mean	1525.686000	27.373220
std	1673.266001	2.885031
min	161.000000	20.900000
25%	580.250000	26.005000
50%	984.000000	26.995000
75%	1467.750000	28.667500
max	11206.000000	34.650000

```
from scipy.stats import spearmanr
spearmanr(mb2)
```

```
SpearmanrResult(correlation=0.5109687757474499, pvalue=1.3234763750800547e-34)
```

Fig. 6 Spearman correlation–Mumbai city

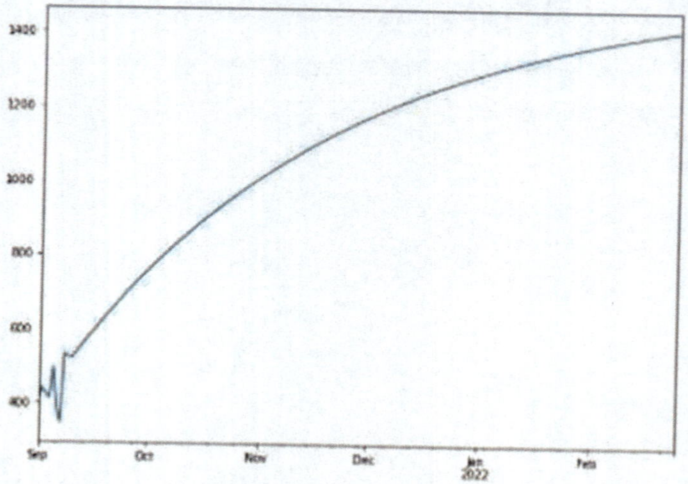

Fig. 7 Output (Mumbai)

found. The below graph shows the forecasted value of corona cases from September 2021 to February 2022 (Figs. 8 and 9).

4.3.3 Bengaluru Urban

The first step in model building is to find the important variable that has some impact on the dependent variable. In this paper, the dependent variable is cases per day. Therefore, a correlation of all the factors with the target variable should be found. The below graph shows the forecasted value of corona cases from September 2021 to February 2022 (Figs. 10 and 11).

4.3.4 Findings

From this study, it is found that the average temperature in Mumbai has a positive correlation with the corona cases per day in Mumbai. This means that with an increase in temperature, the corona cases are also increasing. In Chennai, the daily corona cases show a positive correlation with the maximum and minimum temperatures in the city. In Bengaluru urban, the weather factors have no significant relation with the daily cases of COVID-19 in the city.

```
mb1.corr(method="spearman")
```

	cases_per_day	T2MDEW	T2M_MAX	T2M_MIN	T2M	QV2M	RH2M	PRECTOTCORR	PS
cases_per_day	1.000000	0.002048	0.328064	0.281577	0.510969	0.007494	-0.165322	-0.016271	-0.110775
T2MDEW	0.002048	1.000000	-0.585207	0.616577	0.114745	0.999473	0.861965	0.856312	-0.824766
T2M_MAX	0.328064	-0.585207	1.000000	0.068091	0.612855	-0.582181	-0.877785	-0.637207	0.427621
T2M_MIN	0.281577	0.616577	0.068091	1.000000	0.728567	0.623957	0.258318	0.572070	-0.702826
T2M	0.510969	0.114745	0.612855	0.728567	1.000000	0.123811	-0.300037	0.086269	-0.312683
QV2M	0.007494	0.999473	-0.582181	0.623957	0.123811	1.000000	0.860341	0.861535	-0.835763
RH2M	-0.165322	0.861965	-0.877785	0.258318	-0.300037	0.860341	1.000000	0.813305	-0.679923
PRECTOTCORR	-0.016271	0.856312	-0.637207	0.572070	0.086269	0.861535	0.813305	1.000000	-0.836508
PS	-0.110775	-0.824766	0.427621	-0.702826	-0.312683	-0.835763	-0.679923	-0.836508	1.000000

```
mb2 = mb1[['cases_per_day','T2M']].dropna()
```

```
mb2.describe()
```

	cases_per_day	T2M
count	500.000000	500.000000
mean	1525.686000	27.373220
std	1873.266001	2.885031
min	161.000000	20.800000
25%	580.250000	26.005000
50%	884.000000	26.985000
75%	1467.750000	28.667500
max	11206.000000	34.850000

```
from scipy.stats import spearmanr
spearmanr(mb2)
```

```
SpearmanrResult(correlation=0.5109687757474499, pvalue=1.3234763750800547e-34)
```

Fig. 8 Correlation between variables in heatmap

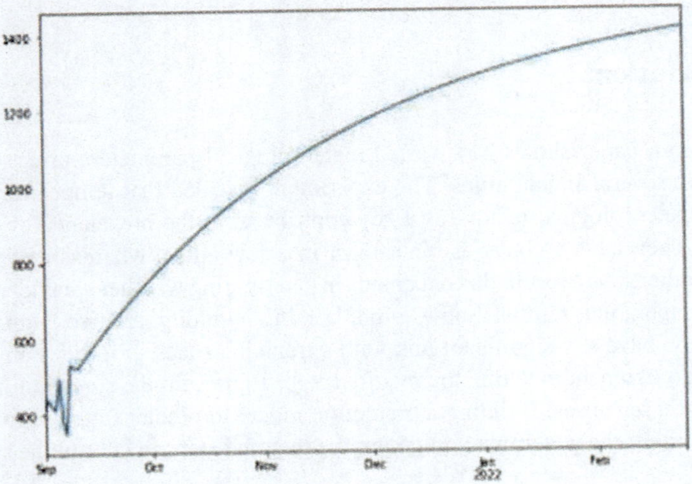

Fig. 9 Line graph (Data Vs cases per day)

```
blr1.corr(method="spearman")
```

	Cases_per_day	T2MDEW	T2M_MAX	T2M_MIN	T2M	QV2M	RH2M	PRECTOTCORR	PS
Cases_per_day	1.000000	0.230285	-0.061089	0.185472	0.125207	0.231215	0.147848	0.164998	-0.169863
T2MDEW	0.230285	1.000000	-0.412719	0.472880	0.041171	0.999519	0.698667	0.743066	-0.552138
T2M_MAX	-0.061089	-0.412719	1.000000	0.342121	0.794592	-0.405440	-0.881376	-0.403528	0.099858
T2M_MIN	0.185472	0.472880	0.342121	1.000000	0.760054	0.482667	-0.115577	0.353572	-0.519269
T2M	0.125207	0.041171	0.794592	0.760054	1.000000	0.051474	-0.584257	-0.020198	-0.276366
QV2M	0.231215	0.999519	-0.405440	0.482667	0.051474	1.000000	0.691640	0.743956	-0.566241
RH2M	0.147848	0.698667	-0.881376	-0.115577	-0.584257	0.691640	1.000000	0.607800	-0.275881
PRECTOTCORR	0.164998	0.743066	-0.403528	0.353572	-0.020198	0.743956	0.607800	1.000000	-0.462114
PS	-0.169863	-0.552138	0.099858	-0.519269	-0.276366	-0.566241	-0.275881	-0.462114	1.000000

Fig. 10 Spearman correlation

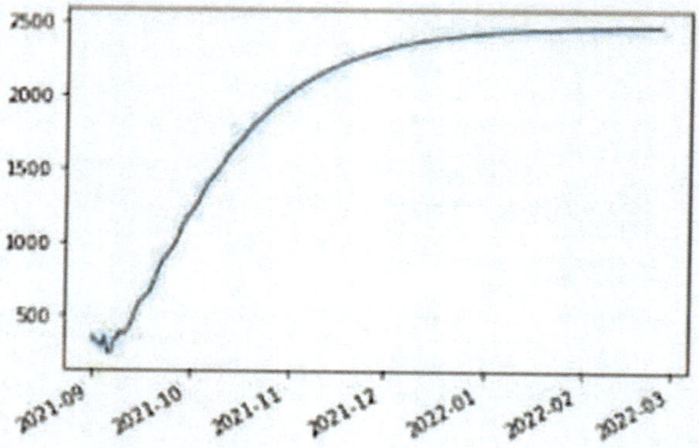

Fig. 11 Output (Bengaluru)

5 Conclusion

This research paper shows how varied meteorological parameters affect COVID-19 cases in several Indian cities. The experiment revealed that temperature, above all other meteorological factors, is a key contributor to the prevalence of COVID-19 cases. They have a considerable impact in a few cities, whereas they have no impact in others. As a result, this paper aids in determining whether weather elements such as temperature, rainfall, snow, wind, specific humidity, relative humidity, and precipitation have any significant link with corona instances. If there is any, it may assist the government in taking the required steps to prevent the spread of instances caused by these causes. Building a prediction model to predict future corona cases could also assist the government in taking the required steps to control the spread of cases.

References

1. Harmooshi N, Shirbandi K, Rahim F (2020) Environmental concern regarding the effect of humidity and temperature on 2019-nCoV survival: fact or fiction. Environ Sci Pollut Res Int
2. Das M, Ghosh A, Ghosh S (2021) Does climate variability impact COVID-19 outbreak? An enhanced semantics-driven theory-guided model. SN Comput Sci
3. Santiago G-H, Erik SJ, Sandra ME-D, André HR (2021) Distinct weather conditions and human mobility impacts on the SARS-CoV-2 outbreak in Colombia: application of an artificial neural network approach. Int J Hyg Environ Health
4. Bolaño-Ortiz TR, Camargo-Caicedo Y, Puliafito SE, Ruggeri MF, Bolaño-Diaz S, PascualFlores R, Cereceda-Balic F (2020) Spread of SARS-CoV-2 through Latin America and the Caribbean region: a look from its economic conditions, climate and air pollution indicators. Environ Res
5. Rosario D, Mutz Y, Bernardes P, Conte-Junior C (2020) Relationship between COVID-19 and weather: case study in a tropical country. Int J Hyg Environ Health
6. Auler AC, Cássaro FA, Da Silva VO, Pires LF (2020) Evidence that high temperatures and intermediate relative humidity might favor the spread of COVID-19 in tropical climate: a case study for the most affected Brazilian cities. Sci Total Environ
7. Supari S, Danang EN, Amsari MS, Furqon A, Ardhasena S, Wido H, Yunus S (2021) The association between initial COVID-19 spread and meteorological factors in Indonesia. Environ Sustain
8. Ran X, Rahmandad H, Marichi G, Catherine D, Navid G, Heresh A, Mohammad SJ (2021) Weather, air pollution, and SARS-CoV-2 transmission. Lancet Planet Health
9. Zarei M, Rahimi K, Hassanzadeh K, Abdi M, Hosseini V, Fathi A, Kakaei K (2021) From the environment to the cells: an overview on pivotal factors which affect spreading and infection in COVID-19 pandemic. Environ Res.
10. Karimi SM, Majbouri MP, DuPré NS, White KB, Little BB, McKinney WP (2021. Weather and COVID-19 deaths during the stay-at-home order in the United States. J Occup Environ Med
11. Shiv TS, Justin DS, Douglas JW, Shelby F, Joshua FB (2020) Maximum daily temperature, precipitation, ultraviolet light, and rates of transmission of severe acute respiratory syndrome coronavirus 2 in the United States. Clin Infect Dis
12. Muhammad R, Wan Mohd RI (2021) New COVID-19 variant (B.1.1.7): forecasting the occasion of virus and the related meteorological factors. J Infect Public Health 1320–1327

A Review on Scheduling in Cloud Fog Computing Environments

M. Santhosh Kumar and Ganesh Reddy Karri

1 Introduction

Cloud computing is the transfer of calculating organizations, for example, waiters, gathering, informational indexes, arranging, programming, examination, and comprehension, over the Internet ("the cloud") to offer quicker turn of events, adaptable resources, and economies of scale [1]. What's more, Cloud figuring is a development that involves the web placing ceaselessly, managing d accessing data through the web.

Fog computing is a decentralized figuring establishment where data, registers, amassing, and applications are seen in some place near the data source and the cloud. The dimness portrayal comes from the meteorological term for a cloud close to the ground, also as haze centers around the edge of the network [2]. Fog Computing additionally fosters efficiency and diminishes how much data is transported to the cloud for dealing with, assessment and limiting. This is much of the time done to additionally foster efficiency, but it could similarly be used for security and consistency reasons.

Cisco proposed the Fog figuring worldview concept. When the fog is a cloud close to the ground or the edge, a fog-figure relationship exists [2]. Fog is used as a contraction for "From core to edge" in some works [3]. The estimation and limit capacities are communicated in frameworks, organization gadgets, and at various levels of the association design in fog registering. These devices have schedulers that determine whether a request should be served at Fog center points or Cloud Datacenters (DCs). This type of decision is usually made on distant servers, based on a variety of elements related to sales and available resources in the Fog and Cloud. Fog enables Internet of Things (IoT) applications by moving computational and organizational resources closer to edge devices. As a result, organizations and

M. S. Kumar (✉) · G. R. Karri
VIT-AP University Amaravathi, Vijayawada, A.P 522237, India
e-mail: santhosh.21phd7113@vitap.ac.in

A. Ramdane-Cherif et al. (eds.), *Machine Intelligence and Data Science Applications*, Algorithms for Intelligent Systems, https://doi.org/10.1007/978-981-99-1620-7_3

forecasts can stay as near to end users and IoT devices that necessitate constant computational constraints as possible while remaining sluggish.

People and associations are a lot dependent on PCs and astute handheld gadgets to manage their everyday tasks [4]. These gadgets produce a gigantic heft of information and are put away by the association for future investigation. Because of the quick expansion in the IoT area, information produced by the sensors and applications has greatly expanded. These are the cloud's primary impediments. In 2012, investigations in contrast to Cisco provided another model, named haze registering [5].

Fog computing is a type of long-term distributed computing characterized as "it is a colossally virtualized stage that upholds capacity, information, process, and systems administration between the customary cloud server and edge gadgets" [6]. "Haze registering is additionally portrayed as countless remote and an autonomous inescapable and decentralized gadget to speak with the organization to play out the activity of putting away and change undertakings without the cooperation of the outsider".

The basic purpose of fog computing is to reduce the amount of data used in computing information handling, investigation, and stockpiling while increasing productivity. The fundamental model of haze registering is depicted in Fig. 1. To see the edge device, mist hub, and cloud are the critical processing elements in a mist registering model. Any device with capacity, measurement, and system administration capacities can function as a haze hub. Switches, switches, mobiles, base stations, and other processing devices are examples of end devices. The server in the cloud manages haze hubs, which are in charge of deciphering administration between edge gadgets and the cloud [7].

In cloud computing, scheduling refers to the process of allocating several jobs to several virtual machines (VMs) or designating VMs to execute on available assets to satisfy the needs of clients [8]. The purpose of adopting booking mechanisms in cloud climates is to improve framework throughput and burden balance, increase asset use, save energy, reduce costs, and shorten total handling time. As a result, to achieve effective matching between positions and assets, the scheduler should take into account the virtualized assets as well as the clients' expected limitations [9]. At least one system should be the foundation of each planning procedure. The

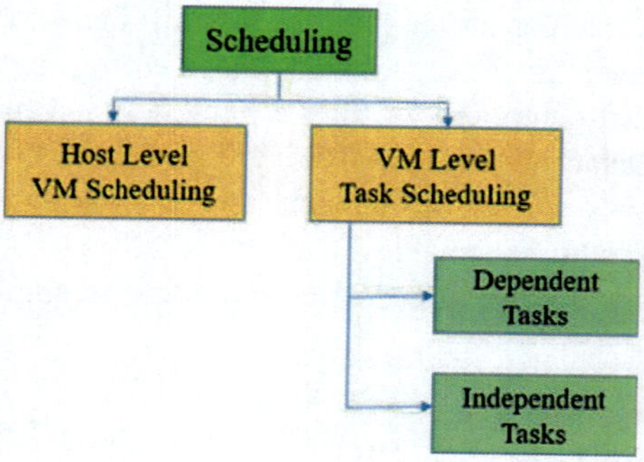

Fig. 1 Scheduling levels

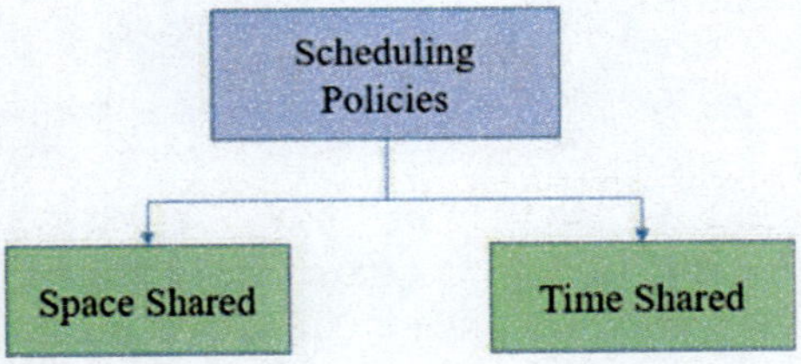

Fig. 2 Scheduling policies

most commonly used systems or goals are time, cost, energy, QoS, and adaptation to non-critical failure [10].

Cloud computing resources as scheduled at two levels, to see in the diagram below [11]: VM-level and Host-level. At the VM-level, a task/job scheduler is utilized to map tasks for execution to the allotted VMs. It is referred to as Task Scheduling. At the host level, a VM scheduler is used to allocate VMs to physical hardware. This form of interaction is referred to as VM Scheduling.

As depicted in Fig. 2. For VM planning and Task Scheduling, the two layouts are used. In a space-shared booking method, only one virtual machine is allowed to run on a given host/VM at any given time. Numerous VMs/tasks can do multiple tasks and operate simultaneously utilizing the Time-Shared planning technique within a host it allotted time to all VMs and tasks [12, 13].

In this work, to study the characteristics of fog computing and problems, as well as the examination table, we compare the exhibition metric and conclusion.

The fundamental target of this paper is to introduce an outline of planning instruments in the cloud and haze processing framework. It will examine attributes of scheduling, procedures, and types. The remainder of the paper is coordinated as follows, Sect. 2 will explain the characteristics of fog computing. Section 3 will discuss the challenges of cloud computing. Section 4 will define the obstacles and problems that exist and these are looked at in fog computing. Section 5 will explain the techniques and types of scheduling. Section 6 will explain literature survey. Section 7 will explain the comparative study of various existing scheduling algorithms. Finally, Sect. 8 will explain the conclusion.

2 Fog Computing Characteristics

Fog computing is a type of cloud computing that is beneficial to items that are linked to IoT data, and it is also forming a cloud computing square [14]. Figure 3 depicts the evident amount of fog processing applied perspective. A three-layered architecture is created by combining fog cloud and edge computing. The three-layered are shown below in Fig. 3

1. Edge layer: It is made up of terminal hubs, implanted frameworks, sensors, and actuators that have extremely limited calculation, energy, and transmission capability.

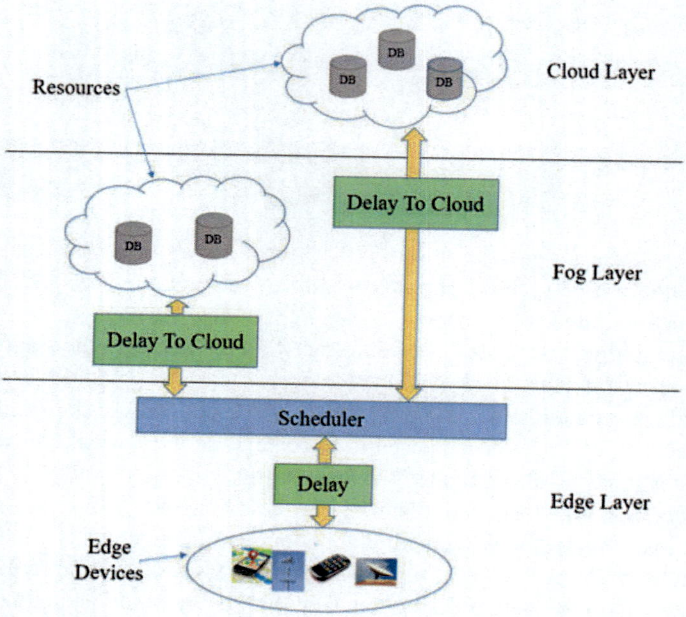

Fig. 3 Cloud-fog computing high-level architecture

2. Fog layer: It has middle-of-the-road organizing devices like entryways, switches, and routes that work on a variety of networks including 3G, 4G, LTE, and Wi-Fi. The computational and limit capacities of these devices are intact.

3. Cloud Layer: this layer incorporates a cloud server cultivated with extraordinarily rich virtual capacities for managing limits and impacts.

The following are some of the key properties of fog computing.

2.1 *An Area on the Outskirts*

The fog processing is transmitted at the breaking point, so region care is entwined it retains end-contraption with the rich assistance of an association consolidating the application with low-idleness requirements, such as gaming.

2.2 *Heterogenetic*

To enroll and grant accumulating, network connectivity between that the end client and the cloud server, haze figuring is frequently virtualized. Both fog and cloud are traditionally constructed using network resources [15].

2.3 *Topographical Dissemination*

The hazy need is widespread and is created by concentrating on applications and organizations. The darkness is critical in delivering tremendous streams of information from fog to moving cars, i.e., in following the information [15].

2.4 *Sensor Network on a Large Scale*

Different models in the suitable system [15] maintain track of and monitor the climate and brilliant matrix's capacity assets.

2.5 *Assistance with Mobility*

An application must have convenience sponsorship to communicate with end devices that use the LISP show and have a different host than the region character [16].

3 Challenges of Cloud Computing

3.1 *Information Security and Privacy*

Information security is one of the most severe challenges in the migration to distributed computing. Client or authoritative data storage in the cloud is simple and safe. It is your responsibility to complete customer verification and approval, personality the board, data encryption, and access control, regardless of whether the cloud specialised organization ensures data respectability. Concerns about cloud security include fraud, data breaches, malware infections, and other issues that lead people to lose trust in your company. As a result, there is a possibility of a drop in income near notoriety and height. Similarly, controlling distributed computing necessitates delivering and receiving massive amounts of data at breakneck speeds, leaving it vulnerable to data leaks.

3.2 *Management of Costs*

Even though the majority of cloud-specific cooperatives follow the "Pay More Only When Fees Emerge" concept, which reduces the overall cost of the resources employed, there are moments when distributed computing incurs large costs. Hidden costs are incurred when assets are not exploited to their full potential, such as when servers are underutilized. If there is poor application execution, or if there are sudden spikes or overages in utilization, it adds to the overall cost. Underutilized assets are another major cause of cost increases. If you enable administrations or a cloud, for example, and fail to switch it off at the end of the week or when not in use, the expense will climb despite the assets not being employed.

3.3 Environments with Multiple Clouds

Because of the increased possibilities available to them, organizations use a single cloud as well as rely on several cloud-specialized cooperatives. A sizable proportion of these firms use hybrid cloud techniques, with more than 84 percent relying on different mists. For the most part, this is complicated to handle for the foundation group. The cycle is frequently quite hard for the IT department due to the variances between different cloud providers.

3.4 Execution Challenges

Execution is critical when it comes to cloud-based solutions. Customers may be turned off and their benefits may be limited if the cloud's presentation is bad. Even a minor lapse in concentration while stacking an application or a website page can lead to a considerable drop in customer satisfaction. This sluggishness could be caused by inefficient load balancing, which shows that the server is unable to split approaching traffic properly to deliver the best client experience. Challenges can also develop as a result of internal failure adaptability, which implies that activities continue as needed even when at least one of the components fails.

3.5 Interoperability and Flexibility

When a corporation uses a cloud-specific co-op and needs to switch to another cloud-based arrangement, the process can be time-consuming because apps created for one cloud using the application stack must be re-composed for the other. Because of the difficulties involved, there is a lack of adaptability when shifting from one cloud to the next. Dealing with information development, establishing security without prior planning, and organization all contribute to the difficulties encountered while altering cloud arrangements, and reducing adaptability.

3.6 *High Dependence on Network*

Cloud computing manages enormous quantities of data going to and from computers since asset provisioning is controlled incrementally. This is simply made possible by the high-speed organization's accessibility. Even though this information and assets are traded around the organization, it may become incredibly helpless in the event of constrained data transport or an unforeseen interruption. In any case, when companies reduce their equipment expenses, they must ensure that the online data transmission capacity is high ensuring that there are no organizational blackouts, or else money would be lost. As a result of this, it is difficult for smaller enterprises to keep up with network data transmission, which is costly.

4 Challenges of Fog Computing

4.1 *Programming Models*

Versatile Edge Computing (VEC) offloads tasks to edge devices, allowing for the creation of useful, flexible, and versatile edge-based applications. Because the application is broken, the offloading option has been altered. The offloading foundation's objectives are to (1) achieve incredible processing speed, (2) save energy, (3) save bandwidth, and (4) achieve low torpidity [16].

4.2 *Security and Reliability*

The primary test is to keep security visible in Fog while calculating an appropriate system like Fog. Confirmation at discrete levels of fog center locations poses a security issue as well. Another test addressed by the use of a sandbox is Trust and Privacy. In fog, the application licenses pariah hardware/software to protect the client data [16].

4.3 *Resource Management*

It forecasts end-device asset usage and allocates resources based on client behaviour and the likelihood of usage when the asset is utilized, allowing for increased efficiency and decency. The mist is supported by a flexible administrative component. A billion little gadgets must be arranged [16].

4.4 Job Scheduling

A basic errand should deal with the structure's enduring quality and versatility in fog, according to task scheduling. The main reason for assigning a position to the resource with a certain time limit is to find the best ideal course of action by carrying out the numerous scenarios with complete and optimal groups According to the job preparing for fog management, determines the best available resource for task execution or PC allocation [16].

5 Scheduling Techniques and Types

5.1 Scheduling Techniques

In general, scheduling is defined as the assignment of errands to available resources at a specific time. Typically, a booking issue is dependent upon numerous requirements and targets that should be satisfied. The booking issue can be planned as an advancement issue where the planning issue goals can be, for example, limiting dormancy and augmenting asset usage. The work in [17] indicates that the majority of planning issues comprise four fundamental components:

1. Assets: are physical/sensible objects capable of carrying out or managing tasks.

2. Tasks: These are physical/sensible tasks that the assets must complete.

3. Constraints: These are the prerequisites for organizing tasks into resources. They can be based on mobility, tasks, resources, or a mix of the three. They could also be hard needs, suggesting that goals must be completed before weak constraints are lifted.

4. Objectives: these are the evaluation criteria that should be utilized to assess the presence of the timetable.

To observe an answer for a booking issue, there are two general classes of strategies: precise techniques and heuristic strategies. Careful techniques track down indisputably the ideal answer for the planning issue. Instances of accurate calculations incorporate Simplex [18] and Branch-and-Bound [19, 20]. Then again, heuristic strategies don't ensure tracking down the ideal arrangement. Regardless, they can observe a solution with when compared to the time expected by an accurate approach, some level of optimality in a tolerable computation time Heuristic computations include repeated reinforcement [21, 22], genetic algorithms [23–25], and ant colony algorithms [26].

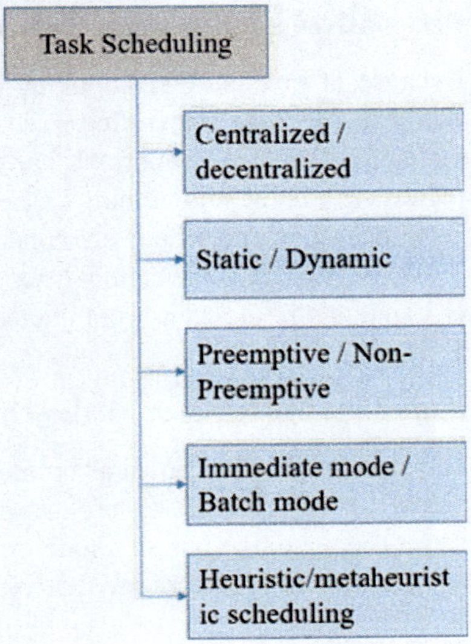

Fig. 4 Scheduling types

5.2 *Scheduling Types*

Figure 4 depicts the many forms and categories of scheduling in a cloud computing system Scheduling Types.

These are the types described in the following [27, 28].

Centralized/Decentralized Scheduling

Choices are made in a focal hub when scheduling is centralized. It ensures the efficiency and simplicity of asset checking. It's a lack of adaptability and adaptation to internal failure in any scenario. Despite the lack of productivity, decentralized or communicated planning is becoming prevalent in actual cloud environments.

Static/Dynamic Scheduling

All arranging information about the jobs is available before executing any effort in static scheduling, thus the execution timetable of each work is enlisted. It is effective in applications with predictable demands. Furthermore, in static booking, the customer chooses simultaneity for organizations with the cloud provider, and the cloud provider prepares standard resources before the start of required support [29].

The situation information about the tasks isn't known at runtime in the powerful booking. As a result, the undertaking's execution timetable may alter depending on the client's preferences. Dynamic booking outperforms static planning in terms of runtime. The cloud provider cannot plan ahead of time due to exceptional planning. It distributes and kills resources as needed [30].

Preemptive/Non-preemptive Scheduling

Because of precautionary booking licenses, every task is disrupted during execution, causing the endeavor to be redirected to yet another source For example when one activity is more important than another and must be prioritized and completed regardless of how it is running in the virtual machine. Similarly, when goals such as need, deadline, and money are constrained, this type of planning is essential.

The virtual machine cannot be withdrawn in non-precautionary booking until the task that is running on it is completed. It prevents interruptions to the task while it is in process.

Immediate Mode/Batch Mode Scheduling

In quick mode, also known as online mode, chores are designed to be completed as quickly as possible. Tasks are assigned for a specific period and cannot be modified. Of course, the bundle mode groups tasks into a set and examine them for the organization at predetermined intervals. It's also known as an unconnected mode or detached mode.

Heuristic/Metaheuristic Scheduling

Heuristic scheduling techniques are issue subordinated that can take care of explicit issues. Metaheuristic methods are undeniable level issue-free procedures that give ace methodologies to take care of general issues and can be applied to a wide scope of issues. Both planning strategies don't ensure that they will track down the ideal arrangement, yet they furnish great arrangement when contrasted and the time spent on calculation.

6 Literature Survey

Work planning based on the Berger model estimation was presented by Baomin Xu et al. (2011) to reduce finish time and boost bandwidth. The suggested approach directs resources to the task is completed in the shortest amount of time possible on the virtual machine [4].

Celal Ozturk and colleagues (2015) as a high-level variation of the non-continuous binary artificial colony technique, developed a differential development-based computation. In rundown quality states, this estimate has been attempted as a benchmark. The proposed structure aims to chip away at the show by differentiating the precision of the inherited computation and the proposed approach [5].

Ranieri Baraglia et al. (2010) examined the value to determine if the machine and work were compatible. They presented a straightforward and adaptable topping-off estimate. The scheduler will most likely game plan showing up positions based on the submission deadline, work on Hardware and programming resources, as well as the associated costs They reduced the execution time to save energy in half and used a less powerful machine [6].

Salim Bitam et al. (2017) proposed to split the difference between CPU execution time and RAM distribution, and the Bees Life Algorithm (BLA) is used. The proposed system was able to achieve with a limited execution time, differentiated and particle swarm optimization (PSO) and genetic algorithms were used to achieve the target genetic algorithm (GA) [7].

A hereditary calculation was introduced by Bin Charm et al. (2017). To limit the makespan, the essential is not permanently set up by arranging position as well as the amount and location of RAM on the same computer. They were able to attain reaction time and handled the booking algorithm's presentation [31].

The creator, Subhadeep et al. (2015), researched the haze viewpoint and the aim, the sales are serto be gradual, with little dormancy, and submit the IoT application. The appropriateness of dimness figuring in terms of genuine time interest and the presentation against the required scattered handling was examined, and power utilization, organization inaction, and cost [14] were all assessed.

Ruilong Deng et al. (2016) The fog contraption was used to investigate the relationship between transmission latency and power consumption in delivered calculations, the cloudiness configuration supports flexibility, geo-scattering, region, and sustained client convey ability. The geo flow has the potential to provide high association setting information, with the eventual goal of being used in haze applications [32].

Fruity fly computation was proposed by Lawanyashri et al. (2017). This strategy emphasized the issue of pile shifting while simultaneously emphasizing the server ranches' diminishing energy and expense. The proposed technique resulted in optimal resource utilization while lowering energy consumption and costs [12].

Mayank Tiwari et al. (2018) provide a game mode where the player's outcome is enhanced, resulting in a limited response time This project concentrated on the verification of flexibility, which was assigned as a computational work, and it kept past data to aggregate device availability. In light of gadget availability, they've additionally encouraged shop transportation [13].

MahendraBhuta et al. (2018) developed a BAT + BAR streamlining technique for task booking and asset allocation, in which the aid is distributed and handling is compensated more as expenses arise. On a virtual machine, the proposed framework placed the endeavors and distributed them based on the weight and information move limits. They looked at both the estimation and the show metric to see if it had improved [33].

Energy-execution Trade-off Multi-Resource Cloud Task Planning was proposed by Li Mao et al. (2018). Its purpose is to reduce energy consumption. Similarly, the parameters influencing the virtual machine, host, and server's show and energy use were isolated. The task plan was created based on the client's specifications to maximize energy efficiency while saving the client time [34].

To address the issue of multi-asset planning, Mehdi Sheikhalishahi et al. (2015) presented a multi-limit mindful asset booking. Following that, with the open system and a specified extreme aim to handle resource use in a multi-resource scenario, the booking plan allowed any work to be selected while taking into account similar resource requirements. Execution improvement, wait-time, and log jam measures all yielded identical results [35].

For the competing needs, Fan Zhan et al. (2014) presented the goal of multi-objective scheduling (MOS) is to redesign the runtime conditions. The MOS approach resulted in faster vertical booking and better execution. It could deal with a large amount of search space to solve a variety of task work process planning issues [36].

Yasmina Sedaka et al. (2017) presented disconnected staggered planning to address the issue of the criticality level of each occupation based on necessity and interest. They investigated how to recall a specific goal to that weighted number of completed occupations while still meeting the thriving requirement. To achieve higher criticality occupations, the proposed technique necessitates more dealing time at run time [37].

Prasanna Kumar et al. (2018) investigated the distributed computing situation, which provides the client with organized resources by delivering various help models. They presented a booking framework that considers server ranch the board when making arrangements for prosperity-related tasks. It was able to manage a group of jobs that can recover from a natural disaster [38].

The authors of [39] tended to regard Cloudlet as an unusual instance of Fog enlistment. It's a temporary layer that sits between each cell and the Cloud server ranches. Rather than traveling to the remote Cloud server ranch, the edge devices partner with the closest Cloudlet from afar. Fog figuring was also used in clever organizations, where each smaller than expected grid might be portrayed as a fog enrolling center point. Instead of communicating with the Cloud, clients communicate with local fog centers. To convey power organizations, murkiness contraptions collaborate with Cloud server homesteads and customers.

The work in [40] provided a major-level programming model for future Internet applications that is geospatially appropriated, has a large extension, and is dormancy sensitive. Fog, its resources are dedicated to assisting low-idleness enterprises, while permissive higher degree assistance is distributed in that Cloud data center.

Furthermore, the work in [41] suggested a four-layer clinical comparison model: a distinct layer that permits different real sensors to collect data, a sorting out layer that allows putting together sponsorship and data movement in wired and remote associations, and an organization layer that creates and manages a diverse range of organizations to meet client needs, and an association point layer that engages the correspondence between cl The work also clarified the framework's organization layer, which is divided into two sub-layers: the Fog layer, which maintains area buffering and other accessibility needs for IoT devices, and that the Cloud layer, which manages the Fog layer, client/contraption/data the board, and application organizations, such as dashboards, rule engines, and big data analysis, and coordination structure.

In the study [42] End users are connected to the Internet via Fog servers, which ensure that all web requests are routed through the Fog servers before reaching the Cloud's web servers. There are two web request association streams: (1) propelling site page for hidden sales and (2) an upgrading page for upcoming requests, where all essential records of the web objects are recorded and processed locally within the Fog server.

7 Comparative Study

In our comparative study, we have analyzed various cloud fog scheduling algorithms. With respect to their merits and de merits, we also analyzed that each algorithm with different performance merits at different levels such as high, low and average. Hybrid fruit fly optimization [12] algorithm is shown that optimized memory utilization. Easy and flexible backfilling [6] algorithm attains high accuracy in intrusion detection. The Sleeping Barber algorithm [33] achieves a good response time and average turnaround time. Dynamic programming Branch-and-bound, CPU intensive scheduling Data-intensive Scheduling [37, 38] both algorithms effectively utilization of resources during the scheduling. The Rest of the algorithms are not achieved high performance in various parameters such as Memory utilization, Bandwidth, Processing time, Transmission delay, Energy consumption and Cost, CPU utilization, Memory utilization, Waiting time, and optimal disk capacity (Table 1).

8 Conclusion

In this paper, we have discussed the challenges in both cloud and fog computing. First, we study the features of cloud and fog computing, and then we study the security requirements of both cloud and fog computing. In both cloud and fog computing, Quality of Service (QoS) depends on how well our existing scheduling algorithms schedule the IoT tasks. We have studied various cloud and fog scheduling algorithms, however, these algorithms fail to classify the tasks to be executed in cloud and fog environments. As a result, even though we had sufficient resources, the quality of service was not maintained. To overcome these limitations, we need to correctly classify the tasks to be executed at cloud and fog levels.

Table 1 Comparison of performance metrics-based scheduling algorithms

Refs. no	Input parameters	Algorithm	Merits	Demerits	Performances metric		
					High	Average	Low
[12]	Arrival time size of task	Hybrid fruit fly optimization	Improves availability, scalability	Imbalance in resource usage	Memory utilization	–	Energy consumption and cost
[13]	Execution time Arrival time	Non cooperative game model	Improve response time	Limited portable size	Response time	Memory utilization	Latency
[14]	Bandwidth delay time	Fog computing paradigm	Improve latency-sensitive Application	Large data traffic	–	Power consumption	Service latency
[4]	Task length file size computation time	Berger model	Improved efficiency	Lack of quality of service	–	Bandwidth	Complete time
[5]	Number of population Iteration criteria	Artificial bees colony optimization	It overcomes the local minimal problem	Lack of information and overlapping cluster problem	Intrusion detection	GA accuracy	k-mean accuracy
[6]	Number of CPU License ratio	Easy and Flexible backfilling	Improved quality of service and less cost	Resource utilization is not efficient	QoS	Execution time	Waiting time
[31]	CPU frequency length of jobs	Genetic algorithm	Less execution time	Poor makespan	-	Processing time	Makespa n
[32]	Delay Traffic arrival CPU frequency	Fog-cloud architecture	Improves mobility, geo distribution, and location	Response latency leads to loss of service	Power consumption	–	Transmission delay
[33]	Task length delay time	Sleeping barber algorithm	No difficulty in renting the cloud resource	Scheduling and allocation of resources are hurdles	Response time	Turnaround time	-

(continued)

Table 1 (continued)

Refs. no	Input parameters	Algorithm	Merits	Demerits	Performances metric		
					High	Average	Low
[34]	Frequency Length	Time aware energy-aware	Improves control of the energy and performance of the system	Load balance in a task cannot be reflected in a virtual machine	–	Energy consumption	Cost
[35]	Number of instruction Start time End time	Multi-capacity aware resource scheduling	Improves resource utilization, and energy efficiency	Utilizes less CPU	–	Waiting time	Slow down
[36]	Execution time length of instruction	Multi-objective scheduling and ordinal optimization	It reduces overhead time	Random workload distribution	Probability of job	Makespan	Cost
[37]	Number of critical levels Processing time Job slack	Dynamic programming Branch-and-bound	Improves resource utilization	Complex processing time of a job	Resource utilization	-	Waiting time
[38]	Number of CPU Finish time delay time	CPU intensive scheduling Data-intensive Scheduling	Improves server and network infrastructure	Unavailability of resources and Poor QOS	CPU resource	–	Elapse time
[39]	Current/start CPU Map task population size	Energy-aware multi job scheduling	It improves the energy efficiency of the server in allocating data	Huge task scheduling and optimization problem	–	Optimal disk capacity	CPU utilization

References

1. Antonio S (2017) Cisco delivers vision of fog computing to accelerate value from Billions of connected devices. https://newsroom.cisco.com/press-release-content?type=webcontent&articleId=1334100. Accessed 29 Jan 2014 [Jan. 24, 2017]
2. Bonomi F, Milito R, Zhu J, Addepalli S (2012) Fog computing and its role in the internet of things. In: Proceeding of 1st Edition of MCC workshop on mobile cloud Computing, 2012, pp 13–16
3. Sarkar S, Chatterjee S, Misra S (2015) Assessment of the suitability of fog computing in the context of internet of things. IEEE Trans Cloud Comput 99:1–1
4. Xu B, Zhao C, Hu E, Hu B (2011) Job scheduling algorithm based on Berger model in the cloud environment. Adv Eng Softw 42(7):419–425
5. Ozturk C, Hancer E, Karaboga D (2015) Dynamic clustering with improved binary artificial bee colony algorithm. Appl Soft Comput J 28:69–80
6. Baraglia R, Capannini G, Dazzi P, Pagano G (2013) A multi-criteria job scheduling framework for large computing farms. J Comput Syst Sci 79(2):230–244
7. Bitam S, Zeadally S, Mellouk A (2018) Fog computing job scheduling optimization based on bees swarm. Enterp Inf Syst 12(4):373–397
8. Manglani V, Jain A, Prasad V (2017) Task scheduling in cloud computing. Int J Adv Res Comput Sci 8(3)
9. Liu L, Qiu Z (2016) A survey on virtual machine scheduling in cloud computing. In 2016 2nd IEEE international conference on computer and communications (ICCC), 2016, pp 2717–2721
10. Chandrashekar DP (2015) Robust and fault-tolerant scheduling for scientific workflows in cloud computing environments. University of Melbourne, Australia
11. Pacini E, Mateos C, Garino CG (2016) Multi-objective swarm intelligence schedulers for online scientific clouds. Computing 98(5):495–522
12. Lawanyashri M, Balusamy B, Subha S (2016) Energy-aware hybrid fruit fly optimization for load balancing in cloud environments for EHR applications. Inform Med Unlocked 8(March):42–50
13. Puthal D, Sahoo B, Yang LT, Tiwary M, Sahoo KS (2018) Response time optimization for cloudlets in mobile edge computing. J Parallel Distrib Comput 119:81–91
14. Sarkar S, Chatterjee S, Misra S (2018) Assessment of the suitability of fog computing in the context of internet of things. IEEE Trans Cloud Comput 6(1):46–59
15. Yi S, Li C, Li Q (2015) A survey of fog computing: concepts, applications, and issues. In: Proceedings of 2015 workshop on mobile big data—Mobidata'15, pp 37–42
16. Hu P, Dhelim S, Ning H, Qiu T (2017) Survey on fog computing: architecture, key technologies, applications, and open issues. J Netw Comput Appl 98:27–42
17. Aburukba R, Ghenniwa H, Shen W (2006) Agent-based approach for dynamic scheduling in content-based networks. In: IEEE international conference on e-business engineering, 2006, pp 425–432
18. Maros I (2002) Computational techniques of the simplex method, vol 61, 1st ed. Springer, New York
19. Clausen J (1999) Branch and bound algorithms-principles and examples. In: Department of computer science in University of Copenhagen, 1999, pp 1–30
20. Lawler E, Wood D (1966) Branch-and-bound methods: a survey. Oper Res 14:699–719
21. Mansour N, El-Fakih K (1999) Simulated annealing and genetic algorithms for optimal regression testing. J Softw Maint 11:19–34
22. Kirkpatrick S, Gelatt CD, Vecchi MP (1983) Optimization by simulated annealing. Science 220:671–680
23. Goldberg DE, Holland JH (1988) Genetic algorithms and machine learning. Mach Learn 3:95–99
24. Croce FD, Tadei R, Volta G (1995) A genetic algorithm for the job shop problem. Comput Oper Res 22:15–24

25. Yu J, Buyya R (2006) A budget constrained scheduling of workflow applications on utility grids using genetic algorithms. In: Workflows in support of large-scale science workshop, pp 1–10
26. Dorigo M, Maniezzo V, Colorni A (1991) Positive feedback as a search strategy. In: Technical Report, Politecnico di Milano, pp 91–016
27. Tikar AP, Jaybhaye SM, Pathak GR (2015) A systematic review on scheduling types, methods and simulators in a cloud computing system. In: 2015 international conference on applied and theoretical computing and communication technology (iCATccT), 2015, pp 382–388
28. Banga P, Rana S (2017) Heuristic-based independent task scheduling techniques in cloud computing: a review. Int J Comput Appl 166(1):27–32
29. Chawla Y, Bhonsle M (2012) A study on scheduling methods in cloud computing. Int J Emerg Trends Technol Comput Sci 1(3):12–17
30. Ma T, Chu Y, Zhao L, Ankhbayar O (2014) Resource allocation and scheduling in cloud computing: policy and algorithm. IETE Tech Rev 31(1):4–16
31. Bin Woo Y, Jung S, Kim BS (2017) A rule-based genetic algorithm with an improvement heuristic for an unrelated parallel machine scheduling problem with time-dependent deterioration and multiple rate-modifying activities. Comput Ind Eng 109:179–190
32. Lai C, Luan TH, Lu R, Liang H, Deng R (2016) Optimal workload allocation in fog-cloud computing towards balanced delay and power consumption. IEEE Internet Things J X(X):1–1
33. Gawali MB, Shinde SK (2018) Task scheduling and resource allocation in cloud computing using a heuristic approach. J Cloud Comput 7(1)
34. Mao L, Li Y, Peng G, Xu X, Lin W A multi-resource task scheduling algorithm for energy-performance tradeoffs in green clouds.pdf
35. Sheikhalishahi M, Grandinetti L, Guerriero F, Wallace RM, Vazquez-Poletti JL (2015) A multi-dimensional job scheduling. Futur Gener Comput Syst 54:123–131
36. Zhang F, Cao J, Li K, Khan SU, Hwang K (2014) Multi-objective scheduling of many tasks in cloud platforms. Futur Gener Comput Syst 37:309–320
37. Seddik Y, Hanzálek Z (2017) Match-up scheduling of mixed-criticality jobs: maximizing the probability of jobs execution. Eur J Oper Res 262(1):46–59
38. Sahoo PK, Dehury CK (2018) Efficient data and CPU- intensive job scheduling algorithms for healthcare cloud. Comput Electr Eng 68(March):119–139
39. Stojmenovic I (2014) Fog computing: a cloud to the ground support for smart things and machine-to-machine networks. In: Telecommunication networks and application Conferences, 2014, pp 117–122
40. Hong K, Lillethun D, Ramachandran U, Ottenwälder B, Koldehofe B (2013) Mobile fog: a programming model for large-scale applications on the internet of things. In: Proceedings of the second ACM SIGCOMM workshop on Mobile cloud computing, 2013, pp 15–20
41. Farahani B, Firouzi F, Chang V, Badaroglu M, Constant N, Mankodiya K (2018) Towards fog-driven IoT eHealth: promises and challenges of IoT in medicine and healthcare. Futur Gener Comput Syst 78:659–676
42. Zhu J, Chan D, Prabhu M, Natarajan P, Hu H, Bonomi F (2013) Improving web sites performance using edge servers in fog computing architecture. In: IEEE 7th international symposium on service-oriented system engineering, 2013, pp 320– 323

Simplifying Handwritten Medical Prescription: OCR Approach

Parminder Singh Sethi, Manish Gupta, Praveen Kumar, and Gurleen Kaur

1 Introduction

Doctors' prescriptions are critical in the treatment of any patient. Misunderstood or wrongly read drug names in prescriptions may severely affect patients. It may lead not only worsen the illness but it can also lead to major threats such as opposite effects and demise [1]. A doctor's prescription is one of the toughest tasks to understand. It is triggered by either incomprehensible handwriting or the inability of pharmacists to recognize medicine names/qty/related details written in the medical prescription. With this research, we have tried to solve this problem. In our study, we use OCR (i.e., Optical Character Recognition) to convert scanned copies of handwritten prescriptions into a list of medicines in readable digital text format. This research will help pharmacists in reducing their efforts and will help in selling the correct medicine to patients. One important fact about India is that here around 90% of medical prescriptions are handwritten.

OCR has been a subject of research for a long. As per our study, handwriting recognition is a classification problem. OCR is a used case of pattern recognition using artificial intelligence (AI). In OCR usually, the basic flow of the process starts with scans and finally ends with identifying stationary images of the characters. It refers to the digital translation of images of handwritten characters into pixel-based machine code.

P. S. Sethi
Dell EMC Bangalore, Bangalore, India

M. Gupta · P. Kumar (✉)
Dell International Services Pvt. Ltd., Gurugram, India
e-mail: praveen24_kumar@rediffmail.com

G. Kaur
Enterprise SoftLabs Pvt. Ltd., Bangalore, India

A. Ramdane-Cherif et al. (eds.), *Machine Intelligence and Data Science Applications*, Algorithms for Intelligent Systems, https://doi.org/10.1007/978-981-99-1620-7_4

2 Literature Review on Optical Character Recognition

For the literature review, we surveyed around seven research papers written in the last two decades for understanding the core of the problem and various approaches taken by others to solve it. On reviewing those research articles, we concluded that every individual has a specific style of writing thus, understanding the handwriting of different people is difficult at certain times [1]. Having illegible handwriting often standardizes the Doctors. Old and ongoing studies have proven that medical errors can be led by Doctor's poor handwriting [2], [3]. The reasons that make it difficult to read doctors' handwriting are as under:

(1) Some similar medicine names,
(2) Use of abbreviations, and
(3) If written unreadably then it becomes confusing for chemists.

Character recognition systems are comprehended as pattern recognition systems that work as learning models for the classification of characters. One distinguishing feature of Character recognition systems is "the type of characters".

Most of the research work is divided into two approaches, one is ***Optimization approach***, and the other is ***Segmentation approach***. Example of Optimization approach used was PSO and BFO. One of the studies was proposing a heuristic solution approach, ***Particle Swarm Optimization (PSO)***. PSO is a bio-inspired algorithm, which is influenced by the movement in a school of fish or a flock of birds. Another algorithm was ***Bacterial Foraging Optimization (BFO)***, which is known for its global minima/maxima optimization approach. In comparison, BFO performed better than PSO. For both models, the loss function used was SRPN.

Another widely used approach was the ***Segmentation approach***. Character segmentation is a crucial method in this approach. Various algorithms used are

(a) Pixel Counting Approach,
(b) Histogram Approach,
(c) Smearing Approach,
(d) Stochastic Approach,
(e) Water Flow Approach, and
(f) Mixed Approach.

Also, considered papers are suggesting various ways to identify handwritten running hand-written characters. Our investigational results prove the higher degree of performance of our projected approaches. A study done by us defines that the identification of characters significantly depends upon feature engineering. The system was trained and tested with representative samples of handwritten prescriptions. Assessment of ranking and accuracy of the model was based on hyperparameters and feature selection for those cases. Selected features give average accuracy of around 78.37%. We would like to honestly accept that, there is further scope for enhancement in it.

3 Solution Design and Methodology

It includes the following steps:

- Pre-processing of data,
- Segmentation,
- Feature extraction,
- Classification (using AI/Deep learning technique), and
- Recognition (Mapping with recognized characters).

All the above-listed steps are dependent and iterative. The removal of noise and variation in handwritten is related to the task of preprocessing. Please find below the process flow for the same (Fig. 1):

Digitization:

The conversion of handwritten documents into electronic format is termed Digitalization. Here, each document contains only one character. A method in which a document is scanned and a pixel-based interpretation of the original document is produced, which accomplishes the electronic conversion [4]. Various scanners are used for digitization, and the digital image goes to the next step which is the preprocessing phase. In OCR, Preprocessing and Segmentation are the most important phases.

Pre-processing

The success of any machine learning algorithm is very much dependent on preprocessing of data. Under preprocessing, we followed the following steps:

- **Binarization:** Converted scanned input images into black and white images. As it is known white pixels are 0 and black are 255. Using the Local Maxima and

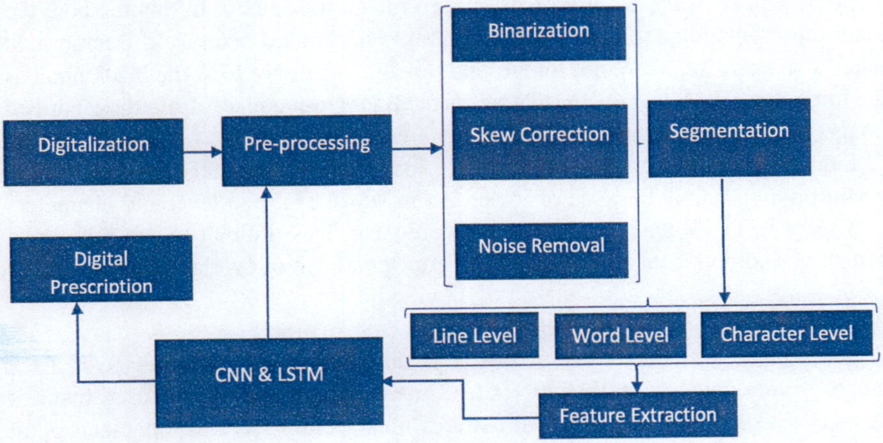

Fig. 1 Solution Architecture

Fig. 2 Binarization

$$C(i,j) = \frac{I_{max} - I_{min}}{I_{\max} - I_{min} + \varepsilon}$$

Minima Method, we calculated the threshold for the binarization of the image. Anything below that threshold is white and above the threshold is black (Fig. 2).

- **Skew Correction:** While scanning it may be the possible image is skewed (i.e., aligned at an angle with horizontal). Used Projection profile method for skew correction.
- **Noise Removal:** The main objective of the Noise removal phase is to remove bias from the image by removing high-intensity small areas. Noise removal can be performed for both colored as well as binary images.

In this stage, there are multiple sequential operations performed on scanned input images. It converts the images into numerical matrix data suitable for segmentation, data are normalized. After pre-processing and variable reduction process to make features independent, a bitmap image is produced, which is then transformed into a diluted image.

Segmentation:

Segmentation is one of the important phases. The process of breaking the whole image into subparts, to further process them, is defined as Segmentation [4]. We segmented handwritten prescriptions at the following levels:

- Line level,
- Word level, and
- Character level,

For this segmentation, we used Histogram Projection Method. In this method, the pixel representation of images is divided into Foreground pixels and Background pixels (The pixels that are not foreground pixels). Where we took the black pixel as the foreground pixel and the black pixel as the Background pixels from the binarized image.

Segmentation becomes a bit difficult for handwritten images (i.e., medical prescriptions) mainly because of variations in words at a line level and characters of a word level, skewed words, curved words, etc. It is difficult to segment when sometimes components of two adjacent characters touch or overlap each other [4].

Feature Extraction:

Features of each character are extracted in this phase. The performance of each model widely depends on the correct feature extraction process. We used features extraction techniques like diagonal features, intersection features, open-end point features, transition features, zoning features, directional features, parabola curve fitting-based features, and power curve fitting-based features to find the feature set for a given character have been used by us.

Modeling:

Training Data:

Our model-building process started with a selection of a learning algorithm. The training data set was built up of 1634 customer prescriptions from one of the pharmacists and its labels were the name given to those medicine in receipt. It was completely a manual process. Understanding previous studies, we were clear that we would have to apply a Deep learning algorithm [5], [6]. We used Deep Convolutional Neural Network, and the rationale behind using it was the sequential nature of data. On top of that, we used Recurrent Neural Network to strengthen our model.

We divided our data into training and test set. The training set was containing 70% of the data whereas the test set was containing the remaining 30% of the data. It was based on random sampling.

The researchers proceeded to the training, after the preceding steps. The threshold is set to 10, and the epoch size is set to 50. Since we observed that the training set was not improving after the 35th epoch, stopped early. The decoded text file is displayed when researchers run the testing.py.

(a) **Deep Convolutional Recurrent Neural Network (CRNN):** Convolutional Neural Network (CNN) is considered to be a good algorithm since it requires less pre-processing, which is likely for other classification models. It takes an input image and helps in the identification of important features. It applies filters to capture the Longitudinal and Sequential Dependencies in an image [7]. For character recognition, this was an important factor since characters are in sequential formation. CNN takes care of Feature reduction as well. Resultantly, it includes reduced and independent parameters and reuses weights, and achieves better fitting to the image dataset. Whereas the LSTM model of RNN accesses long-range context of it. Based on this study LSTM gives better results in handwriting recognition.

(b) **Connectionist Temporal Classification:** In this study, we have used Connectionist and Temporal Classification because in this system we can train our model end-to-end. In this model, the back-propagation procedure, the convolutional filters, and the LSTM unit weights are learned together [8], [9]. We combined the backward outputs and forward outputs, at the BLSTM end. Added Batch normalization after each convolutional layer, to speed up the training process. The network was trained in an end-to-end way together with the Connectionist Temporal Classification loss function. Adam Optimizer was used with an initial learning rate of 0.001, in the training of the network. Batch normalization helped in accelerating the training. Also, simply normalized each batch by both mean and variance, after each convolutional layer.

4 Conclusion

This is based on the detailed study of handwritten medical prescriptions where character recognition technology is used to boost further advances in this area. The accuracy of the model is depending on the nature of the features used. We have seen that feature engineering played an important role in the correct recognition of characters. The second most important part was the identification of relevant features. The third but equally important step was the model selection for classification. This hypothesis suggests a holistic system that can convert scanned/ photo images of handwritten content to text output/documents. Papers with different algorithms have been studied in this paper.

References

1. Brits H, Botha A, Niksch L, Terblanché R, Venter K, Joubert G (2017) Illegible Handwriting and Other Prescription Errors on Prescriptions at National District Hospital, Bloemfontein. South African Fam. https://doi.org/10.1080/20786190.2016.1254932
2. Saha S (2018) A comprehensive guide to convolutional neural networks? The ELI5 way. https://towardsdatascience.com/a-comprehensive-guide-to-convolutional-neural-networks-the-eli5-way-3bd2b1164a53
3. Hochuli AG, Oliveira LS, de Souza Britto Jr A, Sabourin R Segmentation-free approaches for handwritten numeral string recognition. https://www.researchgate.net/publication/323831922_Segmentation-Free_Approaches_For_Handwritten_Numeral_String_Recognition
4. Chen YK, Wang JF (2000) Segmentation of single or multiple touching handwritten numeral string using background and foreground analysis. IEEE Trans Pattern Anal Mach Intell 22(11):1304–1317
5. Deep learning using python by Dr. S Lovely Rose, Dr. L Ashok Kumar, Dr. D Karthika Renuka, Wiley Publication
6. Pattern recognition and machine learning by Christopher M. Bishop, Springer Publication
7. Pattern classification by Richard O. Duda, Peter E. Hart, David G. Stork, Wiley Publication
8. Romero DJ, Seijas LM, Ruedin AM (2007) Directional continuous wavelet transform applied to handwritten numerals recognition using neural networks. JCS&T 7(1)
9. Britto AS, Sabourin R, Bortolozzi F, Suen CY (2003) The recognition of handwritten numeral strings using a two-stage HMM-based method. IJDAR 5:102–117

Profiling the IPL Players—Sports Analytics Through Clustering Algorithms

A. Mansurali, V. Harish, Sherin Hussain, and T. Choudhury

1 Introduction

Sports as an industry has always linearly progressed with the help of numbers and data. Data capturing is always a habit in sports. The explosion of data has affected sports and every industry, from driving cars to managing stock portfolios and even

A. Mansurali (✉)
Department of Management, Central University of Tamil Nadu, Thiruvarur, Tamil Nadu 610001, India
e-mail: mansurali@cutn.ac.in

V. Harish
PSG Institute of Management, PSG College of Technology, Coimbatore, Tamil Nadu 641004, India
e-mail: harish@psgim.ac.in

S. Hussain
MBA Business Analytics, School of Business and Management, Christ University, Bengaluru, Karnataka 560076, India
e-mail: sherin.hussain@mba.christuniversity.in

T. Choudhury
CSE Department, Symbiosis Institute of Technology, Symbiosis International University, Pune, Maharashtra 412115, India

Informatics Cluster, School of Computer Science, University of Petroleum and Energy Studies (UPES), Dehradun, Uttarakhand 248007, India

CSE Department, Daffodil International University, Daffodil Smart City, Birulia 1216, Bangladesh

Director Research (Honorary), The AI University, Cutbank Montana 59427, US

T. Choudhury
e-mail: tanupriya.choudhury@sitpune.edu.in; tanupriya@ddn.upes.ac.in; tanupriya1986@gmail.com

A. Ramdane-Cherif et al. (eds.), *Machine Intelligence and Data Science Applications*, Algorithms for Intelligent Systems, https://doi.org/10.1007/978-981-99-1620-7_5

fundraising for political campaigns. This impact of data has been driven by new sources, increased availability, new technologies, and advances in machine learning, all of which allow for deeper insights in real-time. The sports industry, both on the field and the business side, now has a data-driven decision-making approach. General Managers and team coaches use data to gain an edge over competitors. These could be strategic decisions, better player evaluation, or targeted player development. Cricket as a sport is a great hotspot for analysis and prediction, especially in south-east Asian countries, authors of [5] in their paper did a polarity assertation of opinion on Bangladeshi cricket data using machine learning approach. Machine learning approaches in general are known to be very robust in achieving tasks, [17] paper proposes using machine learning approach for Real-Time Multimodal Biometric Authentication of Humans and achieved 99.32% accuracy.

Sports Analytics is a trending research area with several applications. By using sports analytics, any player's talent or market value can be predicted. Sports Analytics comprises of two categories. In the on-field approach, analysts gather data from the game and the players to improve performance. The off-field approach considers the business performance of teams, such as finances, salaries, social media support of fans and so on. Sports Analytics has become essential for both the team and players as it helps improve the performance on-field, enhances business performance, and predicts player injury risk.

One of the most popular sports arguably in India is cricket. Team selection based on the strength of each player has become of utmost necessity since the introduction of limited-overs cricket. IPL has had over ten seasons, comprising teams with players from all over the world. Cricket is highly statistics-driven, with comprehensive data which can be analysed. The likelihood of winning by a team can be determined using bowling measures, career average runs and so on. Several other factors also contribute to determining the outcome of a match. Regression has proved factors such as weather, toss result, lighting, and condition of the pitch all play a role in determining the outcome of a match.

Players' performance is analysed with the help of measures such as strike rate, average runs per wicket etc. Real-time data can be used to classify players by performance. Clustering algorithms are known to handle complex tasks, authors of [4] in their paper suggested the analysis of kidney disease using clustering algorithm and [10] suggested the use of clustering algorithms in terrorist attack analysis. Clustering algorithms can also be applied in sports. It can be used to cluster players, which will help develop players in a focused manner. Using clustering algorithms, separate clusters are formed which differentiate players based on performances and can be used to categorise each player.

2 Literature Review

The teams for IPL are formed based on players bought by bidding. Huge franchises focus on getting the best proportion of players of every type. (Sankaran, 2014). In any format of cricket, performance assessment is crucial, and every game produces a variety of data. This data is performance related. As compared to other sports, in cricket, individual performance contributes to the overall team performance is notably more significant. Several research studies have been done on predicting the results of a match. Bowling statistics and career average runs can calculate a team's winning probability. (Cohen 2002) [3]. However, there is limited research on classifying players based on performance (Saikia et al. 2012) [12–14]. Using historical data, researchers focused on auction and modelling future performance. By using playing factors such as runs scored, age and wickets, Karnik (2010), developed hedonic regression models to estimate the bid price for an IPL player. Factors such as toss outcome, venue, teams competing, can be considered and supervised learning techniques such as Naïve Bayes and SVM can be applied for predictive analysis in ODI matches (Khan et al. 2019) [7].

By utilising harmonic mean, traditional bowling statistics such as bowling average, bowling economy rate and bowling strike rate can be combined; Lemmer (2002) [8] created the combined bowling rate, which can be used to evaluate bowlers' performance. In limited over cricket, the batsman's performance was assessed by a measure created by Lemmer (2004) [9], which was a product of exponentially weighted batting average, standardised strike rate, and standardised coefficient of consistency of the batters. An alternative measure which is a weighted product of batting average and strike rate, for choosing batsmen in limited-overs cricket was proposed by Barr and Kantor (2004) [1]. All-rounders of IPL can be classified based on their performance using naive Bayesian classification and multinomial logistic regression (Saikia et al. 2011). Batting, bowling or wicket-keeping performances of cricketers can be modelled using the data from the scorecard. Lemmer (2011) developed a performance measure for wicket-keepers which combined the dismissal rate.

IPL has become a primary focus and hotspot for researchers from various fields including economics, finance, decision science, etc. Auction-determined salaries of players have created a rare opportunity to determine the monetary worth of players. Using data from IPL, players' batting and all-around performance were assessed by utilising contour plots (Van Staden, 2009) [15, 19]. Additional research on the valuation of players and the factors that affected the valuations in IPL was presented by Parker (2008) [18] and Rastogi (2009).

Research has also been done on the likelihood of selection in a team. Neural networks and historical data were used to predict the performance of 2007 World Cup players (Iyer et al. 2009) [16]. Players were categorised as performers, moderate performers or failures for this classification exercise based solely on strike rate and average; Saikia (2011) classified all-rounders who were participants of the first three seasons of IPL based on performance.

Saikia et al. (2013) [11] attempted to analyse the performance of cricket players based on their pay package. This was done by normalising, combining and stabilising variables related to performance. In addition to generating new performance indicators, the data mining technique K-means cluster analysis was applied to determine four groups of bowlers based on the effectiveness of their performance. The study indicated that the group Good Performing Fast Bowlers and Strong Performing Spinners should be retained. Expensive Underperformers were identified as the group that requires extra mentoring. Expensive Poor Performers were identified as the group which must be traded. Clusters on team rankings were superimposed to validate the cluster analysis. Significant disparities in the team composition between the high-performance and low-performance groups were discovered. Tomar et al. (2020) [20] propose a parameterized approach for comparative study in an intercommunication-based application.

The study can help team owners determine which players should be traded, which players require further coaching, and which should be retained for the next season. Players can be categorised according to the role they are more suited for in a team (Bose et al. 2021) [2]. In the field of sports, clustering algorithms are preferred to detect the development of players. Agglomerative clustering, K-means, and Fuzzy-C-means are all various clustering algorithms which can be used to form separate clusters, which can then be used to differentiate players by performance (Jain et al. 2019) [6].

3 Objective

- Clustering of players based on their performance using Hierarchical clustering
- To understand the profile of players for the focused development and engagement

4 Methodology

The research design is exploratory in nature as it tries to unveil the clusters of players amongst the IPL players. Since this study's target variable, i.e., clusters, are unknown, it falls under the unsupervised learning technique. Data used for the analysis are secondary data from ESPN Cricinfo. The dataset contains the date for the period from 2008 to 2019 and has records of 50 players. The dataset has the basic details about a player, matches, innings, not out, runs, highest score, average, balls faced, strike rate, number of hundreds, number of fifties, number of fours and number of six. R-studio tool has been used for performing clustering algorithms. Hierarchical algorithm has been used for clustering the players using the two functions such as Agnes and Fvizcluster.

Table 1 IPL player data description

Variable	Description	Data type
Player	Name of the player	FACTOR
Mat	Number of matches played	INT
Inns	Number of Innings played	INT
NO	Number of times a player is Not Out	INT
Runs	Runs scored by the player	INT
HS	Highest score	INT
Ave	Average	NUM
BF	Balls faced	INT
SR	Strike rate	NUM
X100	Number of centuries	INT
X50	Number of Half Centuries	INT
X0	Zero run	INT
X4s	Number of boundaries	INT
X6s	Number of sixes	INT

4.1 Data Description

See Table 1.

4.2 Process Flow

5 Analysis

Knowing the IPL Players—Exploratory Data Analysis

IPL players' data considered for the profiling of players through clustering algorithms had been subjected to exploratory data analysis in a journey to understand the data well and also to validate further for transformation and pre-processing. The dataset has 50 rows which is the number of observations and 13 dimensions represent the variables included for the analysis. Missing values were identified with the help of summary commands in strike rate and other columns which have been treated with the help of mean imputation methods (Fig. 1).

The Performance Wave—Distribution of Data

Most of the variables plotted using histogram in an attempt to understand the distribution of data clearly depicts that the data is right skewed other than the strike rate. Right skewness observed in the parameters such as 50's and 100's makes it pretty clear that the winning strategy and player's performance do not just depend on 50's or 100’s or average, it’s all about hitting the runs when needed like 10 runs in 2 balls but it’s not about scoring centuries or half centuries revealing the scarious nature of short format cricket (Figs. 2, 3, 4 and 5).

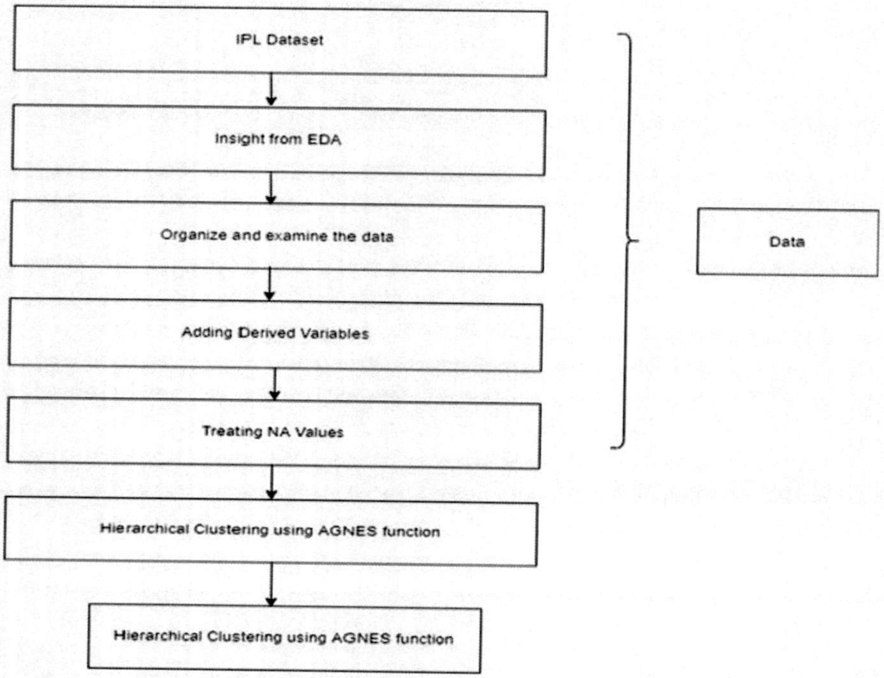

Fig. 1 Process flow diagram for hierarchical clustering

Out of the Boundary Line—Boxplot

As it could be seen in the above box plot of a number of innings and not outs, there are a significant number of outliers can be spotted. Tradition suggests the removal of outlier and handling of outlier treatment and the outliers are the usual courses in the

Fig. 2 Fifty’s

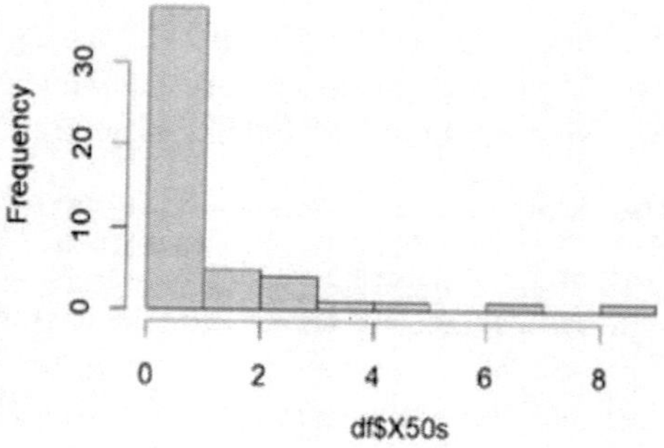

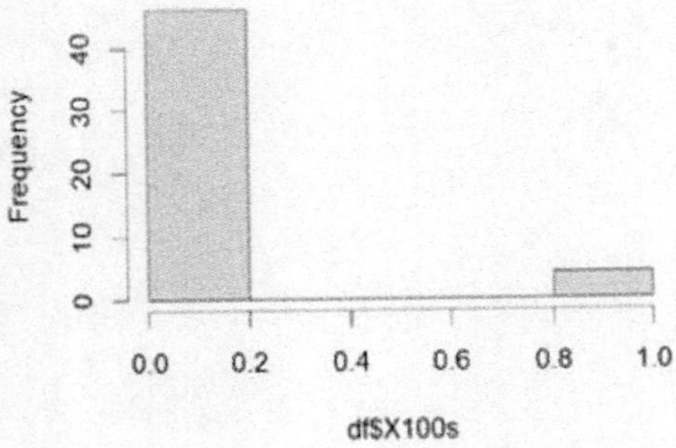

Fig. 3 Hundreds's

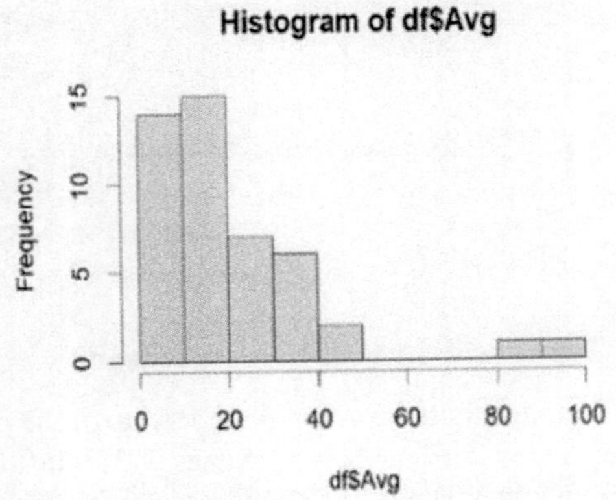

Fig. 4 Average

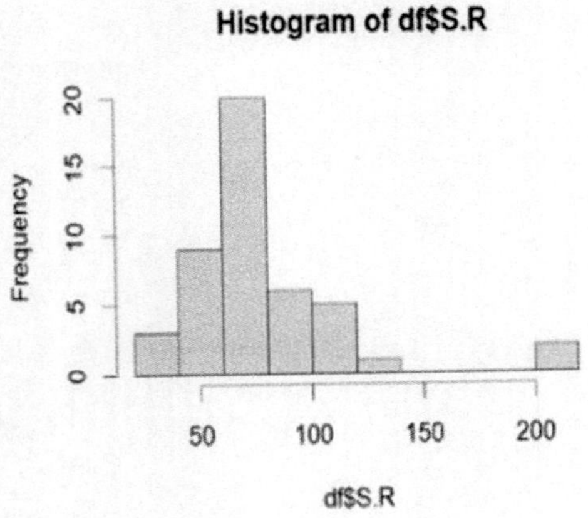

Fig. 5 Strike rate

game of cricket as we see openers go for duck out and the tail ender hits consecutive sixes or four. Here in the cases of Figs. 6 and 7, there are a greater number of players who were not outs, but still, they can't win the games for their teams. Outliers have been treated using capping and flooring methods as its appropriate for the nature of IPL data with the help of players' information.

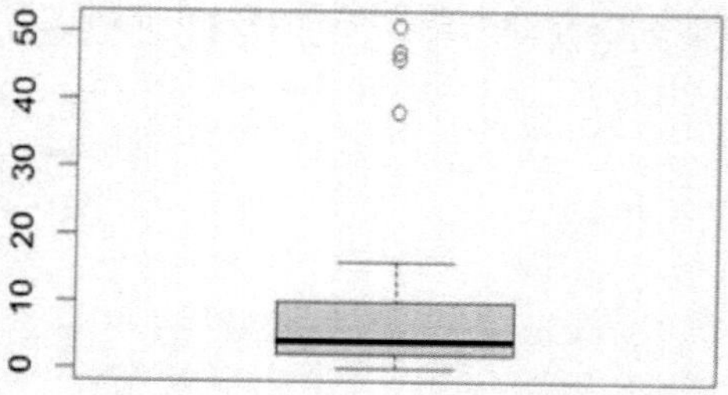

Fig. 6 Innings

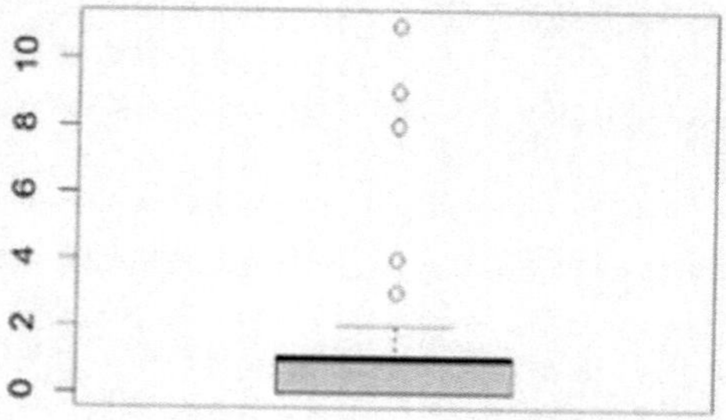

Fig. 7 Not out's

The Hierarchy in IPL—Dendogram

To understand the group membership of the players, a clustering algorithm had been executed with the help of Agnes and Hclust functions and the groups can be visibly seen with the help of a dendrogram using Fviz, dividing data into four groups (Figs. 8 and 9).

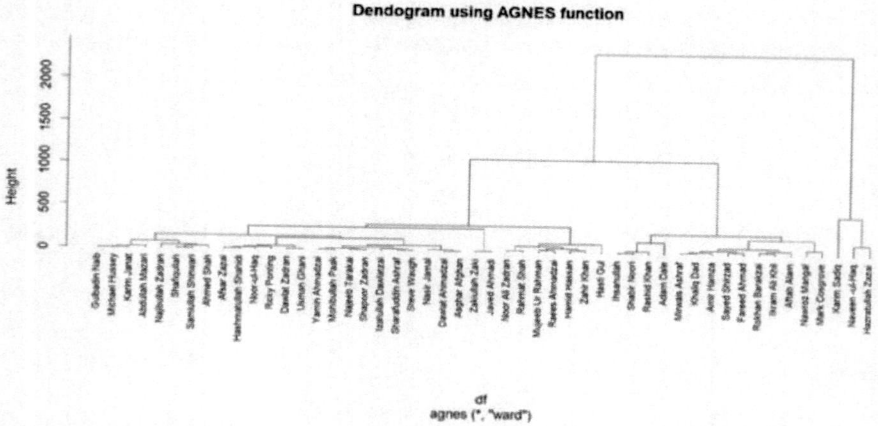

Fig. 8 Clustering of players

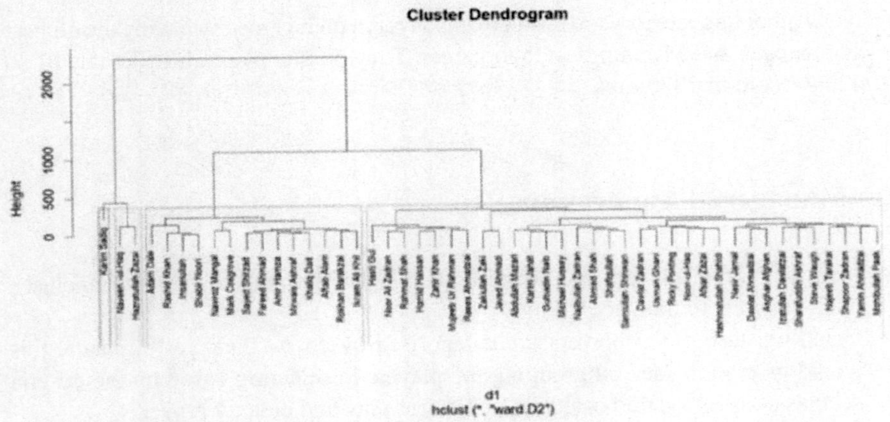

Fig. 9 Clustering of players into four groups

6 Results

The ultimate aim of the clustering algorithm is to understand the profile of the players that falls under each category and also label the players in each cluster. Profiling is the most important outcome of clustering and will help team owners and coaches in understanding what the clusters actually are and they can use this information during the auction to check which cluster a new player might fit into. It also helps to take better decisions in terms of selecting players for a team and to derive strategies for each of the clusters.

Profiling

Cluster 1: Early Entry

Players under this category are early entry, who are just starting their career in cricket and has played the highest of 12 matches and 11 innings in their career. They maintain a high strike rate of 209 and high score of 139 runs.

Cluster 2: Mid Category

Players under this category are mid category, who are had started their careers and playing consistently. This category of players compiles the highest of 46 matches and 38 innings in their career. They maintain a high strike rate of 118 and high score of 387 runs.

Cluster 3: Consistent play

Players under this category are slow in scoring but consistent in their career by having highest of 55 matches and 46 innings in their career. They maintain a high strike rate of 71 and high score of 830 runs.

Cluster 4: Creamy player

Players under this category are most creamy layer in their career by having the highest of 56 matches and 51 innings in their career. They maintain a high strike rate of 90 and high score of 1270 runs.

7 Managerial Applications

- Identification of players and profiling the players help the coach and franchisee team to invest to train accordingly
- Using clustering, the players are categorised based on their performance, this would give a fair idea on grouping the players in selecting based on the criteria such as early entry, mid category, consistent play and creamy player.
- Profiling the players would give a broad idea of selection just based on the group name or category, such as creamy players who hit highest runs, and fast play players who have highest strike rate.
- Sports analytics apart from being seen as a game, it has more business opportunities hidden in it. Selection of the right players for each match plays a significant role in a team's victory.

8 Methodological Limitations

- It is not possible to get precise information regarding data sorting and the output since data used in unsupervised learning is not labelled and is unknown. The results are less accurate because the input data is not known and not labelled in advance.
- Groups are decided based on the input value of K and so to decide the optimal value of K is dynamic to the domain of cricket
- Hierarchical clustering is not very effective when dealing with large sets of data. Once clusters are combined, it is impossible to separate them again, and it can be challenging to manage clusters with distinct characteristics. Additionally, the resulting outcome is affected by the order of the data and can be easily distorted by noise and outliers.
- Using different distance metrics to measure the distances between clusters can provide different results. To make sure the results are accurate, it is important to compare them in order to confirm their veracity.

9 Conclusion

The main objective is to cluster or classify batters based on yearly performance. The clusters are grouped depending on the K value. This can be used later to predict the top batsmen in the upcoming year by using predictive techniques. Clustering is used to determine the profile of the players falling under each category and label the players in each cluster. The players are then classified as Early Entry, Mid Category, Consistent play or Creamy player. The Indian Premier League, a franchise-based tournament, is formed by competitive bidding. Auctions determine the salaries of players. The media and the team owners keep tabs on each player's performance. A franchise has limited resources which need to be used effectively. Hence, using analytics for decision-making has generated interest amongst researchers. By using clustering, team owners can make decisions such as which players can be retained, which can be traded, and which players require additional coaching.

References

1. Barr G, Kantor B (2004) A criterion for comparing and selecting batsmen in limited overs cricket. J Oper Res Soc 55(12):1266–1274
2. Bose A, Mitra S, Ghosh S et al (2021) Unsupervised learning based evaluation of player performances. Innov Syst Softw Eng 17:121–130. https://doi.org/10.1007/s11334-020-00374-3
3. Cohen GL (2002) Cricketing chances. In: Abstracts from the 6th Australian conference on mathematics and computers in sport, 1–3 July 2002, pp 1–14
4. Kunwar V, Sabitha AS, Choudhury T, Aggarwal A (2019) Chronic kidney disease using fuzzy c-means clustering analysis. Int J Bus Anal (IJBAN) 6(3):43–64
5. Faruque MA, Rahman S, Chakraborty P et al (2022) Ascertaining polarity of public opinions on Bangladesh cricket using machine learning techniques. Spat Inf Res 30:1–8. https://doi.org/10.1007/s41324-021-00403-8
6. Jain SS, Gupta R, Tiwari C, Kaur N (2019) Behaviour of players on IPL based on fuzzy C means. Int J Innov Technol Explor Eng (IJITEE) 8(9):150–154
7. Khan JR, Biswas RK, Kabir E (2019) A quantitative approach to influential factors in One Day International cricket: analysis based on Bangladesh. J Sports Anal 5:57–63
8. Lemmer H (2002) The combined bowling rate as a measure of bowling performance in cricket. S Afr J Res Sport Phys Educ Recreat 24(2):37–44
9. Lemmer H (2004) A measure for the batting performance of cricket players. S Afr J Res Sport Phys Educ Recreat 26(1):55–64
10. Gupta P, Sabitha AS, Choudhury T, Bansal A (2018) Terrorist attacks analysis using clustering algorithm. In: Satapathy S, Bhateja V, Das S. (eds) Smart computing and informatics. Smart innovation, systems and technologies, vol 78. Springer, Singapore. https://doi.org/10.1007/978-981-10-5547-8_33
11. Saikia H, Bhattacharjee D, Bhattacharjee A (2013) Performance based market valuation of cricketers in IPL. Sport Bus Manag: Int J 3(2):127–146
12. Saikia H, Bhattacharjee D, Bhattacharjee A (2012) Is IPL responsible for cricketer's performance in Twenty20 world cup? Int J Sport Sci Eng 6(2):96–110
13. Saikia H, Bhattacharjee D, Lemmer HH (2012) Predicting the performance of bowlers in IPL: an application of artificial neural network. Int J Perform Anal Sport 12(1):75–89

14. Saikia H, Bhattacharjee D (2012) A fairer metric for the performance measurement of bowlers in cricket. Gurukul Bus Rev 8:1–5
15. Staden PJ (2009) Comparison of cricketers' bowling and batting performances using graphical displays. Curr Sci 96:764–766
16. Iyer SR, Sharda R (2009) Prediction of athletes performance using neural networks: an application in cricket team selection. Expert Syst Appl 36:5510–5522
17. Srivastava R, TomarR, Sharma A, Dhiman G, Chilamkurti N, Kim B-G (2021) Real-time multimodal biometric authentication of human using face feature analysis. Comput Mater Contin 49(1):1–19. https://doi.org/10.32604/cmc.2021.015466
18. Parker D, Burns P, Natarajan H (2008) Player valuations in the Indian premier league. www.frontier-economics.com/_y/Frontier%20paper%20-%20IPL%20. Accessed 17 June 2010
19. Van Staden PJ (2009) Comparison of cricketers' bowling and batting performances using graphical displays. Curr Sci 96(5):764–766
20. Tomar R, Sastry H, Prateek M (2020) Establishing parameters for comparative analysis of V2V communication in VANET. J Sci Ind Res (JSIR) 79(01):26–29

Data Analytics in Operation Management

V. Harish, A. Mansurali, and Tanupriya Choudhury

1 Introduction

The field of operations has seen significant growth since the eighteenth century, several scientific inventions came into existence and transformed the surface of operation by the introduction of larger machines, which were powered by steam and electric power. Perhaps the most important of these inventions was the steam engine which had the potential to empower the use of large machinery in factories. Newer machines, made of steel, were much stronger and more durable than lightweight wood-changing machines. From the late seventeenth century (1770) to the early eighteenth century, a series of events took place in England when they came together called the industrial revolution. The industrial revolution resulted in

V. Harish (✉)
PSG Institute of Management, Coimbatore, India
e-mail: harish@psgim.ac.in

A. Mansurali
Department of Management, Central University of Tamil Nadu, Thiruvarur, Tamil Nadu 610001, India
e-mail: mansurali@cutn.ac.in

T. Choudhury
CSE Department, Symbiosis Institute of Technology, Symbiosis International University, Pune, Maharashtra 412115, India

Informatics Cluster, School of Computer Science, University of Petroleum and Energy Studies (UPES), Dehradun, Uttarakhand 248007, India

CSE Department, Daffodil International University, Daffodil Smart City Birulia, 1216, Bangladesh

Director Research (Honorary), The AI University, Cutbank Montana 59427, US

T. Choudhury
e-mail: tanupriya.choudhury@sitpune.edu.in; tanupriya@ddn.upes.ac.in; tanupriya1986@gmail.com

A. Ramdane-Cherif et al. (eds.), *Machine Intelligence and Data Science Applications*, Algorithms for Intelligent Systems, https://doi.org/10.1007/978-981-99-1620-7_6

two major processes, the widespread transfer of mechanical power to the use of human power and the establishment of an organized production system known as the industrial system. The acquisition of mechanical power has greatly aided the recruitment of workers in the machinery industry. The original production system had been replaced by a larger industrial system by the middle of the eighteenth century. As time passed, production improved, money demand climbed, and workers became more reliant on jobs and urbanization. One of the most important aspects of operation management is the adoption of technology and one of the most critical aspects of technology influencing operation management is the area of data analytics [28]. Data analytics as a tool has high potential in the area of operation management and its verticals [19].

2 Literature Review

Operation Management

Nowadays in the competitive world, especially in industrial enterprises, operation management is a critical component of management [15]. In the 1950s, operation management (OM) was originally introduced into business school curricula as part of the management department called production management [18]. Because of its significant importance in management success, operation management is a critical factor for operational success of a firm [20]. Operation management is concerned with converting inputs into outputs using a variety of resources [21]. The development of entrepreneurship theory has been aided by operation management, which focuses on how operation management deals with management under uncertainty [8]. OM encompasses all of the tasks required to improve the process in a business that is involved in the manufacture of a product or providing a service [2]. The efficiency and effectiveness of operation are given special attention, as they have a significant impact on how operations can support the firm's strategic goals [26]. It's also worth noting that, in the current atmosphere of intense global competition, operation managers are faced with the problem of enhancing quality while cutting costs and retaining social responsibility [23].

Decisions of Operation Management

Decision-making plays a major role in the field of operation management. Both strategic and logical ways of handling management are essential parts of operation management. Decisions of management may differ according to the type of organization, but its general operation management decisions are classified into the following major categories.

2.1 Supply Chain Management (SCM)

Supply chains are an important core of ambitious pros for businesses. SCM is a group of value-inherited processes that connect a management's vendors and customers [30]. SCM refers to the process of efficiently designing, implementing, and controlling supply chain operations. SCM includes all raw materials and inventory, from origin to consumption, as well as the transfer and storage of finished products. Many manufacturers and retailers rely on the concept of supply chain management to improve efficiency and effectiveness [28]. Vonderembse et al. [33] emphasized three different forms of the supply chain required for three different product types: standard, innovation, and hybrid. They show that simple, poorly differentiated standard products must be created through the same.

2.2 Revenue Management

The process of gathering the maximum potential yield through the process of selling at a firm's maximum capacity is known as revenue management. Revenue management practitioners employ strategies like constructed pricing, breakouts in the market, and forecasting is being sold at the maximum price. Under developing technical conditions, revenue management is defined as one of the most creative and productive systems. Academicians and industry executives have done additional studies aimed at increasing the amount of output generated by this beneficial effect and improving the system. As a result, there is a growing demand for experienced revenue managers [34]. Applications in manufacturing, advertising, energy, and financial services are on the rise. With each new industry, application comes new modeling, forecasting, and optimization issues [12].

2.3 Inventory Management

Inventory management plays a major role in helping companies to estimate what goods to be ordered and when to be ordered. It tracks goods and products from the day when the company purchase to the day when the company sells them. This method monitors and responds to trends so that inventories are ready to meet customer needs early [19]. Even though inventory management is not new, not every organization employs it to cut inventory expenditures [25]. Inventory management has an impact on every area of your organization, from warehousing expenses to the accuracy and timeliness with which you can fulfill orders. From raw supplies to completed items, a firm has to keep track of everything failing which could have long-term consequences for your company.

2.4 Quality Management

In today's highly competitive market, quality has become an existential concern for businesses. Providing quality products and services is the key to gaining and maintaining a competitive advantage. Gunasekaran et al. [14] states that management gains the vitality of the company by meeting the demands of customers, which is the basic principle of TQM, and ensuring the survival of the organization. Organizations that follow TQM have various benefits, rich in quality, satisfied customers, low in costs, higher finance, quality, innovative performance, and higher employee satisfaction. In addition, a successful TQM application creates a competitive advantage [22].

2.5 Goods and Services

Choi et al. [6] asserted within side the eighteenth century that merchandise has exchangeable value and that one of the traits of a very good is that its possession rights can be shaped and transferred. Choi et al. summarized the principal characteristics of products with the aid of using describing them as an entity that exists independently of its proprietor and continue its identification throughout time. As entrepreneurs have become greater aware of the importance of services, they started to sell them [10]. The carrier region of industrialized countries has been at the upward thrust for almost three-quarters of a century, and carrier industries are awesome from production due to their immediacy, and accordingly balancing carrier deliver and call for is difficult.

(a) Process design

Product design is the technique via way of means by which designers integrate client desires with industrial targets as a way to help manufacturers in developing constantly a hit good. Product designers try to enhance the consumer's enjoyment with the answers they offer for his or her clients whilst additionally supporting their corporations via way of means of making sure that their merchandise is long-lasting and meet long-time period commercial enterprise goals [29]. Product design is the technique of making a useable product that satisfies the desires of clients via way of means of figuring out users' issues and devising innovative answers. The period is likewise carried out to the cease manufactured from this procedure, the layout features of a formerly synthetic product.

(b) Location

Both massive and small companies need to make important region considerations. The region of a corporation has an instantaneous effect on its charges in addition to its capability to serve clients and, as a result, its revenues. The charges of moving a corporation are once in a while substantial, and there may be a threat of inconvenience to clients and employees. It's constantly optimal to get the spot properly the primary time. The placement of the facility of the distant place is one of the maximum crucial factors. Firms try and take gain in their area to advantage an aggressive edge and that they cautiously find centers to make the maximum in their area [5].

2.6 Scheduling

Scheduling is genuinely a manufacturing-making plan model's short-time period execution plan. The operations undertaken in a production agency to manipulate and modify the execution of a manufacturing technique are called manufacturing scheduling. A timetable is a venture hassle that specifies which obligations need to be finished and the way the factory's sources should be used to meet the necessities of the plan. Within an organization, it refers to figuring out the timing and utilization of resources. Scheduling is associated with using gadgets and facilities, human pastime scheduling, and fabric reception beneath the operation characteristic of each manufacturing and service [6]. This purpose necessitates the scheduling of operations.

2.7 Maintenance

The significance of the maintenance has grown over time. Despite the increased capital invested in manufacturing equipment and civil buildings due to the large-scale use of mechanization and automation, there has been a discount on production personnel [11]. Maintenance has historically been crucial to business corporations due to its various tasks, resources, assessment, and management. However, because of converting working technology and the converting organizational characteristic of maintenance, the need to deal with the various factors of maintenance extra successfully has won multiplied relevance in current years. Maintenance has a broader point of view in contemporary open device production enterprises. Maintenance has modified from a tightly described operational angle to an organizational approach angle in such enterprises [31].

2.8 HR and Job Design

The education of setting up work into the responsibilities required to finish a positive process is referred to as job design. Job design is the planned try and organizes responsibilities, activities, and duties right into a unit of labor on the way to reap unique goals [4]. The design and specification of precise responsibilities inside the organization need to pique the core of an HR manager. OM researchers and directors are nicely conscious that their structure consists of simplified representations of human behavior.

Data Analytics

It is an area focused on extracting information from large amounts of data. It includes procedures, tools, and methods for analyzing and managing data, as well as collecting, organizing, and storing data. The main goal is to use statistical analysis and the technology of data to identify trends and solve problems. According to [6], data analytics is described as revealing insights using an algorithmic or mechanical approach and screening a large number of datasets to find relevant correlations is all about data analysis. It is implemented in various industries to help companies and companies which follow data analytics make better decisions, test and refurbish the pre-existed structures. The reasoning is the major part that is on focus [9]. The reasoning is a process of reasoning based on what the researcher already knows. The simplest definition of analysis is to use the data to build a model that leads to value-added decision-making [32].

(I) Applications of data analytics in the decisions of operation management

The following are the application of data analytics in operation management

i. **SCM**

The process of the supply chain, understandably, creates a huge amount of data on daily basis. Researchers and experts believe that this data has its very own untold efficiency that might open about the potential of data which about SCM strategy and productivity [28]. But most of the data during the process are unstructured. As the final output, complicated design techniques are needed to make valuable meaning of the data and clear the unknown patterns to obtain useful information. Data analytics in SCM also promotes the usage of technologies like AI and machine learning.

3 Location

The merging of a new era of e-com channel and a connection of bam stores are referred to as Omni channel. This structure enables a retail owner to perform cross-channel with fulfillment, or the fulfilling of orders in which are from various store locations, resulting in a unique location and inventory management issues. Glaeser et al. [13] look at a current world buy from online, pick up through store shop that fulfills online orders via vehicles stationed in conveniently accessible areas. Retailers were looking into a variety of data sets that include data, which are obtained from the point to point sale, regionalized, current situation, and for decades to know more about their receivers. Retailers may now obtain a far better insight into consumers and marketplaces because of the source of a new era of data. Through electronic programs and anonymized mobile location data, retailers now have more customer behavior data sets than they have earlier [24].

4 Inventory Management

Analysis of data plays a major role in the inventory management sector in that case [1] looks at inventory management from a data-driven perspective. When demand measurements are combined with elements that can be based on the prediction of demand through forecasting and by the indicators of economic condition. Analyzing the data will help to identify sources of patterns and current trends that can be uplifted to develop operations of the business by managing inventory at the proper level [3]. The major reason for analyzing the data is stated by Choi et al. [6] was one of the most popular motivations for inventory analytics monitoring is to better understand and improve inventory performance. Inventory management key performance indicators (KPIs) are monitored by an inventory manager to identify areas that are functioning well and processes that require attention. Many businesses will set KPI targets and track their success regularly. Analytics in inventory management results in improved efficiency, predicting needs, optimized replenishment faster recalls, low prevention.

5 Revenue Management

Analyzing the data is creating a significant influence on how businesses decide how to offer their products to customers with that being set, the application of data analytics in revenue management is classified into three major categories, which are structured optimization of the whole process. The amount of data allows a transition from qualitative to quantitative design techniques and how companies may use analytics to increase revenues in both their offline and online channels. Finding the proper path necessitates a high-level overview of each relevant KPI that must be comprehended, as well as a technique to express the total outcomes of all relevant KPIs. By data analytics, the consumer population as a probability distribution over product rankings, [7] suggested a vital statics structure for observing the choice of the customer. Any decision structure that is based on maximum utilization, which covers the maximum source of common interest structures, can be represented by such a model.

5.1 *Maintenance*

Maintenance aims to reduce and eliminate the number of errors that occur during production as a failure of machinery or equipment can disrupt the supply chain. The recent use of big data and related technologies for preventive maintenance significantly improves visibility into system health and improves the speed and accuracy of maintenance decisions [16]. Predictive maintenance is a natural extension of data analysis in conservation. By using machine data in conjunction with other domain data, companies can truly understand the state and performance of their machines. This is done by installing a sensor and using the data to model the performance of the device. Once the model is built, the facility can use real-time data to predict when the machine will fail [17] V2v.

6 Analysis

Approaches of data analytics used to resolve the challenges faced by the decisions of operation management.

6.1 Supply Chain Management

The proper handling of data related to supply chain activities is the subject of using data analytics in SCM. Supply chains, understandably, generate massive volumes of data daily. Experts believe that these data have an unknown worth, which may show the wealth of information, which is related to the current SCM strategy and its efficiency. However, most of the data in SCM are not integrated. As a result, you need advanced software tools that understand these data and uncover hidden patterns to gain useful information. Data analytics in SCM also leverages technologies such as AI and machine learning. Through the following process of data analytics such as planning of capacity, planning of sales and operation, business intelligence, forecasting of demand in the current environment knowing the bottom line through analyzing the data, we can able to rectify the challenges faced by the supply chain sector.

6.2 Revenue Management

Revenue management is the one management sector that requires a huge number of skills sets and updated technology, which makes the major challenge for the revenue management sector. Apart from that measuring the value cost that should be distributed requires a great number of strategic thinking, which is also lagging in the field of revenue management. To tackle the various challenges analyzing the data will help us to improve the efficiency of the company. In revenue management, data analytics allows you to see in real time how many people are looking for your destination, from which country they are coming, and how much money they have. This is critical information to have to establish the proper pricing and attract the right consumer; but, manually capturing and analyzing these data is unfeasible without arms with a big infrastructure and complex analytical models to accurately process the data.

6.3 *Inventory Management*

To run a company or a business, inventory management is one crucial part to handle. Effective inventory software assists in analyzing a huge amount of data from the sales, which was happened in the previous, and analyzing the demand of future for inventory while also considering the seasons and lead times of your organization. Both product performance and channel performance are important. In the modern era, different data analysis tools are used to capture and analyze data. Some of them are barcode data collection, cycle counting to improve accuracy, Pareto analytics for prioritization, integration planning, data tracking. The advantage of data analytics in the field of inventory management is that it increases operational efficiency, improves revenue and profit, reduces costs, and more.

6.4 *Quality Management*

The lack of a genuine culture of quality poses a threat to the QM team in terms of resistance to change as they are reluctant to embrace practices that require changes in current working styles and also threaten the QM team in terms of resistance to change. All information must be communicated to the organization at the right time and in the appropriate format to ensure that no useful results are missed. However, the shortcomings of the communication channel are an obstacle to obtaining high-quality products. Some businesses view quality certification as an autocratic effort that allows them to meet customer needs or contractual commitments while gaining a competitive advantage in the marketplace. Therefore, firms use data analysis tools to analyze quality control problems to get solutions. Recently, process data analysis is applied to quality control to achieve the maximum root solution, improving the system's integrated structure.

6.5 *Process Design*

As the amount of readily available data has grown, organizations have been under pressure to exploit their data collections. This pressure is transferred onto businesses' hired agencies, and designers must back up their qualitative results with a quantitative analysis of the client's data. Too many projects are under pressure to collect quantitative data for no reason, resulting in data that is unusable in terms of design, even though data cannot simply back every design decision. Many individuals make the mistake of relying on statistics based on past behavior to guide future product development. The data-based intuition was superior to the designer's intuition about what should be produced.

6.6 Location

Locating your firm in an area where the majority of businesses are keen to create a presentation is one of the most efficient strategies to increase business success. As a result, the ideal option is to choose an office address in a major metropolis. Geospatial data analysis aids comprehension, insight, decision-making, and prediction. By integrating geographic data such as demographics, traffic, and weather onto a smart map or dashboard, organizations may determine where an event has occurred, analyze why it is occurring, and get insight into what caused it. As part of their digital transformation, many companies are turning to GIS technology to create location intelligence.

6.7 Scheduling

Employees quit for several reasons, but schedule conflicts are one of the most prevalent. In truth, few things cause so much angst among managers and business owners. If you want your company to run smoothly and economically; however, proactively resolving your scheduling issues should be a top priority. Some of the major challenges faced by scheduling are employee shortage, overscheduling of process, problems in the organization, last-minute absences, turnover of an employee lack of availability, etc. To analyze the scheduling, certain analytics tools are used in the record, some of them are constraint summary, criticality summary, time-phased, early finishes, activity duration analysis, activity variance, activity summary, detailed breakdown of activities deleted, etc.

6.8 Maintenance

Maintenance is one of the crucial departments of management. Some challenges which is a threat to the management are unscheduled maintenance is not only detrimental to the business, it also occurs when an unexpected breakdown occurs. Informal and expeditious actions are taken to address and resolve issues. In the process, they are inspected, repaired, old or burnt parts removed and new spare parts installed. These activities are performed in this process. Unplanned downtime is the result of improper expectations or unexpected events. Part accuracy is very important in maintenance, and even a small mistake can cause a machine to fail. It can also affect the effectiveness of staff and equipment. Advances in technology can make management ineffective in analyzing the situation. One of the main tools used for maintenance is predictive analytics, where sensors generate, analytics and monitoring tools help clean up and analyze this data, and planning tools coordinate teams. In this area, maintenance guidance is a key aspect of predictive analytics tools used to improve the effectiveness of the maintenance sector.

6.9 HR and Job Design

In recent years, many modern problems have arisen in the design of work. Challenges include telecommuting, alternate work schedules, technical stress, job redefinition, and skill development. Companies are using this work-planning approach as a way to manage a growing business while facilitating a better work–family balance for employees. The challenges facing the work plan include work hardening, work break, lack of training, frustrated with technical issues. Data-driven ideas will help to rectify the issues faced by the management of HR and job design. Sourcing, monitoring, labeling, analyzing are the major tools that are done with data-driven analysis to improve the process of HR and job design in the management.

6.10 Goods and Services

In the era of globalization, there are several problems in the field of goods production and distribution. Since the recent day's consumers are choosing multichannel for their buying has become the major threat for goods sector. To attain maximum reach expensive marketing is required in general. Field service management is a complex task that requires multiple levels of collaboration. Beyond the technical and managerial aspects of field service management, businesses need effective ways to exceed customer expectations, balance their workforce and remain profitable. Data storage, cleansing, profiling, discovery, mapping, analysis, visualization, and monitoring are some of the major tools that are used to analyze the data and identify its solutions. In the case of the service team, various technical approaches are data-driven to improve consumer satisfaction in another end.

7 Consolidated Application of Data Analytics in Operation Management

Data analytics based on literature review can be classified into three broad categories, namely, data visualization and content analysis at the foundation level, using statistical analysis at the second level and using Artificial Intelligence and Machine learning level at the third stage (Fig. 1).

Operational aspects can be classified into basic operational aspects, process optimization aspects, total productive maintenance, and performance improvement

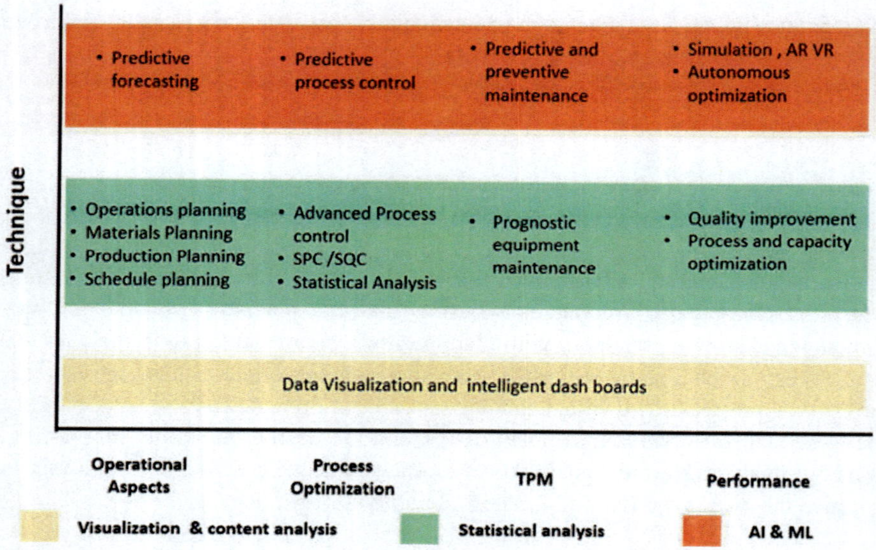

Fig. 1 Application of analytics in operation management

measures. A mapping of the analytics aspects and the operational aspects as indicated in the above-given figure indicates that data visualization and intelligent dashboards can assist in all the operational dimensions such as operational aspects, process optimization, TPM, and performance improvement aspects. Statistical analysis can assist in various aspects such as operational planning, material planning and schedule planning and in SPC/SQC in process optimization, quality improvement and process improvement in performance measures. High level of analytics such as AI and Machine learning can assist in predictive forecasting, predictive process control, predictive maintenance, and autonomous optimization.

Future scope of study:

There are numerous scopes for future research in the subject area such as

- There can be a detailed meta-analysis on the mentioned area and an inclusion of all recent titles can be incorporated.
- A study on the impact of data analytics in various dimensions such as supply chain management and quality management can be done as a separate study.
- An empirical study on the same would provide significant insights on the role of analytics in the area of operation management.

8 Conclusion

Data analytics is a major skill in new era of OM and can be extended to operational scopes such as forecasting, inventory management, logistics management, supply chain management, sales management, and risk analysis through different approaches to big data such as methods and strategies. Data analytics can help uncover hidden variables that are causing non-existent production bottlenecks. After identifying the cause of the problem, manufacturers use targeted data analysis tools to better understand the root cause of bottleneck variables. This allows manufacturers to increase productivity while reducing costs and eliminating waste. Companies that analyze and act on customer data in real time can make faster decisions. Traditionally, there is always a chance that businesses will only be aware of serious business problems on a quarterly or annual basis and will not be able to attend to them on time until they change their operations. On the other hand, businesses using operational analytics can make the necessary adjustments to their processes and workflows in near real time or near real time, increasing profitability and reducing waste. It also helps to quickly identify and respond to problems and defects.

References

1. Ban GY, Rudin C (2019) The big data newsvendor: practical insights from machine learning. Oper Res 67(1):90–108
2. Battistoni E, Bonacelli A, Fronzetti Colladon A, Schiraldi MM (2013) An analysis of the effect of operations management practices on performance. Int J Eng Bus Manag 5(Godište 2013):5–43
3. Bertsimas D, Kallus N (2020) From predictive to prescriptive analytics. Manag Sci 66(3):1025–1044
4. Boudreau J, Hopp W, McClain JO, Thomas LJ (2003) On the interface between operations and human resources management. Manuf Serv Oper Manag 5(3):179–202
5. Butta TL, Abegaz MB (2016) Challenges in the operation of multimodal transport system: the case of Ethiopian shipping and logistics services enterprise. Int J Appl Res 2(7):927–932
6. Choi TM, Kumar S, Yue X, Chan HL (2022) Disruptive technologies and operations management in the industry 4.0 era and beyond. Prod Oper Manag 31(1):9–31
7. Chen Y, Farias VF, Trichakis N (2019) On the efficacy of static prices for revenue management in the face of strategic customers. Manag Sci 65(12):5535–5555
8. Fleury A, Shi Y, Junior SF, Cordeiro JHD, Fleury MTL (2015) Developing an analytical framework for study of emerging country multinationals' operations management. Int J Prod Res 53(18):5418–5436
9. Satpathy R, Choudhury T, Satpathy S, Mohanty SN, Zhang X (2021) Data analytics in bioinformatics: a machine learning perspective. John Wiley & Sons
10. Fisk RP, Brown SW, Bitner MJ (1993) Tracking the evolution of the services marketing literature. J Retail 69(1):61–103
11. Garg A, Deshmukh SG (2006) Maintenance management: literature review and directions. J Qual Maint Eng 12(3):205–238
12. Garrett J, Kalyan TVR, Talluri (2014) An introduction to revenue management. Informs Tutor Oper Res 142–194
13. Glaeser CK, Fisher M, Su X (2018) Optimal retail location: empirical methodology and application to practice. Manuf Service Oper Manag 21(1):86–102

14. Gunasekaran A, Subramanian N, Ngai WTE (2019) Quality management in the 21st century enterprises: research pathway towards industry 4.0. Int J Prod Econ 207:125–129
15. Knod EM, Schonberger RJ (2000) Operation management meeting customers demand. McGraw Hill Irwin, New York
16. Lee CKM, Cao Y, Ng KH (2017) Big data analytics for predictive maintenance strategies. In: Supply chain management in the big data era. IGI Global, pp 50–74
17. Khullar R, Sharma T, Choudhury T, Mittal R (2018)Addressing challenges of Hadoop for big data analysis. In: International conference on communication, computing and internet of things (IC3IoT), Chennai, India
18. Meredith JR (2001) Hopes for the future of operations management. J Oper Manag 19(4):397–402
19. Munyaka JB, Yadavalli VSS (2022) Inventory management concepts and implementations: a systematic review. S Afr J Ind Eng 33(2):15–36
20. Niall P (2012) Business history and operations management. Bus Hist 54(2):154–178
21. Pacana A, Gazda A, Bednarova L (2014) The impact of quality information on the innovatory environment of the public administration. Int J Interdiscip Theory Pract 4:25–26
22. Permana A, Purba HH, Rizkiyah ND (2021) A systematic literature review of Total Quality Management (TQM) implementation in the organization. Int J Prod Manag Eng 9(1):25–36
23. Phan P, Chambers C (2013) Advancing theory in entrepreneurship from the lens of operations management. Prod Oper Manag 22(6):1423–1428
24. Podreciks A, Uhlenbrock N, Ungerman K (2018) Who's shopping where? The power of geospatial analytics in omnichannel retail
25. Rajendran S, Ravindran AR (2019) Inventory management of platelets along blood supply chain to minimize wastage and shortage. Comput Ind Eng 130:714–730
26. Raouf A (1998) Development of operations management in Pakistan. Int J Oper Prod Manag 18(7):649–650
27. Sasser EW (1976) Match supply and demand in service industries. Hardward Business Review pp 5–17
28. Shukla RK, Garg D, Agarwal A (2011) Understanding of supply chain: a literature review. Int J Eng Sci Technol 3(3):2059–2072
29. Singhal J, Singhal K (2002) Supply chains and compatibility among components in product design. J Oper Manag 20(3):289–302
30. Sukati I, Hamida AB, Baharuna R (2012) The study of supply chain management strategy and practices on supply chain performance. Procedia Soc Behav Sci 40:225–233
31. Swanson L (2001) Linking maintenance strategies to performance. Int J Prod Econ 70:237–244
32. Velibor BOŽIĆ (2019) Smart hospital–our experience. Orase Inteligente si dezvoltare regionala 3(2):95–99
33. Vonderembse MA, Uppal M, Huang SH, Dismuke's JP (2006) Designing supply chains: towards theory development. Int J Prod Econ 100(2):223–238
34. Wang XL (2012) Relationship or revenue: Potential management conflicts between customer relationship management and hotel revenue management. Int J Hosp Manag 31(3):864–874

Image Retrieval Technique Using Visual Features in Content-Based Image Retrieval System

Surbhi Tyagi, Vivek Kumar, Sudeshna Chakraborty, and Praveen Shukla

1 Introduction

The main source of generating massive amount of image databases is education, industry, medical, social, and other life facilities. Here living in the information age, when the amount of data available from science and culture is nearly endless. It has raised the issue of how to handle these massive datasets. This required a perfect system that is able to retrieve the desired image from big collection of images. That means, identifying an item of interest is becoming increasingly difficult, like finding a digital needle in a haystack. One of the key successful elements for all efforts to solving this challenge appears to include the use of effective but simple retrieval techniques. Text-Based IR is the first method out of two common methods of searching the images which is based on keywords that is used to tag images [1] There are many of retrieving images using text [2] and to improve multi-label image annotation. Using image information and Jaccard similarities, research creates communities of comparable images [3]. There are two major problems in this research. The user's inability to convey the anticipated visual content while utilizing the current query that is intention gap, and semantic gap, which is the user's inability to express relatively complex semantic notions using basic visual features. So the focus is to reduce the

S. Tyagi (✉) · V. Kumar
Quantum University, Roorkee, Uttarakhand, India
e-mail: surbhityagi37@gmail.com

V. Kumar
e-mail: vicechancellor@quantumuniversity.edu.in

S. Chakraborty
Galgotias University, Greater Noida, India

P. Shukla
Manipal University Jaipur, Dehmi Kalan, Off Jaipur-Ajmer Expressway, Jaipur, Rajasthan, India
e-mail: praveen.shukla@jaipur.manipal.edu

A. Ramdane-Cherif et al. (eds.), *Machine Intelligence and Data Science Applications*, Algorithms for Intelligent Systems, https://doi.org/10.1007/978-981-99-1620-7_7

gap between the images and worked on the CBIR algorithm. The following three concerns must be addressed while creating a CBIR algorithm:

i. Representing an image
ii. Organization of images
iii. Additionally, comparing how similar two images are

Methods for visual retrieval that use text search techniques may suffer from inconsistency between text and visuals due to the lack of visual content as a ranking clue. Because of this intention gap and semantic gap issues, this challenge is difficult [4]. In direction of these issues, research is based on the visuals of the content. The goal of visual information retrieval (VIR) is to develop an open prototype that will be the foundation for upcoming research on key VIR challenges. The goals of the VizIRproject are to investigate the framework's intended architecture, the ui, and the integrated structure difficulty. The querying framework contains modules for extracting features, similarity testing, media processing, and database access [5].

A visual search algorithm's flowchart of a general CBIR algorithm is depicted using Fig. 1, in which we have two sections. First one is the query section, in which the user must have to provide a query image for search. In the second section, similar images are retrieved after applying the proposed algorithm.

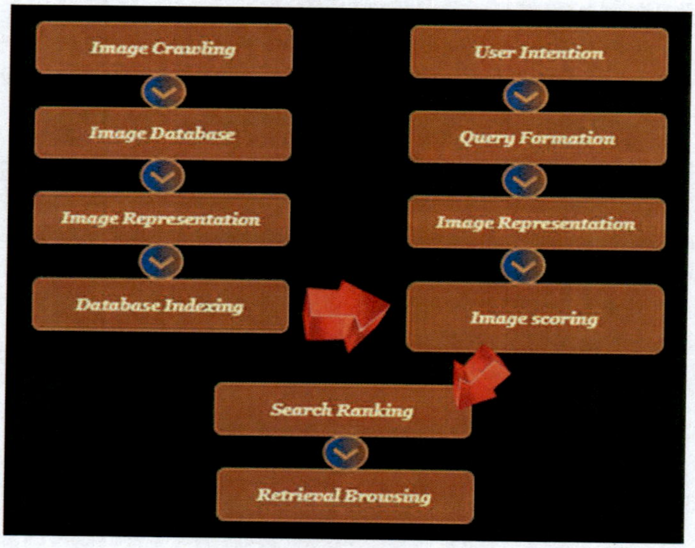

Fig. 1 Architecture of the proposed model CBVIR

1.1 Literature Review

For many years, image retrieval has been a significant research subject, and it has grown more and more exciting in recent years. The task is to build a visual classification model on a large collection of data content without label. The research shows that, despite each image's text being frequently noisy and having only a tenuous link to its data, it may be used to train a deep CNN that accurately predicts the polarity of previously unobserved images [6]. Color, texture, shape, and semantic are the key aspects of CBIRs. The CBIR framework is two stages, mandatory and optional stages. An optional preprocessing stage might include size changing, segmentation, de-noising, and re-scaling. The feature extraction is the stage, in which conversion of visual concept has done into numeric representation [1].

Extraction and reduction of features are the two primary components of visual-based model. When the user's high-level conceptions are difficult to define in terms of low-level features, the results obtained using low-level features are typically appropriate [7]. An image's behavior is described by its features. The CBIR method's effectiveness is mostly determined by extraction. In this paper, the edge of histogram and discrete wavelet transform are combined to improve the performance of an IR system. Artificial Neural Networks (ANN) were used as primary datasets [8]. Different techniques extracted low-level image features, which were later reduced from a larger collection of image data. These algorithms, however, are incapable of expressing high-level features, which are routinely used by humans during analysis. Researchers have devised solutions for dealing with higher quantities of rich content images in which the first one is the ontology, which defined ontology high-level theories, secondly is ML tools trying to equate low-level characteristics with query concepts, and lastly which is RF which is used to enhance lifelong improvement of user interests [9]. This study extracts high-level features from low-level data and uses those features that improve the quality of queries in order to close the semantic gap. According to the findings, using these high-level features improves result set reliability by optimizing recall and precision [10]. From the supplied images, using a query keyword, high-level feature has been formed. The images discovered using low and high-level qualities are then analyzed to determine which are visually and conceptually comparable [11].

Region-based retrieval is also helpful in this issue. In this, a region of any image or particular location has been used to work. Classification, extraction of features, labeling, and retrieving images are all required processes in retrieving photographs that typically contain regions as specified in a query [12]. The problem of estimating region connectivity turned into an assigned similarity measure problem. The researcher suggests that the image region's basic spatial arrangement and significance be kept while re-estimating area features throughout the relevant feedback [13]. For improving results of internet, scholars have long shown an interest in Query Expansion as a part of information retrieval [14]. To improve CBIR accuracy, a novel Dominant Color Descriptor (DCD) approach is developed. At first, Canny edges of images are extracted and then broadened using morphological techniques in the following stage. Finally, pixels that seem to be away from edges are valued less than pixels that are close to the edges. Similarly, pixels in low color variability regions are less valued in constructing DCD attributes, whereas pixels in far more relevant regions are more weighted [15]. The CBIR technique was used to aggregate low-level characteristics. To obtain texture details of the image, the Color Histogram, Discrete Wavelet Transform, and Edge Histogram Descriptor have been employed [16]. In creating a strong algorithm, similarity measurement between images is also essential. Human Activities Recognition is a method of ordering an individual's action by utilizing reactive different types of sensors that are impacted by human activity. This study focuses on the detection of human activity using smartphone sensors and some ML and deep learning characterization methodologies [17]. Brain-computer interface system communicates with the outside environment by using electroencephalography (EEG) signals [18]. The classification problem of right hand and right foot motor images has been explored in this study. And, based on three wavelet types, determine the optimal wavelet and decomposition level, as well as the best classifier perspective of accuracy and receiver operating curve [19]. The proposed system can detect the subject's IC and NC states. As an RCP, a novel 23 matrix made up of household appliances has been presented. For healthy and paralyzed subjects, the average accuracy rate in synchronous mode was 92.44% and 89.33%, respectively [20].

This paper has been divided into different sections. In which Sect. 2 describes the proposed methodology, Sect. 2.1 describes dataset and method used, Sect. 3 presents the experimental result, and Sect. 4 concludes the whole paper.

2 Methodology

2.1 Dataset

A IR algorithm will be constructed in this work utilizing MATLAB as a framework. Using this model user has to provide a input and a similar image has been reverted. The Indian Sign Language database, which contains 1000 images, will be used to get the outcome [21]. The similarity criteria might vary significantly; however, in the current project, the picture characteristics retrieved will be color-based, and the similarities will be using histogram similarity. The produced code, as well as testing and analyzing results, will be thoroughly explained, as will comparisons of the current task to competing systems and methods.

2.2 Visual Search Algorithm

As per the model, the color histogram feature will be used to display the user input, with the HSV region selected. To work on this model recovering 32 dimensions, each Hue, Saturation, and Value component is uniformly quantized into 8, 2, and 2 bins, respectively. HSV has the angular dimension Hue and is a cylindrical geometry. At zero degree having red, moving through green primary at 120 degrees, the blue primary at 240 degrees, and the black primary at 360 degrees. The vertical axis is represented by neutral or grey colors, from black to white values (Figs. 2, 3, and 4).

The issues of image organization and image similarity measurement were discussed. The relevance score was used in the results of this analysis, which is defined by calculating the distance between the collected feature vector of the visual representation or by choosing from image feature matches. The distance measurement method has been selected. And algorithm converts image into

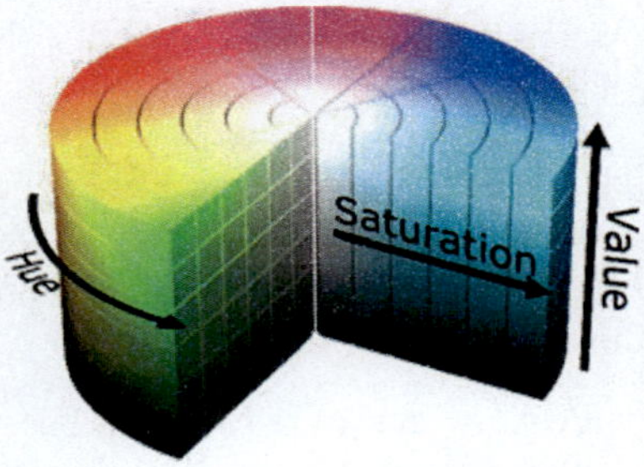

Fig. 2 HSV space representation

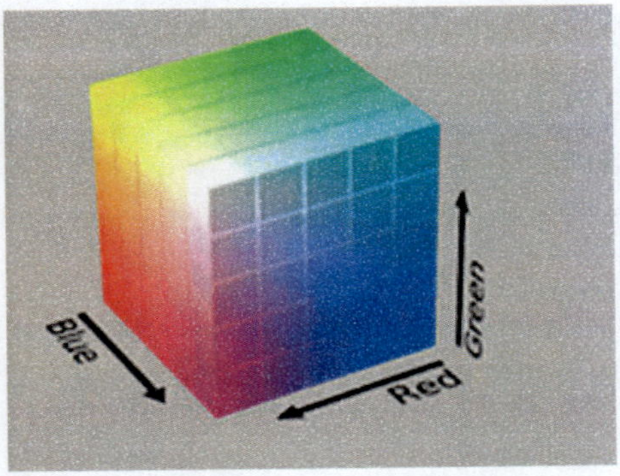

Fig. 3 HSV representation of RGB

Fig. 4 An illustration of the Hue dimension

$$\sqrt{\sum_{i=1}^{N} (X_i - Y_i)^2} \tag{1}$$

where N is defining the feature's number and both X and Y are comparison between them.

Algorithm: The algorithmic approach is divided into two phases:

- Feature Extraction
- Applying CBIR

In the first phase images are passed through feature extraction technique that is color histogram is used to extracting the feature from the images. The second phase analyzes the similarities between the training and query image.

It is reasonable to expect that the proposed strategy of just employing the color histogram for image representation will not yield a working solid algorithm. If two distinct photos share a same color palette, they may have a comparable representation. As a result, there is a need to move beyond the color histogram and incorporate other elements for display.

The combined features will generate a 190-dimensional feature vector, significantly increasing retrieval efficiency. The total of the characteristics produces a 190-dimensional feature vector, which considerably improves retrieval effectiveness. Other changes can close the previously mentioned the semantic gap, there are three basic tiers of questionnaires:

- Primary level of retrieving is based on our algorithm's creation of basic attributes like color, texture, or areas. At this level, the user types in a query, for instance, and anticipates that the program will identify images that match it.
- The retrieval of items of a particular kind that have been identified by derived features constitutes the second level and calls for some logical implication.

Retrieval using abstract qualities, the third and final stage, necessitates sophisticated thinking. To demonstrate our point, let's say the consumer expects more specialized results and searches for both a dog and a cat image.

3 Experimental Results

3.1 Parameters

To obtain precision finds an image from the database. The feature vectors for the query and images presented in the dataset has been extracted and accuracy is then evaluated using algorithm. Lastly, first T visuals in the collection that are closest to the query image are regarded to be images comparable to the query. Some of these T photos (say N) belong to the query image class, whereas the others do not. As a

Table 1 Proposed model's precision, recall, and F1 score on ISL dataset [21]

Sign	Precision (%)	Recall (%)	F1 Score (%)
D	**86.07**	**27.40**	**41.57**
A	**87.57**	**26.14**	40.27
B	**84.94**	**25.59**	39.35
F	**88.02**	**26.40**	40.62

result, the precision and recall of these results are as follows:

$$\text{Precision} = \frac{\text{N}}{\text{T}} = \frac{\text{Number of similar Image retrieved}}{\text{Total Number of Images Retrieved}}$$

$$\text{Recall} = \frac{\text{N}}{\text{K}} = \frac{\text{Number of similar Image retrieved}}{\text{Total Number of Images in Dataset}}$$

where

N = number of similar photos recovered,

T = defines the total retrieved images number, and

K = defines the total of similar images number of the dataset

Next parameter is the F1 score, this metric is calculated as

$$\text{F1 Score} = 2 * \frac{\text{Precision} * \text{Recall}}{\text{Precision} + \text{Recall}}$$

3.2 *Proposed Methodology Result*

The precision, recall, and accuracy of the proposed approach have been shared in Table 1 using Corel-1k and ISL. It exhibit the precision and recall of the proposed technique and literature in the Corel-1k dataset and ISL. So as per the results of the proposed methodology, got the better results with maximum precision in ISL are 86.07%, 87.57%, 84.94%, 88.02% respectively and in Corel-1k are 62.07%, 94.86%, 92%, 52.02%, 72.56%, respectively.

3.3 *Comparison with Other Models*

In this experiment, the proposed method's recall and precision are measured against those of earlier approaches on the ISL and Corel-1k datasets (Fig. 5). Figures 6, 7, and 8 display the suggested method's precision and recall on the Corel-1k dataset in comparison to the literature [15, 16, 18]. The accuracy of the proposed and alternative methods has been shown in Figs. 7 and 9.

Figures 6 and 7 present the suggested model that outperforms competing models in terms of precision in fig out of ten classes. The average precision for all classes in Corel-1k for all methods is shown in Figs. 8 and 9. The suggested technique outperforms literature [8, 15, 16] by 62.07%, 94.86%, 92%, 52.02%, and 72.56%. For more than 1000 images dataset the model doesn't have any effect on it because we are working on image quality not on number of images. We are using feature segmentation technique in which we are searching for similar intensity images.

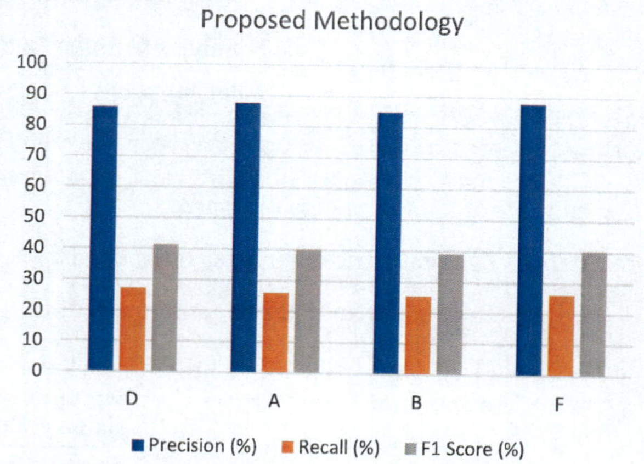

Fig. 5 Proposed method's precision, recall, and f1 score on ISL dataset

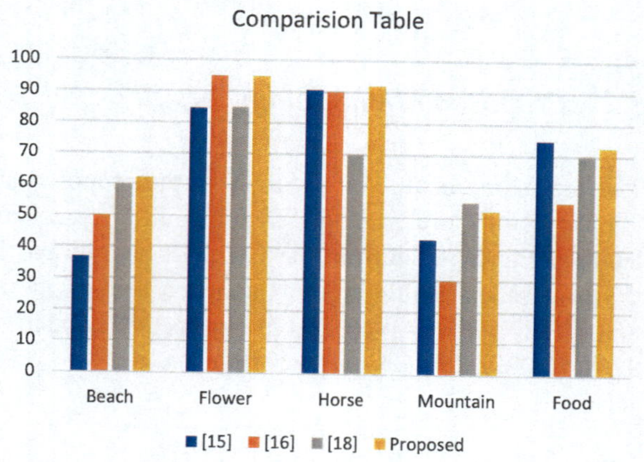

Fig. 6 Precision of the proposed model and literature review [15, 16, 18] in Corel-1 k dataset

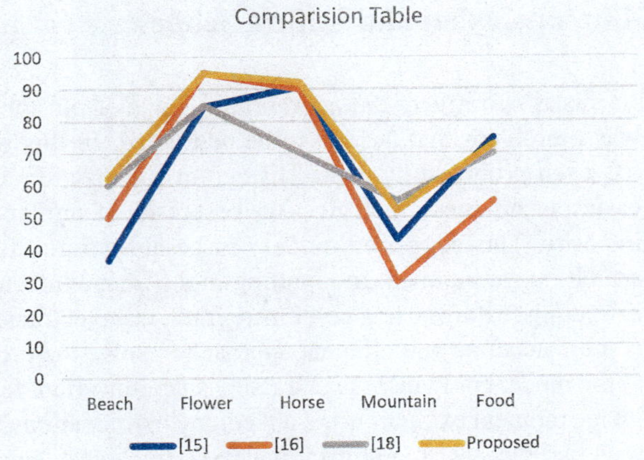

Fig. 7 Precision of the proposed method and literature [15, 16, 18] in Corel-1 k dataset

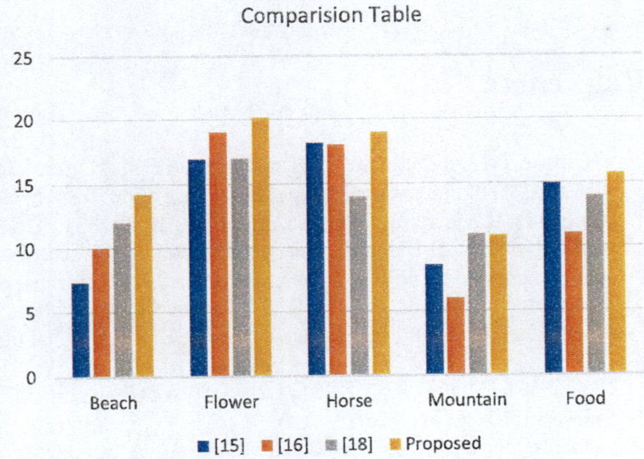

Fig. 8 Recall of the proposed method and literature [15, 16, 18] in Corel-1 k dataset

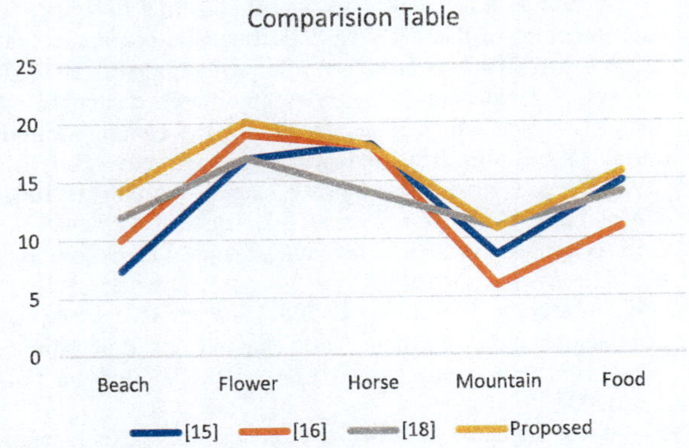

Fig. 9 Recall of the proposed method and literature [15, 16, 18] in Corel-1 k dataset

4 Conclusion and Future Scope

In this is to retrieve the appropriate images from the collection of images, there was a problem that needed to be addressed. In this paper CBVIR technique that has been proposed by using Histogram equalization technique. This is one of the easiest techniques which gives the best result of similar images on the basis of image intensity. This suggested model can be able to get the desired images as per the search image inputted. In inputting in the image information retrieval challenge by extracting features using color histogram, wavelet transform, and Gabor filter which is more accurate and efficient approach to solve the problem. Future improvements to the model could include extracting shape-invariant features using approaches like image moment extraction and ant colony optimization. The RF classifier can be used with particle swarm optimization (PSO) to choose features and train features.

References

1. Hameed IM, Abdulhussain SH, Mahmmod BM, Hameed IM, Abdulhussain SH, Mahmmod BM (2021) Content-based image retrieval: a review of recent trends computer science. Cogent Eng 8(1). Review article. https://doi.org/10.1080/23311916.2021.1927469
2. Li X, Uricchio T, Ballan L, Bertini M, Snoek C, Del Bimbo A (2015) Image tag assignment, refinement and retrieval (ACM Multimedia 2015 Tutorial), pp 1325–1326
3. Johnson J, Ballan L, Fei-Fei L (2015) Love thy neighbors: image annotation by exploiting image metadata. In: Proceedings of the IEEE international conference on computer vision, vol 2015 Inter, pp 4624–4632. https://doi.org/10.1109/ICCV.2015.525
4. Zhou W, Li H, Fellow QT (2017) Recent advance in content-based image retrieval: a literature survey, pp 1–22
5. Eidenberger H, Breiteneder C (2003) VizIR—A framework for visual information retrieval. J Vis Lang Comput 14(5):443–469. https://doi.org/10.1016/S1045-926X(03)00035-1
6. Vadicamo L et al (2018) Cross-media learning for image sentiment analysis in the wild. In: Proceedings of the 2017 IEEE international conference on computer vision work. ICCVW 2017, vol 2018-Jan, pp 308–317. https://doi.org/10.1109/ICCVW.2017.45
7. Goyal N, Singh N (2014) A review on different content based image retrieval techniques using high level semantic features. Int J Innov Res Comput Commun Eng 2(7):4933–4938
8. Ashraf R et al (2018) Content based image retrieval by using color descriptor and discrete wavelet transform. J Med Syst 42(3). https://doi.org/10.1007/s10916-017-0880-7
9. Mohd Zin NA, Yusof R, Lashari SA, Mustapha A, Senan N, Ibrahim R (2018) Content-based image retrieval in medical domain: a review. J Phys Conf Ser 1019(1). https://doi.org/10.1088/1742-6596/1019/1/012044
10. Eidenberger H, Breiteneder C (2002) Semantic feature layers in content-based image retrieval: implementation of human world features. In: Proceedings of 7th international conference control, automation robotic vision, ICARCV 2002, pp 174–179. https://doi.org/10.1109/icarcv.2002.1234816
11. Singh S, Sontakke T (2014) An effective mechanism to neutralize the semantic gap in content based image retrieval (CBIR). Int Arab J Inf Technol 11(2):124–133
12. Jan MM, Zainal N, Jamaludin S (2020) Region of interest-based image retrieval techniques: a review. IAES Int J Artif Intell 9(3):520–528. https://doi.org/10.11591/ijai.v9.i3.pp520-528
13. Li CY, Hsu CT (2008) Image retrieval with relevance feedback based on graph-theoretic region correspondence estimation. IEEE Trans Multimed 10(3):447–456. https://doi.org/10.1109/TMM.2008.917421

14. Azad HK, Deepak A (2019) Query expansion techniques for information retrieval: a survey. Inf Process Manag 56(5):1698–1735. https://doi.org/10.1016/j.ipm.2019.05.009
15. Fadaei S (2022) New dominant color descriptor features based on weighting of more informative pixels using suitable masks for content-based image retrieval 35(8):1457–1467
16. Nazir A, Ashraf R, Hamdani T, Ali N (2018) Content based image retrieval system by using HSV color histogram, discrete wavelet transform and edge histogram descriptor. In: 2018 international conference computing mathematics and engineering technologies invent, innovative integration socioeconomics development. iCoMET 2018—Proceedings, vol 2018-Jan, pp 1–6. https://doi.org/10.1109/ICOMET.2018.8346343
17. Khimraj P, Shukla K, Vijayvargiya A, Kumar R (2020) Human activity recognition using accelerometer and gyroscope data from smartphones. In: Proceedings of the 2020 international conference emerging trends community control computing ICONC3 2020. https://doi.org/10.1109/ICONC345789.2020.9117456
18. Shukla PK, Chaurasiya RK, Verma S (2020) Single-channel region-based speller for controlling home appliances. Int J E-Health Med Commun 11(4):65–89. https://doi.org/10.4018/IJEHMC.2020100105
19. Shukla PK, Kumar Chaurasiya R (2019) An experimental analysis of motor imagery EEG signals using feature extraction and classification methodologies. In: 2018 international conference on computing power communication technology GUCON 2018, no dataset III, pp 796–799. https://doi.org/10.1109/GUCON.2018.8675032
20. Shukla PK, Chaurasiya RK, Verma S, Sinha GR (2021) A thresholding-free state detection approach for home appliance control using P300-based BCI. IEEE Sens J 21(15):16927–16936. https://doi.org/10.1109/JSEN.2021.3078512
21. Tyagi A, Bansal S (2022) Sign language recognition using hand mark analysis for vision-based system (HMASL). In: Emergent converging technologies and biomedical systems. Springer, pp 431–445

Intelligent Hybrid VLC/RF Communication Protocol for Train Data Transfer

Hongyu Guan, Zhuosha Guo, Amar Ramdane-Cherif, Abderraouf Khezaz, Manolo Dulva Hina, and Luc Chassagne

1 Introduction

For security and traffic management purposes in a train system, a general train can be equipped with up to 20 cameras [1]. Standard cameras used for on-board CCTV, such as the one in the Montreal subway system [2], have a resolution of 800*600. According to the APTA (American Public Transportation Association) guidelines, a CCTV camera covering a passenger area must be able to record at a frequency of 5 to 15 fps. With an H.264 coding, this would correspond to around 1.1 Mb/s of data generated by one camera. For a 2.5 hours trip, each camera in a train would generate around 10 Gb of data during the trip, so about 200 Gb of data for a train equipped with 20 cameras.

For security and performance reasons, this important amount of data needs to be stored, and because of spatial constraints in the train, the train station is the natural

H. Guan · A. Ramdane-Cherif · L. Chassagne
Université Paris-Saclay, UVSQ, Versailles, France
e-mail: hongyu.guan@uvsq.fr

A. Ramdane-Cherif
e-mail: amar.ramdane-cherif@uvsq.fr

L. Chassagne
e-mail: luc.chassagne@uvsq.fr

Z. Guo (✉)
OLEDCOMM, Vélizy-Villacoublay, France
e-mail: zhuosha.guo@oledcomm.net

A. Khezaz · M. D. Hina
ECE Paris, Paris, France
e-mail: abderraouf.khezaz@ece.fr

M. D. Hina
e-mail: manolo-dulva.hina@ece.fr

A. Ramdane-Cherif et al. (eds.), *Machine Intelligence and Data Science Applications*, Algorithms for Intelligent Systems, https://doi.org/10.1007/978-981-99-1620-7_8

place to do so. This implies the need to transfer such important amount of data from the train to the station. There are multiple ways of doing so, for example manually having a human agent swapping hard-drives and transporting them to the station, or the use of an RF communication between the train and the station. However this RF communication solution depends on the architecture of the station : concrete and iron can interfere with the electromagnetic waves. This issue can be overcome by the use of a complementary mean of wireless communication, such as VLC (Visible Light Communication).

The structure of the rest of this work is organized as follow. Section 2 presents the literature review of hybrid VLC protocols and VLC applications for trains. Section 3 describes the concept of our intelligent hybrid approach, and Sect. 4 is the conclusion.

2 State of the Art

It is necessary that the communication between two or more intelligent units be reliable. This reliability in communication is hampered by the fact that the most common communication technologies are rather slow. On the other hand, the fastest technologies, such as the VLC, are easily disturbed by the environment and therefore not very reliable in an uncontrolled space. These drawbacks can be addressed by using 5G but then its implementation is costly. In addition, its throughput is dependent upon the number of its users. It is important to note that its frequency range (between 700 MHz and 28 GHz) is highly regulated.

In lieu of these constraints in communication, various researchers find what alternative means can be exploited. For example, various studies were conducted on how VLC can be used for train applications, in particular, on the use of in-board wireless communication. For instance, Vijayalakshmi and Nesasudha [3] experimented on the use of VLC within a wagon in order to provide internet access to its passengers. As a result, it was noted that wireless traffic has increased exponentially. Hence, VLC is considered as a suitable alternative communication protocol to use in an enclosed environment, one that is equipped with LED in dimmed lights. This approach is similar to the one proposed by Oledcomm [4]. The deployment of VLC for data transfer as well as internet access is also contained in the studies conducted by Lamas et al. [5].

Recently, some experiments were conducted on the use of VLC in a train and its communication with its surroundings. For instance, Dagli [6] proposed V2X (Vehicle to Everything) in a train using VLC, making use of tunnel lights and signal lights in order to control trains.

The key advantage of using VLC over other communication protocols is that it is immune to RF (radio frequency) interference. That being said, it is still not immune to other sources of noise. Given that it is working within visible light bandwidth, it follows then that anything that the human eyes can see is a potential source of noise.

In the field of hybrid communication, the VLC communication protocol is generally considered as an alternative solution. According to [7], RF communication is susceptible to interference and jamming attacks. The use of cognitive radio (intelligent detection of unused transmission channels) is generally perceived as less risky option. Nonetheless, the use of VLC communication is suggested to strengthen the security need. In addition, [8], proposed a 5G/VLC joint prototype. In such experiment, sensors data from smart city are collected and transmitted to the road infrastructure (traffic lights) using 5G and then transmitted to vehicles using VLC. Such ingenious hybrid solution makes data reach the intended vehicle destination, ensuring that the wireless network is not saturated. In a similar study, Rahaim et al. [9], used VLC as a substitute communication bus whenever WiFi reaches maximum capacity. In another related undertaking, Rakia et al. [10] proposed a dual-hop data transmission system. In the first hop, VLC transmits data to a relay node, therein a RF protocol takes over. For energy optimization purposes, the DC component of the received optical signal is collected to be used as power to the RF communication. Notwithstanding the optimal throughput results obtained from the experiment, the authors still acknowledge that their DC bias and power-harvesting component performance can still be improved.

Another related work is that of Pan et al. [11] on energy-harvesting within a hybrid RF/VLC set-up. This work focuses on data privacy. Such RF/VLC approach allows that the designated data receiver acquires the message as intended, without eavesdropping.

In addition, the Acemind European Collaborative Project [12] manifested the benefit of a hybrid communication protocol and VLC set-up. This project intends to provide a consistent solution for the management of local networks. In this work, VLC is considered as a complementary solution to the current RF system, such as 4G, 5G, and WiFi.

Having used VLC as a component of a hybrid system, they have listed its benefits as follows:

- Having multiple and complementary technologies ensures that none of them gets congested. It also guarantees that a back-up is available if one technology shuts down.
- Using the existing lighting systems, the VLC technology optimizes that infrastructure of the communication system.
- The VLC technology provides an energy-efficient communication system given that the use of such technology minimizes energy consumption.

In their experiments, the VLC technology was used in streaming a video. No problems were encountered, and the overall delay was smaller than 1 ms.

In 2016, Ahamed [13] studied the footage inside the train requirements of the UK Network Rail, whose operation is similar to APTA. Such footage needs to be downloaded into a server whenever a train is entering a depot/siting [14], VLC technology is the proposed methodology.

It is interesting to note that most of hybrid VLC systems are focused on Data Link Layer and Network Layer of the OSI model. If applied on a higher level, such as in the Application Layer, then the use of a hybrid communication system could also bring added benefits, such as less complex implementation.

As previously stated, a hybrid VLC/RF provides desirable performance in data transfer between a train and a train station. In relation to previous works, our work provides a comprehensive study on the Application Layer of a hybrid VLC/RF communication in train data transfer applications. In this paper, the train uses its headlights to communicate with photo-sensors located in its front. As soon as the communication gets established, video data, after having parsed into smaller packets, are transmitted into the station using VLC. In parallel, the hash of each packet is transmitted to the receiving server using RF means. The receiving server compares the received hash against the one it just calculated based on the received packets. If there is an error between the received hash and the calculated hash, the receiver can ask for a re-transmission of the previous packet.

This approach provides few advantages, as listed below: There are a few advantages to this approach :

- **Speed**: The VLC technology provides high-speed communication with a data rate reaching up to 220 GB/s throughput [15].
- **Interference immunity**: The VLC technology requires a LoS (line-of-sight) condition yet it is immune to electromagnetic interference.
- **Optimal context**: As stated, the VLC technology is sensitive to sunlight and other light noise. This specific context (data communication within the train application) is therefore an ideal application.
- **Logistic**: Unlike other communication protocols, the VLC technology does not need to use an antenna. In contrast, it does require a dedicated set of LED, photodiode and other components that make up the transmission system. In our domain of application, a train already comes with its headlights. This only must be modulated to be used in transmitting data, providing lots of savings on implementation time. The only required physical changes would be on the receiving ends of the train station.
- **Quality Of Service**: In contrast to most works (focused on improvements in the Logical Link Control Layer of the OSI Model) found in the literature, the intelligent hybrid solution presented in this paper focuses on the quality of service (QoS) consideration in the Application Layer.

As outlined in the literature review, the VLC technology shows strong potential when paired with another communication protocol. There have also been some works done on VLC applications for trains, including data transfer from trains to stations.

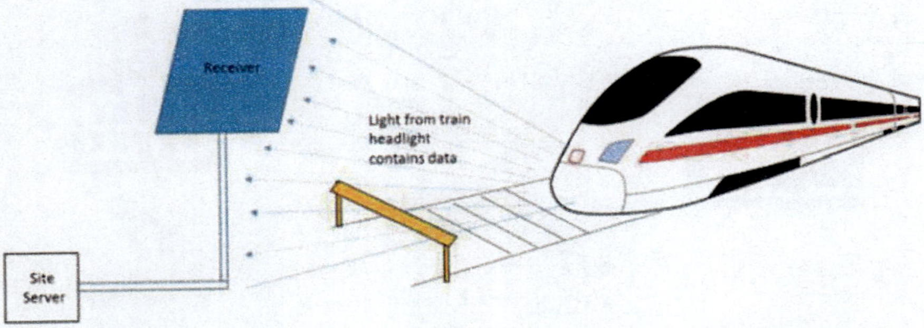

Fig. 1 Illustration of the use case [13] for the train data transfer using an hybrid VLC/RF approach

3 Common use Case Application

This paper proposes the use of an intelligent hybrid VLC/RF communication protocol in order to ensure a correct transmission of data between 2 agents. The use case chosen here is *a train transmitting an important amount of data to a train station.* The intelligent hybrid VLC/RF approach is suitable in this situation because:

- The train is immobile and usually sheltered from light noise sources (the sun)
- Trains already possess light bulbs and can modulate the output in order to communicate
- Train station infrastructures are usually made of metal and concrete [16], making it hard for RF waves to communicate [17].

As stated by that last point, the architectural infrastructure of a train station can make RF communication difficult, so it could be interesting to pair this protocol with a complementary channel of wireless communication.

The concept proposed in this paper is illustrated in Fig. 2. The train uses its headlights in order to communicate with stationary photo-sensors located at the end of the line in its front, as illustrated by Ahamed in Fig. 1. Once the communication is established, the system parses the video data in smaller packets, these smaller packets are transmitted via the headlights of the train into the station. In parallel, the hash of each packet is transmitted via RF communication to the receiving server of the station, which will compare it to the one that is computed from the transferred data. If an error occurs, a re-sent of the previous packet will be asked from the receiver.

3.1 Hashing Algorithms

Hashing is the process of transforming a data by passing it through a function to produce a fixed-size string of characters. This is usually represented by a shorter,

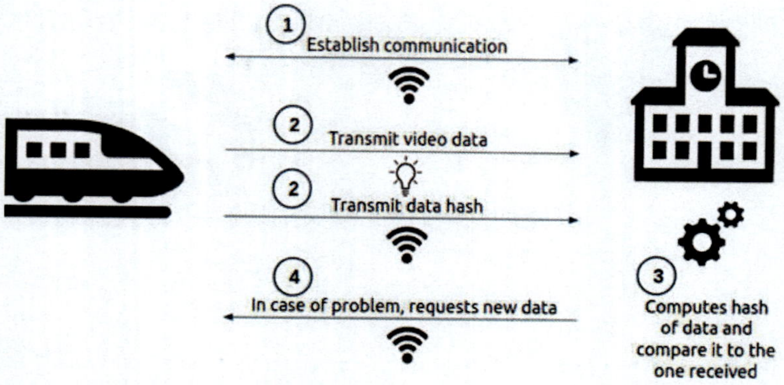

Fig. 2 Illustration of the proposed model of this work in the application layer

fixed-length value or key that represents and makes it easier to find or employ the original string. The most popular use for hashing is the implementation of hash tables.

Hashing is important as it is a useful tool to manage data security and integrity. This is due to the fact that given the same frame of data, it will always produce the same string of characters as output. In other words, if an initial data is corrupted or compromised, even with a few bits, then the resulting hash output will not be the same as what is expected.

There are various types of hash functions available; each one provides its own level of security guarantees. In this paper, we are focusing on guarantee that some data are not lost during data transmission. Some hash functions algorithms, such as MD5, are acceptable [18]. There were also studies related to the optimization of video hash. For example, Coskun et al. [19] converted a video into a 3D matrix. Every video frame is positioned into the third axis, and the hash value is based on Discrete Cosine Transform function. In addition, Liong et al. [20] developed a deep neural network in order to hash a video frame-by-frame. Pamula and Ziebinski [21], on the other hand, proposed hashing of a buffered video livestream in real time using FPGA (field programmable gate array). In this study, a signature is generated for each video frame, producing a hashing throughput of more than 400 Mb/s for blocks of 512 bits due to hardware acceleration manifested by FPGA.

If such an approach is used in our case, a single 800×600 frame would be hashed in 2 ns. In the same manner, a video record of 200 Gb at a frequency of 15 fps would be hashed in about 13 s.

An FPGA set up at the receiving end of the transmission system would then be able to generate the hash of a 1 Gb of data in roughly 65 ms. It would be able to quicky demand for a re-transmission request should an error get detected.

Table 1 Transmission speed of some VLC studies

Study	Speed	Transmission time (for 200 Gb)
Haigh et al. 2013 [23]	3 Mb/s	18 hours
Haigh et al. 2016 [24]	170 Mb/s	19 min
Shi et al. 2019 [25]	5 Gb/s	40 s

3.2 VLC Transmission Speed

As stated in the previous section, VLC also presents an advantageous transmission speed. This would mean that, for a similar Round-Trip Delay (RTD), the size of the data frames would be bigger than when using RF communication. This would allow for the acknowledgement process, and the hashing data check to happen more regularly too, enabling a faster transmission of the data.

Depending on the type of LED and receivers used, the transmission data rate can greatly vary: Table 1 offers an illustration of some of the throughput reached by other works, as well as the average unloading time of 200 Gb of data. The first study used of OLED lighting in order to reach a speed of 3 Mb/s, and the same team made an Artificial Neural Network as an equalizer and high-speed receivers in order to improve the speed to 170 Mb/s. Another study made use of FPGA and 64 QAM (Quadrature Amplitude Modulation [22]) to reach a VLC throughput of 5 Gb/s.

With the use of VLC the data can then be transmitted at an extremely high-speed. This system ensures a good transmission of the video recording. Most digests generated by hashing functions are only a few hundred bits long, making their transmission very easily via RF. In addition, as stated above, the computing of a new hash can be reduced to a few nanoseconds with the use of FPGA technology, guaranteeing a fast verification of the received data. In the optimal settings, the whole operation can be done in a few minutes.

4 Experimentations

The VLC communication setup used for making tests was based on the OpenVLC project.[1] It is an open-source, resource constraints VLC platform allowing for the fast prototyping of VLC communication scenarios. The VLC setup can be seen in Fig. 3. There is an emitter system plugged on a car's headlight (cf. Fig. 4) and interfaced with the network commands, while the receiver was positioned at a 20m distance in the line of sight. For the practical reason, in this VLC set up, we use a car's headlight

[1] http://www.openvlc.org/.

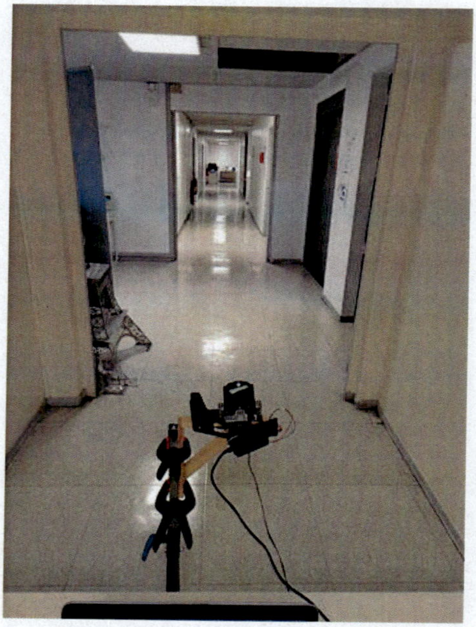

Fig. 3 Setup of the VLC technology testing platform using a headlight and a photo-detector

instead of a locomotive headlight. Because the needed candela value[2] is very similar to the value of a car's headlight. The network payloads are transmitted through the headlight and detected by the receiver. In this set-up, we equip also one WiFi dongle in the headlight to simulate the train side and in the receiver to simulate the train station side in the real condition.

For the basic communication experiments, we use the *iperf* network measurement tool [26] to evaluate our system. Our VLC system send packages at the speed of 100kb/s using an UDP protocol, the speed at which the blinking was undetectable to the human eye. As seen in Fig. 5, the VLC communication works perfectly without any loss. The hash is sent by WiFi dongle equipped in the headlight transmitter side and the receiver side also. When the small data package is received without any error, we don't need to request for re-sending the package by WiFi communication. The result illustrates no interference between the VLC communication and the WiFi. Due to the limit of our platform performance, the test speed of VLC communication is not so high. But this basic experiments show the feasibility of our proposed model in Fig. 2.

[2] https://www.law.cornell.edu/cfr/text/49/229.125.

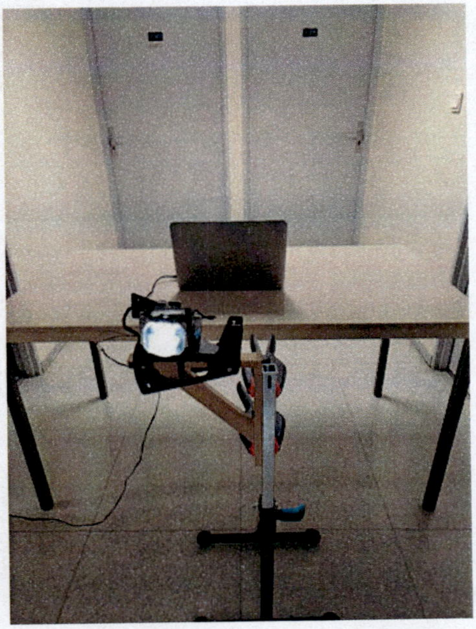

Fig. 4 Car headlight with almost the same characteristics of the headlight of the train using in the experiment

Fig. 5 Testing setup and Iperf results for the VLC receiver equipped in the train station

5 Conclusion

This paper introduced the idea of an intelligent hybrid VLC/RF model in order to ensure no information is lost during data transmission between 2 agents : The “heavy” data is sent through VLC, because of its high-speed and reliability, and the hash of the data (much smaller and used to verify the integrity of the transmitted information) is sent through an RF channel.

One potential application would be in train-to-stations communication. Trains are already equipped with headlights and stations usually offer protection from weather and sunlight. This system could, for example, be used in order to transfer data from on-board CCTV into a storage server.

In this use case of the data transfer between the train and the train station, we achieve five objectives. Firstly, the use of VLC communication can reach **the very high data rate** up to 220 GB/s potential throughput. This is very useful for such big volume video data exchange in our use case. Secondly, we combine the VLC technology with RF technology. This gives **a very good immunity performance**, because of no interference between the VLC technology and the electromagnetic RF technology. In this use case, the train station has no sun light so the noise of light is very limited. Our system has **a very optimal context**. As the train equip already the LED headlight, we have **a very good logistic** for our system. We just need to equip the photo-detector in the train station side for the VLC communication. Finally, our approach propose an intelligent hybrid VLC/RF system in the application layer. Thus we can ensure **a very good quality of service** in the higher layer of communication.

For the same reason, other type of use cases could be for satellite networks or car-to-drone communication. VLC offers great potential in intelligent hybrid communication models.

References

1. Info sûreté du mois : les caméras embarquées (2017). https://malignec.transilien.com/2016/05/12/info-surete-camera-embarquees/
2. On-board cctv surveillance with rear-facing ip camera-asmag.com provide moxa inc. latest press release information. https://www.asmag.com/mobile/suppliers/pressreleases.aspx?
3. Vijayalakshmi BA, Nesasudha M (2020) Li-Fi delivering broad band and multimedia content under dimming in train. Opt Quant Electron 52(12):518. http://link.springer.com/10.1007/s11082-020-02652-7
4. Guan H, Jutras D, Guo Z (2018) " Secured and green data processing and transmission in a human-vehicle interaction adas system. In: Global LIFI Congress (GLC). pp 1–5
5. Fraga-Lamas P, Fernández-Caramés TM, Castedo L (2017) Towards the internet of smart trains: a review on industrial iot-connected railways. Sens 17(6). https://www.mdpi.com/1424-8220/17/6/1457
6. Dagli MA (2020) Visible light communications based train control. pp. 87–94. http://library.witpress.com/viewpaper.asp?pcode=CR20-008-1
7. Nauryzbayev G, Abdallah M, Al-Dhahir N (2020) Outage analysis of cognitive electric vehicular networks over mixed RF/VLC channels. arXiv:2004.11143 [cs, math]
8. Marabissi D, Mucchi L, Caputo S, Nizzi F, Pecorella T, Fantacci R, Nawaz T, Seminara M, Catani J (2020) Experimental measurements of a joint 5G-VLC communication for future vehicular networks. JSAN 9(3):32. https://www.mdpi.com/2224-2708/9/3/32
9. Rahaim MB, Vegni AM, Little TDC (2011) A hybrid radio frequency and broadcast visible light communication system. In: 2011 IEEE GLOBECOM workshops (GC Wkshps). IEEE, Houston, TX, USA, pp 792–796. http://ieeexplore.ieee.org/document/6162563/
10. Rakia T, Yang H-C, Gebali F, Alouini M-S (2016) Optimal design of dual-hop VLC/RF communication system with energy harvesting. IEEE Commun Lett 20(10):1979–1982. http://ieeexplore.ieee.org/document/7524747/

11. Pan G,Ye J, Ding Z (2017) Secure hybrid VLC-RF systems with light energy harvesting. IEEE Trans Commun :1. http://ieeexplore.ieee.org/document/7935472/
12. Bouchet O, Perrufel M, Topsu S, Guan H (2016) Acemind new indoor full duplex optical wireless communication prototype. In: van Eijk AMJ, Davis CC, Hammel SM (eds). San Diego, California, United States, p 99790F. https://doi.org/10.1117/12.2235740
13. Ahamed S (2016) Visible Light Communication in Railways. In: International conference on railway engineering (ICRE 2016). Institution of Engineering and Technology, Brussels, Belgium, pp 7 (5 .)–7 (5 .). https://digital-library.theiet.org/content/conferences/10.1049/cp.2016.0516
14. R. P. . S. I. Group (2015) National rail & underground closed circuit television (cctv)guidance document
15. Gomez A, Shi K, Quintana C, Sato M, Faulkner G, Thomsen BC, O'Brien D (2015) Beyond 100-Gb/s indoor wide field-of-view optical wireless communications. IEEE Photon Technol Lett 27(4):367–370. http://ieeexplore.ieee.org/document/6967750/
16. Perry VH, Zakariasen D (2004) First use of ultra-high performance concrete for an innovative train station canopy. p 8
17. Horikoshi J, Tanaka K, Morinaga T (1986) 1.2 GHz band wave propagation measurements in concrete building for indoor radio communications. IEEE Trans Veh Technol 35(4):146–152. http://ieeexplore.ieee.org/document/1623397/
18. Rivest R, Dusse S (1992) The md5 message-digest algorithm
19. Coskun B, Sankur B, Memon N (2006) Spatio-temporal transform based video hashing. IEEE Trans Multimed 8(6):1190–1208. http://ieeexplore.ieee.org/document/4014210/
20. Liong VE, Lu J, Tan Y-P, Zhou J (2017) Deep video hashing. IEEE Trans Multimed 19(6):1209–1219. http://ieeexplore.ieee.org/document/7797446/
21. Pamułia D, Zi A Securing video stream captured in real time. p 3
22. Webb WT, Hanzo L (1994) Modern quadrature amplitude modulation: principles and applications for fixed and wireless channels: one. IEEE Press-John Wiley
23. Haigh PA, Ghassemlooy Z, Rajbhandari S, Papakonstantinou I (2013) Visible light communications using organic light emitting diodes. IEEE Commun Mag 51(8): 148–154. http://ieeexplore.ieee.org/document/6576353/
24. Haigh PA, Ghassemlooy Z, Rajbhandari S, Papakonstantinou I, Popoola W (2014) Visible light communications: 170 Mb/s using an artificial neural network equalizer in a low bandwidth white light configuration. J Light Technol 32(9):1807–1813. http://ieeexplore.ieee.org/document/6781659/
25. Shi M , Wang C, Li G, Liu Y, Wang K, Chi N A 5Gb/s 2 × 2 MIMO real-time visible light communication system based on silicon substrate. p 5
26. Tirumala A (1999) Iperf: the tcp/udp bandwidth measurement tool. http://dast.nlanr.net/Projects/Iperf/

Unmanned Aerial Vehicles Path Planning Based on an Enhanced Whale Optimization Algorithm

Amylia Ait Saadi, Assia Soukane, Yassine Meraihi, Asma Benmessaoud Gabis, Amar Ramdane-Cherif, and Sylia Mekhmoukh Taleb

1 Introduction

Recently, Unmanned Aerial Vehicles (UAVs) are considered as powerful systems which are used in various military and civil applications. The development of UAVs is continuously growing over the years due to their features such as autonomy, accessibility, low cost, etc. These features allow UAVs to reach difficult places and perform complex missions. In these kinds of missions, the challenging key is the capability of UAVs to determine efficiently its own trajectory which is known as path planning. The UAV path planning process consists of finding the optimal path from starting to ending points under various requirements including the UAV safety from obstacles and environment constraints [1]. Based on this definition, the UAV path planning problem can be solved by an optimization algorithm that minimizes the cost related to the traveled route and the safety from obstacles. In the literature, various methods already exist. Because the UAV path planning problem is an NP-hard problem, we focused on meta-heuristic-based approaches.

In the work of Qu et al. [2], a hybrid algorithm (HSGWO-MSOS) based on the hybridization of Simplified Grey Wolf Optimizer (SGWO) and Modified Symbiotic Organisms search (MSOS) algorithms was introduced for solving the 3D UAV

A. A. Saadi (✉) · Y. Meraihi · S. M. Taleb
LIST Laboratory, University of M'Hamed Bougara Boumerdes, Avenue of Independence, 35000 Boumerdes, Algeria
e-mail: a.aitsaadi@univ-boumerdes.dz

A. A. Saadi · A. Ramdane-Cherif
LISV Laboratory, University of Paris-Saclay, 10-12 Avenue of Europe, 78140 Velizy, France

A. Soukane
ECE Paris School of Engineering, 37 quai de Grenelle, 75015 Paris, France

A. B. Gabis
Ecole nationale Supérieure d'Informatique, Laboratoire des Méthodes de Conception des Systèmes, BP 68M, 16309, Oued-Smar Alger, Algérie

A. Ramdane-Cherif et al. (eds.), *Machine Intelligence and Data Science Applications*, Algorithms for Intelligent Systems, https://doi.org/10.1007/978-981-99-1620-7_9

path planning. The performance of the HSGWO-MSOS algorithm was assessed in 3D different cases using various obstacles. The comparative study showed that HSGWO-MSOS gives competitive results compared to SOS, GWO, and Simulated Annealing (SA) algorithms regarding the fitness values and execution time. Authors in [3] presented an improved version of Particle Swarm Optimization (PSO) algorithm, called taw-PSO, for enhancing the UAV path planning problem. The taw-PSO algorithm integrates a time-varying adaptive inertia weight coefficient to enhance the population diversity of PSO algorithm. The taw-PSO algorithm was tested in one case using static obstacles. The proposed approach shows better performance than other variants of adaptive PSO algorithms [4–7] in terms of fitness and diversity costs. In [8], authors proposed an improved Bat Algorithm, called IBA for solving the UAV path planning problem. In IBA algorithm, Artificial Bee Colony was combined to the original Bat Algorithm to enhance its exploitation capabilities. IBA algorithm was evaluated in a 3D environment using different number of dimensions. IBA algorithm outperforms BA, ABC, PSO, and BAM algorithms in terms of path cost and execution time. Pan et al. [9] developed an enhanced version of Golden Eagle Optimization (GEO) algorithm which integrates the Double Learning Strategy (DLS) for solving the UAV path planning problem. The efficiency of the GEO-DLS algorithm was validated in 2 different scenarios using various obstacles. Simulation results demonstrated the superiority of the GEO-DSL algorithm in terms of fitness value compared to PSO, GEO [10], GWO, SOGWO [11], and LWOA [12] algorithms. In this work [13], an Improved Hybrid Salp Swarm Algorithm with Aquila Optimizer algorithm (IHSSAO) was proposed for UAV path planning. In the IHSSAO algorithm, a Tent chaotic map is introduced to generate solutions and the Opposition-based Learning was applied to enhance the population diversity of the hybrid SSA with AO algorithm. The IHSSAO algorithm was assessed in one scenario using 3 obstacles. The IHSSAO outperforms SSA and AO algorithms in terms of fitness value. In [14], authors proposed an enhanced Genetic algorithm, called (GAF) based on the integration of Fuzzy Logic strategy to the original Genetic Algorithm for solving the UAV path planning problem. The proposed GAF algorithm was evaluated in one scenario in complex terrain. The GAF algorithm gives good performance compared to the original GA algorithm regarding the fitness value. Jiang et al. [15] proposed an improved version of Grey Wolf Optimizer algorithm, called (GWOLC) based on the integration of the Partially observable Markov decision process (POMDP) to GWO algorithm. The effectiveness of the proposed approach was validated in a complex scenario using various obstacles. The GWOLC method outperforms state-of-the-art methods [16–18] in terms of fitness value.

The majority of earlier researches on the UAV path planning problem focused on Swarm Intelligence methods due to their good performance in this subject [1]. Whale Optimizer Algorithm (WOA) is one of the recent swarm intelligence approach characterized by its low complexity, easy implementation, etc. However, as with the other meta-heuristics, the WOA algorithm presents some weaknessess, such as trapping into local optima and lack of exploitation. Motivated by that, this paper proposes an enhanced version of the WOA algorithm, called QRWOA, based on the incorporation of the Quasi Reflection Opposition-based scheme for solving the UAV path

Table 1 The main notations used in this paper

Parameter	Description
$[X_m, Y_m, Z_m]$	Map limits
S	Source point
E	End point
L	Total path
n	Number of waypoints
$Cost$	Total cost
C_{Path}	Path cost
C_{Obs}	Obstacles' cost
w	Weight coefficient
$\overrightarrow{L_iL_{i+1}}$	Line segment between ith and $(i+1)th$ points
Obs_j	The jth obstacle
D	Distance to obstacle
R_j	Obstacle radius
m	Number of obstacles
OC_j	The jth obstacle cost

planning problem. The proposed QRWOA is tested using 6 cases with avarious number of obstacles and dimensions in comparison with the classical WOA, PSO, and BA algorithms.
The remaining of this paper is organized as follows. Section 2 gives the formulation of the UAV path planning problem. Section 3 gives the description of the WOA algorithm and QROBL strategy. The research explains the structure of the proposed QRWOA algorithm for solving the UAV path planning issue. Section 5 discusses the simulation findings. Finally, Sect. 5 shows the concluding remarks and future works.

2 UAV Path Planning Problem Formulation

As mentioned before, the objective of the UAV path planning is to optimize the UAV flight between source and destination nodes. To begin with, a mathematical model to represent the environment is required. Then, the requirements of the UAV path planning are transformed into an objective function that reflects the solution's quality given by the algorithms. For better readability, the main notation used in this paper are summarized in Table 1.

2.1 Environment Representation

We consider a 3D environment where the UAV begins its mission from a start point S located at (x_s, y_s, z_s) to the end point E located at (x_e, y_e, z_e). Between these two points, the UAV passes through n adjacent points. In this area, various obstacles can be presented. In our environment, the obstacles are modeled as cylinders where their radius represents the obstacles range. The objective of the UAV path planning algorithm is to minimize the distance between the n points while maintaining a safe distance between them and the obstacles as shown in Fig. 1.

2.2 Objective Function

Our objective in this work is to ensure that the UAV takes the shortest distance to reach the destination and to stay as far as possible from obstacles. Based on these two objectives, we defined the objective function as the following equation:

$$Cost = w_1 \cdot C_{Path} + w_2 \cdot C_{Obs} \tag{1}$$

where C_{Path} and C_{Obs} represent the cost related to the path and obstacles, respectively. w_1 and w_2 are the weight coefficient expressed as follow:

$$\sum_{i=1}^{2} w_i = 1 \tag{2}$$

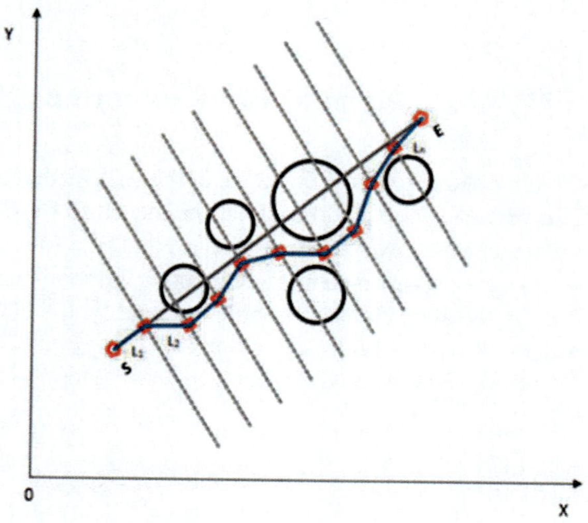

Fig. 1 Environment modeling of the UAV path planning

2.2.1 Path Cost

$$C_{Path} = \sum_{i=0}^{n} L_i \tag{3}$$

$$L_i = \sqrt{(x_{i+1} - x_i)^2 + (y_{i+1} - y_i)^2 + (z_{i+1} - z_i)^2} \tag{4}$$

2.2.2 Obstacles Cost

$$C_{Obs} = \sum_{j=0}^{m-1} OC_j \tag{5}$$

where OC_j is the cost related to the j-th obstacle which can be calculated as follows:

$$OC_j = \begin{cases} 0, & \text{if } D > R_j \\ \infty, & \text{otherwise} \end{cases} \tag{6}$$

where D denotes the distance between the line segment $\overrightarrow{\mathrm{L_i L_{i+1}}}$ and the center of j-th the obstacle. R_j stands for the j-th the obstacle's radius.

3 Preliminaries

3.1 Whale Optimization Algorithm

Whale Optimization Algorithm (WOA) is a nature-inspired meta-heuristic proposed by Mirjalili et al. [19]. Similar to other meta-heuristics, WOA was widely applied for solving optimization problems in various fields such as feature selection [20], PID controllers[21], energy [22], wireless networks [23], etc. The WOA algorithm is inspired by the behavior of the whale in the hunting process which includes three methods, namely Encircling prey, Bubble-net attacking, and searching for the prey.

3.1.1 Encircling Prey

During this method, the whale considers the current best position as the prey's position and adjusts their locations according to the prey's position. This behaviour is mathematically presented in Eq. (7) and (8).

$$\overrightarrow{D} = |\overrightarrow{C} \cdot \overrightarrow{X^b}(t) - \overrightarrow{X}(t)| \tag{7}$$

$$\overrightarrow{X}(t+1) = \overrightarrow{X^b}(t) - \overrightarrow{A} \cdot \overrightarrow{D} \tag{8}$$

where $\overrightarrow{X^b}(t)$ denotes the best position found so far. $\overrightarrow{X}(t)$ represents the current position. $\overrightarrow{A}$ and $\overrightarrow{C}$ are coefficient parameters formulated as the following equations:

$$\overrightarrow{A} = 2 \cdot \overrightarrow{a} \cdot \overrightarrow{r} - \overrightarrow{a} \tag{9}$$

$$\overrightarrow{C} = 2 \cdot \overrightarrow{r} \tag{10}$$

where $\overrightarrow{a}$ is a control parameter which decreases linearly from 2 to 0. $\overrightarrow{r}$ is a random vector in the range of [0, 1].

3.1.2 Bubble-net Attacking

At this point, the whale creates bubbles in spiral shapes and moves towards the surface. The spiral movement of the whale can be expressed mathematically in Eq. (11)

$$\overrightarrow{X}(t+1) = \overrightarrow{D'} \cdot e^{\delta.l} \cdot cos(2\pi l) + \overrightarrow{X^b}(t) \tag{11}$$

where $\overrightarrow{D'}$ represents the distance between the best position found so far and the current whale location. It can be calculated as in Eq. (12). δ is a fixed value. l stands for a random variable in the range of $[-1, 1]$. $\overrightarrow{X}(t+1)$ is the updated position at the $t+1$ iteration.

$$\overrightarrow{D'} = |\overrightarrow{X^b}(t) - \overrightarrow{X}(t)| \tag{12}$$

During this hunting method, the whale makes simultaneously both shrink circle and spiral movements. These movements are formulated mathematically in Eq. (13).

$$\overrightarrow{X}(t+1) = \begin{cases} \overrightarrow{X^b}(t) - \overrightarrow{A} \cdot \overrightarrow{D}, & \text{if } rand < 0.5 \\ \overrightarrow{D'}.e^{\delta l} \cdot cos(2\pi l) + \overrightarrow{X^b}(t), & rand \geq 0.5 \end{cases} \tag{13}$$

3.1.3 Search for the Prey

In this phase, the whale searches randomly for the prey according to its position. It is mathematically expressed in Eqs. (14) and (15).

$$\vec{D} = |\vec{C} \cdot \overrightarrow{X_{rand}} - \vec{X}| \tag{14}$$

where $\overrightarrow{X_{rand}}$ is a random position.

$$\vec{X}(t+1) = \overrightarrow{X_{rand}} - \vec{A}.\vec{D} \tag{15}$$

3.2 *Quasi Reflection Opposition-Based Learning*

The concept of Opposition Based Leaning was proposed first by Tizhoosh et al. [24]. The Opposition-Based Leaning strategy aims to enhance the exploration search of meta-heuristics by searching and evaluating the opposite of the current candidates. The Quasi Reflection Opposition-based learning is an improved version of OBL proposed by Ergezer et al. [25]. The Quasi Reflection Opposition-based learning (QROBL) stipulates that the opposition solution should be closer to the center of the search area rather than the opposite value. By applying QROBL to a candidate solution X_j, its opposite candidate $\overline{X_j^r}$ can be calculated as the following equation:

$$\overline{X_j^r} = \begin{cases} rand(X_j, M_j), & \text{if } X_j < M_j \\ rand(M_j, X_j), & Otherwise \end{cases} \tag{16}$$

where M_j denotes the interval center which is expressed in Eq. (17). $rand(X_j, M_j)$ is a random value in the range of $[X_j, M_j]$.

$$M_j = \frac{(Lb_j + Ub_j)}{2} \tag{17}$$

4 Quasi Reflection Whale Optimization Algorithm for UAV Path Planning

This section details the implementation steps of the Quasi Reflection Whale Optimization Algorithm (QRWOA) for UAV path planning.

To begin with, the UAV position is initialized randomly in the search area delimited by the lower and upper bounds $[x_{min}, y_{min}, z_{min}]$ and $[x_{max}, y_{max}, z_{max}]$,respectively. The total UAV's position is expressed in the following equation:

$$Positions = \begin{bmatrix} Pos_{1,1} & Pos_{1,2} & \dots & Pos_{1,3n} \\ Pos_{2,1} & Pos_{2,2} & \dots & Pos_{2,3n} \\ \vdots & \vdots & \vdots & \\ Pos_{N,1} & Pos_{N,2} & \dots & Pos_{N,3n} \end{bmatrix} \tag{18}$$

where N is the total population size. n donates the number of waypoints in each solution. The coordinate of each waypoint $(Pos_{i,j}, Pos_{i,j+n}, Pos_{i,j+2n})$ which corresponds to (x, y, z), in the i-th solution are represented in Eq. (19)

$$Pos_{i,j} = \begin{cases} x_{min} \leq Pos_{i,j} \leq x_{max}, \ j = 1, \dots, n \\ y_{min} \leq Pos_{i,j} \leq y_{max}, \ j = n+1, \dots, 2n \\ z_{min} \leq Pos_{i,j} \leq z_{max}, \ j = 2n+1, \dots, 3n \end{cases} \tag{19}$$

where $x_{min}, x_{max}, y_{min}, y_{max}, z_{min}, z_{max}$ represent the $3D$ area limits.
After initialization, the proposed QRWOA algorithm evaluates the candidate solutions according to their fitness value. Since the problem of UAV path planning is a minimization problem, QRWOA selects the least cost position as the best position as formulated in Eq. (20).

$$Pos_B = \arg\min Cost(Pos_i) \tag{20}$$

Before updating UAV's position, QRWOA searches for other potential solutions using Quasi Reflection Opposition based-learning using Eqs. (16) and (17). Then, it evaluates the $(Pos_i \cup \overline{Pos_i})$ solutions and selects the N best positions. After the population's selection, the QRWOA updates its position like the original WOA algorithm. According to the new positions, QRWOA searches for better best position until the maximum number of iterations is reached. The pseudo-code of the Quasi Reflection Whale Optimization Algorithm for UAV path planning is given in 1.

5 Results and Discussions

This section presents the results of applying the Quasi Reflection Whale optimization Algorithm (QRWOA) for the UAV path planning. The effectiveness of the proposed algorithm was evaluated in two different scenarios using multiple obstacles and dimensions. The details of simulation parameters are summarized in Table 2. To demonstrate that the proposed QRWOA algorithm provides optimal results, we applied 3 well-known meta-heuristics: PSO, BA, and WOA algorithms for the UAV

Algorithm 1 Quasi Reflection Whale Optimization Algorithm for UAV path planning

Input : Start and destination points, obstacles' position, and number of waypoints n

Output: The best UAV positions from start to end

```
Initialize QRWOA parameters: Maximum number of iterations T, Population' size N, Dimension Dim, δ, etc.
Initialize the population of QRWOA: Pos_i (i = 1, 2, ..., N)
Calculate the fitness value f(Pos_i)
Determine the best position Pos_b
while (t < T) do
    for i = 1, 2, ..., N do
        Find the N opposite positions based on QROBL using Equations (17) and (16), and select the N fittest positions using Equation
        Update a, A, C, l, and p
        if p < 0.5 then
            if |A| < 1 then
                Update the position using Equation (7)
            else if
                thenUpdate the position using Equation (14)
            end if
        else if p > 0.5 then
            Update the position using Equation (11)
        end if
        Update the best position Pos_b
    end for
    t = t + 1
end while
return The best position Pos_b
```

path planning. Simulations are conducted using MATLAB 2021b running on Core i7 2.90GHz, RAM 32 GB machine.

Table 3 reports the fitness value obtained by QRWOA, PSO, BA, and WOA algorithms in the first case. For different number of n, it can be noticed that the QRWOA algorithm provides the smallest value regarding the worst, mean, and standard deviation of fitness value. In this case, the proposed QRWOA algorithm provides the best solution quality with better stability. These results reflect the good quality of the provided path as shown in Fig. 3. Regarding the best fitness value, we can see that in the case of $n = \{10, 20\}$, the PSO algorithm achieves the best fitness value. QRWOA and PSO algorithms offer the same best fitness value when $n = 30$. In the case of $n = 2$, in some tests, the WOA algorithm fails to provide a safe path which is explained by an infinite cost. Figure 2 illustrates the convergence curve of the algorithms for $n = 30$. We can see that the proposed QRWOA algorithm converges to the optimal value from the 20-th iteration which demonstrates its convergence speed

Table 2 Simulation details

Parameter	Value
Population size	50
Maximum number of iterations	100
Number of runs	50
Start point	[200,100,150]
End point	[800,800,150]
Area dimension	[1000,1000]
Altitude limits	[100,200]
Obstacles position	[400,500,100], [600,200,150] , [500,350,150] [350,200,150], [700,550,150]
Obstacles radius	80, 70, 80 70, 70

Table 3 Fitness results of the first case

Fitness value	QRWOA	PSO	BA	WOA	
Best	313.77	**313.64**	367.65	313.74	$n = 10$
Worst	**315.03**	330.36	530.85	315.04	
Mean	**314.23**	314.78	437.19	314.34	
Std	**0.31**	3.29	35.62	0.33	
Best	313.78	**313.73**	388.2	313.82	$n = 20$
Worst	**316.06**	316.31	501.95	/	
Mean	**314.55**	314.56	440.59	/	
Std	**0.46**	1.01	24.94	/	
Best	**313.87**	**313.87**	365.14	313.89	$n = 30$
Worst	**315.46**	316.62	483.14	315.65	
Mean	**314.65**	314.71	469.35	314.7	
Std	**0.38**	0.96	26.53	0.4	

efficiency. The PSO algorithm converges in the last iterations. As for BA algorithm, we can see that it converges quickly but to a non-optimal value.

Table 4 summarized the fitness results provided by the algorithms in the second case. From the results, we noticed that the proposed QRWOA algorithm provides the best fitness values for different number of dimensions. For this case of obstacles, we can say that the QRWOA algorithm is the most appropriate algorithm. AS for BA and WOA algorithms, we can see from the worst fitness value that at some tests, they both failed to provide a collision-free path. Both of them are not suitable for this number of obstacles. PSO algorithm provides good and acceptable results. Figure 4 represents the convergence curve of the algorithm in the second case for $n = 30$. From the figure, we can see that the proposed QRWOA converges first from the

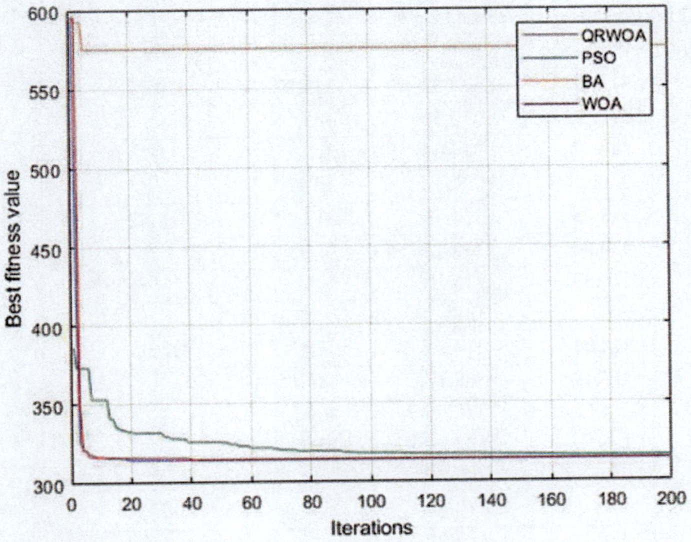

Fig. 2 Comparison of convergence curves of QRWOA and literature algorithms obtained in the first case for $n = 30$

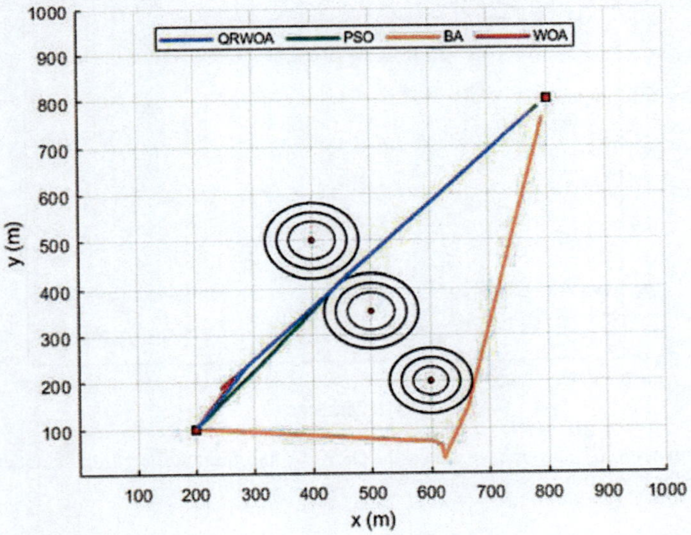

Fig. 3 The path obtained by QRWOA and literature algorithms obtained in the first case for $n = 30$

Table 4 Fitness results of the second case

Fitness value	QRWOA	PSO	BA	WOA	
Best	**314.02**	**314.02**	315.27	**314.02**	$n = 10$
Worst	**315.3**	316.53	485.56	/	
Mean	**314.42**	314.62	375.14	/	
Std	**0.32**	0.83	50.47	/	
Best	**313.78**	314.73	317.07	313.82	$n = 20$
Worst	**314.06**	315.13	/	/	
Mean	**314.58**	315.1	/	/	
Std	**0.41**	1.02	/	/	
Best	**314.04**	314.25	365.14	314.05	$n = 30$
Worst	**315.48**	368.77	483.14	/	
Mean	**314.61**	320.55	469.35	/	
Std	**0.39**	13.98	26.53	/	

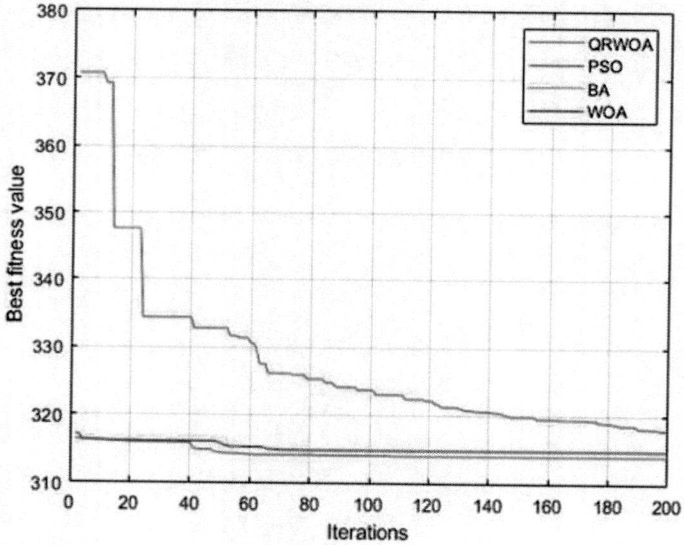

Fig. 4 Comparison of convergence curves of QRWOA and literature algorithms obtained in the second case for $n = 30$

60-th iteration. WOA algorithm starts its convergence starting from $t = 80$. PSO algorithm needs more iteration to converge in this case. The convergence of BA algorithm is not shown because in this example BA fails. Figure 5 illustrates the path obtained by the algorithms. We can see that the path given by QRWOA and PSO is the shortest path towards the destination which respects the obstacle constraint. The path given by BA algorithm, in this case, is optimal and represents its best solution which is not often offered. Regarding the path provided by WOA algorithm, we can

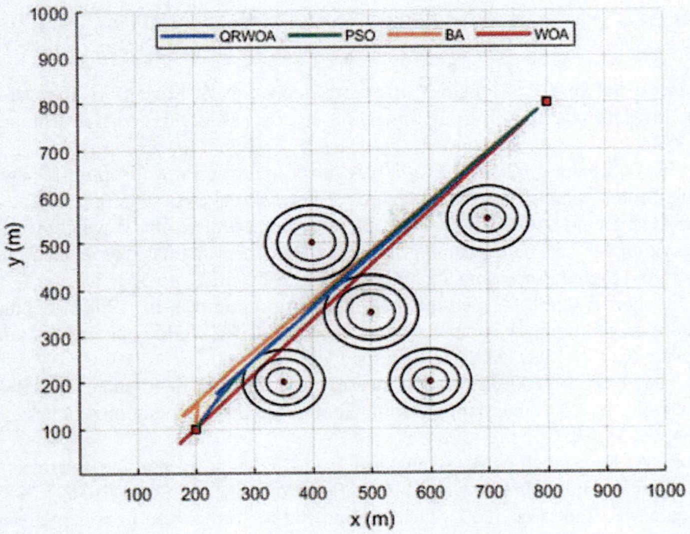

Fig. 5 The path obtained by QRWOA and literature algorithms obtained in the second case for $n = 30$

see that it represents a failure since it collides with two obstacles. For both cases, by comparing the dimension n. We can see that the fitness values are increasing proportionally to n. In fact, the dimension n represents the number of nodes and the problem size simultaneously. So, the complexity of the problem is also increasing with n. For this level of complexity, only QRWOA and PSO algorithms succeeded.

Comparing the two cases, we can conclude that the obstacle number affects the fitness values. Actually, obstacle constraint is part of the environment's complexity that varies proportionally with. This fact explains the rise of the fitness values from one case to another. For complex environments, QRWOA and PSO algorithms are suitable.

6 Conclusion

In this paper, an ameliorated version of the WOA algorithm has been proposed for tackling the UAV path planning issue in a 3D environment. The proposed method called QRWOA employed the Quasi Reflection Opposition-based Learning scheme to guarantee a good balance between exploitation and exploration. The proposed QRWOA algorithm was assessed using six benchmark maps with different dimensions and obstacles considering the fitness metric. Compared to BA, PSO, and WOA algorithms, the proposed QRWOA scheme shows competitive performance by providing optimal UAV path planning.

References

1. Ait Saadi A, Soukane A, Meraihi Y, Benmessaoud Gabis A, Mirjalili S, Ramdane-Cherif A (2022) Uav path planning using optimization approaches: a survey. Arch Comput Methods Eng :1–52
2. Qu C, Gai W, Zhang J, Zhong M (2020) A novel hybrid grey wolf optimizer algorithm for unmanned aerial vehicle (uav) path planning. Knowl Based Syst 194:105530
3. Nayeem GM, Fan M, Akhter Y (2021) A time-varying adaptive inertia weight based modified pso algorithm for uav path planning. In: 2021 2nd International Conference on Robotics, Electrical and Signal Processing Techniques (ICREST). IEEE, pp 573–576
4. Shi Y, Eberhart R (1998) A modified particle swarm optimizer. In: 1998 IEEE international conference on evolutionary computation proceedings. IEEE world congress on computational intelligence (Cat. No. 98TH8360). IEEE, pp 69–73
5. Xin J, Chen G, Hai Y (2009) A particle swarm optimizer with multi-stage linearly-decreasing inertia weight. In: 2009 International Joint Conference on Computational Sciences and Optimization, vol 1. IEEE, pp 505–508
6. Nickabadi A, Ebadzadeh MM, Safabakhsh R (2011) A novel particle swarm optimization algorithm with adaptive inertia weight. Appl Soft Comput 11(4):3658–3670
7. Arumugam MS, Rao MVC (2008) On the improved performances of the particle swarm optimization algorithms with adaptive parameters, cross-over operators and root mean square (rms) variants for computing optimal control of a class of hybrid systems. Appl Soft Comput 8(1):324–336
8. Zhou Xianjin, Gao Fei, Fang Xi, Lan Zehong (2021) Improved bat algorithm for uav path planning in three-dimensional space. IEEE Access 9:20100–20116
9. Pan Jeng-Shyang, Lv Ji-Xiang, Yan Li-Jun, Weng Shao-Wei, Chu Shu-Chuan, Xue Jian-Kai (2022) Golden eagle optimizer with double learning strategies for 3d path planning of uav in power inspection. Math Comput Simul 193:509–532
10. Mohammadi-Balani A, Nayeri MD, Azar A, Taghizadeh-Yazdi M (2021) Golden eagle optimizer: a nature-inspired metaheuristic algorithm. Comput & Ind Eng 152:107050
11. Dhargupta S, Ghosh M, Mirjalili S, Sarkar R (2020) Selective opposition based grey wolf optimization. Expert Syst Appl 151:113389
12. Ling Ying, Zhou Yongquan, Luo Qifang (2017) Lévy flight trajectory-based whale optimization algorithm for global optimization. IEEE access 5:6168–6186
13. Yao Jinyan, Sha Yongbai, Chen Yanli, Zhang Guoqing, Xinyu Hu, Bai Guiqiang, Liu Jun (2022) Ihssao: an improved hybrid salp swarm algorithm and aquila optimizer for uav path planning in complex terrain. Appl Sci 12(11):5634
14. Ntakolia C, Platanitis KS, Kladis GP, Skliros C, Zagorianos AD (2022) A genetic algorithm enhanced with fuzzy-logic for multi-objective unmanned aircraft vehicle path planning missions. In: 2022 International Conference on Unmanned Aircraft Systems (ICUAS). IEEE, pp 114–123
15. Jiang W, Lyu Y, Li Y, Guo Y, Zhang W (2022) Uav path planning and collision avoidance in 3d environments based on pompd and improved grey wolf optimizer. Aerosp Sci Technol 121:107314
16. Zhang D, Tan J, Yao X (2021) Vortex evolution and flame propagation driven by oblique shock wave in supersonic reactive mixing layer. Aerosp Sci Technol 118:106993
17. Ge F, Li K, Xu W et al (2019) Path planning of uav for oilfield inspection based on improved grey wolf optimization algorithm. In: 2019 Chinese Control And Decision Conference (CCDC). IEEE, pp 3666–3671
18. Kharwar PK, Verma RK (2020) Exploration of nature inspired grey wolf algorithm and grey theory in machining of multiwall carbon nanotube/polymer nanocomposites. Eng Comput :1–22
19. Mirjalili Seyedali, Lewis Andrew (2016) The whale optimization algorithm. Adv Eng Softw 95:51–67

20. Hussien AG, Hassanien AE, Houssein EH, Bhattacharyya S, Amin M (2019) S-shaped binary whale optimization algorithm for feature selection. Recent trends in signal and image processing. Springer, pp 79–87
21. Mosaad AM, Attia MA, Abdelaziz AY (2019) Whale optimization algorithm to tune pid and pida controllers on avr system. Ain Shams Eng J 10(4):755–767
22. Wong LA, Ramachandaramurthy VK, Walker SL, Taylor P, Sanjari MJ (2019) Optimal placement and sizing of battery energy storage system for losses reduction using whale optimization algorithm. J Energy Storage 26:100892
23. Pham Quoc-Viet, Mirjalili Seyedali, Kumar Neeraj, Alazab Mamoun, Hwang Won-Joo (2020) Whale optimization algorithm with applications to resource allocation in wireless networks. IEEE Trans Veh Technol 69(4):4285–4297
24. Tizhoosh HR (2005) Opposition-based learning: a new scheme for machine intelligence. In: International Conference on Computational Intelligence for Modelling, Control and Automation and International Conference on Intelligent Agents, Web Technologies and Internet Commerce (CIMCA-IAWTIC'06), vol 1. IEEE, pp 695–701
25. Ergezer M, Simon D, Du D (2009) Oppositional biogeography-based optimization. In: 2009 IEEE International Conference on Systems, Man and Cybernetics. IEEE, pp 1009–1014

Reviewing Scope of Multimodal Approach in Face and Iris Recognition

Sonal, Ajit Singh, and Chander Kant

1 Introduction

Biometric authentication depends on some components of your biological traits to authenticate your identification. An iris scan, retina scan, or fingerprint may all suit the bill here. You could employ merely one quality or a combo of many attributes. Pattern recognition systems are at the core of biometrics. These "unique" features may now be recorded, photographed, or quantified using electrical or optical sensors like cameras and scanners. In addition to security, border control, public aid/social benefits and customs/immigration and passport and healthcare identification verification are some of the uses of these technologies. Most biometric systems uses a single biometric attribute to establish identification (GAR) and the False Acceptance Rate (FAR) of the system is enhanced by noise in the sensed data and non-universality effects the Genuine Acceptance Rate (GAR) (GAR). Biometric security, on the other hand, reduces the system's advantages.

1.1 Biometric

Measurements and computations based on human traits, known as biometrics, are a subset of this field. As a method of identification and restriction of access in the field of computer science, biometric authentication (also known as realistic authentication) is

Sonal (✉)
UTU, Dehradun, India
e-mail: sonalkharb@gmail.com

A. Singh
BTKIT, Dwarahat, Uttarakhand, India

C. Kant
KUK, Kurukshetra, Haryana, India

A. Ramdane-Cherif et al. (eds.), *Machine Intelligence and Data Science Applications*, Algorithms for Intelligent Systems, https://doi.org/10.1007/978-981-99-1620-7_10

often used by researchers. People in groups under observation may also be identified using this method.

To identify and categorize people, biometrics are the unique, quantifiable features that may be employed. In the context of biometric identifiers, the form of a person's body is typically referred to as a physiological feature. All these methods are only a few instances of the various ways to identify a person. Typing rhythm, movement, signature, behavioral profiling, and voice are all examples of behavioral features. Some researchers have used the term "behaviometrics" to describe this kind of biometrics.

1.2 Biometric Authentication

User identification is verified using biometric authentication devices that use physical traits such as fingerprints, face patterns, iris, or retinal patterns, amongst other biometrics. Network login is one of the several uses for biometrics authentication, which is on the rise. An authorized user's biometrics sample must be saved in a database as a template or identifier so the device may compare it with a fresh sample provided during the login procedure. In high-security settings, biometrics are often utilized in combination with smart cards. The following are the most common kinds of biometric devices:

- **Fingerprint scanners**: A broad range of manufacturers provide them for desktop and laptop computers, and they connect through USB or PCMCIA (also known as PC Card) interfaces.
- **Devices that recognize facial patterns**: Facial geometry analysis is used in these devices to authenticate identification.
- **Recognizing** the shape of a person's hands is used in Face-patterning devices and hand-geometry-testing devices that have a common features.
- **Iris scanners** are used to identify people. Iris scanners make advantage of it since it is permanently produced during the 8th month of pregnancy in human beings.
- **Retinal scanners** are used to identify people. The retina's blood vessels may be mapped using retina scanners.

1.3 Biometric Multimodal

Biometric identification systems use various biometric indications to identify people in multimodal biometric systems. A fingerprint, face, palm print, or iris are just a few examples of the types of biometric data that may be used with multimodal authentication. Using the matching score level, a person's fmgerprint and iris may be combined for automated identification. In order to enhance security, fingerprint and iris images are utilized to authenticate a user.

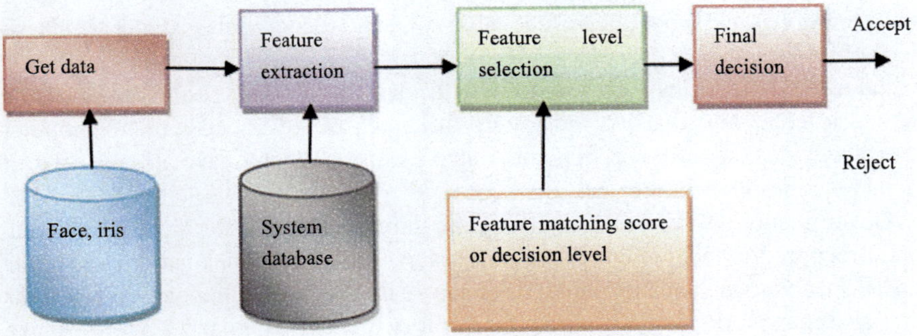

Fig. 1 Multi model system for biometric recognition

Using a person's qualities or attributes, biometrics may be used to identify them scientifically. People in groups under observation may also be identified using this method. Individuals are uniquely identified and described by the use of biometric identifiers. A multimodal biometric system incorporates two or more biometric data recognition results in order to increase authentication security and minimize the false acceptance rate. For the purposes of enrollment, verification, or identification, these systems make use of or are capable of making use of several physiological or behavioral characteristics of an individual. Law enforcement, e-commerce, smart cards, passports, and visas all use multimodal biometric technologies. Problems arise when two or more biometric modalities are combined in the multimodal biometric system. When it comes to biometrics, multimodal biometrics delivers more information than single-modal biometrics systems. Multiple biometric fusion frameworks and recognition algorithms are needed. The most common biometric identifiers include fingerprints, face, voice, iris, and vein patterns on the palm or finger of the hand.

When a biometric system uses a human attribute as an input, it is classified as a biometric modality. Fingerprint, face, iris, voice, hand writer recognition, and hand geometry are examples of biometric modalities. Unimodal biometrics refers to biometrics that employ just one biometric property to identify and verify a person. Multimodal biometrics refers to a kind of biometric identification that relies on more than two distinct features of an individual to confirm their identity (Fig. 1).

1.4 Face Recognition

Frequently, an ID verification service will utilize a facial recognition technology to verify a user's identity by comparing a digital picture or video frame against a database of known faces.

In the 1960s, work on such systems started as a kind of computer program. This is the first time that face recognition technologies have had a larger range of applications on smartphones and other types of technology. Facial recognition systems are classified as biometrics since they employ measurements of a person's physiological traits. Facial recognition technologies are frequently used despite their poorer accuracy as a biometric technology than iris and fingerprint detection. Advanced human–computer interface, video surveillance, and automated picture indexing have all made use of facial recognition technologies.

Governments and commercial enterprises throughout the globe are already using facial recognition technologies. The success of these techniques varies, and some have been abandoned due to their inefficiency in the past. Facial recognition systems have also caused criticism, with concerns that they breach people' privacy, often make inaccurate identifications, foster gender norm and racial profiling and do not preserve sensitive biometric data. Several communities in the United States have enacted bans on face recognition technologies because of these arguments. More than one billion users' face scans will be deleted as a consequence of Meta's announcement that it would shut down Facebook's facial recognition technology. In the history of face recognition, this will be one of the biggest revolutions.

Face recognition systems use computer algorithms to identify a person based on their unique facial traits. These traits, data from other faces, such as the distance between the eyes or the curve of the chin, are then compared to this database. Even while face recognition is most often used to locate and identify the whereabouts of absconders and criminals, its advantages extend well beyond safety concerns. Facial recognition technology may be used instead of cash or credit cards to make purchases at shops.

The iPhone X's security has gone from a one in 50,000 probability of being hacked to a one in a million possibility thanks to Face ID's replacement of Touch ID. When you unlock your phone using Face ID, instead of using a fingerprint, you unlock it using your face. Even the most modern face recognition systems have a greater FAR than the following advanced fingerprint authentication methods. Facial recognition, on the other hand, might be prone to false negatives due to the presence of spectacles, cosmetics, or even altered ambient illumination.

Facial detection encompasses both face recognition and face detection. A picture or video's ability to detect the presence of a person's face is all that the term "facial detection" really signifies. Only facial recognition is one of the many uses of face detection. Cameras that employ autofocus may also benefit from face detection. Facial recognition systems may be used to authenticate a user's identification by comparing a digital photo or video frame to a database of faces, which works by recognizing and quantifying facial traits in the image.

1.5 Iris Recognition

One or both of an individual's irises may be used as a biometric identifying tool using iris recognition, it uses video images of one or both irises to apply mathematical pattern-recognition algorithms. All biometric methods have varying degrees of discriminating capacity based on the amount of entropy present. It is possible for them to encode and utilize in matching. There are no "collisions" (False Matches) when using Iris recognition, even when comparing large populations of people throughout the globe. Image collection across distances of more than a meter or two or without the cooperation of the subject is very challenging.

In contrast to iris identification, which analyses the distinctive patterns on a person's retina blood vessels to identify them, retina scanning is an ocular-based biometric method of its own. An externally viewable, high-detail picture of the iris is captured using video camera technology and near infrared light for iris identification. Using mathematical and statistical methods, these patterns may be encoded into digital templates that can be used to identify a person or someone posing as that individual. Millions of templates per second per single-core CPU are scanned by matcher engines with very low false match rates.

More than one and a half billion individuals have signed up to use iris recognition technologies for everything from national identification to electronic government services and benefits distribution to security and convenience measures like passport-free automated border crossings. Iris recognition is used to authenticate identification in these systems. Additionally, iris recognition has the advantage of being able to quickly match a person based on their iris, which is an internal and protected yet outwardly visible eye organ.

2 Literature Review

See Table 1.

Abdurrahim et al. [1] provided little research on the impact of demographic variables on face recognition algorithm performance. An extensive and concentrated overview of current research on demographic variables was provided in this publication by the authors. To better understand how race and age interact with other factors, researchers drew conclusions from their findings and gave suggestions for the field's future development [1].

De-xin et al. [2] a video search engine that uses facial recognition as an indexing component was proposed in this study. Facial recognition is a natural match for sifting through the massive amounts of video data that have been generated in recent years. However, the quality of the facial photos captured in the video signals affects the search's performance. Face occlusion is fairly prevalent in daily video since security cameras don't capture footage with fixed postures for the subject. FPCA is used to develop a model of occluded faces, which is subsequently used to reconstruct human faces in the suggested method. The results of the experiments reveal that the system processes real-life movies extremely efficiently and is particularly resistant to different types of facial occlusions. As a result, it frees up the monitors' front row for reviewers while also boosting productivity [2].

Li et al. [3] summarizes and categorizes recently announced facial recognition virtual sample generation technologies into three groups: Face structure-based virtual

Table 1 Literature survey

Serial number	Author/year	Title	Objective	Technique	Limitation
[1]	Abdurrahim [1]	Look at how factors like age, gender, and race affect automated facial recognition technology	Face recognition	Facial recognition technology	Scope of this research is very less
[2]	Xin Zhang [2]	Face recognition technology in video surveillance systems is becoming more common	Face recognition	Video surveillance systems	There is no implication in future
[3]	Li [3]	For face recognition, a review of virtual sample creation technologies	Face recognition	Virtual sample creation technologies	There is no security in this system
[4]	Huo [4]	An effective feature descriptor for iris recognition is developed using the Gabor filter and uniform local binary pattern coding	Iris recognition	Gabor filter and uniform local binary pattern coding	Lack of technical work
[5]	Regouid [5]	Based on local descriptors, a multimodal biometric system recognizes ECG, ear, and iris data	E-mail spam	Multimodal biometric system	Performance of this research is very low
[6]	Raja [6]	Biometric features detection based on human face, fingerprint, and iris pictures using SVM classifiers for prognostic purposes	Multimodal biometric, biometric	SVM classifiers	There is no security in this system
[7]	Kaur [7]	Using multimodal biometric keys and fractional transformations, a safe picture encryption technique is created	Multimodal biometric	Multimodal biometric system, encryption technique	Lack of accuracy

(continued)

Table 1 (continued)

Serial number	Author/year	Title	Objective	Technique	Limitation
[8]	Rathgeb [8]	Prerequisites, benefits, and potential enhancements for SIFT-based iris recognition	Iris biometrics	Iris recognition system	Lack of security
[9]	Acien [9]	Face recognition algorithms that include gender and racial biases	Face recognition	Facial recognition technology	Lack of accuracy
[10]	Okokpujie [10]	User authentication may be improved with the use of iris biometrics in automated teller machines	Iris biometrics, biometrics	Iris biometric system	There is lack of performance

portraits; perturbation- and distribution-based portraits; and sample-based portraits are three different approaches to creating virtual portraits of people. An in-depth comparison of several approaches was conducted as part of a comprehensive comparative research. Using virtual sample generation technologies in conjunction with representation-based approaches yields considerable benefits, as shown below [3].

Huo et al. [4] created a serious expression on his face. A 2D Gabor filter is used to extract multi-orientation imaginary (MOI) properties prior to storing area information utilizing a consistent LBP approach in this paper's Iris feature descriptor. Modifications were made by converting MOI's feature-by-point energy into a uniform LBP histogram by block and widening the distributions of both classes in and out of class. This procedure greatly enhances the ability to tell apart MOI features. As a final step in matching, they use the Bhattacharyya distance (BD). This method outperforms the individual functions of MOI features and LBP encoding when tested on CASIA and JLU iris image databases [4].

Regouid et al. [5] expressed that ECG, ear, and iris biometrics are used in this study to create novel multimodal biometric systems. Normalization and segmentation were used to process the ECG, ear, and iris biometrics. Techniques such as 1D-LBP (One D Local Binary Patterns) and Shifted-1D-LBP are used to extract the most essential aspects from an ECG signal and to transform pictures of the ear and iris into 1D signals utilizing local texture descriptors. Unknown users may be classified as either real community members or imposters using KNN and RBF. ID-ECG, AMI ear, and CASIA v1 iris databases were used, created system was tested. Proposed biometric system outperforms unimodal biometric system in experiments. It possible to attain a 100% correct recognition rate (CRR) with an EER of 0.5% [5].

Raja et al. [6] expressed that the problem was addressed by presenting the ESVM-KM technique for multimodal biometric identification. The ESVM-KM technique has been created to increase the accuracy of multimodal biometric identification utilizing human face, fingerprint, and iris images. The image quality is enhanced for human recognition during the preprocessing stage by eliminating noise and increasing contrast. Using the ESVM-KM methodology with the Gabor wavelet transformation, an excellent feature extraction method for human face, fingerprint, and iris images was developed. The ESVM-KM technique used an ensemble SVM classifier to boost recognition rates for the multimodal biometric system. The ESVM-KM technique's simulations employ a variety of measures, including computation time, recognition rate, and the percentage of true positives. Multimodal biometric identification systems may benefit from ESVM-faster KM's processing times while improving recognition rates, according to the simulation results. Storage of ESVM-KM findings was made possible by storing them in a cloud-based environment [6].

Kaur et al. [7] study proposes a picture encryption technique using a fractional transform and a scarmbled biometric key. The biometric keys are very difficult for unauthorized parties to get. XORing both the iris and fingerprint binary codes yields the original picture. Fractional order serves as a key to this randomized image's security measures. Fractional transforms have a unique characteristic that benefits from the additional latitude supplied by their fractional orders. The iris key was used to determine the fractional order. Scrambling was used to rearrange the pixels, thus complicating the encryption process. Using histogram analysis and correlation analysis, as well as peak signal-to-noise ratio, mean square error, and structural similarity index metrics, spoofing and heterogeneous attacks may be identified. Verify the efficacy of the provided methodology [7].

Rathgeb et al. [8] looked SIFT-based iris recognition has been proposed in this study, which addresses the shortcomings of prior systems. Biometric data were used to design and evaluate an improved SIFT-based baseline system that works on normalized enhanced iris textures derived from near-infrared iris pictures. An array of binary SIFT-based feature vector extraction approaches, each with a different approach to identifying important landmarks, will be shown and compared here. The suggested solution maintains the performance of several conventional systems in terms of identification accuracy and authentication speed on the CASIAv1, CASIAv4-Interval, and Bio Secure iris databases. A multi-algorithm fusion at the score level, on the other hand, produces considerable gains in recognition accuracy since SIFT-based characteristics complement those derived by standard techniques [8].

Acien et al. [9] expressed that sensitive data like as gender and ethnicity were examined in this paper's studies to see whether the most widely used facial recognition networks include any of those information. VGGFace and Resnet50, two well-known pre-trained networks, are employed in our research. For both pre-trained models, gender and ethnicity categorization is straightforward even when the embedding layer neurons are 80% suppressed. More than 13,000 photographs of faces of diverse ages, races, and nations were evaluated on a publicly available database called Labeled Faces in the Wild [9].

Okokpujie et al. [10] reviewed the world's most widely used ATM that was presently plagued by a slew of security issues. Personal Identification Number (PIN)-based automated teller machines were at the top of the list of these problems. There have been a slew of ideas floated recently for combating the problem of automated teller machine fraud. As a means of preventing fraud at ATMs, biometrics is one of the most promising technologies available. A microprocessor, an iris scanner, and a strong database form the basis of this paper's iris biometric-based automated teller machine. The prototype, which has been built and constructed, is capable of detecting ATM fraud and is also simple to adopt in poor countries [10].

Tiong et al. [11] created MDLN (Multi-Feature Deep Learning Network), to increase identification performance, this research proposed using modalities from the face and particular places together with texture descriptors. Biometric data from various sources are combined with a texture description in this feature-level fusion technique to build a new feature representation. As a result, the MLDN model is projected to outperform the existing unethical deep learning approaches in terms of information delivery via feature representation. Their research shows that the suggested MDLN has improved biometric identification in challenging settings, such as light fluctuations, appearances, and posture misalignment, utilizing numerous public datasets and our own proposed MDLN [11].

Abozaid et al. [12] Compared the results of Eigen face and Principle Component Analysis (PCA) in extracting facial recognition characteristics. GMM (Gaussian Mixture Model), ANN (Artificial Neural Network), SVM are some of the several classifiers that may be used to recognize voice and face (SVM). Speech and face biometrics are combined into a single multimodal biometric system through the fusion of features and scores. It has been shown via computer simulation studies that using the campestral and statistical coefficients for sound recognition yields better results than doing the same thing using Eigen face and SVM for face recognition. in the suggested multimodal biometrics system, fusion of score data was superior than other alternatives [12].

Vasavi et al. [13] presented that classification methods were used in the biometric authentication procedure known as iris scanning. Humans are concerned primarily with their own well-being. As a result of its individuality, biometric identification offers a higher level of security. To identify people, iris recognition employs the use of eye patterns. The goal of this research was to build an iris recognition system utilizing Machine Learning. For biometric identification and verification, the major objective is to improve iris image capture and processing (Machine learning strategies that have recently been developed to improve the performance of iris recognition systems were discussed in detail in this study). It was necessary to examine the fundamental principles and connections among diverse approaches in order to get an overview and understanding of the evolution of iris recognition [13].

Vasanthi et al. [14] looked at the relevance of the ACCs, an autocorrelation test was performed. Assumptions were made that if the ACC passes; the little image area (the window) reflects the texture and is processed accordingly. It is referred to as "structure" since it is a feature of the shape rather than a structure. In certain cases, it is feasible to combine the texture and form qualities into one feature vector

(FV). In this way, all color submodels are vetted. This holistic vector, which serves as the FV for the primary face picture, may be used to produce the FVs for the submodels. Twenty features make up the core of the FV. If the key and target FVs are same, multivariate parametric statistical tests are used to ascertain this. Key and target photos are tested to see if their FVs are the same; if not, they are considered to be different. A number of datasets will be employed in the research: GT, FSW, Pointing'04, and Bio ID. Additionally, we have used Google images and photos shot with digital cameras from throughout the globe as supplementary datasets. The experiment is also carried out on it. According to the recognition results, the suggested ARBBPNN approach beats the current methods [14].

Wati et al. [15] presented many benefits of documenting attendance by scanning a person's face. It was the elimination of the need to establish direct eye contact with the scanning device. It is necessary to do face detection before completing face recognition in order to identify whether or not the face image contains the eyes, nose, mouth, and other facial traits. Viola-Jones facial recognition, Gabor Wavelet feature extraction, and Template Matching are all used in this work. Individual faces may be recorded individually or in groups, in which case all the faces are recorded simultaneously and then each one is extracted and recognized. This research achieved an accuracy of 75%, a recall rate of 64%, and a precision rate of 88% for individual attendance identification. Using simultaneous/group facial recognition, the research achieved 88% accuracy, a recall rate of 75%, and an accuracy score of 97% [15].

3 Problem Statement

There have been many researches in area of biometric application. These researches have used unimodal as well as multimodal approaches. But the limitation of the existing model is their performance issue and accuracy issue. There is a need to do more work in area of biometrics to improve the reliability, flexibility, and scalability. The accuracy factors considered in previous research are recall value, precision, and F1-score that are required to be improved.

4 Need of Research

Instead of a physical ID card, citizens may show their true identity using a biometric document. The government is able to provide a secure environment and guarantee that residents have access to the services and benefits they need by giving fingerprint-based identification cards. After gaining traction in other industries, it is now making inroads into the financial sector. Biometric payment cards are already on the market, and they're gaining more and more attention from consumers. When making a payment, customers are more likely to mistype or forget their PINs, and they are more likely to be wary about using it. Because the user simply has to use their biometric

Table 2 Comparison of features in existing research

References	Face recognition	Iris recognition	Multi-model	Deep learning	Security
[13]	No	Yes	No	Yes	No
[14]	Yes	No	Yes	Yes	No
[15]	Yes	No	No	No	Yes
[16]	Yes	No	Yes	No	No
[17]	Yes	No	No	Yes	No
[18]	Yes	No	No	No	No
[19]	No	Yes	Yes	No	No
[20]	No	Yes	Yes	No	N
[21]	Yes	Yes	Yes	No	No
[22]	Yes	Yes	No	No	No
[23]	Yes	Yes	No	No	No
[24]	Yes	No	Yes	Yes	No

traits as their identity using fingerprint identification, this problem is eliminated. Biometric authentication has become a need because of the high level of security it provides. Using biometric authentication, you can prevent card skimming, chip swapping, and shoulder surfing, as well as other types of payment fraud (Table 2).

5 Scope of Research

Biometric identification is very instantaneous; however, PIN authentications might take a few seconds or more if you use the incorrect one. Customers and companies alike will benefit greatly from this. PINs and passwords are expensive to maintain, whereas biometric-enabled payment cards are less expensive. Fingerprint sensor technology suppliers are already working on a rapid and cost-effective enrollment solution for their customers as demand for biometric credit and debit payment cards develops. When it comes to the design of biometric payment cards, they are thin, tiny, and flexible. It may be carried in a wallet or purse like any other credit card because of its unique design. A biometric payment card may sound outlandish at first, but the procedure of enrolling couldn't be easier. Enrollment is as simple as putting one's fingerprint on the card's sensor. There is no need to contact a bank or submit biometric data to any one central database. In future, further researches are expected to enhance the performance of biometric authentication.

References

1. Abdurrahim SH, Samad SA, Huddin AB (2018) Review on the effects of age, gender, and race demographics on automatic face recognition. Vis Comput 34(11):1617–1630. https://doi.org/10.1007/s00371-017-1428-z
2. Xin Zhang D, An P, Xiang Zhang H (2018) Application of robust face recognition in video surveillance systems. Optoelectron Lett 14(2):152–155. https://doi.org/10.1007/s11801-018-7199-6
3. Li L, Peng Y, Qiu G, Sun Z, Liu S (2018) A survey of virtual sample generation technology for face recognition. Artif Intell Rev 50(1):1–20. https://doi.org/10.1007/s10462-016-9537-z
4. Huo G, Guo H, Zhang Y, Zhang Q, Li W, Li B (2019) An effective feature descriptor with gabor filter and uniform local binary pattern transcoding for iris recognition. Pattern Recognit Image Anal 29(4):688–694. https://doi.org/10.1134/S1054661819040059
5. Regouid M, Touahria M, Benouis M, Costen N (2019) Multimodal biometric system for ECG, ear and iris recognition based on local descriptors. Multimed Tools Appl 78(16):22509–22535. https://doi.org/10.1007/s11042-019-7467-x
6. Raja J, Gunasekaran K, Pitchai R (2019) Prognostic evaluation of multimodal biometric traits recognition based human face, finger print and iris images using ensembled SVM classifier. Cluster Comput 22:215–228. https://doi.org/10.1007/s10586-018-2649-2
7. Kaur J, Jindal N (2019) A secure image encryption algorithm based on fractional transforms and scrambling in combination with multimodal biometric keys. Multimed Tools Appl 78(9):11585–11606. https://doi.org/10.1007/s11042-018-6701-2
8. Rathgeb C, Wagner J, Busch C (2019) SIFT-based iris recognition revisited: prerequisites, advantages and improvements. Pattern Anal Appl 22(3):889–906. https://doi.org/10.1007/s10044-018-0719-y
9. Acien A, Morales A, Vera-Rodriguez R, Bartolome I, Fierrez J (2019) Measuring the gender and ethnicity bias in deep models for face recognition, vol 11401. Springer International Publishing, LNCS
10. Okokpujie K et al (2019) Integration of iris biometrics in automated teller machines for enhanced user authentication, vol 514. Springer, Singapore
11. Tiong LCO, Kim ST, Ro YM (2019) Implementation of multimodal biometric recognition via multi-feature deep learning networks and feature fusion. Multimed Tools Appl 78(16):22743–22772. https://doi.org/10.1007/s11042-019-7618-0
12. Abozaid A, Haggag A, Kasban H, Eltokhy M (2019) Multimodal biometric scheme for human authentication technique based on voice and face recognition fusion. Multimed Tools Appl 78(12):16345–16361. https://doi.org/10.1007/s11042-018-7012-3
13. Abirami MS, Vasavi J (2020) A qualitative performance comparison of supervised machine learning algorithms for iris recognition. Eur J Mol Clin Med 7(6):1937–1946
14. Vasanthi M, Seetharaman K (2021) A hybrid method for biometric authentication-oriented face detection using autoregressive model with Bayes backpropagation neural network. Soft Comput 25(2):1659–1680. https://doi.org/10.1007/s00500-020-05500-8
15. Wati V, Kusrini K, Al Fatta H, Kapoor N (2021) Security of facial biometric authentication for attendance system. Multimed Tools Appl 80(15):23625–23646. https://doi.org/10.1007/s11042-020-10246-4
16. Vorakulpipat C, Pichetjamroen S, Polprasert C (2021) Interacting face detection-based access control with various authentication factors. In: ACM international conference proceedings series, pp 65–69. https://doi.org/10.1145/3453800.3453813
17. Teoh KH, Ismail RC, Naziri SZM, Hussin R, Isa MNM, Basir MSSM (2021) Face recognition and identification using deep learning approach. J Phys Conf Ser 1755(1). https://doi.org/10.1088/1742-6596/1755/1/012006
18. Imaoka H et al (2021) The future of biometrics technology: from face recognition to related applications. APSIPA Trans Signal Inf Process 10:1–13. https://doi.org/10.1017/ATSIP.2021.8

19. Vyas R, Kanumuri T, Sheoran G, Dubey P (2021) Accurate feature extraction for multimodal biometrics combining iris and palmprint. J Ambient Intell Humaniz Comput 0123456789. https://doi.org/10.1007/s12652-021-03190-0
20. Vijayakumar T (2021) Synthesis of palm print in feature fusion techniques for multimodal biometric recognition system online signature. J Innov Image Process 3(2):131–143. https://doi.org/10.36548/jiip.2021.2.005
21. Purohit H, Ajmera PK (2021) Optimal feature level fusion for secured human authentication in multimodal biometric system. Mach Vis Appl 32(1). https://doi.org/10.1007/s00138-020-01146-6
22. Sarangi PP, Nayak DR, Panda M, Majhi B (2022) A feature-level fusion based improved multimodal biometric recognition system using ear and profile face, vol 13, no 4. Springer Berlin Heidelberg
23. Xiao K, Tian Y, Lu Y, Lai Y, Wang X (2022) Quality assessment-based iris and face fusion recognition with dynamic weight. Vis Comput 38(5):1631–1643. https://doi.org/10.1007/s00371-021-02093-7
24. Yang Q, Chen X, He Z, Chang L (2022) Survey on deep learning based fusion recognition of multimodal biometrics. In: Biometric recognition. CCBR 2022. Lecture Notes in Computer Science, vol 13628. Springer, Cham. https://doi.org/10.1007/978-3-031-20233-9_52

Applications of Deep Reinforcement Learning for Drug Discovery

Parminder Singh Sethi, Gurleen Kaur, D. S. Vasanth, Malathi Ramakrishnan, and Nithish Kote

1 Introduction

Drug discovery is a time-consuming process, and it takes 10–15 years and up to $200 million to bring a drug to market. Regardless of the huge investment and time commitment, not all drugs discovered using traditional methods are approved for public usage. This is because of the efficacy of drugs and their adverse side effects. Furthermore, the traditional methods contrast with the "one-size-fits-all" approach. In other words, a medicine is created only with typical person in mind, irrespective of individual differences.

The drawbacks and regulations in the drug discovery process have influenced the healthcare industry to focus more on computational models to accelerate the drug discovery process. This is augmented by the exponential growth in the amount of chemical and biological data available to design, analyze, and apply novel learning methodologies. A significant amount of biological data has been compiled and generated in recent years from a variety of sources, including tiny individual laboratories and huge companies [17]. The scientific community uses biomedical data (genomic, proteomic, metabolomic, and pharmacogenomic) to support drug discovery through computational models [1].

The recent advancements in information technology and the ability to expedite and simplify the analysis of a large amount of information enabled us to find a new computational model for screening extensive drug libraries for drug discovery [1]. On the other hand, the capacity of systems (storage capacity and compute power) was drastically improved to process a large amount of biomedical data. With these computational advancements, machine learning (ML) techniques are gaining more

P. S. Sethi · D. S. Vasanth · M. Ramakrishnan (✉) · N. Kote
Dell EMC Bangalore, Bangalore, India
e-mail: rmalathimdu@gmail.com

G. Kaur
Enterprise SoftLabs Pvt. Ltd., Bangalore, India

A. Ramdane-Cherif et al. (eds.), *Machine Intelligence and Data Science Applications*, Algorithms for Intelligent Systems, https://doi.org/10.1007/978-981-99-1620-7_11

prominence due to their ability to predict and recommend drugs. The use of machine learning models significantly reduces the cost and time for discovering new drugs. Currently, machine learning and a large amount of data on drugs and diseases hold promise for potentially improving the drug discovery process and accelerating the discovery of innovative treatments [7]. Furthermore, the ability to predict the association between therapeutic indicators and side effects using ML helps researchers identify drug indications and associated side effects.

The following are the machine learning algorithms widely used in the pharmaceutical industry for drug discovery [10].

- Deep Reinforcement Learning
- Naive Bayes Classifier (NB)
- Support Vector Machine (SVM)
- Random Forest (RF)
- Artificial Neural Network (ANN)

The treatment of a disease does not just end with the discovery of a drug. It is very important to test the efficiency and side effects of drugs before administration. According to the precise medicine initiative, treatment and side effects differ depending on the individual's lifestyle, gene, and community. The current commercial drug development model does not serve this purpose because drug efficacy and side effects can only be tested in a limited number of patents or regions. Therefore, it is wise to gather data through crowdsourcing [3, 15].

Through crowdsourcing, molecular designs of a drug are shared to test its efficiency with various research organizations and institutions, biotech and pharma companies, and contract research organizations (CROs). Based on the data received through crowdsourcing, ML techniques can be further applied to rapidly validate the administration of drugs to a large number of causalities.

2 Literature Review

The process of finding new drugs has undergone a fundamental transformation due to development of big data, artificial intelligence, and machine learning technologies. Machine learning techniques are used in pharmaceutical industry applications such as hit discoveries and de novo drug architectures. Deep reinforcement learning models are preferred over supervised learning as they can better understand the chemical space of the environment in comparison to supervised learning methods. Deep learning has transformed various technologies, including drug discovery. The drug discovery techniques take a long time, are not affordable, and cannot supply drugs quickly for emerging pathogenic microorganisms. Big data sets created by large-scale testing are utilized in the most recent alternative drug development techniques. These techniques enable more precise target bioactivities and molecular characteristics prediction (Stephenson et al. 2019). Dara et a. [4] deliver workable research on the use of machine learning strategies and tools in all stages of drug development in order to speed up research and lower the risk and cost of clinical trials. Jose et al. [5] use a reinforcement learning-based approach to identify the potential binding site and the correct ligan pose. Zhang et al. [18] explain how the identification of numerous possible SARS-CoV-2 medicines has been aided by deep learning algorithms. Popova et al. [12] propose a novel computational strategy for the de-nova design of molecules with desired properties termed ReLeaSE (Reinforcement Learning for Structural Evolution). Tan et al. [14] discuss the application of reinforcement learning for systems pharmacology-oriented and personalized drug design. Ståhl et al. [13] present a fragment-based reinforcement learning approach to generate novel molecules with optimal properties. Nantasenamat [9] provides step-by-step instructions on how to build a simple regression model in Python for predicting the solubility of molecules. Mach [6] leverages reinforcement learning to enhance structure-based drug design models by optimizing generated molecules to demonstrate desired properties. Pereira et al. [11] explore deep learning methods for identifying potential drugs with molecular properties. Rakant [2] the tutorial provides the basics of reinforcement learning and how to use neural networks within it. Maragakis et al. [8] present DESMILES, an advanced deep neural network used to design molecules. [16] provides a general introduction to deep reinforcement learning.

3 Drug Discovery Process

The drug development process involves identifying the targets in a biological molecule, identifying drugs among the millions of drug molecules that will bind to the target, and making the required biological impact. Following drug identification, the drugs are optimized to assess toxicity. Then it goes under a certain preclinical test, followed by animal trials, clinical trials, and human trials.

Though drugs have been identified for the target, there are some concerns about the drug's effectiveness on the target. This is because it is not sure how the identified drug would produce a biological activity on the target molecule, effectively bind to the target, or cause drug toxicity.

4 Deep Reinforcement Learning Model for Drug Discovery

The machine learning framework discussed in this work has two parts. Part one consists of two deep neural network models: the generative and the QSAR models. The generative model builds rules to generate new molecules by using SMILES [11]. The QSAR model predicts the specific biological activities of molecules. With Long Short-Term Memory (LSTM) and Gated Recurrent Unit (GRU) cells, both generative and QSAR models are created.

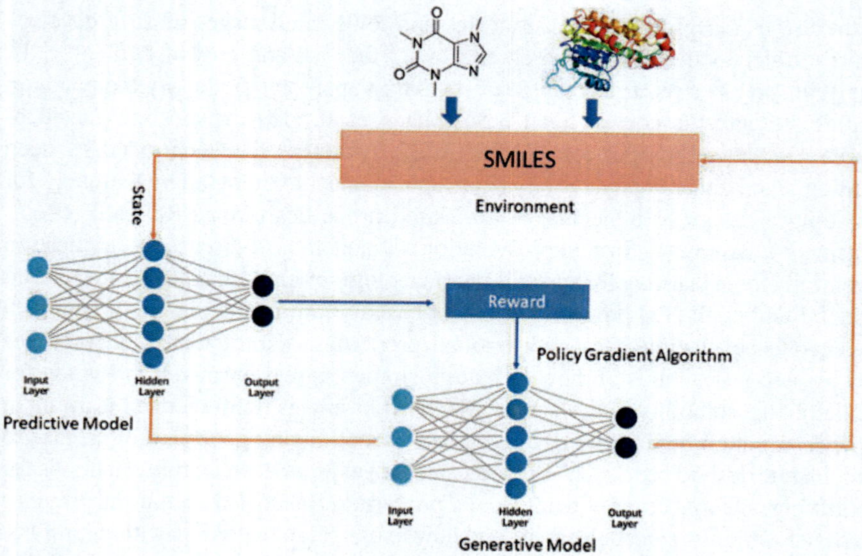

Fig.1 Deep reinforcement learning framework

In Part 2, Reinforcement Learning employs the predictor model to predict the affinity of the newly generated molecule for the desired target and train the generative model, resulting in molecules with optimized desired properties [11].

Figure 1 illustrates the Deep Reinforcement learning framework discussed in this paper.

5 Generative Model

The generative model uses SMILE data to generate new molecules. However, the SMILE cannot be provided as input data. The SMILE has to be converted into tokens for this model to process. In this framework, all the input SMILES strings are converted into numerical values and used as input for the generative model. The SMILES strings are tokenized, padded, and finally transformed into encoding vectors [11].

The generative model consists of the following layers:

- A layer of neurons with the softmax function (1)
- Densely connected layer (1)
- LSTM layers (2)
- Hot layer (1).

Each token array is used in the training process to determine the output based on the probability distribution of all the potential tokens. After multiple training cycles, the

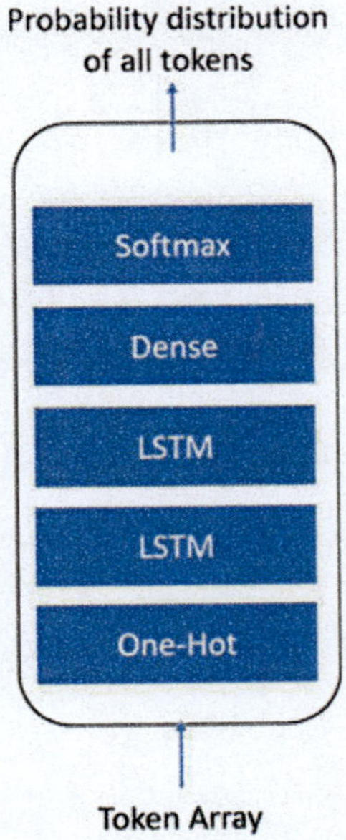

Fig. 2 Generative model

model awards a token with the highest probability [11]. Figure 2 shows the layers used in the generative model.

6 Predictive Model

The predictor model is a Quantitative structure–activity relationship (QSAR) model that maps the molecules' structure and their capacity for bonding to the target [11]. Due to the logarithmic nature of potency values, the value of p IC50 is equal to log(IC50). Higher p IC50 values represent exponentially more potent inhibitors [11].

The model consists of the following layers [11]. The SMILES strings are the input data for the predictive model. Figure 3 shows the layers used in the predictive model.

- Layer of embedding that transforms each token into a vector.
- Gated recurrent unit (2).
- Dense layer (1) activated by a Rectified Linear Unit (ReLU) function.
- Output layer with a linear activation function.

The pre-processing of the predictive model starts with converting SMILES to tokens and then to numeric descriptors [11]. Each numeric descriptor that represents a token is transformed into a continuous vector space. The tokens are passed to the two layers of GRU cells, which produce a vector of 128 units, after converting them into continuous value vectors [11]. To transfer this 128-unit vector into a single value,

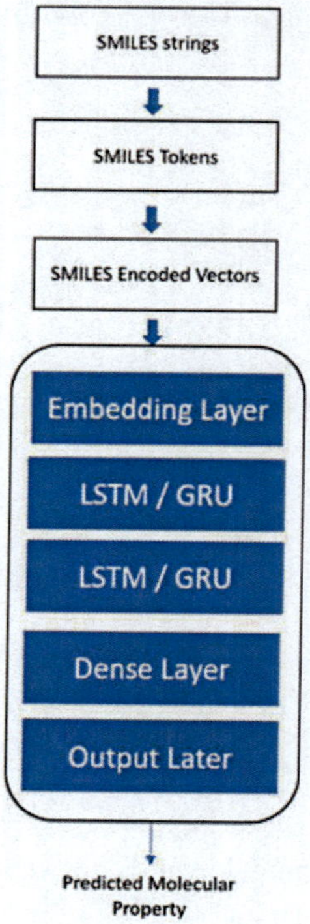

Fig. 3 Predictive model

it is routed to a densely connected layer. Lastly, the biological activity is estimated using a linear activation [11].

7 Reinforcement Learning Process

The aim of the reinforcement learning model used in the framework is to optimize the generative model. As a result, the generative model can generate molecules with significant binding and inhibitory properties for the target. The generative model uses the experience-driven approach of RL to learn [11]. The agent in the generative model interacts with the environment, learns how to map states into action, and maximizes the reward. Based on the status of the environment, the agent chooses the optimum course of action [11]. Based on the outcome of the previous action and state, a reward is assigned, the environment is updated, and the new state is presented to the agent to repeat the process [11].

The RL model starts the learning process with the new compounds. In other words, the SMILES that correspond to new molecules are created and are assigned with a reward from the predictor model. The loss function is then calculated after the tokens used to represent the molecules have been "decomposed" and "reconstructed" yet again. This results in the initial incentive being assigned in a discounted form, and the agent is then shown the new status. The intersection of the rebuilt molecule and the token selected in the previous phase is represented by this state. In order to forecast the following token or action, the newly produced state is employed [11].

8 Policy Gradient

The policy function used to output the probability distribution for all possible actions uses a SoftMax function to transform the raw numbers outputted by the policy function, called logits, into probabilities between 0 and 1. All the probabilities add up to 1 (this is the reason why SoftMax is used instead of a sigmoid activation function [9].

The policy gradient loss function is defined as

R is the reward at a given state after taking a given action.
y (gamma) is the discount factor and has a value between 0 and 1.
The gamma parameter represents how much future reward can be expected.

To find the optimal policy that will lead to the greatest reward, the policy gradient loss is maximized instead of minimizing it as in gradient descent. The negative sign in the policy gradient loss function is important as it serves the purpose of maximization [9].

Logarithms are used because they are more numerically stable, and therefore perform better. Backpropagation is also less complicated [9].

In simple terms, if the value of the action taken is better than the expected value of the action taken, then the probability of that specific action increases. Next time, the agent will be way more likely to take that specific action again. In contrast, if the value of the action taken is worse than the expected value, then the probability of that specific action decreases. Therefore, to maximize the reward, the loss function is maximized [13].

9 Reward Function

It is important to design reward functions well because this can dictate how well the reinforcement learning model trains [9].

In this case, the reward function is designed so that the agent (the generative model) learns to optimize the generated molecules so that they demonstrate a potent inhibitory effect against the target. Essentially, the goal of the reward function is to entice the agent to maximize the pIC-50 values for the target [9].

The reward function is defined as in Eq. 1:

$$R(S) = exp\left(\frac{pp(s)}{3}\right) \quad (1)$$

S is the current state of the agent.

pp is the property prediction model.

The property prediction model takes in input **S**, which is the current state the agent is in or the molecule produced by the agent, and outputs the corresponding pIC-50 value for JAK2. The agent gets rewarded more if the molecules it generates have a higher pIC-50 value in comparison to a lower pIC-50 value [9].

10 Feature Dataset

The model uses the already experimentally validated drug-target complexes available in the public repositories. The drug-target complex data that can be downloaded from the public repositories includes information about the type of drug-target interaction, special coordinates, and other information [11].

The collected drug-target complex data is converted into SMILE strings. The SMILES strings are the input data for this model. Each sequential or structural character in SMILE strings is transformed into a numerical value so that it can be used in the model [11].

11 Conclusion

In this paper, we discussed about the drawback of the traditional drug discovery process and availability of various computation models that are used to expedite the drug discovery process. Furthermore, we discussed about the application of a machine learning framework that is used to generate new molecules and estimate the compounds' biological affinity for the intended target.

References

1. Carracedo-Reboredo P, Liñares-Blanco J, Rodríguez-Fernández N et al (2021) A review on machine learning approaches and trends in drug discovery. Computat Struct Biotechnol J 19:4538-4558. https://doi.org/10.1016/j.csbj.2021.08.011
2. Rakant CK (2020) Reinforcement learning with neural network. Baeldung on Computer Science. www.baeldung.com. Published 30 September. https://www.baeldung.com/cs/reinforcement-learning-neural-network
3. Chodera J, Lee AA, London N, von Delft F (2020) Crowdsourcing drug discovery for pandemics. Nat Chem 12(7):581–581. https://doi.org/10.1038/s41557-020-0496-2

4. Dara S, Dhamercherla S, Jadav SS, Babu CM, Ahsan MJ (2021) Machine learning in drug discovery: a review. Artificial Intelligence Review. Published online 11 Aug. https://doi.org/10.1007/s10462-021-10058-4
5. Jose J, Gupta K, Alam U, Jatana, N, Arora P (2021) Reinforcement learning based approach for ligand pose prediction. www.biorxiv.org. Published 11 April. https://www.biorxiv.org/content/10.1101/2021.04.10.438538v1.abstract
6. Mach J (2020) Drug design made fun using reinforcement learning. Medium. Published 4 March. https://medium.datadriveninvestor.com/drug-design-made-fun-using-reinforcement-learning-212a4f867f33. Accessed 14 Sep 2022
7. Machine learning models for drug discovery. IBM Research Blog. Published 7 April 2017. https://www.ibm.com/blogs/research/2017/04/machine-learning-models-drug-discovery/
8. Maragakis P, Nisonoff H, Cole B, Shaw DE (2020) A deep-learning view of chemical space designed to facilitate drug discovery. J Chem Inf Model 60(10):4487–4496. https://doi.org/10.1021/acs.jcim.0c00321
9. Nantasenamat C (2020) How to use machine learning for drug discovery. Medium. Published 24 August. https://towardsdatascience.com/how-to-use-machine-learning-for-drug-discovery-1ccb5fdf81ad
10. Patel L, Shukla T, Huang X, Ussery DW, Wang S (2020) machine learning methods in drug discovery. Molecules 25(22):5277. https://doi.org/10.3390/molecules25225277
11. Pereira T, Abbasi M, Ribeiro B, Arrais JP (2021) Diversity oriented deep reinforcement learning for targeted molecule generation. J Cheminf 13(1). https://doi.org/10.1186/s13321-021-00498-z
12. Popova M, Isayev O, Tropsha A (2018) Deep reinforcement learning for de novo drug design. Sci Adv 4(7):eaap7885. https://doi.org/10.1126/sciadv.aap7885
13. Ståhl N, Falkman G, Karlsson A, Mathiason G, Boström J (2019) Deep reinforcement learning for multiparameter optimization in de novo drug design. J Chem Inf Model 59(7):3166–3176. https://doi.org/10.1021/acs.jcim.9b00325
14. Tan RK, Liu Y, Xie L (2022) Reinforcement learning for systems pharmacology-oriented and personalized drug design. Expert Opin Drug Discov 17(8):849–863. https://doi.org/10.1080/17460441.2022.2072288
15. The Moonshot: Crowdsourcing To Develop The First Open-Source, Generic COVID-19 Antiviral Pill—Health Policy Watch. Published 24 Dec 2021. https://healthpolicy-watch.news/the-moonshot/
16. TORRES.AI J. DRL 01: A gentle introduction to Deep Reinforcement Learning. Medium. Published 10 May 2021. https://towardsdatascience.com/drl-01-a-gentle-introduction-to-deep-reinforcement-learning-405b79866bf4
17. White House Precision Medicine Initiative. The White House. Published 2015. https://obamawhitehouse.archives.gov/precision-medicine
18. Zhang Y, Ye T, Xi H, Juhas M, Li J (2021) Deep learning driven drug discovery: tackling severe acute respiratory syndrome coronavirus 2. www.frontiersin.org. Published 28 Oct 2021. https://www.frontiersin.org/articles/10.3389/fmicb.2021.739684/full#:~:text=In%20parasite%20research%2C%20deep%20learning,ChEMBL%20database%20(Table%201)

Deep Learning-Based Framework for URL Phishing Detection

Meet Kava, Rajeev Kumar Gupta, Santosh Bharti, and Abhijit Kumar

1 Introduction

Phishing is a sort of attack or deception in which users are approached by various communication channels, such as phone, text message, or e-mail. In this attack, the attackers masquerade as a reputable and trustworthy entity and try to persuade consumers to give over or leak sensitive information such as passwords, banking credentials, identification information, or other personal information. The victim's submitted or leaked information is then utilized to access the crucial documents and accounts, allowing the attackers to profit financially or use their personal information for other purposes.

Among all the cyber-attacks, phishing is the most common. Phishing can be termed as the trapping of users to get the information in form of credentials or documents. Phishing attacks can be done via various means like SMS, URLs (websites), voice calls, emails, etc. Each term has different types of phishing methods. The various types of phishing methods are described briefly below.

Spear phishing: It is a type of phishing attack in which the target is specified and then the attack is designed and carried out accordingly. It targets some specific groups or individuals such as the company's network administrator or member of the board of that company. The term spear phishing is derived from the term spear, i.e. is a sharp side-edged weapon and phishing is the term that is used to manipulate the target to

M. Kava · R. K. Gupta · S. Bharti
Department of Computer Science and Engineering, Pandit Deendayal Energy University, Gandhinagar, Gujarat 382421, India
e-mail: santosh.bharti@sot.pdpu.ac.in

A. Kumar (✉)
School of Computer Science, University of Petroleum and Energy Studies Dehradun, Dehradun 248001, India
e-mail: abhijitkaran@hotmail.com

A. Ramdane-Cherif et al. (eds.), *Machine Intelligence and Data Science Applications*, Algorithms for Intelligent Systems, https://doi.org/10.1007/978-981-99-1620-7_12

fall into the trap of the attacker. In general terms Fishing with a spear allows you to target a specific fish. Hence, the name can be related to the term spear phishing.

Vishing: It is the type of phishing attack in which the phishing attack is carried out using a voice call or any audio medium. Hence, the term is called "Vishing". V stands for the voice here. Such attacks are popular among people who are less educated and can be easily manipulated by fraudulent schemes and fraud promises. e.g. A fraud voice call asks you about your bank account details for the deposit of the winning lottery amount. If the user/victim falls into the trap and shares the details then the sharing of account details can harm the user monetarily.

Whaling: It is a type of phishing attack in which the target chosen is very rich with the wealth or holds an extremely high position in any organization. He/she may be the CEO, CFO, or CMO of the company, which can provide a high amount of money in exchange for not leaking the acquired credentials of the company or using their position to cause harm to the company by any means by use of the resources stolen from him/her. Here, the term whale is used because the persons in the largest designation of organizations are targeted and referred to as whales of their organizations as the whale is the largest mammal among all the mammals.

Smishing/Email phishing: It is the type of phishing attack in which the phishing attack is carried out using SMS, email, or any text medium. Hence, the term is called "smishing". "sm" stands for the SMS here, i.e. the short message service. It is the most commonly used type of phishing since mobile services were invented. Attackers send these SMS/emails to any email addresses they can obtain.

Sextortion: It is a type of phishing attack in which nude images, videos, or audio are being used for phishing and blackmailing the victim for sharing or harming them in exchange for the monetary benefits, any personal rage, or any other reasons. In most cases, the demand is that you give them money, usually in Bitcoin, or the attackers will release the clip in publicly.

Search Engine Phishing: In this type of phishing attack, the attackers work on search engine optimization in which the top search is hit in the search engine. Clicking on that link displayed within the search engine redirects the victim to the attacker's website. After visiting the website, the attacker has access to the credentials of the victim, which can be used for various harmful purposed.

Pharming: In this type of attack, the website looks fake or cloned the same as the user thinks is as the legitimate one and the same user is manipulated to visit that URL and gains the information and credentials of that user.

Angler Phishing: It is a type of phishing attack in which the users targeted are specifically from the social media accounts of the organizations or corporate companies. It means the targeted users visit these fake social media accounts of organizations thinking that these are legitimate and share their information and credentials with them.

Pop-up phishing: It is a type of attack in which pop-up ads are used for targeting the users to install the malware into their systems. These pop-ups are either in form of the taking permission of the device or in form of advertisements. Usually, the popups are designed to be attractive and that it convinces the users to click on them, e.g. A pop up for enhancing the security of your device free.

Clone phishing: It is a type of attack in which the existing or previously used email is contained with any attachments or links. The cloned version elements are then replaced by the malware, which can be used for various purposes like flooding or encrypting for the ransom demand. The main aim of this technique is the trust of the user as it comes from the verified source, and the victim is easily allowed to enter into his/her system.

Evil twinning: It is a type of attack in which a Fake Wi-Fi access point is set up and it is designed in such a way that it looks like a legitimate access point of the network purposively the stealing the user's sensitive data. Such attacks can also be used as a man-in-the-middle attack. The fake Wi-Fi access point is used to eavesdrop on the user's devices and gain sensitive credentials like login details, banking passwords etc. As the network used on the device is of the attacker, the victim is unaware of what is being shared via that network.

1.1 Various Approaches Used for the Detection of the Phishing Websites

Blacklist and Whitelist Filtering Approach: In this strategy, URLs are filtered by whitelisting benign links and blacklisting phishing links, which means benign links are permitted to be viewed in the future but phishing links are not.

User Behaviour Prediction Approach: In this approach, the filtering is done using the behaviour of the URL, which means how the URL is performing when it is executed. For example, if a URL is benign then it will function as intended otherwise if it is phishing then it will cause some misbehaviour like redirection, error, etc.

Google Page Rank Approach: In this approach, the page rank of Google is identified and the URL having a bad score of page rank is considered a Phishing URL, and pages with a good score are considered benign URLs.

Machine Learning Approach: In this approach, the features of the URLs are fetched and a dataset is created. Using these datasets of URL features, the models are being trained and tested for the URLs being Phishing URLs or Benign URL.

Image-Based Approach: In this approach, the images of the URLs are being used and the proofs or the characteristics of certain URLs are being identified. In the image, the identification can be done whether the website has certain features such as a legitimate website or certain features a phishing URL has. From identification, we can say that the URL is benign or a phishing one.

This paper is organized as follows: Sect. 2 discusses the comparative analysis of the existing approaches. Section 3 discusses the dataset that we have used for the training and testing of our proposed model, the dataset for the phishing detection websites (URLs) with the description in which the attributes are the URL properties, URL resolving metrics, etc. Section 4 discusses the proposed Deep Learning-based model. We have used the Machine learning approach for phishing detection. Section 5 discusses the proposed model in detail. The models have been trained using eight machine learning and two deep learning methods after the significant features have been chosen. It also discusses the result analysis of different machine learning and deep learning model. Section 6 concludes the work and gives some future direction.

2 Literature Review

When utilizing machine learning to detect phishing, we don't look for their types. The URLs are solely classified as phishing URLs or benign URLs. If we want to classify the many sorts of phishing, we'll need a large dataset with reliable information on the features that the user extracts. For the prevention or detection of phishing links, the literature of methods and techniques used have been well-studied in the academic literature. The following authors have used the different approaches and achieved significant results, which have been used as a part of the literature survey and also after that, the comparative analysis of various work done in this domain is discussed below.

2.1 Literature Survey of Papers

The paper "Towards lightweight URL detection" [1] had used the URL dataset with 10 popular features of URLs like URL length, HTTPS usage, etc., and various classification algorithms like Random forest, SVM, MLP, etc., but the highest accuracy achieved was of RF classifier, which is about 99.29%. Also, the F1 score of the RF algorithm used is 98.26. As a result, the conclusion is that the method employed to construct the model was pretty good and produced good prediction results.

A variant of the CNN-based BAT algorithm was proposed in the paper "A novel phishing website detection based on Swarm intelligence with deep learning" [2], which uses the same dataset as our work, with 11 k data points and 32 features. The algorithm they proposed, i.e. SI-BBA has used the Adam optimizer for optimization purposes and gave an accuracy of 94.85%.

The paper "Performance Analysis of ML Algorithms Used for Phishing Detection" [3] proposed ELM-based Deep learning algorithm for the classification of

Phishing links. The algorithm they used on the particular dataset is SVM, Naïve Bayes, and ELM, whose accuracy was 85.43%, 60%, and 70%, respectively.

The paper "Towards Benchmark Datasets for ML-based website Phishing detection" [4] created the database for the URLs in which the data are extracted from the Yandex and Whois database. After that, they used this dataset for the prediction of URLs using binary and multi-class classification. They had used various algorithms like SVM, DT, NB, LR, and Random Forest, among which the highest sensitivity for the classifiers achieved for the binary classification was using Random Forest, i.e. for Pearson coefficient as 96.65% and for chi-square as 96.83%.

The paper "Towards Detecting and Classifying Malicious URLs Using DL" [5–7] had used filtering and cleaning approaches for remaking the dataset. After that, the dataset is tested for various algorithms. They have trained the algorithms using CPU, GPU, and TPU and compared the results. RF, DT, and K-NN algorithms performed best among all, whose accuracy is 98.68 ± 0.22, 97.63 ± 0.24, and 97.47 ± 0.39, respectively.

For the detection of Phishing websites, the study "Optimized Stacking Ensemble Model for Phishing Websites Detection" [8–11] uses a genetic algorithm-based ensemble classification approach. Their system is divided into three stages: training, ranking, and testing. They used AdaBoost, XGBoost, random forest, and LGBM, among other ensemble-based learning models. The dataset they have used is the merger of three publicly available datasets. On one of the datasets, the stacking ensemble method generates an accuracy of 98.58%, a precision score of 98.50, and a recall score of 98.74%

2.2 *Research Gap Identified from the Literature Survey*

By reading, the literature related to the topic, we came to know that the research gap for the detection of phishing links is the dynamic nature of the URLs.

There are three main points to highlight the research gap:

- Dynamic nature of URLs.
- Short life span of URLs.
- Difficulties in updating the databases of URLs frequently.

3 Dataset Description

The dataset used is taken from the Kaggle website, in which there are 32 features and 11,055 data points. These features are fetched from the Alexa and who is database and converted it into readable form. The dataset referred to from [12] is downloaded from Kaggle. The features of the URL that are used in the database are further classified and illustrated in Fig. 1.

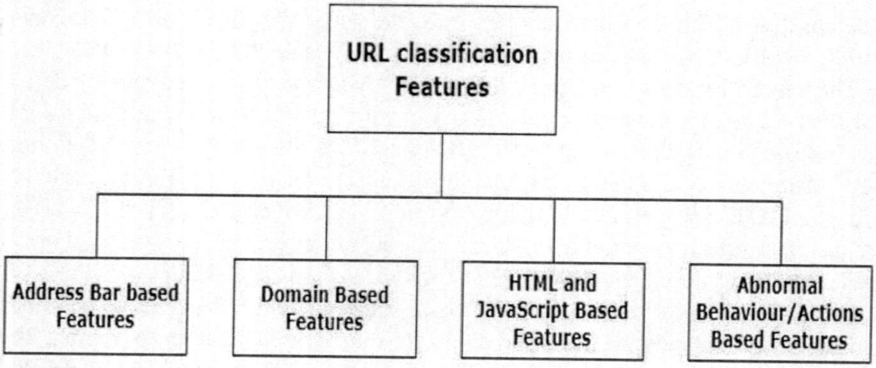

Fig. 1 Types of URL features

URL-based features

These features are extracted and collected using to form a dataset that can be used for purpose of the classification of the URLs as phishing or benign.

4 Working Methodology

- EDA and data pre-processing
- Features extraction and selection
- Splitting of the dataset
- Model construction
- Performance parameters of the models.

4.1 EDA and Data Pre-processing

In exploratory data analysis, the dataset is checked for error values, missing values, null values, and conversion of the values is performed if it is in another datatype like float to int, if required.

Our dataset is checked for all such values and filtered accordingly.

After checking the values, we need to filter the dataset into two categories from three categories.

In our dataset, three categories are mentioned.

- 1 → Legitimate/benign URL
- −1 → Phishing URL
- 0 → Suspicious of being Phishing URL

Therefore, instead of using three classes for classification and due to data insufficiency, we merge two classes by replacing those values to include them into another class. Hence, we make the suspicious/doubtful links as phishing links. After replacement, we now have two classes for the classification. Hence, the problem statement is reduced to a binary classification problem.

Two classes for the prediction with the values assigned are:

1 → Legitimate/benign URL
0 → Phishing URL

4.2 Dataset Balancing Performed

Dataset balancing is the technique in which the number of data points is made equal so that the prediction of the classes could be done better. If the dataset is not balanced, then the learning of the model would be done by taking the bias of the majority class and giving predictions based on the overfitting of the majority class. Hence, balancing the dataset is required to regular fit the data points to the model and gives better prediction accuracy to the model.

SMOTE Sampling (Synthetic Minority Oversampling Technique)

In the SMOTE oversampling technique, the balancing is done the same as the Oversampling technique mentioned above, but the only difference is the sample used for the oversamples, instead of duplicating they are synthetically generated using similar features and, the balancing is done.

We have done balancing using SMOTE, which gives the number of data points for each class as 6157 samples for benign and phishing links. Total data points are 12,314.

4.3 Feature Selection

Since there are 32 features, we could check for all features and whether they are contributing to the prediction of the output or not. For checking, we perform the method feature selection to see whether all the features are contributing to the accurate detection of the phishing links or not.

We have used feature selection methods like Chi-squared Selection, LASSO Regression, and Pearson coefficient Correlation. Out of all three ways, Pearson coefficient Correlation performed the best and gave quite a good selection of required features.

Pearson coefficient Correlation:

The absolute value of Pearson's correlation between the goal and numerical features in our dataset is checked using this method. Based on this criterion, we keep the top n-features.

$$r = \frac{\sum(x - \overline{x})(y - \overline{y})}{\sqrt{\sum(x - \overline{x})^2 \sum(y - \overline{y})^2}} \tag{1}$$

where,

r > 0.8 → features are highly correlated

r = 0 → features are not correlated at all

r < −ve → features are negatively correlated.

Approach for the feature selection using Pearson's coefficient

1. First, we will check the correlation with the Target Variable. If there is no correlation with the Target variable, then we drop those features (r = 0), as these features are not contributing to the prediction (four features removed).
2. Then we check the correlation of the features with each other. If the features are highly correlated (r > 0.8), then we drop those features as we can signify two highly correlated features as one (two features removed).
3. Total selected features = 25.

4.4 Splitting of the Dataset

The purpose of splitting the dataset is to check whether the model on which we are training the algorithms is suitable for the prediction of the test dataset or not. Hence, for the training and testing of models, we split the models into various splits. The most popular split used for the datasets is 80/20, 70/30, and 75/25. We have used two splits in our model, i.e. 80/20 and 70/30. We have displayed the results of the best split, which is an 80/20 split, where 80% is the training dataset, and 20% is the testing dataset.

5 Model Description

5.1 Random Forest (Bagging Based Classification Technique)

Random Forest is an ensemble-based Bagging supervised learning technique suitable for classification and regression. It is utilized for categorization difficulties, although most of the time. To put it another way, since a forest is made up of trees, and there are many trees, it's more robust. Using random forest techniques, the program constructs decision trees on sample data and then uses voting to choose the optimal option. Incorporating the decision tree results into a larger number of ensemble methods lowers the overfitting of individual decision trees, providing a better overall decision outcome.

Decision trees' predictions are used in the random forest algorithm to generate the final result. By averaging or summarizing the output of multiple trees, it predicts the future as the number of trees grows, so does the accuracy of the final product.

Equation of Gini index is to calculate the importance of each node in a decision tree, claiming two root nodes (binary tree):

$$n_k = W_k C_k - W_{l(k)} C_{l(k)} - W_{r(k)} C_{r(k)} \tag{2}$$

where,

n_k = Node k's importance.

w_k = In a weighted manner, the number of variables that reach node k.

c(k) = Impurity value for node k.

l(k) = The left-hand child node divides on node k.

r(k) = on node k, a right split child node.

The relevance of each feature is then determined on a decision tree as follows:

$$fi_i = \frac{\sum_{j:\text{ node } j \text{ splits on feature } i} ni_j}{\sum_{k\in \text{ all nodes}} ni_k} \tag{3}$$

By dividing by the sum of all feature importance values, these can be normalized between 0 and 1:

$$\text{norm } f_i = \frac{fi_i}{\sum_{j\in all\,features} f_i} \tag{4}$$

where,

$\mathbf{f_i}$ = Significance of the feature i.

$\mathbf{n_k}$ = Significance of the node k.

The ultimate measure of a feature's relevance in the Random Forest is the average value of the feature across all trees in the forest. To get the total number of trees, multiply the sum of the significance values given to each characteristic on individual trees by the total number of trees.

$$\boldsymbol{RFf_i} = \frac{\sum_{j \in \text{all trees}} \text{norm}\, \boldsymbol{f_{ij}}}{\boldsymbol{T}} \tag{5}$$

where, RFfi = The weight allocated to feature i as decided by the Random Forest model's whole tree population.

normf_{ij} = The normalized significance of a feature in tree j for i.

T = Number of trees in total.

5.2 *Perceptron*

The perceptron algorithm is the simplest deep learning-based classification technique. It is a binary classification system. It is made up of a single neuron that takes data points as input and predicts a class label as an output. Calculating the weighted sum of inputs and bias is how it's done (default bias is set to 1). The weighted sum of the input is termed the activation function which allows the neuron to get activated or not, i.e. it activates if the input required is significant for changes/modifications in the weights of the neuron.

$$\text{Activation} = \text{weights} * \text{input values} + \text{bias} \tag{6}$$

Activation > 0.0 → Predicts Yes

Activation <= 0.0 → Predicts No

The formula of perceptron algorithm:

$$\text{w}(\text{t} + 1) = \text{w}(\text{t}) + \text{eta} * (\text{y}_\text{i}) * \text{x}_\text{i} \tag{7}$$

where w = weights, eta = learning rate, $\hat{y}_i$ = Predicted value, y_i = Expected value, x_i = input

5.3 ANN (Artificial Neural Network)

The ANN algorithm is a deep learning-based algorithm, which works like a human brain to understand the features and make intelligent decisions based on the learning ability of the neurons. The ANN is designed to behave simply just like interconnected brain cells in the human body. A neural network that consists of a large number of artificial neurons, which are arranged in a sequence of layers with some input feeding to neurons, is called an ANN. Such a Deep Learning Network consists of three layers, i.e. input layer, an output layer, and a hidden layer.

The weighted sum of the inputs is computed by the artificial neural network, which also incorporates a bias. The weighted sum of inputs, coupled with a bias, is computed and fed to each neuron using an artificial neural network. It's time to figure out how to calculate weights and bias.

$$\sum_{i=1}^{n} w_i * x_i + b_0 \tag{8}$$

The weighted total is passed as an input to an activation function, which produces output. A node's activation function determines whether or not it will fire. Output is limited to those who have been fired. To get the desired result, a set of transfer functions is called an activation function. Non-linear and linear activation functions are the most common types of activation functions.

5.4 Experimental Setup

The dataset used for the purpose of URL phishing detection is about 11,055 data points and 31 features. The data used for the training is about 80% and 20% is used for testing. This is one of the most common splits. Other common splits are 70–30, 75–25, and 60–40 into training and testing, respectively. Python 3.10 was used throughout the project's development. The IDE used for the project's implementation is the Jupyter Notebook.

To further enhance our suggested approach's performance had used Python modules such as Keras, tensor flow, and sci-kit-learn to construct, calculate, and evaluate the model's performance and its results. We also display the accuracy table, classification report, and confusion matrix of different classifiers [12].

5.5 Important Terminologies for Deep Neural Networks

Activation Functions used—*Sigmoid Function*: As an input, the function accepts any real value, and as an output, it produces values in the range of 0 to 1. The greater the size of the input (the more positive), the closer the output value will be to 1.0, whereas the smaller the size of the input (the more negative), the closer the output value will be to 0. The function is calculated by the formula as given below:

$$f(x) = \frac{1}{1 + e^{-x}} \tag{9}$$

Rectified Linear Unit (ReLU): The function returns 0 if it receives any negative input, but for any positive value of x, it returns that value as echo. So it can be written as

$$f(x) = max(0, x) \tag{10}$$

Loss Function used: *Binary cross-entropy*:

As the prediction class is binary in nature, we are using the binary cross-entropy. Such type of loss is expected to get output with one of the two potential outcomes just like outcomes in tossing a coin.

The loss is calculated using this formula given below:

$$L = -(y_i \log(\hat{y}_i) + (1 - y_i)\log(1 - \hat{y}_i)) \tag{11}$$

where,

L = loss function

y_i = expected output

$\hat{y}_i$ = predicted output.

Optimizer used:

Optimizers are used to reduce the loss function and to enhance the performance of the model. Hence, termed as an optimizer.

We have explored multiple optimizers during experimentation and found that using Adam Optimizer enhances the accuracy of the classifier. The Adam optimizer takes exponential weighted average of the gradients and is used to accelerate the gradient descent procedure. Using averages accelerates the algorithm's convergence to the minima, allowing it to reach the minimum more quickly.

The equation for Adaptive learning rate used by Adam optimizer:

$$\begin{aligned} m_t &= \beta_1 m_{t-1} + (1 - \beta_1) g_t \\ v_t &= \beta_2 v_{t-1} + (1 - \beta_2) g_t^2 \end{aligned} \tag{11}$$

where,

β_1, β_2—hyper-parameters control the exponential decay rates of the moving average.

m_t = first-moment estimation (mean estimation).

v_t = second-moment estimation (variance estimation).

5.6 Proposed Method

We have proposed two Deep learning models and performed the hyper-parameter tuning in Machine Learning models to enhance the performance of the models. These models are configured based on the hyperparamters like the number of iterations used, the learning rate of the model, number of epochs used by model and the activation function used to activate the neurons of the deep learning model.

Proposed Deep Learning Models Description:

- Perceptron:

  Following are the hyperparameters tuned for the perceptron algorithm:
  Number of the iterations used → 5000
  Learning rate (eta) → 0.001
  Random state → Constant

- Artificial Neural Network:

Configurations of the model:

Number of layers used. = 14 (1 input, 12 Hidden, 1 output)

Types of layers used = Dense Layers (It means that layers are fully interconnected with each neuron of each previous and next layer)

Activation functions used—Sigmoid in input and output layer, ReLU at each Hidden Layer

Number of neurons at each layer.

L1 - 256 Neurons, Activation Function—Sigmoid
L2 - 128 Neurons, Activation Function—ReLU
L3 - 128 Neurons, Activation Function—ReLU
L4 - 64 Neurons, Activation Function—ReLU
L5 - 64 Neurons, Activation Function—ReLU
L6 - 64 Neurons, Activation Function—ReLU
L7 - 32 Neurons, Activation Function—ReLU
L8 - 32 Neurons, Activation Function—ReLU
L9 - 16 Neurons, Activation Function—ReLU
L10 - 16 Neurons, Activation Function—ReLU
L11 - 8 Neurons, Activation Function—ReLU
L12 - 8 Neurons, Activation Function—ReLU
L13 - 4 Neurons, Activation Function—ReLU

L14 - 1 Neuron, Activation Function—Sigmoid

Salient Features of the ANN model:

- Top-down approach—Top-down technique—We selected a top-down strategy because it allows the neurons to learn more effectively and converge the results of the predictions. As a result, it will be more precise.
- Because the prediction is based on two classes, the loss function utilized is binary cross-entropy; it will provide the best results in error detection.
- The **Adam optimizer** was employed since it has the best performance of all the optimizers because it is an ensemble-based optimizer with properties of both rmsprop and adaptive gradient optimizers.
- The output layers were kept between 0 and 1 using a sigmoid activation function.
- At hidden layers, the ReLU activation function is employed to preserve the prediction output linearly connected and prevent the problem of vanishing gradient descent.

5.7 Model Performance Analysis

The experimentation, exploration, and exploitation were performed over across all 10 below-listed models using various performance parameters, and the analyzed results are listed in Table 1. The results are compared using various algorithms before and after the balancing of the datasets.

From the above table of the accuracy scores of the models used, we can conclude that the balancing of the dataset will give a significant improvement in accuracy of the model. Hence, the balancing gave better outputs than the unbalanced dataset. Also, the algorithm, which has performed best in terms of accuracy score, is Random

Table 1 Accuracy of various models before and after balancing

Serial Number	Algorithm For models	Accuracy before balancing	Accuracy after balancing
1	KNN	93.84	93.95
2	Logistic Regression	91.37	92.04
3	Naïve Bayes	89.96	91.06
4	Decision Tree	93.97	94.12
5	SVM	94.93	95.89
6	Adaptive Boost	90.77	91.71
7	Gradient Boost	92.25	92.35
8	Random Forest	94.60	95.18
9	Perceptron	–	90.98
10	ANN	–	94.58

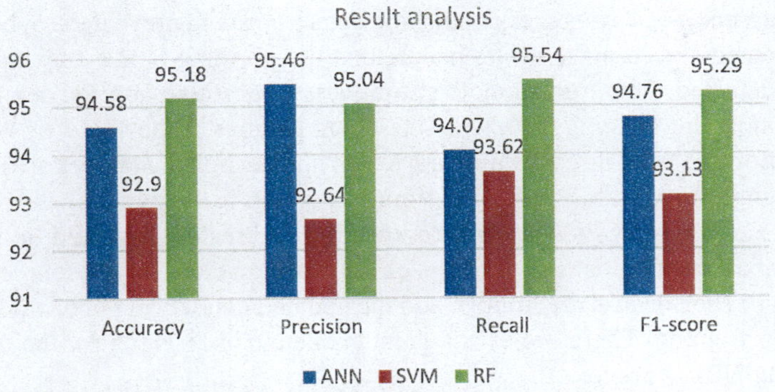

Fig. 2 Experimentation results

Forest Algorithm. As Random Forest Classifier is robust in nature and works on the voting of multiple classifiers, the results of performance parameters are expected to be the best in performance. The second-best algorithm is the proposed model algorithm, i.e. an artificial neural network whose accuracy is the second best among all the other models. Also, the other performance parameters are comparatively best among all the algorithms. Hence, we can conclude that artificial neural network can be used for the purpose of the classification task of the binary data.

From Fig. 2, we can conclude that the best algorithm for the classification of the URLs into Phishing and Benign is Random Forest algorithm. Also, from the above table of performance parameters, we can say that the Precision Score of the ANN algorithm is the best among all the algorithms by which we can state that the Phishing websites are most accurately detected by the ANN as precision detects the total positive predictions among total positive samples.

Also, from the results, we can say that the Random forest algorithm and ANN algorithm work almost similar in terms of performance parameters. Hence, we can say that the proposed model can be significantly used for the purpose of classifying the URLs into Phishing and benign ones.

6 Conclusion and Future Work

The detection of phishing websites should be done with a great percentage of accuracy. Moreover, the false negative cases are of hue concern as if the website is predicted as legitimate but actually it is phishing then this may lead to serious cyber attacks. The approaches used for the detection are the main parameters to be used for classifying the websites as the legitimate ones or the phishing ones. In the proposed work in this thesis, we can conclude that for the classification tasks the bagging-based ensemble technique, i.e. random forest works best. Also, the artificial neural networks can provide a great accuracy and precision for the classification of such textual datasets. The advantages of phishing detection are to prevent the cybercrimes to be executed and to make aware people that the phishing is like trap and how it could be detected. Also, regarding the awareness of phishing attacks, people could make aware by introducing them with the salient features of the URLs by which the identification can be done easily until advanced obfuscation techniques are not used. The challenge faced in this task of phishing detection is the dynamic nature of the websites. Hence, every time when a new website is created, it should be included in the database as it belongs to phishing class or benign class. The resources for the collection of such dataset are limited, and the updated dataset cannot always be kept handily for training. Therefore, it is a great challenge of research for the detection of Phishing URL websites.

6.1 Limitations/Challenges of Detection of Phishing Website Detection

There is not enough data to train the models with the known flaws in the URLs and vulnerabilities in the web technologies are developing and improvising day by day.

The lifespan of the URLs is very short. When once the execution or the time limit set on the URL is executed or expired, it just becomes inactive from the Phishing person's side (Hacker). Hence, the detection of expired URLs is very difficult as the data can be retrieved or fetched dynamically.

The zero-day Phishing URLs. These URLs come into the act when new vulnerabilities in the existing web technologies are discovered. These cannot be detected easily as we don't know for what purpose they are designed so the features of such websites are unknown. Therefore, the detection of such websites whose behaviour is unknown is almost impossible to detect the phishing behaviour.

The limitation of the work we have performed is the use of this model in real-life scenarios. As the dataset used is already old, this model would not detect the new phishing URLs and vulnerabilities after that. Also, the model needs a dynamic approach for feeding the data and high computational resources to make the model work in a real-life scenario.

6.2 Future Work

In future work, we could make the dataset to be updated when a website is registered for its domain to be hosted on the internet. In addition, we can use the dynamic updating of the websites by using the data of the ISP registry. However, this task is very much impossible as after the exponential growth of internet users, the content of websites is almost more than trillions [13].

Another thing in the future we could make an application interface website by which the continuous updating of the websites could be done on real-time cloud storage. Once the database is updated, the extension to the browser should be created dynamically, it could prevent users to visit phishing websites.

References

1. Pavan Kumar P, Jaya T, Rajendran V (2021) SI-BBA—A novel phishing website detection based on Swarm intelligence with deep learning. Mater Today Proc https://doi.org/10.1016/j.matpr.2021.07.178
2. Al-Sarem M et al (2021) An optimized stacking ensemble model for phishing websites detection. Electron 10(11). https://doi.org/10.3390/electronics10111285
3. Tang L, Mahmoud QH (2021) A survey of machine learning-based solutions for phishing website detection. Mach Learn Knowl Extr 3(3):672–694. https://doi.org/10.3390/make3030034
4. Ganesh JS, SwarupV N, KumarR N (2020) Machine learning based malicious website detection. http://www.ijser.org
5. Sameen M, Han K, Hwang SO (2020) PhishHaven—An efficient real-time Al phishing URLs detection system. IEEE Access 8:83425–83443. https://doi.org/10.1109/ACCESS.2020.2991403
6. Sadique F, Kaul R, Badsha S, Sengupta S (2020) An automated framework for real-time phishing URL detection. In: 2020 10th annual computing and communication workshop and conference, CCWC 2020, Jan, pp 335–341. https://doi.org/10.1109/CCWC47524..9031269
7. Kulkarni A, Brown LL (2019) Phishing websites detection using machine learning. www.ijacsa.thesai.org
8. Sharma DK, Singh NK (2016) An approach for collaborative decision in distributed intrusion detection system. Int J Comput Appl 133
9. Butnaru A, Mylonas A, Pitropakis N (2021) Towards lightweight url-based phishing detection. Future Internet 13(6). https://doi.org/10.3390/fi13060154
10. Johnson C, Khadka B, Basnet RB, Doleck T (2020) Towards detecting and classifying malicious urls using deep learning. J Wirel Mob Netw, Ubiquit Comput Depend Appl 11(4):31–48. https://doi.org/10.22667/JOWUA.2020.12.31.031
11. Liew SW, Sani NFM, Abdullah MT, Yaakob R, Sharum MY (2019) An effective security alert mechanism for real-time phishing tweet detection on Twitter. Comput Secur 83:201–207. https://doi.org/10.1016/j.cose.2019.02.004
12. https://www.kaggle.com/code/dnyaneshsatpute/phishing-webiste-detection/data?select=dataset.csv
13. Prakash R, Chithaluru P, Sharma D, Srikanth P (2019) Implementation of trapdoor functionality to two-layer encryption and decryption by using RSA-AES cryptography algorithms. In: Nath V, Mandal J (eds) Nanoelectronics, circuits and communication systems. Lecture Notes in Electrical Engineering, vol 511. Springer, Singapore. https://doi.org/10.1007/978-981-13-0776-8_8
14. Kiruthiga R, Akila D (2019) Phishing websites detection using machine learning. Int J Recent Technol Eng 8(2) Special Issue 11:111–114. https://doi.org/10.35940/ijrte.B1018.0982S1119
15. Jeeva SC, Rajsingh EB (2016) Intelligent phishing url detection using association rule mining. Human-centric Comput Inf Sci 6(1). https://doi.org/10.1186/s13673-016-0064-3
16. Sananse BE (2015) Phishing URL detection: a machine learning and web mining-based approach. www.google.com
17. WhatAPhish: Detecting Phishing Websites. https://towardsdatascience.com/whataphish-detecting-phishing-websites-e5e1f14ef1a9

18. Baliram Torane S, Shekokar N. Performance analysis of machine learning algorithms used for web based phishing detection. https://doi.org/10.51201/JUSST/21/05-187
19. I. International Institute of Information Technology (Pune, Institute of Electrical and Electronics Engineers. Pune Section, and Institute of Electrical and Electronics Engineers, International conference on automatic control & dynamic optimization techniques (ICACDOT 2016) , 9–10 Sep 2016
20. Tomar G et al (2020) 2020 IEEE 9th international conference on communication systems and network technologies: CSNT 2020 : proceedings
21. Khan SA, Khan W, Hussain A. Phishing attacks and websites classification using machine learning and multiple datasets (A Comparative Analysis).
22. Tiwari P, Ratan Singh R. Machine learning based phishing website detection system. www.ijert.org
23. Organizer. Ranganathan Engineering College and Institute of Electrical and Electronics Engineers, proceedings of the international conference on inventive communication and computational technologies: ICICCT 2018 , 20–21 April 2018
24. Annual IEEE Computer Conference, M. and T. 2014. 04. 2.-29 D. N. International Conference on Computing, M. and T. C. 2014. 04. 2.-29 D. N. IEEE Computing, and IEEE ComManTel 2014.04.27–29 Da Nang, 2014 International Conference on Computing, Management and Telecommunications (ComManTel) 27–29 April 2014, Da Nang, Vietnam
25. https://www.itgovernance.co.uk/blog/beware-of-angler-phishing
26. ADSC (2022) Spam Email detection using machine learning classifiers, Naresh Chillur, Debabrata Swain, Meet Kava

Machine Learning-Based Prototype Design for Rainfall Forecasting

Ashish Labade, Bhavana Gupta, Rajeev Kumar Gupta, and Abhijit Kumar

1 Introduction

Forecasting is still a hot topic, with governments, corporations, risk management companies, and among data scientists all interested. Rainfall is a natural variable that influences a wide range of human activities, including agriculture, irrigation, building, power generation, forest, and tourism [1]. Agriculture is among the most important economic elements in the country, and it is completely reliant on rainfall. In this scenario, rainfall forecasting is critical since it has the strongest connection to natural disasters like avalanches, mass movements, flooding, and landslides. These events have left an indelible mark on the world [2]. As an outcome, having a reliable rainfall forecasting tool helps you to take preventative and mitigation measures against natural calamities [3].

Management of water resources, flood warnings, air operations management, transit restrictions, building construction activities, and other vital human factors all gets benefited from rainfall forecasts. Rainfall data is collected for forecasting using weather satellite systems, cable and digital equipment, and high-speed computers. Weather forecasting has been a fascinating and engaging issue since the dawn of civilization, and it represents among the most complex and appealing domains today. Rainfall is forecasted using a range of methodologies and procedures, some more

A. Labade · B. Gupta
Department of Computer Science and Engineering, Sagar Institute of Science and Technology, Bhopal, MP 462030, India
e-mail: bhavanagupta@sistec.ac.in

R. K. Gupta
Department of Computer Science and Engineering, Pandit Deendayal Energy University, Gandhinagar, Gujarat 382421, India

A. Kumar (✉)
School of Computer Science, University of Petroleum and Energy Studies Dehradun, Dehradun 248001, India

A. Ramdane-Cherif et al. (eds.), *Machine Intelligence and Data Science Applications*, Algorithms for Intelligent Systems, https://doi.org/10.1007/978-981-99-1620-7_13

accurate than others. In weather forecasting, atmospheric conditions such as rainfall, temperature, humidity, wind direction and speed, pressure, evaporation, and others are collected.

The most significant feature of most water storage mechanisms across the world right now is rainfall forecasting. The inaccuracy of rainfall statistics is one of the most problematic concerns. Most rainfall forecasting methods are now unable to discern hidden patterns or nonlinear trends of rainfall captured data. This study will aid in the discovery of any nonlinearly separable patterns in data and trends that are required for reliable rainfall forecasting.

To provide complete and rapid predictions, we used numerous machine learning approaches and algorithms, as well as Automated Machine Learning to find one of the best models for the captured data and cross-check the results against other models. The purpose of the proposed research is to provide a complete ML life cycle, from input pre-processing to modeling implementation and monitoring. The data pre-processing process includes operations such as filling in null values, encoding features, feature normalization, and extraction of features. The following algorithms were used: LR (Logistic Regression), K–NN (K Nearest Neighbor), NB (Nave Bayes classifier), RF (Random Forest classifier), DT (Decision Tree), and SVM classifier. For evaluation, criteria model Accuracy, Precision, Recall, and the F1-Scores are measured.

Firstly, we will apply Automated Machine Learning (AutoML) to the dataset. The results obtained have been compared with the below-listed algorithm's outcome:

- Logistic Regression,
- K Nearest Neighbor
- Support Vector Machines
- Naïve Bayes
- Decision Tree
- Random Forest
- XGBoost.

And the result will show the comparative analysis of all the models depending on their Accuracy, Precision, Recall, and F1-Score.

Automated Machine Learning (AutoML):

AutoML (Automated Machine Learning) is concerned with providing Machine Learning solutions to data scientists without having to conduct endless investigations into data preparation, model selection, hyper-parameter tuning, and feature scaling. In other words, AutoML is concerned with method selection, hyper-parameter optimization and tuning, and model validation. It's about using less code and avoiding manual hyper tweaking to make Machine Learning operations easier.

AutoML (Automated Machine Learning) is the process of automation of the entire application of machine learning to real-world and business-critical situations. A data scientist must use data pre-processing, hyper-parameter optimization, feature extraction, and feature selection approaches to prepare the dataset for inference, and hence data analysis.

Following pre-processing processes, they must choose an algorithm and execute hyperparameter optimization to improve the prediction accuracy of their end machine learning model. AutoML was offered as an AI-based approach to the difficulty of implementing machine learning without significant knowledge because many of these stages can only be handled by ML experts. There are so many types of AutoML techniques.

There are four types of models one can build using AutoML:

1. Image Data,
2. Tabular Data,
3. Text Data,
4. Video Data.

There are different types of AutoML libraries available, some of them are listed below:

- Auto-sklearn
- Auto-keras
- H20 AutoML
- TPOT and many more

TPOT (Tree-based Pipeline Optimization Tool):

TPOT will automate the most time-consuming component of machine learning by automatically exploring millions of potential pipelines to find out the best one for the data (Fig. 1).

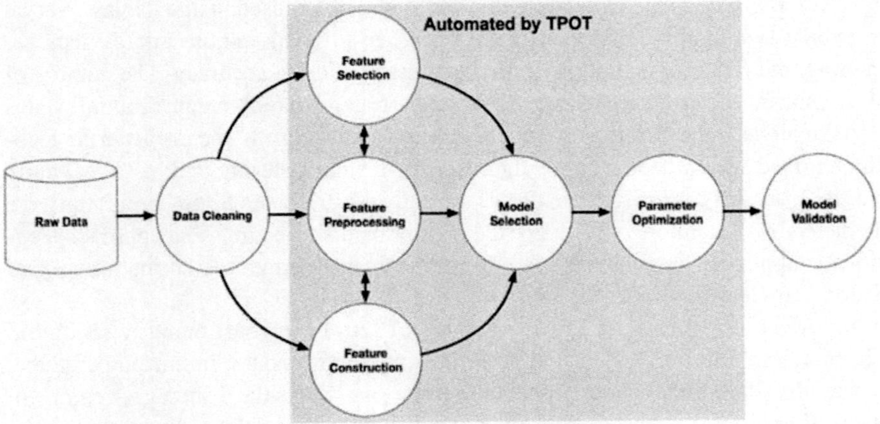

Fig. 1 Flow graph of the proposed model

2 Literature Review

The authors of [19] integrated RNN and TDNN characteristics and found that composite models outperform single models in terms of accuracy. For predicting Indian summer monsoon rainfall, the authors of [18] employed Multilayer Feed Forward Neural Networks (MLFNN). To predict rainfall, the Error Back Propagation (EBP) algorithm is taught and used. We looked at three network architectures with two, three, and ten input parameters. The obtained result from the model has been compared with the statistical models. The authors of [17] employed an error back-propagation technique to estimate India's Summer Monsoon Rainfall on monthly and annual time series. For rainfall prediction, they used data from the previous five years' monthly and annual mean rainfall values. The authors of [16] employed an ABF neural network to anticipate annual rainfall in Kerala. According to their findings, ABFNN outperforms Fourier analysis.

The authors of [7] used an ANN model to predict rainfall in the Hyderabad, INDIA region. They also compared the ARIMA approach to ANN. They fed rainfall data from the previous four months into a neural network model. For rainfall forecasting, the authors of [9] used two parameters: min temp and max temp. The CGD (Conjugate Gradient Descent) and Levenberg–Marquardt (LM) learning algorithms have been adopted by the authors of [13]. In the prediction challenge, both algorithms performed similarly. The authors of [14] applied Modular Artificial Neural Network (MANN) to predict India and China's rains. The performance of MANN was compared with LR, KNN, and ANN.

To forecast rainfall, the authors of [12] used yearly, biannually, quarterly, and monthly rainfall data. They trained four distinct Focused Time Delay Neural Networks to anticipate rainfall (FTDNN). When utilizing annual rainfall data for training, the FTDNN model offers the greatest prediction accuracy. The authors of [11] contributed to the creation of the CART strategy, which creates rainfall states from large-scale meteorological data in a river basin using a K-mean clustering technique mixed with a decision tree algorithm. K-mean clustering and historical daily multisite rainfall data are used to derive the daily rainfall state. Short-term rains were predicted by the authors of [10]. For the prediction task, an empirical approach technique is applied. For each area, data from three separate months during a five-year period is reviewed.

An ANN model was used by the authors of [20] to forecast monthly rainfall in Chennai. The authors of [8] looked at rainfall, maximum and minimum temperatures, and relative humidity. They predicted weekly rainfall in the Pantnagar region. In terms of prediction accuracy, the ANN outperformed the multiple linear regression model. The authors of [7] employed a Time Delay Recurrent Neural Network to predict monthly precipitation over Australia (TDRNN). The authors of [6] predicted average rainfall in the Udipi district of Karnataka. They used ANN models for the rainfall prediction task. Layer recurrent and cascaded backpropagation was shown to be inferior to the Backpropagation Algorithm (BPA). The authors of [5] forecasted the hourly rainfall. CART and C4.5 are used for prediction, and they may reveal hidden important patterns and their causes. Eighteen variables were gathered from the weather station. A tenfold cross-validation technique is used for validation. In this test, CART outperformed C4.5.

2.1 Literature Survey of Papers

Title	Proposed Methodology	Outcome/Remark
Oswal, "Predicting Rainfall using Machine Learning Techniques," 2019, [Online]. [1]	The author of the paper has implemented data over different models like; Logistic Regression, DecisionTree, K–Nearest Neighbor, Random Forest, Ada Boost, and Gradient Boosting and compared the result of both undersampling and oversampling of the data. Moreover, on original data also The evaluation criteria are: Accuracy, Area Under Curve (AUC), Precision, Recall, F1-Score, and Confusion Matrix	Because Australian weather is unpredictable, there is no such link between rainfall and location or time. In the future can apply the Deep learning model and provide a comparative analysis between them
G. Sethupathi M et. al. "Efficient Rainfall Prediction and Analysis using Machine Learning Techniques," Turkish J. Comput. Math. Educ., vol. 12, no. 6, pp. 3467–3474, 2021. [2]	The authors have applied two different machine learning models and shown the comparative analysis between these two models. These two models are: Logistic Regression and Random Forest Classifier	The results are compared using Accuracy score and Confusion matrix Acc (LR): 95.94%, Acc (RF): 94.42%
Sarasa-Cabezuelo, "Prediction of Rainfall in Australia Using Machine Learning," Information, vol. 13, no. 4, p. 163, 2022, https://doi.org/10.3390/info13040163. [5]	The authors of the paper have applied different models on the dataset as per their region and shown how different models vary their accuracy depending upon the different region or location Models used are: KNN, Random Forest, Decision Trees, and Neural Network	Depending upon the different location or region different models vary their output

(continued)

(continued)

Title	Proposed Methodology	Outcome/Remark
Preethi et. al. "Rainfall Prediction using Machine Learning and Deep Learning Algorithms," Int. J. Recent Technol. Eng., vol. 10, no. 4, pp. 251–254, 2021, https://doi.org/10.35940/ijrte.d6611.1110421. 20]	The authors of the paper have worked with four different machine learning models they are: ANN, AdaBoost, GRADIENT BOOST, and XGBOOST, and their evaluation criteria includes Accuracy, Precision, Recall, and F1-core	Compared to the other three boosting algorithms, ANN has performed well Acc (ANN): 88.64%, Acc (XGBOOST): 86.20%, Acc (ADABOOST): 86.38%, Acc (GB): 86.25%

3 Working Methodology

To understand the working flow, first, we need to understand the data. The dataset contains 145,461 records.

3.1 *Preprocessing*

To apply this data to the model, we need to convert categorical data into numerical data. Some features are encoded using the label encoder method and others are by using get_dummies() of pandas.

3.1.1 Null/NA Value Elimination

NaN values are imputed using mean imputation.

3.1.2 Target Variable Analysis

- As we can see the target variable is highly imbalanced.
  - 'No': 110,316,
  - 'Yes': 31,877.
- Value counts:
- If we train our model on this imbalanced data, our model's prediction will be highly biased toward 'No' classification.

  So, before applying this data to the model, we need to balance the dataset (Fig. 2).

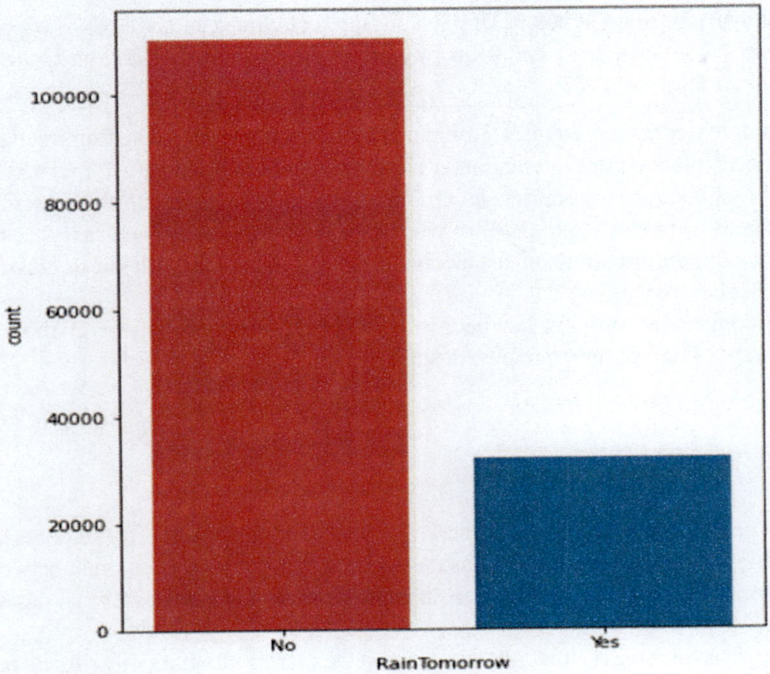

Fig. 2 Rainfall count

Types of Data Balancing Techniques:

- Undersampling,
- Oversampling.

3.2 *Undersampling*

- Undersampling lowers the number of members of the majority class in the training set. As an outcome, the training set's overall record count is drastically lowered. This means that training time is drastically decreased during classification. There is also huge memory savings because we are working with very high-dimensional datasets.
- Nevertheless, because we are removing items from the majority class, it is probable that we may lose a lot of relevant information if we remove documents that could help our classifier develop a more accurate model.

There are so many types of Undersampling techniques available but I have used Random Undersampling. The following are some features of Random Undersampling:

- Random undersampling is a simple resampling technique. In the training dataset, majority class entries are randomly eliminated until the minority to majority class ratio achieves the appropriate level. One of the issues with random undersampling is that it is impossible to monitor what data about the majority class is discarded.
- Important information on the decision-making between both these classes, in particular, may be lost.
- Random undersampling has been scientifically proven to be one of the highly effective resampling strategies, despite its simplicity.

3.3 Oversampling

- The goal of oversampling is to increase the percentage of minority students in the training group. Oversampling has the advantage of preserving all members of the minority and majority classes, ensuring that no data from the original dataset is lost.
- The disadvantage is that the training set expands dramatically. As a result, we increase the amount of memory required to retain the training set and extend the training period. To keep time and memory complexity within acceptable constraints, we must continue with caution because we are working with extraordinarily high-dimensional datasets.
- If the time required to resample is omitted, undersampling outperforms oversampling in terms of time and memory complexity.
- As an outcome, oversampling must perform much superior to undersampling in terms of classification ability to be viable.
- Previous research has yielded no definitive findings on whether undersampling or oversampling is better for classification performance.
- Conflicting results are most likely owing to the use of multiple datasets and classification techniques. Furthermore, the resampling approach chosen is most likely domain and problem specific.

There are many types of Oversampling techniques available but I have used SMOTE (Synthetic Minority Oversampling Technique) for this dataset. The following are some features of SMOTE:

- SMOTE was first introduced by Chawla [2]. SMOTE is a simple yet effective strategy that outperforms random oversampling in a number of low-dimensional scenarios.
- SMOTE is one of the few resampling methods to regularly outperform random undersampling and random oversampling in terms of performance.

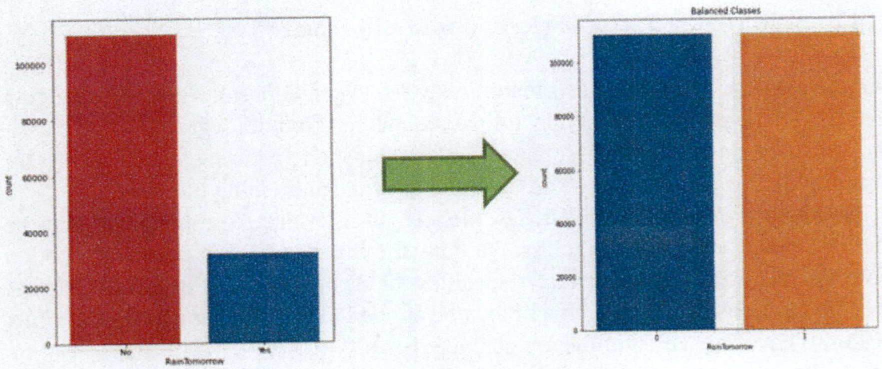

Fig. 3 Rainfall count with balanced class

- For each minority class item in the training set, use the SMOTE algorithm to find the 5 closest neighbors from the other minority class papers in the training set.

We choose one of the five closest neighbors at random to produce a new minority class document. The new minority category record is a point on the hyperplane connecting the two minority class records that were chosen at random (Fig. 3).

3.4 Feature Selection

Now, our dataset has 23 features, but there are some unnecessary features that we need to remove from the data before we train our model. So, we need to apply feature extraction method to get important features. We have applied univariate feature selection method on our data and selected best 16 features from the dataset.

3.4.1 Univariate Feature Selection

The best features are selected using univariate statistical tests in univariate feature selection. Each attribute is compared to the target variable to see if there is a statistically significant link between the two. Analysis of variance is another name for it. We disregard the other features while analyzing the link between a feature and the target variable. That is why it is referred to as "univariate." Each feature has a test score associated with it. Finally, all of the test results are examined, and the features with the highest scores are chosen.

3.4.2 SelectKBest Feature Selection with Chi-Square Test

The Select KBest method selects features based on their highest k scores. By changing the score func parameter, we may utilize the method for both regression and classification data. Selecting the best features is crucial when preparing a large dataset for training. It aids us in deleting non-essential data and shortening training time.

To begin, we'll use the *SelectKBest* model to classify data. The model will then be defined using the *SelectKBest* class. We'll use the chi2 approach as a scoring function for classification. The k parameter specifies the desired number of characteristics. The fit and transform approach will then be applied to the training x and y data. In x data, we have chosen 16 of the top characteristics.

And now our data is ready to be fed into the model.

4 Result Analysis

Output of Oversampling data on different models.

Model	Accuracy	F1-Score	Precision	Recall
Random forest	0.907	0.907	0.913	0.901
K–nearest neighbors	0.849	0.848	0.808	0.918
Decision tree	0.840	0.840	0.836	0.847
Support vector machines	0.830	0.829	0.843	0.812
XGBoost	0.815	0.815	0.818	0.813
Logistic regression	0.780	0.784	0.795	0.769
Naïve bayes	0.750	0.752	0.782	0.704

Output of Undersampling data on different models.

SN	Model	Accuracy	F1 score	Precision	Recall
1	Random forest	0.804	0.804	0.808	0.792
2	Support vector machines	0.790	0.794	0.802	0.776
3	Logistic regression	0.780	0.781	0.791	0.760
4	XGBoost	0.771	0.770	0.785	0.740
5	K–nearest neighbors	0.756	0.756	0.758	0.745
6	Naïve bayes	0.750	0.748	0.787	0.679
7	Decision tree	0.720	0.715	0.711	0.714

5 Conclusion and Future Work

From the results, it is clear that Oversampling performs better than Undersampling, and AutoML technique has provided the best outcome among all the models. But still forecasting of weather is very much a tough task. We require more detailed data, to predict future conditions. But as of now, the Random Forest algorithm outperforms all the models in terms of everything like, accuracy, precision, recall, and F1-score. In future, we can apply different models on the data "Location" wise, because rainfall is highly depending upon the atmosphere of particular locations. So, if we apply models on Location wise data, we may get good result but for that we require lot of data Location wise.

References

1. Oswal N (2019) Predicting rainfall using machine learning techniques. Online Available http://arxiv.org/abs/1910.13827
2. Gowtham Sethupathi M, Ganesh YS, Ali MM (2021) Efficient rainfall prediction and analysis using machine learning techniques. Turkish J Comput Math Educ 12(6): 3467–3474
3. Dutta K, G. P* (2020) Rainfall. Int J Recent Technol Eng 9(1):1954–1961. https://doi.org/10.35940/ijrte.a2747.059120
4. Basha CZ, Bhavana N, Bhavya P, Sowmya V (2020) Rainfall prediction using machine learning deep learning techniques. In: Proc Int Conf Electron Sustain Commun Syst. ICESC 2020, no. Icesc, pp 92–97. https://doi.org/10.1109/ICESC48915.2020.9155896
5. Sarasa-Cabezuelo A (2022) Prediction of rainfall in Australia using machine learning. Information 13(4):163. https://doi.org/10.3390/info13040163
6. W. M. Ridwan, M. Sapitang, A. Aziz, K. F. Kushiar, A. N. Ahmed, and A. El-Shafie, "Rainfall forecasting model using machine learning methods: Case study Terengganu, Malaysia," *Ain Shams Eng. J.*, vol. 12, no. 2, pp. 1651–1663, 2021, doi: https://doi.org/10.1016/j.asej.2020.09.011.
7. Raval M, Sivashanmugam P, Pham V, Gohel H, Kaushik A, Wan Y (2021) Automated predictive analytics tool for rainfall forecasting. Sci Rep 11(1):1–13. https://doi.org/10.1038/s41598-021-95735-8
8. Ji SY, Sharma S, Yu B, Jeong DH (2012) Designing a rule-based hourly rainfall prediction model. In: Proc 2012 IEEE 13th Int Conf Inf Reuse Integr. IRI 2012, pp 303–308. https://doi.org/10.1109/IRI.2012.6303024
9. Abhishek K, Kumar A, Ranjan R, Kumar S (2012) A rainfall prediction model using artificial neural network. In: Proc—2012 IEEE Control Syst Grad Res Colloquium, ICSGRC 2012, no. Icsgrc, pp 82–87. https://doi.org/10.1109/ICSGRC.2012.6287140
10. Parmar MSA, Mistree K (2017) Machine learning techniques for rainfall prediction: A Review. Int Conf Innov Inf Embed Commun Syst (September)
11. Kannan S, Ghosh S (2011) Prediction of daily rainfall state in a river basin using statistical downscaling from GCM output. Stoch Environ Res Risk Assess 25(4):457–474. https://doi.org/10.1007/s00477-010-0415-y
12. Htike KK, Khalifa OO (2010) Rainfall forecasting models using focused time-delay neural networks. Int Conf Comput Commun Eng. ICCCE'10 (May): 11–13. https://doi.org/10.1109/ICCCE.2010.5556806
13. Wu CL, Chau KW, Fan C (2010) Prediction of rainfall time series using modular artificial neural networks coupled with data-preprocessing techniques. J Hydrol 389(1–2):146–167. https://doi.org/10.1016/j.jhydrol.2010.05.040

14. Guhathakurta P (2008) Long lead monsoon rainfall prediction for meteorological sub-divisions of India using deterministic artificial neural network model. Meteorol Atmos Phys 101(1–2):93–108. https://doi.org/10.1007/s00703-008-0335-2
15. Agrawal K (2006) Modelling and prediction of rainfall using artificial neural network and ARIMA techniques. J Ind Geophys Union 10(2):141–151
16. Philip NS, Joseph KB (2003) A neural network tool for analyzing trends in rainfall. Comput Geosci 29(2):215–223. https://doi.org/10.1016/S0098-3004(02)00117-6
17. Sahai AK, Soman MK, Satyan V (2000) All India summer monsoon rainfall prediction using an artificial neural network. Clim Dyn 16(4):291–302. https://doi.org/10.1007/s003820050328
18. Venkatesan C, Raskar SD, Tambe SS, Kulkarni BD, Keshavamurty RN (1997) Prediction of all India summer monsoon rainfall using error-back-propagation neural networks. Meteorol Atmos Phys 62(3–4):225–240. https://doi.org/10.1007/BF01029704
19. Goswami P, Srividya (1996) A novel neural network design for long range prediction of rainfall pattern Curr Sci 70(6), 447–457
20. Preethi BM, Gowtham R, Aishvarya S, Karthick S, Sabareesh DG (2021) Rainfall prediction using machine learning and deep learning algorithms. Int J Recent Technol Eng 10(4):251–254. https://doi.org/10.35940/ijrte.d6611.1110421
21. towardsdatascience.com
22. https://www.kaggle.com/jsphyg/weather-dataset-rattle-package
23. www.analyticsvidhya.com
24. https://github.com/

BERT Based Language Identification in Code-Mixed English-Assamese Social Media Text

Nayan Jyoti Kalita, Pritam Deka, Vijay Chennareddy, and Shikhar Kumar Sarma

1 Introduction

Due to the widespread accessibility of the Internet, social media usage among the general public is constantly expanding. As a result of that, people are using such platforms for daily use, such as texting. In a multilingual society, text exchange using such social media platforms often contains elements of multiple languages resulting in a code-mixed conversation. The term "code mixing" refers to the process of combining words, phrases, and even entire sentences from two distinct languages within the context of a single speech event [5]. Code mixing has developed into its own phenomenon in multilingual societies as a direct result of the proliferation of social media [13].

Identification of different languages within such code-mixed texts has been studied around the world [1, 9, 18, 27, 28, 32, 36, 37]. However, there are far fewer research works in identifying a low resource language from India in a code-mixed environment. The north-eastern part of India is home to a sizable number of low-resource languages [8, 16]. In this paper, we have explored the identification of such

N. J. Kalita (✉)
The Assam Royal Global University, Assam, India
e-mail: nayan.kalita@rgi.edu.in

P. Deka
Queen's University Belfast, Belfast, UK
e-mail: pdeka01@qub.ac.uk

V. Chennareddy
Middlesex University, England, UK
e-mail: vc381@live.mdx.ac.uk

S. K. Sarma
Gauhati University, Assam, India
e-mail: sks@gauhati.ac.in

A. Ramdane-Cherif et al. (eds.), *Machine Intelligence and Data Science Applications*, Algorithms for Intelligent Systems, https://doi.org/10.1007/978-981-99-1620-7_14

a low resource language, Assamese, which is predominantly spoken in the region of Assam, in a code-mixed environment by using state-of-the-art NLP techniques.

A large number of transformer-based language models have emerged as a result of recent developments in natural language processing [34]. One such model, BERT, has achieved state-of-the-art results on various NLP tasks, including text classification, natural language inference, question answering, and textual similarity [14]. These models have been trained on huge amounts of text data such as Wikipedia and BookCorpus, and as a result, can be applied to many downstream tasks. This is the first effort that we are aware of that employs BERT [14] models in conjunction with a neural network in order to detect languages inside a code-mixed English and Assamese language combination. Our results which were evaluated on a dataset shows that BERT [14] achieves better results with a scope for further research using transformer models for these tasks.

2 Relevant Work

We briefly go over the language identification problem for code-mixed language pairings for Indian languages in this section.

Veena et al. [35] devised a technique for identifying the languages spoken in code-mixed language pairings of Tamil and Malayalam with English words at the word level. They used word2vec [25] embeddings as features that were fed into an SVM classifier to identify the languages. The work of [10] also uses word2vec [25] CBOW, as well as the skip-gram architecture, to calculate word embeddings. The embeddings are then used as features in various classifiers such as SVM classifier, Gaussian Naïve Bayes, Logistic Regression, Adaboost and Random Forest for the language identification task. The authors used the ICON 2016 dataset [19] for their work.

Barman et al. [4] used a dictionary-based classifier, a linear SVM and k-nearest neighbors for language identification in a Nepali-English and a Spanish-English dataset. The word length and character n-grams were employed as features in the SVM classifier. Das and Gambäck [12] also used a dictionary-based classifier and a SVM based classifier for language identification from English-Hindi and English-Bengali language pairs. They used features such as weighted n-grams and word context information. The recent work by [3] used factors such as context, capitalization, special characters, character n-grams, and lexicon to correctly identify the language in code-mixed English and Punjabi text using Conditional Random Fields (CRF).

Recent works also include the use of transformer [34] models. The work of [2] includes the use of BERT [14] for language identification in Hindi-English language pairs at the word level. The authors fine-tuned BERT using code-mixed data containing English, Hindi and Urdu data, and then utilised the fine-tuned model to classify Hindi-English code-mix terms. They found that the fine-tuned models give better results.

Chakravarthi et al. [11] did a study on analysing sentiment and detection of objectionable language on a data set gathered from YouTube with over 60,000 comments. Three Dravidian languages-Tamil, Kannada, and Malayalam-were primarily the focus of their attention. The authors used models such as BERT [14], DistilBERT [29], CharacterBERT [7], ALBERT [23], RoBERTa [24], XLM [22] and XLM-R for the purpose of their research.

Another recent study on Tamil code-mixed YouTube comments language recognition was conducted by [33]. Multilingual posts and comments on YouTube were used to compile the dataset. CNN-BiLSTM, DistilBERT, and XLM-R models performed similarly but poorly on this dataset, but ULMFiT outperformed the others due to improved fine-tuning approaches. They presented a code-mixed corpus selective translation and transliteration. The benefit of employing transformer-based models for low resource languages was also demonstrated.

For the Assamese language, there are fewer works in this direction. Sarma et al. [30] used word2vec with triplet loss [31] to calculate word embeddings. These are then used as features in various classifiers such as SVM, Naïve Bayes and a convolutional neural network classifier to identify languages. The authors used a mix of English, Assamese, Hindi and Bengali language for their work. Bora and Kumar et al. [6] experimented with SVM and decision trees for the language identification task involving English, Assamese and Hindi text. The authors used n-grams and word context information as features. Kalita and Saharia [20] used a linear SVM with features such as weighted n-grams and word position for language identification in a data set in the English-Assamese language.

The language identification works for Assamese-English language pairs included here mostly uses basic features with a few using semantic word embeddings. Our method, on the other hand, employs context-based word embeddings in conjunction with neural networks. We demonstrate that this method performs significantly better than simply relying on word-level characteristics or even semantic word embeddings.

3 Approach

We first describe the specifics of the dataset we are using in the following sub-section. We next briefly go through the BERT technique in the sub-section that follows.

3.1 Dataset details

Code-mixed data have elements of different languages within the same conversation environment. As a result, there can be words which are English words, and within the same sentence there can be words which are romanized words of the other language. The number of datasets for the code-mixed English-Assamese language combination is however limited. For our experiments, we have taken the dataset from [20].

Table 1 Dataset details

Tag	Desciption	Count (%)
EN	English words	64.6
AS	Assamese words	16.3
UNIV	Emoticons, universal characters	13.1
NE	Named entities	2.0
HI	Hindi words	3.0
ACRO	Acronyms	0.5
WM	Word mixing	0.2
UNDEF	Tokens that are hard to categorize	0.1

Although the dataset is small, due to the unavailability of a large dataset for this work, we decided to use it for evaluation purposes.

The dataset primarily consists of words written in English and romanized Assamese words with a few romanized Hindi words. There is a total of 26,444 words in the dataset. Tagging has been done at the level of the individual words, utilising eight distinct tags, namely: EN, AS, UNIV, NE, HI, ACRO, WM and UNDEF. The tags EN, AS, and HI are used to identify the terms in English, Assamese, and Hindi, respectively. Emoticons or special characters like ", ', !, @ are denoted by the tag UNIV. Named entities are denoted by the tag NE. The UNDEF tag is used to define the tokens that are hard to categorize. The ACRO tag is used to denote acronyms and WM is used to denote mixed words, which contain words in English but are mixed with certain romanized Assamese elements so that it becomes a separate word. Table 1 provides information about the dataset.

3.2 *BERT Based Neural Network Model*

When it was first released, the transformer [34] language model BERT (bidirectional encoder representations for transformers) [14] caused a stir because it produced cutting-edge results for a variety of NLP applications. BERT can be used to calculate word embeddings. It has unsupervised pre-training on BookCorpus and Wikipedia and excels in producing semantically rich word representations or word vectors that significantly rely on context. Because of the diverse contexts in which the words appear, the word embeddings for "apple" that are generated by BERT will be radically different when applied to the phrases "I ate an apple" and "Apple purchased a startup". Because their embeddings did not constantly alter in accordance with the context of the adjoining vector, older systems like word2vec [25] and GloVe [26] had lower performance.

Our approach is to use a BERT based word vector representation for the tokens present in the corpus and then use these as training features for a neural network. The motivation behind using BERT is the fact that it can learn contexts which can be leveraged for the work of language identification in such a code mix corpus. We describe the details of the experiment in the next section.

4 Experiment Details

In order to identify languages, the baseline method [20] employed a linear SVM and a bag-of-words strategy. However, Weka [15] was used to perform the studies and since there is no dedicated test file for evaluation purposes, we decided to use a linear SVM with a bag-of-words approach with TF-IDF in Python using the scikit-learn library.[1] First, we employed an 80:20 train-test split, where 80% of the data was used for training and the remaining 20% was used for testing. We kept the same split across all our experiments. Due to the class imbalance in the dataset, we decided to conduct all of our tests using the weighted average of precision, recall, and f-scores. We also applied two additional classifiers using the tf-idf approach, multinomial Naïve Bayes and logistic regression for comparison with the linear SVM approach.

We also experimented with word2vec [25] using GloVe [26] word embeddings along with an LSTM [17] neural network. The word embeddings have a dimension of 300 that were generated using the Glove 300 dimension file.[2] The embeddings were then passed as a matrix to an LSTM layer with 128 neurons whose output was passed to another 128-neuron dense neural layer. The activation function used was *relu*. Three more dense layers were used with the same activation function having 128, 128 and 64 neurons respectively. The last dense classification layer employed a *sigmoid* activation function with eight neurons. A *categorical cross entropy* loss function and the Adam [21] optimizer were used to train the neural network over 20 epochs. The *categorical cross entropy* loss has been employed as our loss function because there are more than two categorization labels in our dataset. The following equation describes the categorical cross-entropy function:

$$Loss, \lambda = -\frac{1}{N}\sum_{i=1}^{N} \log p_{model}\left[y_i \in C_{y_i}\right] \tag{1}$$

where each y_i belongs to exactly one class, C_{y_i} and $p_{model}\left[y_i \in C_{y_i}\right]$ is the probability predicted by the model.

For the purposes of the BERT experiment, we have used different BERT base models from Tensorflow Hub.[3] The neural network is built using three dense layers

[1] https://scikit-learn.org/.

[2] http://nlp.stanford.edu/data/glove.6B.zip.

[3] https://tfhub.dev/.

on top of the BERT layer with two layers containing 64 and 32 neurons respectively with *relu* activation function. The final dense classification layer contains 8 neurons with a *softmax* activation funtion. The BERT layer consists of the word embeddings from the BERT base model along with the input word ids and the masked sequence of the words. During the neural network model training, we have used a learning rate of 1e-5, which is taken from the original BERT paper [14] with a *categorical cross entropy* loss function. In order to train the neural network model, we employed a learning rate of 1e-5 with a *categorical cross entropy* loss function, which was adopted from the original BERT paper [14]. A batch size of 32 is used for the training. The model is trained over 4 epochs, and the best model is saved. For optimization, we have used the Adam optimizer [21].

We calculated the weighted average precision, recall and f-1 score on the test set for all experiments. The results are shown in Table 3. We also compared the scores from the different BERT models whose results are shown in Table 4. We also show the precision, recall and f-score for the different class labels in Table 2.

5 Results and Discussion

From Table 3, we can see that linear SVM has the lowest accuracy among the three tf-idf based classifiers. However, the neural network method using word2vec has the lowest accuracy among all the approaches. It could identify words with the EN tag with a precision of 94% and a recall of 78%. This can be attributed to the fact that GloVe embeddings were pre-trained over English words. It could also identify words with the AS tag, however, the precision is low with only 33%. It could not identify words with the other tags. This could be due to the fact that GloVe embeddings fail to learn the context as compared to BERT. In fact, the multinomial Naïve Bayes and logistic regression using TF-IDF performed better.

The logistic regression classifier performed the best using the tf-idf approach. The comparison between this and BERT can be seen for each word tag in Table 2. From Table 3, the best performance is shown by the BERT classifier method with an accuracy of 94%, outperforming all the other approaches in terms of precision, recall and f-score. Although not much difference can be seen in the performance of the different BERT models from Table 4, it can be seen that the BERT base model which is not fine-tuned like the other models, provides the better performance.

From the results we can see that using BERT, identification of languages in a code mix English-Assamese text corpus can be achieved with very good results. Our motivation behind using BERT is justified, as it can learn word contexts and as a result show better results in identifying languages with higher precision and recall.

Table 2 Comparison of Precision, Recall and F-score for different class labels

Tag	Logisitic regression			BERT		
	Precision	Recall	F-1	Precision	Recall	F-1
EN	0.97	0.81	0.88	0.96	0.98	0.97
AS	0.94	0.64	0.76	0.89	0.90	0.89
UNIV	0.39	0.98	0.56	0.97	0.98	0.97
HI	0.87	0.36	0.51	0.71	0.52	0.60
NE	0.72	0.30	0.42	0.63	0.53	0.51
ACRO	0.86	0.54	0.67	0.80	0.57	0.67
WM	0.50	0.20	0.29	1.00	0.13	0.24

Table 3 Comparison of BERT with other methods

	Precision	Recall	F-1	Acc. (%)
LinearSVM	0.59	0.67	0.56	67
MultinomialNB	0.77	0.76	0.68	76
Logistic regression	0.88	0.78	0.80	78
word2vec + LSTM	0.66	0.66	0.63	63
BERT-base + NN	0.93	0.94	0.93	94

Table 4 Comparison of different BERT models

	Precision	Recall	F-1	Acc. (%)
BERT-base	0.93	0.94	0.93	94
BERT_Wiki	0.88	0.92	0.90	92
BERT_Wiki_MNLI	0.88	0.92	0.90	92
BERT_Wiki_SST2	0.93	0.93	0.93	93

6 Conclusion

Automatic language identification in a code mix text has great scope for research. In a multilingual society where speakers exchange information in a mix of languages, it becomes necessary to identify the languages in order to detect hate speeches or misinformation being spread. However, in an environment of low resource language mixed with English, it becomes even more difficult due to shortage of data. In this paper, we have used BERT based word embeddings in a neural network to identify language in an English-Assamese code mix corpus. We have seen significant improvements when using these models paving way for further research in this direction using such models. We intend to expand the corpus in the future by integrating a broader range of material and experimenting with different transformer models. We also plan to use unsupervised learning approaches to see if we can identify the languages without the need for costly training.

References

1. Al-Badrashiny M, Diab M (2016) The george washington university system for the code-switching workshop shared task 2016. In: Proceedings of the second workshop on computational approaches to code switching. pp 108–111
2. Ansari MZ, Beg M, Ahmad T, Khan MJ, Wasim G (2021) Language identification of hindi-english tweets using code-mixed bert. arXiv preprint arXiv:2107.01202
3. Bansal N, Goyal V, Rani S (2020) Language identification and normalization of code-mixed english and punjabi text. In: 17th international conference on natural language processing. p 30
4. Barman U, Wagner J, Chrupała G, Foster J (2014) Dcu-uvt: word-level language classification with code-mixed data. In: Proceedings of the first workshop on computational approaches to code switching. pp 127–132
5. Bokamba EG (1989) Are there syntactic constraints on code-mixing? World Engl 8(3):277–292
6. Bora MJ, Kumar R (2018) Automatic word-level identification of language in assamese english hindi code-mixed data. In: 4th workshop on indian language data and Resources, Proceedings of the Eleventh International Conference on Language Resources and Evaluation (LREC 2018). pp 7–12
7. Boukkouri HE, Ferret O, Lavergne T, Noji H, Zweigenbaum P, Tsujii J (2020) Characterbert: reconciling elmo and bert for word-level open-vocabulary representations from characters. arXiv preprint arXiv:2010.10392
8. Burling R (2003) The tibeto-burman languages of northeastern india. Sino Tibet Lang 3:169–191
9. Carpuat M (2014) Mixed language and code-switching in the canadian hansard. In: Proceedings of the first workshop on computational approaches to code switching. pp 107–115
10. Chaitanya I, Madapakula I, Gupta SK, Thara S (2018) Word level language identification in code-mixed data using word embedding methods for indian languages. In: 2018 international conference on advances in computing, communications and informatics (ICACCI). IEEE, pp 1137–1141
11. Chakravarthi BR, Priyadharshini R, Muralidaran V, Jose N, Suryawanshi S, Sherly E, McCrae JP (2022) Dravidiancodemix: Sentiment analysis and offensive language identification dataset for dravidian languages in code-mixed text. Lang Resour Eval :1–42
12. Das A, Gambäck B (2014) Identifying languages at the word level in code-mixed indian social media text
13. Das A, Gambäck B (2015) Code-mixing in social media text: the last language identification frontier?
14. Devlin J, Chang MW, Lee K, Toutanova K (2018) Bert: pre-training of deep bidirectional transformers for language understanding. arXiv preprint arXiv:1810.04805
15. Eibe F, Hall MA, Witten IH (2016) The weka workbench. online appendix for data mining: practical machine learning tools and techniques. In: Morgan Kaufmann
16. Grierson GA (2006) Languages of North-Eastern India. Gyan Publishing House
17. Hochreiter S, Schmidhuber J (1997) Long short-term memory. Neural Comput 9(8):1735–1780
18. Jaech A, Mulcaire G, Ostendorf M, Smith NA (2016) A neural model for language identification in code-switched tweets. In: Proceedings of the second workshop on computational approaches to code switching. pp 60–64
19. Jamatia A, Das A (2016) Task report: tool contest on pos tagging for code-mixed indian social media (facebook, twitter, and whatsapp) text@ icon 2016. In: Proceedings of ICON

20. Kalita NJ, Saharia N (2018) Language identification on code-mix social text. In: Proceedings of the international conference on computing and communication systems. Springer, pp 433–440
21. Kingma DP, Ba J (2014) Adam: a method for stochastic optimization. arXiv preprint arXiv:1412.6980
22. Lample G, Conneau A (2019) Cross-lingual language model pretraining. arXiv preprint arXiv:1901.07291
23. Lan Z, Chen M, Goodman S, Gimpel K, Sharma P, Soricut R (2019) Albert: a lite bert for self-supervised learning of language representations. arXiv preprint arXiv:1909.11942
24. Liu Y, Ott M, Goyal N, Du J, Joshi M, Chen D, Levy O, Lewis M, Zettlemoyer L, Stoyanov V (2019) Roberta: a robustly optimized bert pretraining approach. arXiv preprint arXiv:1907.11692
25. Mikolov T, Sutskever I, Chen K, Corrado GS, Dean J (2013) Distributed representations of words and phrases and their compositionality. In: Advances in neural information processing systems. pp 3111–3119
26. Pennington J, Socher R, Manning CD (2014) Glove: global vectors for word representation. In: Proceedings of the 2014 conference on empirical methods in natural language processing (EMNLP). pp 1532–1543
27. Piergallini M, Shirvani R, Gautam GS, Chouikha M (2016) Word-level language identification and predicting codeswitching points in swahili-english language data. In: Proceedings of the second workshop on computational approaches to code switching. pp 21–29
28. Samih Y, Maharjan S, Attia M, Kallmeyer L, Solorio T (2016) Multilingual code-switching identification via lstm recurrent neural networks. In: Proceedings of the second workshop on computational approaches to code switching. pp 50–59
29. Sanh V, Debut L, Chaumond J, Wolf T (2019) Distilbert, a distilled version of bert: smaller, faster, cheaper and lighter. arXiv preprint arXiv:1910.01108
30. Sarma N, Singh SR, Goswami D Identifying languages at the word level in assamese-bengali-hindi-english code-mixed social media text
31. Schroff F, Kalenichenko D, Philbin J (2015) Facenet: a unified embedding for face recognition and clustering. In: Proceedings of the IEEE conference on computer vision and pattern recognition. pp 815–823
32. Shirvani R, Piergallini M, Gautam GS, Chouikha M (2016) The howard university system submission for the shared task in language identification in spanish-english codeswitching. In: Proceedings of the second workshop on computational approaches to code switching. pp 116–120
33. Vasantharajan C, Thayasivam U (2022) Towards offensive language identification for tamil code-mixed youtube comments and posts. SN Comput Sci 3(1):1–13
34. Vaswani A, Shazeer N, Parmar N, Uszkoreit J, Jones L, Gomez AN, Kaiser Ł, Polosukhin I (2017) Attention is all you need. In: Advances in neural information processing systems. pp 5998–6008
35. Veena P, Kumar MA, Soman K (2017) An effective way of word-level language identification for code-mixed facebook comments using word-embedding via character-embedding. In: 2017 international conference on advances in computing, communications and informatics (ICACCI). IEEE, pp 1552–1556
36. Volk M, Clematide S (2014) Detecting code-switching in a multilingual alpine heritage corpus. Association for Computational Linguistics
37. Xia MX (2016) Codeswitching language identification using subword information enriched word vectors. In: Proceedings of the second workshop on computational approaches to code switching. pp 132–136

Enhanced White Shark Optimization Algorithm for the Mesh Routers Placement Problem with Service Priority in Wireless Mesh Networks

Sylia Mekhmoukh Taleb, Yassine Meraihi, Amar Ramdane-Cherif, Asma Benmessaoud Gabis, and Dalila Acheli

1 Introduction

WMN, or wireless mesh network, is an emerging communication technology that provides connectivity for regions without infrastructure. It has gained great attention due to its easy implementation at low cost. In this sense, WMN is used in a large variety of applications such as broadband home networking, enterprise networking, metropolitan area networking, transportation systems, building automation, health and medical science, and safety and surveillance systems. WMN is composed of three types of nodes including Mesh clients (MCs), Mesh Routers (MRs), and Mesh Gateways (MGs) as illustrated in Fig. 1.

Despite its benefits, WMN has some limitations, including those related to security, throughput, scalability, and service quality. In actuality, node placement has a big impact on WMN performance. More specifically, poor WMN nodes location results in numerous interference that significantly increase packet loss and congestion. To solve this problem, an effective optimization technique must be applied.

The nodes placement problem belongs to the category of NP-Hard problems, solved successfully using meta-heuristics. In literature several meta-heuristics were

S. M. Taleb (✉) · Y. Meraihi
LIST Laboratory, University of M'Hamed Bougara Boumerdes, Avenue of Independence, 35000 Boumerdes, Algeria
e-mail: s.mekhmoukh@univ-boumerdes.dz

A. Ramdane-Cherif
LISV Laboratory, University of Versailles St-Quentin-en-Yvelines, 10-12 Avenue of Europe, 78140 Velizy, France

A. B. Gabis
Ecole nationale Supérieure d'Informatique, Laboratoire LMCS, BP 68M, 16309, Oued-Smar Alger, Algérie

D. Acheli
LAA Laboratory, University of M'Hamed Bougara Boumerdes, Avenue of Independence, 35000 Boumerdes, Algeria

A. Ramdane-Cherif et al. (eds.), *Machine Intelligence and Data Science Applications*, Algorithms for Intelligent Systems, https://doi.org/10.1007/978-981-99-1620-7_15

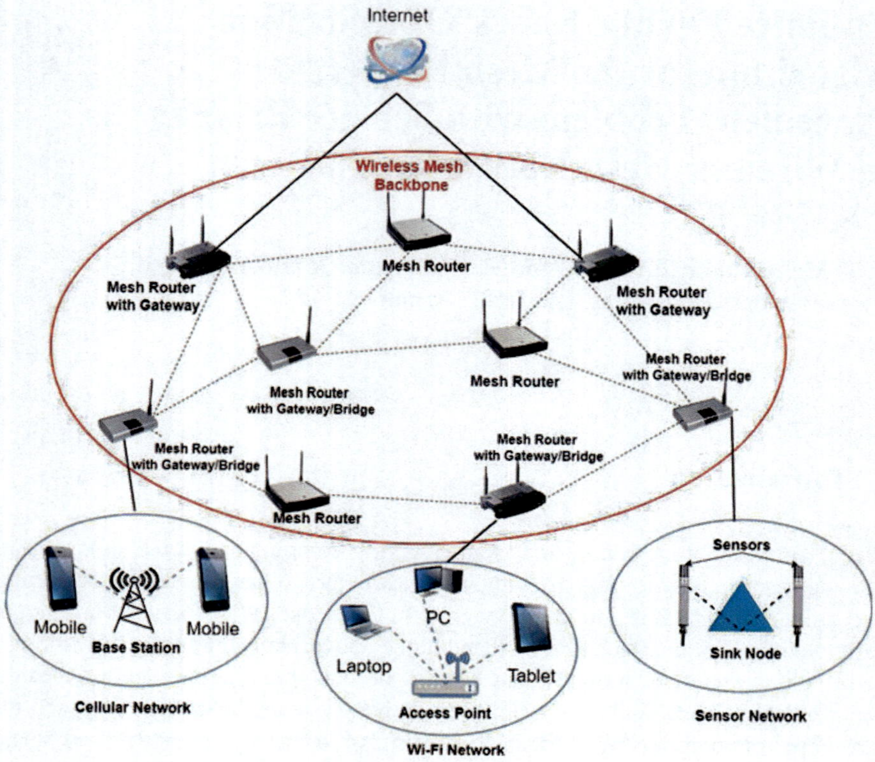

Fig. 1 Wireless Mesh Network design

applied to solve this issue such as Particle Swarm Optimization (PSO) [1], Harris' Hawk Optimization (HHO) [2], Coyote Optimization Algorithm (COA) [3], and Multi-Verse Optimization (MVO) [4].

The majority of earlier investigations on the Mesh Routers Placement (MRP) problem assumed that all mesh clients are treated equally when assessing the performance of a network. In reality, few studies [5, 6] have taken into account weighted clients where mesh clients with greater priorities than a threshold must be served as well. This lack motivates our attempt to address the MRP problem with service priority constraint using the newly White Shark Optimization (WSO) meta-heuristic.

In this paper, we suggest an improved version of WSO, called RFDB-WSO, for addressing the MRP problem with service priority constraint. RFDB-WSO is based on the incorporation of the Roulette Fitness Distance Balance (RFDB) selection method into the original WSO. The RFDB-WSO algorithm was assessed in terms of fitness value under distinct scenarios, taking into account the weights of mesh clients, in comparison with WSO, PSO, Genetic Algorithm (GA), Scine Cosine Algorithm (SCA), and Harmony Search (HS) algorithms.

The remainder of this paper is organized as follows. The formulation of the MRP problem with service priority constraint is described in Sect. 2. Section 3 presents the WSO algorithm and RFDB selection method. Section 4 describes the RFDB-WSO algorithm for solving the addressed problem. Additionally, we assess the effectiveness of RFDB-WSO and present the outcomes in Sect. 5. Finally, Sect. 6 concludes the paper.

2 MRP Problem Formulation

An undirected graph $G = (V, L)$ can be used to mathematically represent WMN, where V represents the set of MRs and L represents the set of links connecting these MRs. A group of unrelated sub-networks come together to form the network G. Let:

- MRS is the set of n of MRs: $MRS = \{mr_1, mr_2.....mr_n\}$. Each mesh router mr_i is equipped with one radio interface having transmission range TR_i. A connection between two mesh routers mr_i and mr_j is possible only if the distance between them does not exceed the sum of their transmission ranges. Each installed MR must covers at least one MC.
- MCS is the set of m MCs: $MCS = \{mc_1, mc_2.....mc_m\}$. MCs are randomly dispersed in the deployment area. We assume that MCs are not equal, a priority is associated with each mesh client mc_j. The priority of the jth client is denoted by wc_j. A mesh client mc_j is said covered by a mesh router mr_i if it is within the transmission range of this router. It can be linked at most with one router. It can be within the transmission range of several MRs but it is associated with the nearest one.

Our work's main goal is to determine where n MRs should be placed in a $2D$ space with dimensions $W \times L$, taking into account the positions of m mesh clients and their priorities, maximizing two metrics such as:

- Weighted clients coverage: It described a sum of priorities of the covered mesh clients by all MRs according to Eq. (1):

$$\Psi(G) = \sum_{i=1}^{m} (max_{j \in \{1,...n\} \sigma_{ij}}) * wc_i \tag{1}$$

where σ_{ij} defines the coverage variable represented as described in Eq. (2):

$$\sigma_{ij} = \begin{cases} 1 \ if \ mc_i \ is \ covered \ by \ mr_j \\ 0 \ Otherwise. \end{cases} \tag{2}$$

- Network connectivity: It is defined as the biggest sub-network among k formed sub-networks with regard to the number of MRs. It is calculated as defined in Eq. (3):

$$\Phi(G) = Max_{i\in\{1,\ldots k\}}|G| \tag{3}$$

where $|G_i|$, $i \in \{1, k\}$ is the size of ith sub-network and $G = G_1 \cup G_2 \cup \ldots \cup G_k$. The following criteria were used to define the objective function as described in Eq. (4):

$$f(X) = \gamma.\frac{\Psi(G)}{m} + (1-\gamma), \frac{\Phi(G)}{n} \tag{4}$$

where G denotes the graph associated with the given solution X. The floating parameter γ, with a range of [0 1], is used to regulate the importance of metrics.

3 Preliminaries

3.1 White Shark Optimization (WSO) Algorithm

White Shark Optimization (WSO) is a new bio-inspired meta-heuristic, developed by Braik et al.[7] in 2022. The WSO concept is based on the strategies used by white sharks to help them survive in the deep ocean, detailed in this section.

(I) **Initialization of WSO**:

The WSO process begins by initializing adjustable parameters including:

- nw: The number of white sharks in the population.
- d: Dimension of the search space.
- *ub*,*lb*: Upper and lower bounds of the search space.
- $fmax$: Maximum and minimum frequencies of the wavy motion.
- $pmax$,$pmin$: The starting and subordinate velocities for white shark motion.
- μ: The constriction factor used to analyze the convergence of WSO
- $a0$, $a1$, $a2$: Th exploration and exploitation control location constants
- $itmax$: The total number of iterations

In a d search domain space, the population of nw WSO denoted by X is initialized, X is presented by the following 2D matrix:

$$X = \begin{bmatrix} X_1^1 & X_2^1 & \ldots\ldots & X_d^1 \\ X_1^2 & X_2^2 & \ldots\ldots & X_d^2 \\ . & . & . \quad . & . \\ . & . & . \quad . & . \\ X_1^{nw} & X_2^{nw} & \ldots\ldots & X_d^{nw} \end{bmatrix}$$

Each white shark hark X^i is characterized by its velocity denoted by V^i. The velocity of the initial population is described as a zeros matrix of dimensions $nw \times d$.

(II) **Speed of movement towards prey**:

At each iteration $t + 1$, the velocity of the ith shark is denoted by V_{t+1}^i and is updated as described in Eq. (5):

$$V_{t+1}^i = \mu\left[V_t^i + p_1\left(X_{gbest-t} - X_t^i\right) \times r_1 + p_2\left(X_{best}^{u_t^i} - X_t^i\right) \times r_2\right] \quad (5)$$

where $X_{gbest-t}$ denotes the global best position vector. u_t^i is the $i^t h$ index vector of the white sharks achieving the best position indicated by Eq. (6). r_1 and r_2 are random numbers in the range [0 1]. p_1 and p_2 stand for the forces of the white sharks that regulate the effect $X_{gbest-t}$ and $X_{best}^{u_t^i}$ on X_t^i, respectively, which are computed as stated in Eqs. (7) and (8).

$$u = \lfloor nw \times \text{rand}(1, nw)\rfloor + 1 \quad (6)$$

$$p_1 = pmax + (pmax - pmin) \times e^{-(4t/itmax)^2} \quad (7)$$

$$p_2 = pmin + (pmax - pmin) \times e^{-(4t/itmax)^2} \quad (8)$$

(III) **Movement towards optimal prey**:
The behavior of the white sharks as they move toward the optimal prey was described in Eq. (9):

$$X_{t+1}^i = \begin{cases} X_{t*}^i \to \oplus X_0 + ub \cdot a + lb.b;\ r_3 < mv \\ X_t^i + V_t^i/f;\ r_3 \geq mv \end{cases} \quad (9)$$

where r_3 is a random number in range [0 1]. a and b present binary vectors defined as illustrated in Eqs. (10) and (11).

$$a = \text{sgn}\left(X_t^i - ub\right) > 0 \quad (10)$$

$$b = \text{sgn}\left(X_t^i - lb\right) < 0 \quad (11)$$

$$X_0 = \oplus(a, b) \quad (12)$$

Equations (13) and (14), respectively, define how often a white shark will move in a wavy pattern and how many times it will attack a target.

$$f = fmin + \frac{fmax - fmin}{fmax + fmin} \quad (13)$$

$$mv = \frac{1}{\left(a0 + e^{(itmax/2-t)/a1}\right)} \quad (14)$$

(IV) **Movement towards the best white shark**:

Great white sharks can keep their position toward the finest one that is near to the prey. Equation (15) describes this conduct.

$$X_{t+1}^{\prime i} = X_{gbest-t} + r_4 \vec{D}_x \operatorname{sgn}(r_5 - 0.5)\, r_6 < S_s \tag{15}$$

$X_{t+1}^{\prime i}$ denotes the updated shark's position. r_4, r_5, and r_6 are random number in range [0 1]. D_x denotes the target's and the shark's lengths defined as illustrated in Eq. (16). S_s is parameter has been suggested to represent the power of white shark as given in Eq. (17).

$$\vec{D}_x = \mid r_7 \times \left(X_{gbest} - X_t^i\right) \mid \tag{16}$$

$$S_s = \left|1 - e^{(-a2 \times t / itmax)}\right| \tag{17}$$

(V) Fish school behavior:

The white shark fish school behavior is formulated as illustrated in Eq. (18):

$$X_{t+1}^i = \frac{X_t^i + X_{t+1}^{\prime i}}{2 \times r_7} \tag{18}$$

where r_7 is random number in range [0 1].

3.2 *Roulette Fitness Balance Distance (RFDB)*

Kahraman et al. [8] introduced the selection approach known as Fitness Distance Balance (FDB) in 2020, for solving the premature convergence issue in the meta-heuristics. The objective is to identify the most influential people who can successfully lead the search strategy. The following list contains the key FDB steps:

- The distance between candidates' solutions and the great solution ($Pbest$) is calculated in the first step of the FDB selection method. The predefined objective function is used to determine ($Pbest$). The distance between each individual P_i in the population P and $Pbest$ is calculated as given in Eq. (19). Let P the population of m solutions and N is the number of variables in each potential solution, respectively. The DP distance vector for the candidates' solution is given as specified in Eq. (20).

$$_{i=1}^{m}, \forall P_i, D_{P_i} = \sqrt{\left(P_{1[i]} - P_{1[\text{best}]}\right)^2 + \left(P_{2[i]} - P_{2[\text{best}]}\right)^2 + \cdots + \left(P_{N[i]} - P_{N[\text{best}]}\right)^2} \tag{19}$$

$$D_P \equiv \begin{bmatrix} D_{P_1} \\ . \\ . \\ D_{P_m} \end{bmatrix}_{m \times 1} \tag{20}$$

- The candidates' scores in the second stage of FDB are determined by the normalized fitness values ($normFV$) and normalized distance values ($normDP$) from the candidates, as stated in Eq. 21:

$$_{i=1}^{m} \forall P_i, \quad FDB_{P_i} = w * \text{norm}FV_i + (1 - w) * \text{norm}D_{P_i} \tag{21}$$

where w is the weighted factor.
- Finally, the score vector FDB_p obtained from the P population is reported in the form of Eq. (22):

$$FDB_p \equiv \begin{bmatrix} FDB_{p_1} \\ . \\ . \\ . \\ FDB_{p_m} \end{bmatrix}_{m \times 1} \tag{22}$$

The search process's leading candidate solution, FDB_{best}, can be chosen using random, greedy, or probabilistic approaches (roulette and wheel). Readers who are interested can learn more about the FDB approach in [8]. In our study article, the best candidate, FDB_{best}, was chosen using the roulette technique. The selection method used in this paper will be denoted by RFDB.

Algorithm 1 describes the steps of the RFDB selection method.

Algorithm 1 The pseudo-code of RFDB selection method

Input:
w: Weighting coefficient
P: Population of solutions
Output:
The leading candidate solution FDB_{best}

```
Determine P_best using objective function
for i ← 1 to m do
    Calculate the distance between P_i and P_best using Eq.(19)
    Update the D_p distance vector using Eq.(20)
for i ← 1 to m do
    Normalizing the fitness and distance vector in the range [0 1]
    Update the FDB-based score vector using Eq.(21)
Select the candidate FDB_best using roulette selection method
Return the best candidate FDB_best
```

4 Roulette Fitness Distance Balance White Shark Optimization (RFDB-WSO) for the MRP Problem with Service Priority in WMNs

The structure of the suggested RFDB-WSO, which is based on the incorporation of the RFDB method in WSO, will be discussed in this section. To resolve the premature convergence of WSO, RFDB was employed. RFDB was implemented in two stages: (1) movement toward the best white shark and (2) fish school behavior. The behavior based on the movement towards the best white shark is formulated in Eq. (23). On the other hand, the fish school behavior is described in Eq. (24).

$$X'^{i}_{t+1} = RFDB - X_{gbest-t} + r_4 \overrightarrow{RFDB - D_x}\ \mathrm{sgn}\,(r_5 - 0.5)\, r_6 < S_s \tag{23}$$

$$X^{i}_{t+1} = \frac{X^{i}_{t} + RFDB - X'^{i}_{t+1}}{2 \times r_7} \tag{24}$$

$RFDB - X_{gbest-t}$ is the leading solution obtained using the RFDB selection method. $RFDB - D_x$ is determined as described in Eq. 25.

$$\overrightarrow{RFDB - D_x} = \mid r_7 \times \left(RFDB - X_{gbest-t} - X^{i}_{t}\right) \mid \tag{25}$$

The main steps of RFDB-WSO are given in algorithm 2

Algorithm 2 The pseudo code of RFDB WSO algorithm for solving the MRP problem

Input:
n: Number of MRs
m,wc: Number of MCs and theirs priorities
nw: Number of white sharks in the population
ub,lb: Upper and upper bounds of the deployment area
$fmax$: Maximum and minimum frequencies of the wavy motion
$pmax$,$pmin$: The starting and subordinate velocities for white shark motion
μ: The constriction factor used to analyze the convergence of WSO
$a0$, $a1$, $a2$: Th exploration and exploitation control location constants
$itmax$: The total number of iterations
Output:
Return the best solution X_{gbest} and its fitness value

```
for i ← 1 to nw do
    Initialize the positions and transmission ranges of mesh routers of the i^th
    white shark
    Initialize the velocity of the i^th white shark
Initialize randomly the positions of MCs in the deployment area
Evaluate the initial population and identify the X_gbest-0
while t < itmax do
    Update the parameters u, p1, p2, a, b, wo, f , mv and S_s using Eqs. (6), (7),
    (8), (10), (11), (12), (13), (14) and (17)
    for i ← 1 to nw do
        Update V_t+1 shark using Equation 5
    for i ← 1 to nw do
        Update X_t+1 using Equation 9
    for i ← 1 to nw do
        if r_6 < S_s then
            Determine RFDB−X_gbest-t using Algorithm 1 Update RFDB − D_s (vector)
            using Equation (25)
            if t==1 then
                Update X^1_t+1 using Equation 23
            Update X^n_t+1 using Equation 23
            Update X^1_t+1 using Equation 24
    Evaluate the population, update the best positions, and determine X_gbest-t
    t = t + 1
Return the best solution
```

5 Simulation Results

In this section, we will evaluate the RFDB-WSO algorithm's performance in comparison to the WSO, PSO, GA, SCA, and HS algorithms. All of the algorithms were coded in Matlab. A Core i7 2.5 GHz computer is used to run the simulations. The total number of iterations is set to 2000. All simulations took into account a population of 30 solutions. The results displayed here are the average of an average of 15 trials.

The performance of the RFDB-WSO was analysed in terms of fitness value considering the effect of varying two parameters such as (1) number of mesh clients (2) number of mesh routers. A weighted mesh clients randomly dispersed in an area of $1500m \times 1500m$ were considered in all scenarios. Weights are integer numbers in range [1 5]. The weighting coefficient γ is set to 0.7 [9]. A maximum transmission range of $200m$ is considered in all simulations.

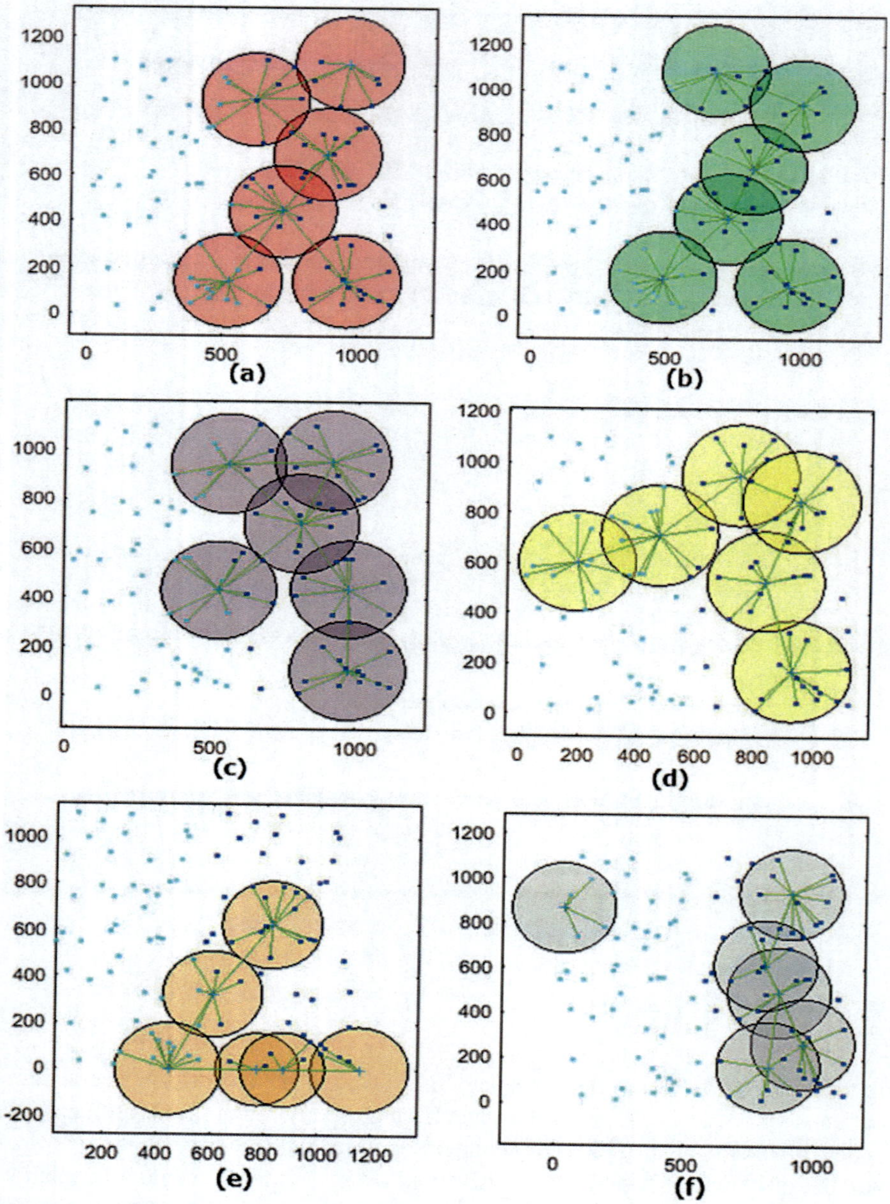

Fig. 2 Obtained placement using: **a** RFDB-WSO **b** WSO **c** PSO **d** GA **e** SCA **f** HS

Figures 2a, 2b, 2c, 2d, 2e, and 2f report examples of a planned network using RFDB-WSO, WSO, PSO, GA, SCA and HS, respectively. The planned network is a solution of network instance with 6 MRs and 100 MCs. Light blue points represent MCS having priority equal to 1 (there are 50 MCS distributed in the left half part of

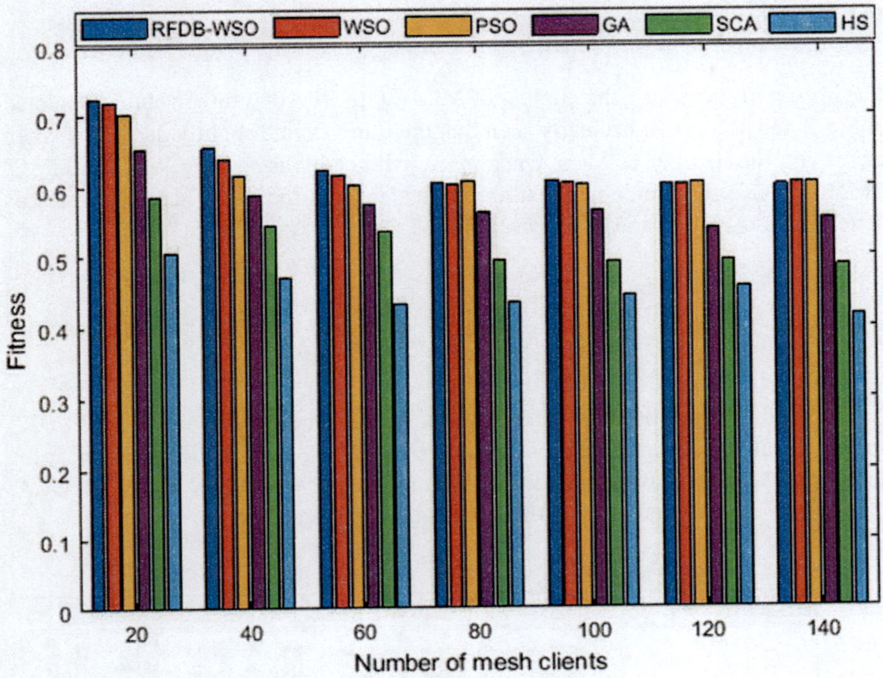

Fig. 3 Fitness under various number of mesh clients

Table 1 Fitness under various number of MRs

	RFDB-WSO	WSO	PSO	GA	SCA	HS
m = 20	**0.724**	0.720	0.704	0.655	0.586	0.505
m = 40	**0.656**	0.640	0.617	0.590	0.544	0.470
m = 60	**0.625**	0.617	0.604	0.576	0.536	0.433
m = 80	0.606	0.603	**0.608**	0.563	0.493	0.434
m = 100	**0.608**	0.605	0.602	0.565	0.491	0.445
m = 120	0.602	0.603	**0.605**	0.540	0.494	0.456
m = 140	0.602	**0.606**	0.606	0.554	0.486	0.417

the deployment space). Dark blue points represent MCS having priority equal to 5 (there are 50 MCS distributed in the right half part of the deployment space). Each installed MR is represented by light blue plus sign. It is clearly seen from this figures that MCs with greater priorities are the first served. Again, RFDB-WSO, WSO, PSO, and GA are more appropriate for solving the MRP problem with service priority.

5.1 *Effect of Varying the Number of MCs*

The impact of increasing the number of MCs (20 to 140) on fitness value is depicted in Fig. 3 and Table 1. It is clearly seen that the fitness value slightly decreases when increasing the number of MCs. Once more, when contrasted with WSO and PSO, RFDB-WSO offers outcomes that are equivalent. On the other hand, It performs better than GA, SCA, and HS algorithms.

5.2 *Effect of Varying the Number of MRs*

Figure 4 and Table 2 describe the impact of varying the number of MRs (2 to 22) on fitness value. The presented results showed that the fitness value increases as the number of MRs increases. In addition, in comparison with WSO, PSO, GA, SCA, and HS, the effectiveness of RFDB-WSO is demonstrated.

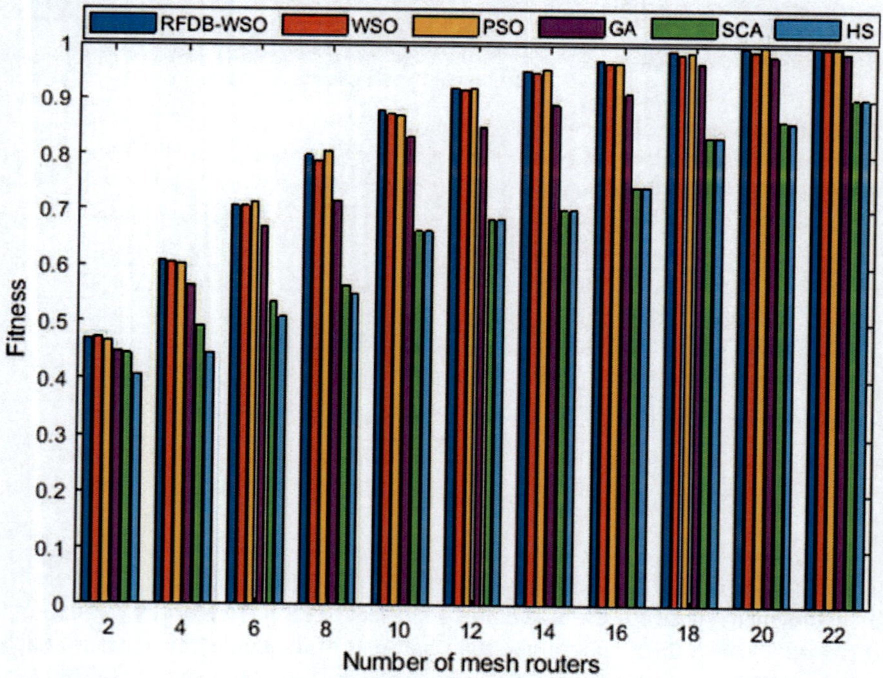

Fig. 4 Fitness under various number of mesh routers

Table 2 Fitness under various number of MRs

	RFDB-WSO	WSO	PSO	GA	SCA	HS
n = 2	0.469	**0.470**	0.466	0.447	0.444	0.407
n = 4	**0.608**	0.605	0.602	0.565	0.491	0.445
n = 6	0.706	0.706	**0.713**	0.668	0.536	0.508
n = 8	0.799	0.789	**0.807**	0.717	0.565	0.550
n = 10	**0.880**	0.873	0.871	0.831	0.662	0.662
n = 12	**0.920**	0.918	0.920	0.850	0.684	0.683
n = 14	0.953	0.948	**0.956**	0.890	0.700	0.700
n = 16	**0.971**	0.967	0.967	0.910	0.743	0.743
n = 18	**0.991**	0.983	0.987	0.967	0.833	0.833
n = 20	0.996	0.990	**0.998**	0.980	0.861	0.860
n = 22	**0.998**	0.996	0.997	0.987	0.901	0.901

6 Conclusion

In order to solve the mesh router placement (MRP) problem with service priority, this study suggests an enhanced version of WSO, called RFDB-WSO, based on the integration of RFDB selection method into the original WSO. Depending on a set of client weights and client locations, the performance of the RFDB-WSO algorithm is tested in terms of fitness value maximizing both coverage and connectivity. The simulation trials showed that RFDB-WSO produces comparable results to those of WSO, PSO, GA, SCA, and HS algorithms.

References

1. Le VT, Dinh NH, Nguyen NG (2011) A novel pso-based algorithm for gateway placement in wireless mesh networks. In: 2011 IEEE 3rd International Conference on Communication Software and Networks. IEEE, pp 41–45
2. Abdulrab HQA, Hussin FA, Abd Aziz A, Awang A, Ismail I, Saat MSM, Shutari H (2022) Optimal coverage and connectivity in industrial wireless mesh networks based on harris' hawk optimization algorithm. IEEE Access 10:51048–51061
3. Taleb SM, Meraihi Y, Gabis AB, Mirjalili S, Zaguia A, Ramdane-Cherif A (2022) Solving the mesh router nodes placement in wireless mesh networks using coyote optimization algorithm. IEEE Access
4. Binh LH, Truong TK et al (2022) An efficient method for solving router placement problem in wireless mesh networks using multi-verse optimizer algorithm. Sensors 22(15):5494
5. Lin CC, Li YS, Deng DJ (2014) A bat-inspired algorithm for router node placement with weighted clients in wireless mesh networks. In: 9th International Conference on Communications and Networking in China. IEEE, pp 139–143

6. Lin Chun-Cheng, Shu Lei, Deng Der-Jiunn (2014) Router node placement with service priority in wireless mesh networks using simulated annealing with momentum terms. IEEE Syst J 10(4):1402–1411
7. Braik M, Hammouri A, Atwan J, Al-Betar MA, Awadallah MA (2022) White shark optimizer: a novel bio-inspired meta-heuristic algorithm for global optimization problems. Knowl Based Syst 243:108457
8. Kahraman HT, Aras S, Gedikli E (2020) Fitness-distance balance (fdb): a new selection method for meta-heuristic search algorithms. Knowl Based Syst 190:105169
9. Ohara S, Durresi H, Barolli A, Sakamoto S, Barolli L (2019) A hybrid intelligent simulation system for node placement in wmns considering load balancing: a comparison study for exponential and normal distribution of mesh clients. In: International Conference on Broadband and Wireless Computing, Communication and Applications. Springer, pp 555–569

Wearable Technologies for Health: Investigating Behavioral Intention to Adopt Cloud-Based Smartwatch

Ashok Singh Malhi, Raj K. Kovid, and Kanika

1 Introduction

In recent years so many Bollywood celebrities were passed away due to heart attack like Krishnakumar Kunnat (KK), Praveen Patwardhan, Puneeth Rajkumar, Sidharth Shukla, Brahma Swaroop Mishra, Mithilesh Chaturvedi, Praveen Kumar Sobti, Surekha Sikri, Arvind Trivedi [1]. As per The Indian Express, six-fold rise in heart attacks in Mumbai was reported in the first half of 2021 [2] and more shocking fact is that normal citizens are also being victim of this life-threatening disease details are given in Fig. 1. Major reasons of the heart attacks are coronary artery disease and what are the different causes is given in Fig. 2, like complete blockage/ partial blockage in heart artery. But this is the reason of heart attack occurred to an individual that can be measured but why this blockage happens is revealed by so many studies [3, 4]. Wearing and using cloud-based smartwatches can to the extent enough pre-empt the occurrence of heart attack through early indicators of the incident. These watches are of many types as given below.

1.1 Cloud-Based Smartwatches (C-BSW)

C-BSWs are more advanced versions of smart bands. They have an embedded operating system (OS), which allows the installation of third-party or in-house cloud-based applications. A smartwatch is a wristwatch-sized wearable touchscreen computer device meant to facilitate physical activity. A smartwatch combines the

A. S. Malhi (✉) · R. K. Kovid
School of Business Studies, Sharda University, Greater Noida, Uttar Pradesh 201310, India

Kanika
Management, Amity Global Business School, Amity University, Noida, India

A. Ramdane-Cherif et al. (eds.), *Machine Intelligence and Data Science Applications*, Algorithms for Intelligent Systems, https://doi.org/10.1007/978-981-99-1620-7_16

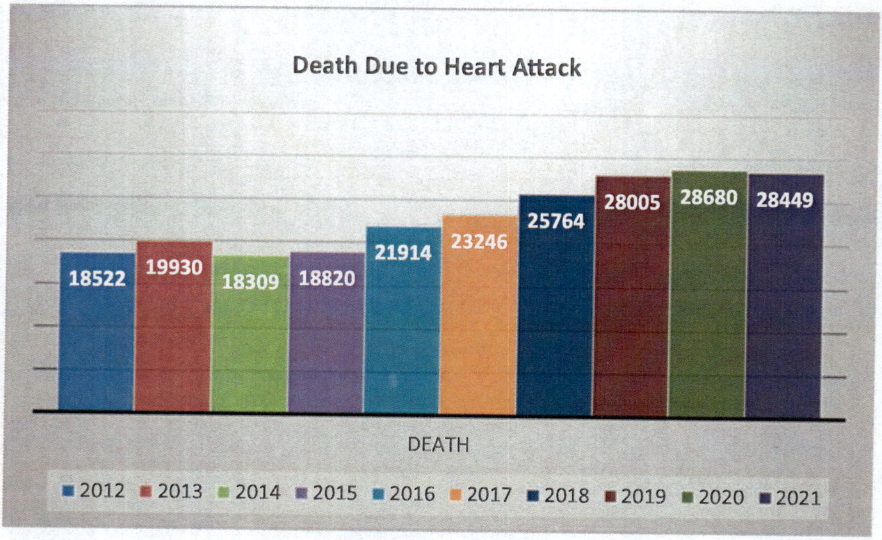

Fig. 1 The number of heart attack deaths from 2012 to 2021. *Source* (NCRB)

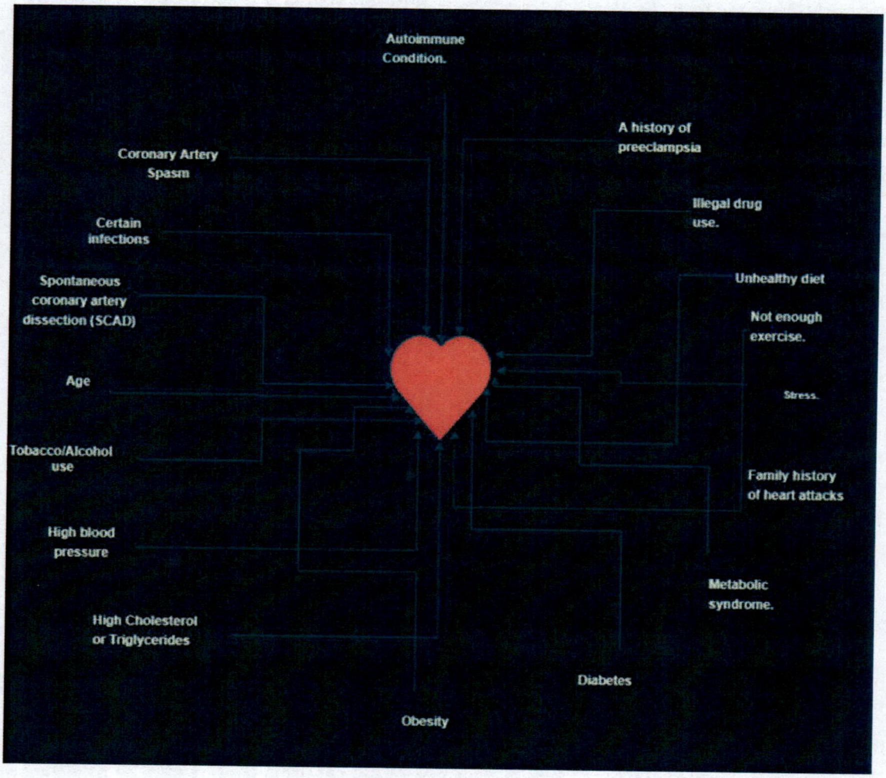

Fig. 2 Reasons of heart attack for young generation of India

functionalities of a traditional watch with more sophisticated smart features. Smartwatches, like smartphones, include touchscreens and app support, and can monitor the user's heart rate and other vital indications. Smartwatches have been heralded as the next generation of electronics that will revolutionize the lives of customers. It is planned to expand the capabilities of smartwatches in the next years by adding features and functionality. It was earlier anticipated that smartwatch shipments will reach 70 million devices by 2021. Smartwatches are on their way to replacing cell phones. One of the most significant advantages of smartwatches is that they assist the user create healthy habits and stay on time and up to date by utilizing a variety of various behavioral and physiological sensors. Apple, Samsung, Sony, Fitbit, and LG are among the smartwatch manufacturers.

1.2 Cloud-Based Smart Bands

These do not have all of the same features as smartwatches. In other words, they have only limited features in comparison with smartwatches. The main purpose of a smart band is to monitor the sleep pattern and the physical activity of the user like different exercise modes, calories burn, and heart rate. As per McIntyre "Smartwatch is a multi-functional device that appeals to a broad range of user interests, including not only fitness, health monitoring, and location tracking but also extended communication and smart features" [5]. As per Kim "Smart watches serve mostly as satellite devices for amassing useful data from a paired smartphone via wireless Bluetooth connection and providing more convenient, faster, and substitutable access to information, especially as its information processing is less demanding and using a smartphone is sometimes impractical" [6] and "A wrist-worn device with computational power that can connect to other devices via short-range wireless connectivity; provides alert notifications; collects personal data through a range of sensors and stores them; and has an integrated clock" [7] some other dimension also include in this device as per Choi [8] "A smartwatch is a unique form of information technology in that its shape resembles an item that has been a close companion to us humans for many centuries, namely the 'wristwatch'" and Hsiao [9] "Smartwatch is devices that can connect with smartphones and receive a lot of information, such as time, text messages, schedules, and GPS data. While it can perform basic data and communications tasks, it is also capable of running mobile applications". This article does not include all factors that cause heart attack but focuses on that factor that helps to monitor by the majority of these smartwatches, like daily step count, calories burnt, live heart rate count, breathing speed, blood oxygen level, sleeping Monitor, BMI, Water Intake Reminder, Handwash Reminder, 100 Different Exercise Modes, Stopwatch, Compass, music, daily target goal index. In the twenty-first-century people, especially young generation are trapped unhealthy lifestyle, fast food consumption, not proper sleeping, avoid exercise, consuming alcohol-based drink, and tobacco-based products (sometimes, restricted medication). This smartwatch has appeared to be like a virtue to humankind, that helps to live a balanced lifestyle with proper monitor individuals' daily activities.

Cloud-based smartwatches use tangible aspects but its effectiveness more increase when it combines its respective application like Google Health, Apple Healthkit, MI-Fit Apps, and Noise Fit App. They increase its acceptability and accuracy in analyzing our daily workout data, body-weight ratio, total calories burnt, average heart rate, breath rate, and average sleeping hours. That lead to any individual to adopt a healthy lifestyle and minimize the occurrence of heart attack.

1.3 Heartbeat Rate

An optical heart rate sensor is included with a smartwatch. It aids in the monitoring of blood flow from your heart. It computes heartbeats per minute because as blood flows from your heart, it reaches your artery and is registered on your wrist. The speed of your blood flow is measured on one's wrist to ensure accuracy. Heart-related problems are more severe. As a result, in the event of a medical emergency, a qualified doctor or cardiologist should be consulted. These smartwatch metrics should only be used as a guideline and not as medical data.

1.4 Calorie Counter

A wristwatch can measure the motions and acceleration of a body. The BMR, or basal metabolic rate, is used by the sensors inside the wristwatch to deliver precise statistics. The calorie counter restarts at midnight each day to give you accurate statistics. The more you can push yourself, the more calories you will burn. If you want to monitor your health while working out.

1.5 Sleep Monitor

Some people often experience sleep unrest or insomnia. A smartwatch can help you keep track of this. When you are in an inactive state, the smartwatch assumes you are sleeping. The number of times you awoke during the night will also be tracked.

This information is really useful and can assist professionals in dealing with your health problem more effectively. However, abnormal sleep patterns can be caused by a variety of different circumstances.

1.6 Revolutionary Pedometer

Some individuals mistake a smartwatch for a fitness tracker. The explanation for this might be that the pedometer in a smartwatch is still the most popular and used function. This tool allows you to calculate your daily step count, which assists you in remaining as fit as possible. It is now feasible to count your steps and create objectives for living a better life thanks to motion sensors. Along with the steps you take, it is critical that you carefully monitor your nutrition to stay fit.

1.7 Blood Oxygen Level

SpO2 sensors detect your blood oxygen saturation the quantity of oxygen in your blood. In reality, most current smartwatches and fitness watches integrate these sensors and analytics. Smartwatch SpO2 sensors function similarly to photoplethysmography (PPG) sensors used to assess heart rate. PPG—or optical heart rate—sensors function by beaming a green light into your skin and measuring the light reflected to establish your heart rate. The difference is that wristwatch SpO2 sensors beam red and infrared light onto your skin instead of green light. The gadgets then employ an algorithm to evaluate your blood oxygen levels depending on how light bounces back.

1.8 Breathing Monitor

The smartwatches have a blood oxygen meter, a heart rate monitor, and meditative breathing functions. The meditative breathing feature is designed to assist the wearer in engaging in breathing exercises for overall well-being. Like Yoga and Meditation practicing people also use it for better result.

1.9 Drinking Water and Handwash Reminder

It is just a customized feature that helps the end users to create alert based on time like hours to consume water and same alert created for hand washing. That helps to create proper motion of the body, decrease the chance of water deficiency, and cleaning hands consider a more common but serious issue because the maximum time an individual occupied with his/her mobile phone, laptop, desktop, elevator, common doors for either pull or push instruction that has to be major involvement of hands, and this might be a major issue of exposure of bacteria. So, handwash reminder is common but really a best feature.

It is the latest trend to have Smartwatch not only it helps individual to get several notifications of different applications like calls, SMS, Twitter, Facebook, LinkedIn, and Email but also it has also some advanced feature that helps to monitor health-related issues like heart rate, calories, footstep count, GPS helps to move outdoor activity like cycling, running and due to water-proof ability make able the consumer to use this watch while swimming. So in short health is a prime concern of every person and this gadget helps to maintain a healthy lifestyle that reduces the chances of getting heart disease to some extent This study integrates the extended technology acceptance model TAM" [12] to examine how consumers evaluate Cloud-BSW on perceived usefulness, Cloud-BSW perceived ease of use and Health Benefits in determining their attitude and behavioral intentions. The data was collected, through an online structured questionnaire, from Delhi NCR consumers who either had used or were likely to use the Cloud-BSW. The research questions include as how do perceived usefulness, perceived ease of use and health benefits influence consumers behavioural attitude and intention to use Cloud-BSW.

2 Literature Review and Hypotheses

2.1 *Technology Acceptance Model (TAM)*

Theory of rational action [13] and planned behaviour [14] are the foundations of TAM [12]. TAM has been extensively used in the study of how individuals and organizations embrace new technologies [8–10]. It has also been used to discover the reasons why consumers are willing to accept new technologies or the motives that drive them. Additionally, TAM is essential in identifying or predicting the system-specific advantages that consumers want in order to maintain a long-term working relationship with a new technology. This is an area in which TAM plays a vital role.

As a result, TAM is well versed in a wide range of technical adoptions in several marketing and retail environments. Examples include internet banking and mobile banking as well as mobile commerce and RFID as well as the usage of augmented reality and electronic labeling. It is thus possible to analyze customer acceptance of the Cloud-BSW by using TAM which has been the prevailing paradigm in many different contexts in recent decades for user acceptance procedures in information systems. Customers' adoption of a new technology is mostly influenced by two factors, namely the perceived utility and perceived ease of use of the new technology, according to TAM, a simple and robust methodology [4]. There are several ways that the Cloud-BSW might be considered valuable by consumers. Cloud-BSW's perceived ease of use relates to how easy it is for consumers to utilize Cloud-BSW technology while using a smartwatch. Research shows that consumers' behaviors are influenced by perceived ease of use directly and indirectly through perceived usefulness [10, 11, 14].

2.2 *Perceived Usefulness*

The term "perceived usefulness" (PU) refers to the degree to which a person feels that he/she would be better off by making use of Cloud-BSW because of the close connection between health app and smart watch. It implies that if the user considers the service to be beneficial, then the service has the potential to improve health performance and productivity, as well as increase the less chance of heart attacks. [24–26]. Related research may be found here: According to Kim, people evaluate the outcomes of their actions and then settle on a decision based on the importance they place on the perceived usefulness of those outcomes [27]. The quantity of information that we get via health application is a significant component that plays a very "significant" role in determining whether a user would accept this service or not. It is a reference to the criteria that consumers might use to determine whether they would get advantages from using smart health watch services in their day-to-day life. People use cloud based technology for a variety of reasons, one of which is that they find this system beneficial for their health data visualization. A system is seen to be helpful only adopters think that there is a positive link between usage and performance on the one hand, and effectiveness and satisfaction on the other side. This is because perceived usefulness plays an important part in adoption of Cloud-BSW. Thus, the authors hypothesize that:

H1: Perceived usefulness (Cloud-BSW-PU) has positive associated with consumer's attitude (Cloud-BSW-ATT) towards Cloud-BSW.

2.3 *Perceived Ease of Use*

Perceived ease of use refers to freedom from difficulty and "effort while using Cloud-BSW services" [29], and complexity were used to develop this definition [30]. When individuals talk about "perceived ease of use," they're talking about how easy it is to utilize a particular piece of technology. Users are able to accept hassle-free technology-based health-related data and related services in daily usage if they have a positive perceived ease of use, also known as PEU [31]. According to the findings of [24], the initial inclination to accept new technology in the smart wearables industry is contingent upon accessibility. When clients are supplied with knowledge about the specifics of the goods or services, the advantages they bring, and the use of advice information, it makes it simpler for them to embrace personalized Cloud-BSW. Users are more likely to gladly embrace cloud-based application mobile services that have a reputation for being simple and straightforward to use. This results in a positive association between the user's desire to use something and their perception of how easy it will be to utilize it [32]. As a result of continued practice, users eventually acquire skill. It is hypothesized as follows:

H2: Perceived ease of use (Cloud-BSW-PEU) has positive associated with consumers' attitude (Cloud-BSW-ATT) towards Cloud-BSW through perceived usefulness (Cloud-BSW PU).

H3: Perceived ease of use (Cloud-BSW-PEU) has positive associated with consumers' attitude (Cloud-BSW-ATT) towards Cloud-BSW.

2.4 Attitude and Behavioral Intention

Attitude is the composite of a consumer's beliefs, feelings, and behavioral intentions towards some product/service within the context of marketing [33]. An individual's general propensity to engage in a particular behavior is also referred to as attitude. This attitude is the product of people's views about their behavior, the consequences of that behavior, and the importance they place on those beliefs. An individual's behavioral intention is defined as the subjective probability that he or she will perform a particular behavior [34], and it is more predictive of actual behavior than attitude [35]. Also, earlier studies have used Behavioral Intention (BI) as the most operated endogenous variable [36]. Behavioral Intention to use a product or a system is influenced by an individual's attitude towards the behavior [37]. Based on the above statements, the study hypothesizes:

H4: Attitude (Cloud-BSW-ATT) has positive associated with consumers' behavioral intentions (Cloud-BSW-BHE) towards Cloud-BSW.

2.5 Health Benefits

In the context of smartwatches, health benefits refer to the degree every individual uses this device to monitor his/her activities. If Cloud-BSW provides a wholesome analytical report that helps the end-user to take appropriate action to reduce that phenomenon, for example, keep an eye on his weight reduction program. Cloud-BSW empowered the individual to take action because they have zero excuses that they do not have their health-related data and it is like a motivation factor also. This motivation leads to a healthy wellness lifestyle. in this research, we look at how consumer acceptance of Cloud-BSW is influenced by health benefits. Customers' perceptions of health benefits are referred to as their "positive effect on his health status" [13]. To put it another way, it symbolizes the health benefits capacity to provide valuable outcomes to its clients and other stakeholders [14]. The significance of health benefit variables in new technology uptake has been studied very little. We hypothesize in this research that people's attitudes and behaviors about Cloud-BSW will be more positive if they interact with Cloud-BSW application with an accurate analysis of their health data. [15].

H5: Health Benefits (Cloud-BSW-HB) positively affect customers' attitude (Cloud-BSW-ATT) towards Cloud-BSW.

2.6 Habit

When an individual does the same activity more frequently that can be observed by the frequency of that activity done by that individual consider habit. It is the pattern of behavior that is considered as a habit. In the context of Cloud-BSW, the individual has to tie this gadget in his hand so the frequency is increased to observe the different ridings provided by that watch. So in this model, hypothetical representation of this latent variable is given below.

H6: Habit (Cloud-BSW-HBIT) positively affects customers' attitude (Cloud-BSW-ATT) towards Cloud-BSW.

2.7 Social Influence

This construct is used in this conceptualized model to provide some of the evidence that how an individual gets influenced by its peer groups that might be his family person, friends, and co-workers, use this Cloud-BSW and peer groups adoption of Smartwatches.

H7: Social Influence (Cloud-BSW-SI) positively affects customers' attitude (Cloud-BSW-ATT) towards Cloud-BSW.

A detailed pictorial representation of hypotheses in the purposed research model is presented in Fig. 3.

3 Methodology

3.1 Sample and Data Collection Procedure

We received a total of 200 legitimate replies from those who responded online. Using the sample size calculator located at [18], it was determined "that the total number of answers" needed for "the model structure was enough" ("the minimum sample recommended for structural equation modeling with 7 latent variables, 18 observed variables, p-level, 0.05, and anticipated size effect,0.3 is 150"). Descriptive findings are given below. In the survey, 70% participants were male and majority of the participants i.e., 80% belonged to the 21–35 age group with 65% graduate degree holder.

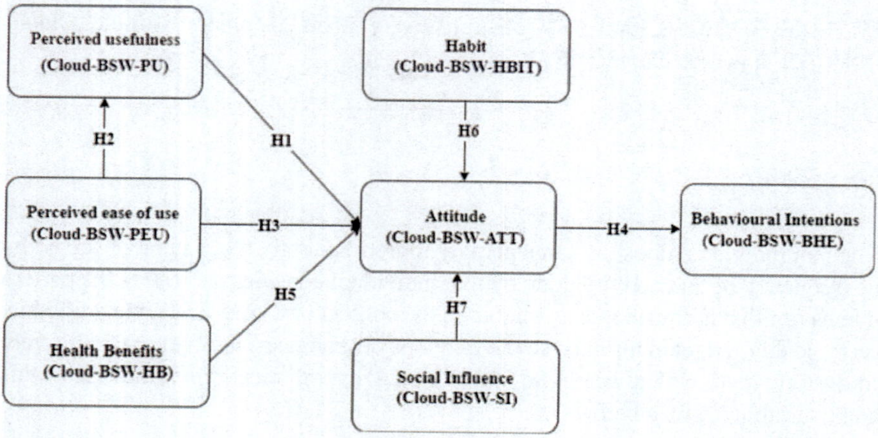

Fig. 3 Purposed research model

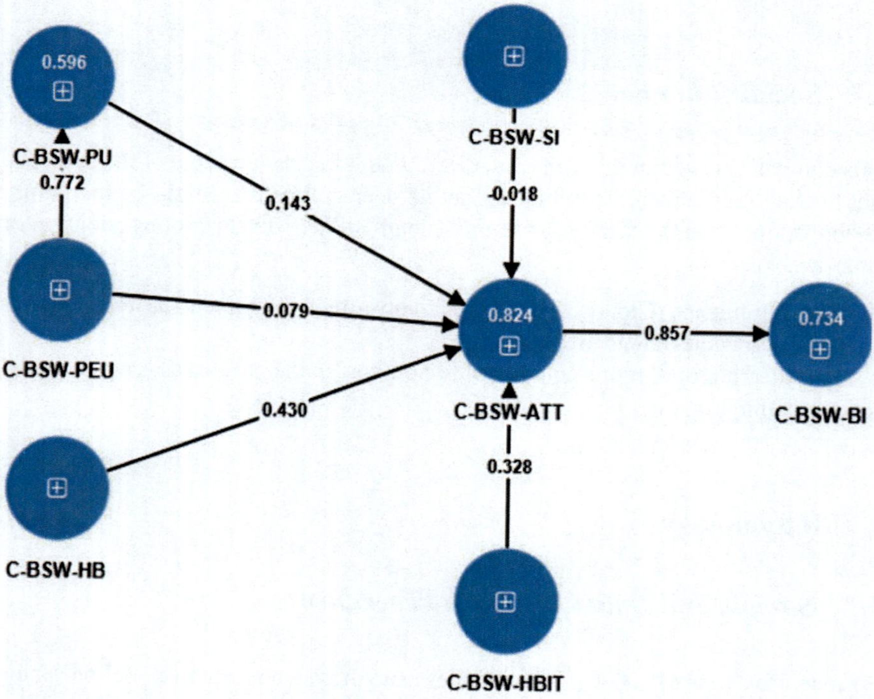

Fig. 4 SEM diagram with path co-efficient and t-values

3.2 Variable Measures and Method of Analysis

The study has used variables and measurement scales as used in TAM framework. Perceived Usefulness, Perceived Ease of Use [16] and Behavioral intentions [11] were measured using 4 item for each construct while three items were used for measuring each of Attitude [17] and Bank Reputation [13] construct. A 5-point Likert scale was used to record respondents' reaction, i.e., 1 is strongly disagree and 5 is strongly agree. It was necessary to employ a multivariate analysis tool since this study dealt with a high number of variables at the same time. In multivariate analysis, a large number of variables are examined concurrently using statistical approaches. These variables are often used to measure various aspects of a person's or a company's actions, as well as many aspects of a specific circumstance. Surveys and observations are widely used to gather primary data for measures. A multivariate methodology called Structural Equation Modeling (SEM) is utilized in this study to integrate factors that cannot be directly seen. Latent variables or constructs fill this void, since they cannot be assessed directly but rather via the use of indicator variables or statements, which are abstract and hence cannot be measured directly. It was necessary to employ a variance-based SEM approach known as PLS-SEM because of the small sample size (Partial Least Square-Structural Equation Modeling). The PLS-SEM is able to handle both reflective and formative measurement models, as well as single-item and multi-item measures. Because the findings of PLS-SEM and CB-SEM tend to be similar, PLS-SEM estimates may serve as a viable substitute for CB-SEM outputs. Non-normality, a small sample size, and the employment of several scaling types are among the most often cited reasons for using PLS-SEM [19, 20, 29]. Using SmartPLS 3, which was available for a 30-day free trial on the official website, the PLS analysis was carried out using the software.

4 Results and Discussion

A two-step reporting method is used in this investigation. First, the measurement model's suitability will be evaluated and second the structural model would be used.

4.1 Assessment of Measurement Model

The model's adequacy is proved by comparing its reliability and convergent and discriminant validity. A reasonable level of reliability is shown in Table 1 by alpha coefficients and composite reliabilities over the bottom limit of 0.7 for all constructs except Behavioral (Cloud-BSW-BHE) and Health Benefits (Cloud-BSW-HB), which do not exhibit any value.

Table 1 Composite reliability, rho AVE & Cronbach alpha

	Cronbach's alpha	rho_A	Composite reliability	Average variance extracted (AVE)
C-BSW-ATT	0.883	0.889	0.927	0.81
C-BSW-BI	0.883	0.89	0.928	0.811
C-BSW-HB	0.907	0.918	0.936	0.786
C-BSW-HBIT	0.868	0.875	0.919	0.79
C-BSW-PEU	0.934	0.934	0.953	0.834
C-BSW-PU	0.944	0.944	0.964	0.9
C-BSW-SI	0.887	0.89	0.922	0.748

Source Output from smart PLS v4.0

An indication of convergent validity is the square root of the average variance extracted (AVE), which measures the shared variance between a concept and its indicators. Because each concept has an AVE of at least 0.5, convergent validity may be demonstrated. Each item inside a construct may be evaluated for its discriminant validity by comparing the high loadings it has on the construct it aims to measure with the cross-loadings, which must be lower. Table 2 displays the factors' cross-loading. The Path Coefficient and R-squared are shown in Table 3.

Several major results from this research support the suggested SRT integrated framework. As part of this study, researchers used PLS-SEM to examine the major impacts and configurational combination of variables that affect customer approval of the C-BSW. Customers' attitudes and behavior towards C-BSW are influenced by their perceptions of the ease of use, usefulness, and reputation of the service. The summary of results of hypotheses is provided in Table 4.

5 Managerial Implications

The study provides with several ramifications for retail branch banking managers and operators. When it comes to consumer attitudes and behavior, TAM characteristics like perceived ease of use and perceived usefulness play a crucial role. So, health benefits should focus on adopting user-friendly and simple-to-use smart technologies. Customers' irritation and pain may be minimized and their demands be met more effectively through smart technologies which are easy to use and have useful functions. For retail customers/users, Cloud-BSW technology enables them to personalize their interactions, which results in better health monitoring that increases individual performance to achieve their personal journey for betterment of health [21–23]. Cloud-BSW should be adopted more easily if retail outlets give the required assistance to alleviate pain and uneasiness and improve the degree of optimism. Customer perceptions of Cloud-BSW application that capture the periodic records of an individual.

Table 2 Factor cross-loadings

	C-BSW-ATT	C-BSW-BI	C-BSW-HB	C-BSW-HBIT	C-BSW-PEU	C-BSW-PU	C-BSW-SI
C-BSW-ATT1	0.896	0.756	0.724	0.749	0.709	0.681	0.679
C-BSW-ATT2	0.892	0.712	0.745	0.726	0.631	0.716	0.671
C-BSW-ATT3	0.911	0.836	0.891	0.836	0.814	0.765	0.797
C-BSW-BI1	0.734	0.867	0.694	0.766	0.695	0.690	0.716
C-BSW-BI2	0.839	0.938	0.833	0.837	0.850	0.717	0.779
C-BSW-BI3	0.736	0.896	0.773	0.781	0.803	0.734	0.770
C-BSW-HB1	0.834	0.789	0.934	0.829	0.730	0.750	0.794
C-BSW-HB2	0.801	0.782	0.904	0.782	0.764	0.802	0.764
C-BSW-HB3	0.813	0.782	0.929	0.771	0.736	0.794	0.771
C-BSW-HB4	0.658	0.668	0.769	0.649	0.721	0.558	0.664
C-BSW-HBIT1	0.765	0.860	0.787	**0.879**	0.797	0.706	0.784
C-BSW-HBIT2	0.829	0.772	0.800	**0.905**	0.710	0.763	0.771
C-BSW-HBIT3	0.686	0.719	0.691	**0.883**	0.704	0.617	0.697
C-BSW-PEU1	0.700	0.774	0.743	0.758	0.902	0.698	0.742
C-BSW-PEU2	0.736	0.761	0.753	0.733	0.919	0.728	0.735

(continued)

Table 2 (continued)

	C-BSW-ATT	C-BSW-BI	C-BSW-HB	C-BSW-HBIT	C-BSW-PEU	C-BSW-PU	C-BSW-SI
C-BSW-PEU3	0.751	0.825	0.743	0.764	0.915	0.667	0.730
C-BSW-PEU4	0.745	0.823	0.791	0.774	0.918	0.727	0.799
C-BSW-PU1	0.762	0.742	0.794	0.738	0.721	0.943	0.765
C-BSW-PU2	0.756	0.768	0.772	0.744	0.732	0.957	0.727
C-BSW-PU3	0.765	0.743	0.781	0.756	0.744	0.946	0.723
C-BSW-SI1	0.642	0.673	0.650	0.670	0.700	0.698	0.807
C-BSW-SI2	0.679	0.710	0.733	0.705	0.625	0.603	0.840
C-BSW-SI3	0.732	0.757	0.766	0.781	0.771	0.692	0.903
C-BSW-SI4	0.709	0.756	0.771	0.766	0.745	0.700	0.905

Source Output from smart PLS v4.0

Table 3 Path co-efficient and R-squared

	Original sample (O)	T statistics (|O/STDEV|)	P values		R-square adjusted
C-BSW-ATT - > C-BSW-BI	0.857	53.875	0.000	C-BSW-ATT	0.820
C-BSW-HB - > C-BSW-ATT	0.430	6.130	0.000	C-BSW-BI	0.733
C-BSW-HBIT - > C-BSW-ATT	0.328	4.743	0.000	C-BSW-PU	0.595
C-BSW-PEU- > C-BSW-ATT	0.079	1.422	0.155		
C-BSW-PEU - > C-BSW-PU	0.772	31.370	0.000		
C-BSW-PU- > C-BSW-ATT	0.143	2.747	0.006		
C-BSW-SI- > C-BSW-ATT	-0.018	0.292	0.770		

Source Output from Smart PLS v4.0

6 Limitations and Scope for Further Research

There are several drawbacks to this study. First, because the structural model's strongest coefficients are rather low, additional latent factors such as social influence is not contribute significant impact, which was not incorporated in the model and may affect continuous usage intention of smartwatches, need to be examined. Other constructs can also be included such as price, hedonic motivation, flow, and facilitating condition. As a result, such aspects will be studied further. Second, we only evaluated the model with smartwatches; thus, future study might test the model with additional wearable devices (e.g., smart glasses) or even mobile devices (e.g., smartphones) [30–34] to assess device continuous usage intention and boost generalizability. Furthermore, future research should look at the growing importance of smartwatches in gaming and entertainment, as well as methods for this omnipresent media to be used more effectively as a new platform in recreational elements of everyday life. Third, the study does not give empirical insights into smartwatch design requirements. As a result, further study into the size, distinctiveness, and design of smartwatches from both present and prospective user viewpoints is required.

Table 4 Summary of Hypotheses Testing

Symbol		Hypotheses	Result
H1	C-BSW-PU - > C-BSW-ATT	Perceived usefulness (Cloud-BSW-PU) has positive associated with consumer's attitude (Cloud-BSW-ATT) towards Cloud-BSW."	Supported
H2	C-BSW-PEU - > C-BSW-PU	Perceived ease of use (Cloud-BSW-PEU) has positive associated with consumers' attitude (Cloud-BSW-ATT) towards Cloud-BSW through perceived usefulness (Cloud-BSW PU)	Supported
H3	**C-BSW-PEU - > C-BSW-ATT**	Perceived ease of use (Cloud-BSW-PEU) has positive associated with consumers' attitude (Cloud-BSW-ATT) towards Cloud-BSW	Rejected
H4	C-BSW-ATT - > C-BSW-BI	Attitude (Cloud-BSW-ATT) has positive associated with consumers' behavioural intentions (Cloud-BSW-BHE) towards Cloud-BSW	Supported
H5	C-BSW-HB - > C-BSW-ATT	Health Benefits (Cloud-BSW-HB) positively affects customers' attitude (Cloud-BSW-ATT) towards Cloud-BSW	Supported
H6	C-BSW-HBIT - > C-BSW-ATT	Habit (Cloud-BSW-HBIT) positively affects customers' attitude (Cloud-BSW-ATT) towards Cloud-BSW	Supported
H7	**C-BSW-SI - > C-BSW-ATT**	Social Influence (Cloud-BSW-SI) positively affects customers' attitude (Cloud-BSW-ATT) towards Cloud-BSW	Rejected

7 Conclusion

Using smartwatches for health and fitness monitoring has the potential to drive people to live a healthier lifestyle, diagnose chronic illness in its early stages, and eventually enhance their overall health. By using perceived vulnerability and severity as moderators, this study fully analyzed factors that impact consumers' propensity to use smartwatches for personal fitness and health monitoring reasons [35–40]. The findings revealed that behavioral intention to use smartwatches for health and fitness monitoring is positively influenced by performance expectation, effort expectancy, enabling conditions, and hedonic incentive. Furthermore, the findings show that perceived vulnerability mitigates the influence of effort expectation on behavioral intention. Perceived severity reduces the impact of social influence while increasing the impact of hedonic incentive on behavioral intention to use smartwatches for health and fitness monitoring. This study's findings not only add to the literature on smartwatch adoption but also give significant information for smartwatch technology developers, marketers, and managers to understand the elements that may inspire users to embrace smartwatches for health and fitness monitoring reasons.

References

1. https://www.hindustantimes.com/lifestyle/health/sidharth-shukla-to-kk-10-celebs-who-died-of-heart-attack-101660140071975.html
2. https://indianexpress.com/article/explained/mumbai-heart-attack-rise-explained-7984371
3. Tsao CW, Aday AW, Almarzooq ZI, Beaton AZ, Bittencourt MS, Boehme AK et al (2022) Heart disease and stroke statistics—2022 update: A report from the American heart association. Circulation 145(8):e153–e639
4. Fryar CD, Chen T-C, Li X (2012) Prevalence of uncontrolled risk factors for cardiovascular disease: United States, 1999–2010 [PDF-494K]. NCHS data brief, no. 103. Hyattsville, MD: National Center for Health Statistics. Accessed May 9 2019
5. McIntyre A (2014) Wearable computing in the workplace to be dependent on apps and services. Forbes, pp 19–24. Available at www.forbes.com/sites/gartnergroup/2014/03/06/wearable-computing-in-the-workplace-to-be-dependent-on-apps-and-services. Accessed 15 Mar 2015
6. Kim KJ, Shin DH (2015) An acceptance model for smart watches: implications for the adoption of future wearable technology. Int Res 25(4):527–541
7. Choi J, Kim S (2016) Is the smartwatch an IT product or a fashion product? A study on factors affecting the intention to use smartwatches. Comput Hum Behav 63:777–786
8. Chuah SHW, Rauschnabel PA, Krey N, Nguyen B, Ramayah T, Lade S (2016) Wearable technologies: the role of usefulness and visibility in smartwatch adoption. Comput Hum Behav 65:276–284
9. Hsiao KL (2017) What drives smartwatch adoption intention? Comparing Apple and non-Apple watches. Library Hi Tech 35(1):186–206
10. Chae JM (2009) Consumer acceptance model of smart clothing according to innovation. Int J Hum Ecol 10(1):23–33
11. Kalantari M Consumers' adoption of wearable technologies: literature review, synthesis, and future research agenda. Forthcom Int J Technol Mark
12. Brodski S, Desmangles L, Fanfarillo S, Khodabandeh S, Palumbo S, Santinelli M (2020) What does personalization in banking really mean? BCG Global. https://www.bcg.com/publications/2019/what-does-personalization-banking-really-mean. Accessed 14 Apr 2022
13. Ghosh I, Deshmukh A, Conte N (2020) AIoT: When artificial intelligence meets internet of things. Visual Capitalist. https://www.visualcapitalist.com/aiot-when-ai-meets-iot-technology/ . Accessed 15 Apr 2022
14. Lin Y, Chuang C, Yen C, Huang S, Huang P, Chen J, Lee S (2019) Artificial intelligence of things wearable system for cardiac disease detection. In: 2019 IEEE International Conference on Artificial Intelligence Circuits and Systems (AICAS)
15. Davis F (1989) Perceived usefulness, perceived ease of use, and user acceptance of information technology. MIS Quarterly 13(3):319. Available https://doi.org/10.2307/249008
16. Damanpour F (1991) Organizational innovation: a meta-analysis of effects of determinants and moderators. Acad Manag J 34(3):555–590. Available https://doi.org/10.5465/256406
17. Fishbein M, James J, Andrew R, Ajzen I, Barbara L (1980) Predicting and understanding family planning behaviors. In: Understanding attitudes and predicting social behavior. Accessed 15 Apr 2022

18. Sok J, Borges J, Schmidt P, Ajzen I (2020) Farmer behaviour as reasoned action: a critical review of research with the theory of planned behaviour. J Agric Econ 72(2): 388–412. Available https://doi.org/10.1111/1477-9552.12408
19. Legris P, Ingham J, Collerette P (2003) Why do people use information technology? A critical review of the technology acceptance model. Inf Manag 40(3):191–204
20. Reinders MJ, Frambach R, Kleijnen M (2015) Mandatory use of technology-based selfservice: does expertise help or hurt? Eur J Mark 49(1/2):190–211
21. Hernandez B, Jimenez J, Jose Martin M (2009) Adoption vs acceptance of e-commerce: two different decisions. Eur J Mark 43(9/10):1232–1245
22. Gao L, Bai X (2014) A unified perspective on the factors influencing consumer acceptance of internet of things technology. Asia Pac J Mark Logist 26(2):211–231
23. Garaus M, Wolfsteiner E, Wagner U (2016) Shoppers' acceptance and perceptions of electronic shelf labels. J Bus Res 69(9):3687–3692
24. Bao Y, Bao Y, Sheng S (2011) Motivating purchase of private brands: effects of store image, product signatureness, and quality variation. J Bus Res 64(2):220–226
25. Lee K, Shavitt S (2006) The use of cues depends on goals: store reputation affects product judgments when social identity goals are salient. J Consum Psychol 16(3):260–271
26. Boulding W, Kirmani A (1993) A consumer-side experimental examination of signalling theory: do consumers perceive warranties as signals of quality? J Consum Res 20(1):111–123
27. Venkatesh V, Davis FD (2000) A theoretical extension of the technology acceptance model: four longitudinal field studies. Manag Sci 46(2):186–204
28. Weijters B, Rangarajan D, Falk T, Schillewaert N (2007) Determinants and outcomes of customers' use of self-service technology in a retail setting. J Serv Res 10(1):3–21
29. Soper DS (2014) A-priori sample size calculator for structural equation models. Available from http://www.danielsoper.com/statcalc
30. Hair JF, Sarstedt M, Pieper T, Ringle CM (2012) The use of partial least squares structural equation modeling in strategic management research: A review of past practices and recommendations for future applications. Long Range Plan 45:320–340
31. Henseler J, Ringle CM, Sinkovics RR (2009) The use of partial least squares path modeling in international marketing. Adv Int Mark 20:277–320
32. Inman JJ, Nikolova H (2017) Shopper-facing retail technology: a retailer adoption decision framework incorporating shopper attitudes and privacy concerns. J Retail 93(1):7–28
33. Pantano E, Rese A, Baier D (2017) Enhancing the online decision-making process by using augmented reality: a two-country comparison of youth markets. J Retail Consum Serv 38:81–95
34. Wünderlich NV, Heinonen K, Ostrom AL, Patricio L, Sousa R, Voss C, Lemmink JG (2015) "Futurizing" smart service: implications for service researchers and managers. J Serv Mark 29(6/7):442–447
35. Ramdhony D, Munien S (2013) An investigation on mobile banking adoption and usage: A case study of Mauritius Ramdhony & Munien. World J Soc Sci 3(3):197–217
36. Rao S, Troshani I (2007) A conceptual framework and propositions for the acceptance of mobile services. J Theor Appl Electron Com-Merce Resarch 2(2):61–73
37. Sangle PS, Awasthi P (2011) Consumer's expectations from mobile CRM services: a banking context. Bus Process Manag J 17(6):898–918
38. Kim HW, Xu Y, Koh J (2004) Comparison of online trust building factors between potential customers and repeat customers. J Assoc Inf Syst 5(10):392–420
39. Jeong BK, Yoon TE (2013) An empirical investigation on consumer acceptance of mobile banking services. Bus Manag Res 2(1):31–40
40. Davis FD (1989) Perceived usefulness, perceived ease of use, and user acceptance of information technology. MIS Q 13(3):319–340

Comparison of ML Algorithms for Heart Disease Detection

Abhinav Kulshreshth, Mahima Yadav, and Ganga Sharma

1 Introduction

Heart is one of the most complicated organs of a living being. It is very crucial for the cardiovascular system to work properly so that our body and mind could work appropriately. Therefore, it is important for a heart diagnosis to be accurate. While doctors are accurate most of the time, patients still go for a second opinion because a false positive may not just cause monetary issue but may also result in some kind of side effect. Using Machine Learning may solve the issue as it may become the easier method to ascertain a heart disease in a patient. Machine Learning has already become a very important part in healthcare sector with algorithms for cancer detection, medical diagnostic and even drug development.

Prediction of a cardiovascular disease can be done using variables like Cholesterol, Blood Pressure (before and after exercise), Exercise induced angina and even X-rays. Algorithms such as Random Forest, Support Vector Machines, Neural Networks, Naïve Bayes, Logistic Regression and Multi-Layer Perceptron will be used for classification on a set of variables some of which have already been stated. Waikato Environment for Knowledge Analysis (WEKA) software can also be used to perform Data Mining techniques on the data [1].

1.1 Heart

Heart diseases are one of the most prominent causes of death. According to the Indian Medical Association (IMA), 30 out of 50 deaths are caused due to heart diseases.

A. Kulshreshth (✉) · M. Yadav · G. Sharma
SCSE, Galgotias University, Greater Noida 201310, India
e-mail: kulshreshthabhinav@gmail.com

A. Ramdane-Cherif et al. (eds.), *Machine Intelligence and Data Science Applications*, Algorithms for Intelligent Systems, https://doi.org/10.1007/978-981-99-1620-7_17

There are variety of reasons which can cause a heart ailment. Some of the causes are as follows:

1. High Cholesterol
2. High Blood Pressure
3. Diabetes
4. Obesity

Medical Associations such as WHO and IMA collect data on different health issues such as heart attack and cancer. These datasets are then used to study the issue and used for further studies pertaining to the specific illness.

1.2 *Machine Learning*

Machine Learning is a computer-based technique which can be used to teach pattern recognition to a brainless device. This is used to create an intelligent system and is used in various industries such as Robotics, Data Science and Analysis, Computer Vision, Natural Language Processing (NLP) and Recommendation systems. The objective of a machine learning model is to minimize the error rate while predicting or classifying for a problem. A model can be created by using a dataset and running it through different algorithms and tuning the parameters to achieve the objective.

Machine Learning techniques can be classified into three types.

Supervised Machine Learning. When the model is trained on data which is labelled, it is termed as Supervised Machine Learning[2]. Classification and Regression models can be trained using supervised learning. Random Forest [3], Support Vector Machines (SVM) [4], Linear Regression and Logistic Regression are some of the algorithms which can be used for the same purpose.

Unsupervised Machine Learning. When the model is trained on a data which does not have labels, it is called Unsupervised Machine Learning [5]. Machines have to learn to categorize the data based on patterns. Clustering is one of the techniques performed in unsupervised machine learning. It can be used for fraud detection and recommendation system.

Reinforcement Machine Learning. When the model does not use labelled dataset and it does not use data to train but learns using experience, it is called Reinforcement Learning [6]. This type of technique is used for automating games and robots along with creation of self-driving cars and industry automation.

For the purpose of Heart Disease Detection, classification algorithms have to be used which are a part of supervised machine learning.

Section 2 of this paper focuses on the Datasets and the variables used for the training and testing purposes. Section 3 focuses on the techniques and algorithms used by the researchers along with a comparison. Section 4 focuses on the results such as accuracy and precision obtained by the techniques.

2 Datasets Used

Databases are the heart of a machine learning model. For a heart disease detection model, database can be simple tabular data and also an image database such as ECG reports. The datasets used for the research are the Cleveland Dataset [7], People's Hospital Dataset [8], South African Heart Disease Dataset [9] and a custom Indian Dataset [10].

Figure 1 shows that Cleveland Dataset is the most popular dataset. It is a part of the UCI Heart Disease Repository which was originally created in 1988 by creators from Hungary, Switzerland, and the USA. The UCI Repository contains 303 records along with 75 attributes while the Cleveland Dataset contains only 14 of the 75 attributes with 303 records. Table 1 shows the attributes of the Cleveland Dataset.

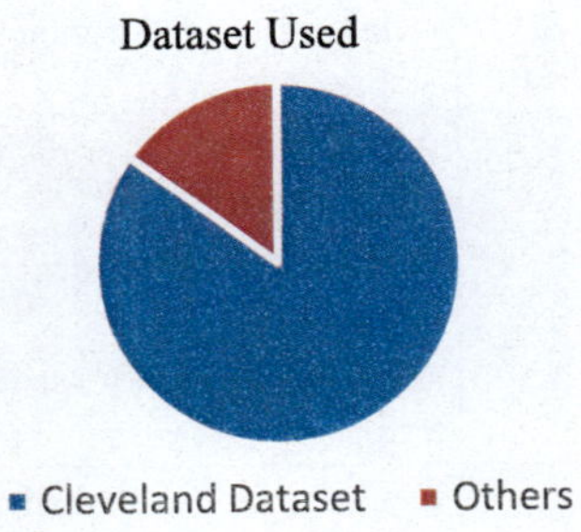

Fig.1 Pie chart showing comparison between cleveland dataset and others'

Table 1 Attributes for cleveland dataset

Sr. no	Attributes
1	Age
2	Resting blood pressure
3	Cholesterol
4	Maximum heart rate achieved
5	Depression
6	Sex
7	Chest pain type
8	Fasting blood sugar
9	Rest ECG
10	Slope
11	Number of major vessel
12	Thalassemia
13	Exercise induces angina
14	Condition (Target)

The PKU People's Hospital Dataset contains more than a million records of inpatients. It contains different types of data ranging from the Basic information to physical diagnosis to medical records, medical images, and diagnosis. Table 2 describes the PKU People's Hospital Dataset.

The South African Heart Dataset is a subset of the dataset collected by Rossouw et al. [9] in 1983. It was further replicated by author[11] in 2016 and it contains 9 attributes. Table 3 shows the attributes and their descriptions for the SA Heart dataset.

The Indian Heart Dataset was created by collecting data from various hospitals and medical practitioners. A total of 214 records were collected with 19 attributes having symptoms and test records. Table 4 shows the attributes of the Indian Heart Dataset.

Table 2 Attributes for PKU people's hospital dataset

Sr. no	Data type	Attributes
1	Basic information	User ID, Gender, Age
2	Physical information	Physical testing, Timestamp
3	Medical testing	389 Attributes such as blood glucose and calcium
4	Medical imaging	47 Attributes such as ECG and Ultrasound cardiogram
5	Diagnosis	729 Attributes containing diseases such as hypertension, diabetes and insomnia

Table 3 Attributes for South Africa heart disease dataset

Sr. no	Attributes
1	Sbp: Systolic blood pressure
2	Tobacco: Cumulative tobacco (kgs)
3	Ldl: Low density lipoprotein cholesterol
4	Adiposity
5	Famhist: Family history of heart disease
6	Typea: Type A behaviour
7	Obesity
8	Alcohol: Alcohol consumption
9	Age
10	Label

Table 4 Attributes for Indian heart dataset

Sr. no	Data type	Attributes
1	Symptoms	Such as Chest pain, Arm pain and Breathlessness
2	Test results	Such as ST Elevation, ST Depression and T-Elevation
3	Target	Label

3 Related Work

Methodologies refer to the methods and algorithms used by the researchers. Feature Engineering, Feature Selection are also part of this section.

Ghumbre et al. [10] did not use any of the popular datasets but used a dataset that they curated on their own. Their dataset consisted of 19 attributes and 214 records. Out of the 214 records, 78 were normal while the rest had some kind of heart disease. They used multiple algorithms, namely, Support Vector Machines (SVM), Multilayer Perceptron (MLP), Radial Basis Function Neural Network (RBF) [12], BayesNet, J48 and Rule algorithm [13]. They validated all the algorithms with fivefold and tenfold cross validations. They concluded that SVM was the overall best algorithm on validation though it can be seen that while most of the algorithms performed better with tenfold cross validation, SVM's performance dipped slightly with the change.

Patel et al. [14] used the complete UCI Heart Disease Dataset with 76 attributes. They implemented multiple tree-based algorithms such as J48 [15], Logistic Model Tree and Random Forest algorithm using WEKA tool. Their research shows that the Linear Tree Model (LTM) provided 0 (zero) as training error but the J48 algorithm provided the least testing error of 0.166. Thus, the J48 algorithm was the most accurate of the bunch with 56.76% and was chosen to be the best.

Chaki et al. [16] selected the Cleveland Dataset for training and testing purpose of their model. They also used the WEKA software to train the machine learning model such as C4.5, Naïve Bayes and Support Vector Machine (SVM). Using the tenfold cross validation, they trained all the models and SVM achieved the best accuracy of 84.12% followed by Naïve Bayes at 83.5% and finally the C4.5 algorithm with 77.56%.

Xu et al. [8] used the Cleveland dataset as well as the People's Hospital dataset to train and test their algorithms. They first performed feature engineering like data normalization with z-score and processed the text for the People's Hospital dataset. Then they performed feature selection on the engineered data. They combined the Best-First Search algorithm [17] with the CFS Subset evaluation [18] to choose the best features for training the model. Finally, they used Random Forest technique for classification and found out that it was the most accurate with 91.6% on the Cleveland Dataset while it scored 97% on People's Hospital dataset but was topped by SVM at 98.9%.

Arabasadi et al. [19] worked with the widely used Cleveland Dataset and applied the Neural Networks along with Genetic Algorithm. They applied four algorithms such as Gini index[20], information gain [20], weight by SVM [21], Principal

Component Analysis (PCA) [22], for selecting the best features. They proposed a method which used Genetic algorithm for weight initiation of Neural Network. They then tested with multiple number of hidden neurons to find the best hyperparameter. Neural Network with 5 hidden neurons performed the best with minimum RMSE score for the train set with 0.257. With the genetic algorithm they were able to score 93.85% in accuracy which was much more than the classic Neural Net.

Saboji et al. [23] used Apache Spark to create the pipeline and the model for classification. They optimized a Random Forest with depth as 4, Gini for removing impurity and number of trees as 3. They observed that a maximum accuracy of 98% can be achieved with 600 records.

Gavhane et al. [24] proposed a machine learning application which trained on the Cleveland dataset. They created a simple Multilayer Perceptron (MLP) which resulted in a 91% precision.

Senthilkumar et al. [25] proposed a method of Hybrid Random Forest with Linear Model (HRFLM) along with algorithms for feature selection. On comparing the proposed algorithm with Naïve Bayes, Generalized Linear Model, Random Forest and Support Vector Machines, HRFLM was found to be working more accurately at 88.4%.

Author Santhan Krishnan and Geetha [26] tested two supervised machine learning algorithms, Decision Trees and Naïve Bayes for the classification problem. Their tests resulted in Decision Trees achieving the better accuracy between the two with 91% while the Naïve Bayes scored 87%.

Authors Kannan and Vasanthi [27] examined four supervised machine learning algorithms to predict the presence of a heart disease. The algorithms, Random Forest, Logistic Regression, Gradient Boosting and SVMs were trained and tested on the Cleveland dataset. Their evaluation of the algorithms resulted in Logistic Regression achieving the best accuracy of 86.5% followed by Gradient Boosting, Random Forest and SVM.

Gonsalves et al. [28] during their research tried three supervised algorithms namely, Naïve Bayes, C4.5 or the J48 algorithm and Support Vector Machine on the South African Heart Disease Dataset. They evaluated their algorithms using confusion matrix which resulted in Naïve Bayes achieving the best accuracy of 71.6% while the SVM achieved 71% and the C4.5 algorithm achieved 70.7%.

Authors Buettner and Schunter [29] created a Random Forest for the prediction of heart disease. They trained and tested the model with a tenfold cross validation and also without it to reach a better conclusion. The algorithm with the tenfold achieved 84.448% accuracy while without the tenfold, it achieved 82.895% which is 1.553% lower.

Rahma Atallah and Amjed Al-Mousa [30] created a Hard Voting Ensemble Method [31] consisting of four machine learning algorithms, namely, Stochastic Gradient Descent (SGD), K Nearest Neighbour (KNN), Random Forests and Logistic Regression. They used GridSearchCV [32] on all the algorithms for attaining the best result. The SGD predicted with 88% accuracy, KNN, Random Forest and Logistic with 87%. On creating an ensemble method with hard voting, they were able to predict with 90% accuracy.

Shah et al. [33] compared and tested four algorithms such as Naïve Bayes, KNN, Decision Tree and Random Forest. On testing these algorithms, he concluded that KNN performed better than other algorithms. It scored 90.789% with Naïve Bayes scoring 88.157% and Decision Tree and Random Forest following at 80.263 and 86.84%.

Rajdhan [34] in his paper trained four algorithms on the Cleveland dataset. The algorithms were Random Forest, Logistic Regression, Decision Tree and Naïve Bayes. They were trained and tested against the Cleveland Dataset. Random Forest recorded the best accuracy with 90.16% followed by Logistic Regression and Naïve Bayes at 85.25% and Decision Tree with 81.97%.

Li et al. [35] proposed a system consisting of Feature selection, hyperparameter tuner and a model for predicting the presence of a heart disease. They used multiple state-of-the-art feature selector such as Relief [36], mRMR [37], LASSO [38] and LLBFS [39] and also proposed a new feature selector Fast Conditional Mutual Information (FCMIM). They tested the different algorithms such as Logistic Regression, SVM, KNN, Artificial Neural Network (ANN), Naïve Bayes and Decision Trees. Logistic Regression and SVM produced the best accuracies with and without feature selection with SVM scoring 92.37% with the proposed FCMIM selection.

Yadav [40] used the Cleveland dataset to predict the presence of heart disease. They tested K-Means, KNN, Naïve Bayes, Logistic Regression, Fuzzy KNN, Back Propagated Neural Network and an ensemble of K-Means and Naïve Bayes. K-Means scored the least with 50.83% while Back Propagated Neural Net and Fuzzy KNN scoring the most with 98 and 94.19%, respectively.

Bharti [41] performed three different testing with six algorithms namely, Logistic Regression, Decision Tree, SVM, Random Forest, K Neighbours and Deep Learning approach. They first tested these algorithms without any feature selection or outlier detection which resulted in SVM achieving the highest accuracy of 84.09%. Then the algorithms were tested with feature selection but no outlier detection. This resulted in Random Forest securing the best accuracy of 88%. The final approach had feature selection and outlier detection done before training the algorithms. Deep Learning was the most accurate algorithm with 94.2% accuracy.

4 Proposed Model

The proposed model uses the Cleveland Dataset [7] with 303 records and an 80:20 split for training and testing purposes. Multiple feature selection algorithms have been applied as a voting ensemble which produced a list of top 8 features. New-age technology of H2O.ai [42] has been used for automating the model selection process which used the top 8 features selected. The resulting model, Gradient Boosting Classifier [43], was selected as the final model for the project. On testing the model, the accuracy was 93%, recall was 94.2% while the precision was 94%.

5 Result and Conclusion

Multiple research projects have been performed in order to create or find the best algorithm that can be used to predict the presence of a heart disease more accurately. This has been done using various datasets such as the Cleveland dataset, South African heart health dataset. Feature selection algorithms have been performed by some, while others relied on just data cleaning and classic algorithms. Researchers have created new algorithms and used software like WEKA for data mining and used Python [44] or R [45] language for creating models.

Table 5 summarizes the algorithms created by researchers for their papers along with the accuracies for their models. It can be seen that SVM algorithm created by Shan Xu [8] scored the highest accuracy of all the algorithms when trained on People's Hospital dataset. Random Forest by Saboji [23] and BP-NN by Samir Yadav [40] scored the highest accuracies when trained on the Cleveland Dataset.

On comparison of accuracy of popular algorithms with and without Feature Selection, it can be seen that only Logistic Regression and Support Vector Machines were the only ones to get a better accuracy when used with a Feature Selection algorithm.

Table 5 Best algorithms and their accuracies

Ghumbre et al. [10]	Custom dataset	SVM	85.05
Jaymin Patel et al. [14]	Cleveland dataset	J48	56.76
Chaki et al. [16]	Cleveland dataset	SVM	84.12
Xu et al. [8]	Cleveland dataset	Random forest	91.6
	People's hospital dataset	SVM	98.9
Arabasadi et al. [19]	Cleveland dataset	Genetic algorithm + Neural network	93.85
Saboji et al. [23]	Cleveland dataset	Random forest	98
Senthilkumar et al. [25]	Cleveland dataset	HRFLM	88.4
Santhana Krishnan [26]	Cleveland dataset	Decision tree	91
Kannan and Vasanthi[27]	Cleveland dataset	Logistic regression	87
Gonsalves et al. [28]	South African heart disease dataset	Naïve bayes	71.6
Buettner and Schunter[29]	Cleveland dataset	Random forest	84.4
Rahma Atallah and AmjedAl-Mousa[30]	Cleveland dataset	Hard voting ensemble method	90
Shah et al. [33]	Cleveland dataset	KNN	90.79
Rajdhan et al. [34]	Cleveland dataset	Random forest	90.16
Li et al. [35]	Cleveland dataset	SVM	92.37
Yadav et al. [40]	Cleveland dataset	BP-NN	98
Bharti et al. [41]	Cleveland dataset	Deep learning	94.2

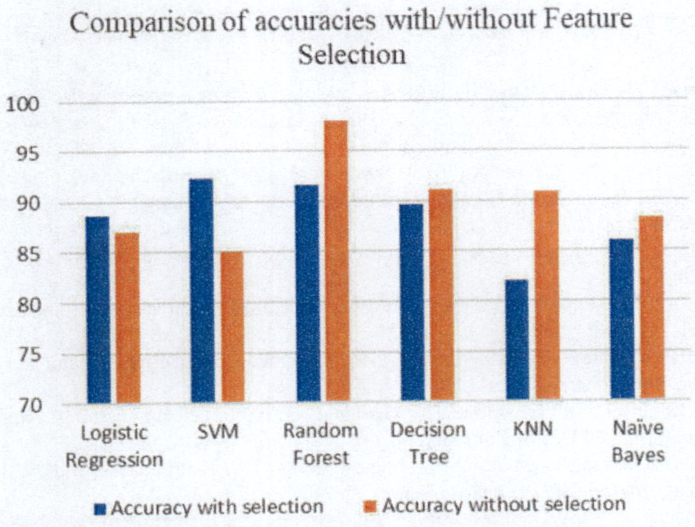

Fig. 2 Comparison of accuracies of algorithms with/without feature selection

Rest of the algorithms scored less when used with a feature selection algorithm. Figure 2 shows the comparison of the algorithms with/without feature selection.

There have been many attempts at creating a good and a useful machine learning model. Many algorithms with different techniques for feature engineering and feature selection have been used for predicting the presence of a heart disease. Back-propagated Neural Network without any feature selection was the most accurate model of all scoring 98%. It can also be seen Neural Networks and Deep Learning algorithms scored between 85 and 90% but it can be increased to more than 95% if used with feature engineering, selection, a larger dataset and hyperparameter tuning. New-age software such as H2O.ai [42] and Google's AutoML [46] can also be used for model selection and creation purposes.

In the future, other technologies such as Google AutoML could be used to enhance the model selection process for the proposed model. Exploration of feature selection algorithms is also an action that can be used to enhance the overall quality of the proposed model. Future scope will also include increasing the size of the dataset used for training and testing purposes.

Acknowledgements We would like to thank God for their blessings throughout our research to complete our paper. Authors Abhinav Kulshreshth and Mahima Yadav would like to extend their gratitude towards their guide Dr. Ganga Sharma, Associate Professor, School of Computer Science Engineering, Galgotias University, Greater Noida, for her invaluable insights and guidance for this research. We would also extend our heartful gratitude to our family and friends for their support and love during the complete period.

References

1. Holmes G, Donkin A, Witten IH WEKA: a machine learning workbench. In: Proceedings of ANZIIS '94—Australian New Zealnd Intelligent Information Systems Conference. IEEE, pp 357–361
2. Singh A, Thakur N, Sharma A (2016) A review of supervised machine learning algorithms. In: 2016 3rd International Conference on Computing for Sustainable Global Development (INDIACom). pp 1310–1315
3. Breiman L (2001) Random forests. Mach Learn 45:5–32. https://doi.org/10.1023/A:1010933404324
4. Zhang X (2017) Support vector machines. Encyclopedia of machine learning and data mining. Springer, US, Boston, MA, pp 1214–1220
5. Gentleman R, Carey VJ (2008) Unsupervised machine learning. Bioconductor Case Studies. Springer, New York, New York, NY, pp 137–157
6. Sutton RS (1992) Introduction: The challenge of reinforcement learning. reinforcement Learning. Springer, US, Boston, MA, pp 1–3
7. Janosi A, Steinbrunn W, Matthias P, Detrano R (1988) UCI Heart Disease Dataset. https://archive.ics.uci.edu/ml/datasets/heart+disease
8. Xu S, Zhang Z, Wang D, et al. (2017) Cardiovascular risk prediction method based on CFS subset evaluation and random forest classification framework. In: 2017 IEEE 2nd International Conference on Big Data Analysis (ICBDA). IEEE, pp 228–232
9. Rossouw JE, Du Plessis JP, Benadé AJ et al (1983) Coronary risk factor screening in three rural communities. The CORIS baseline study. S Afr Med J 64:430–436
10. Ghumbre SU, Ghatol AA (2012) Heart disease diagnosis using machine learning algorithm. pp 217–225
11. Bartley C (2016) Replication data for: South African heart disease
12. Orr MJL, others (1996) Introduction to radial basis function networks
13. Zhao S, Tsang ECC, Chen D, Wang X (2010) Building a rule-based classifier—a fuzzy-rough set approach. IEEE Trans Knowl Data Eng 22:624–638. https://doi.org/10.1109/TKDE.2009.118
14. Patel J, Upadhyay T, Patel S (2015) Heart disease prediction using machine learning and data mining technique. Int J Comput Sci Commun 7:129–137.
15. Mathuria M (2013) Decision tree analysis on j48 algorithm for data mining. Int J Adv Res Comput Sci Soft Eng 3
16. Chaki D, Das A, Zaber M (2015) A comparison of three discrete methods for classification of heart disease data. Bangladesh J Sci Ind Res 50:293–296. https://doi.org/10.3329/bjsir.v50i4.25839
17. Dechter R, Pearl J (1985) Generalized best-first search strategies and the optimality of A*. J ACM 32:505–536. https://doi.org/10.1145/3828.3830
18. Kowshalya AM, Madhumathi R, Gopika N (2019) Correlation based feature selection algorithms for varying datasets of different dimensionality. Wirel Pers Commun 108:1977–1993. https://doi.org/10.1007/s11277-019-06504-w
19. Arabasadi Z, Alizadehsani R, Roshanzamir M, et al (2017) Computer aided decision making for heart disease detection using hybrid neural network-Genetic algorithm. Comput Methods Programs Biomed 141:19–26. https://doi.org/10.1016/j.cmpb.2017.01.004
20. Pang-Ning T, Michael S, Vipin K (2006) Introduction to data mining. EQ 1:337–341
21. Alizadehsani R, Zangooei MH, Hosseini MJ et al (2016) Coronary artery disease detection using computational intelligence methods. Knowledge-Based Syst 109:187–197. https://doi.org/10.1016/j.knosys.2016.07.004
22. Fukunaga K (2013) Introduction to statistical pattern recognition. Elsevier
23. Saboji RG (2017) A scalable solution for heart disease prediction using classification mining technique. In: 2017 International Conference on Energy, Communication, Data Analytics and Soft Computing (ICECDS). IEEE, pp 1780–1785

24. Gavhane A, Kokkula G, Pandya I, Devadkar K (2018) Prediction of heart disease using machine learning. In: 2018 Second International conference on electronics, communication and aerospace technology (ICECA). IEEE, pp 1275–1278
25. Mohan S, Thirumalai C, Srivastava G (2019) Effective heart disease prediction using hybrid machine learning techniques. IEEE Access 7:81542–81554. https://doi.org/10.1109/ACCESS.2019.2923707
26. J. SK, S. G (2019) Prediction of heart disease using machine learning algorithms. In: 2019 1st International conference on innovations in information and communication technology (ICIICT). IEEE, pp 1–5
27. Kannan R, Vasanthi V (2019) Machine learning algorithms with roc curve for predicting and diagnosing the heart disease. pp 63–72
28. Gonsalves AH, Thabtah F, Mohammad RMA, Singh G (2019) Prediction of coronary heart disease using machine learning. In: Proceedings of the 2019 3rd International conference on deep learning technologies—ICDLT 2019. ACM Press, New York, New York, USA, pp 51–56
29. Buettner R, Schunter M (2019) Efficient machine learning based detection of heart disease. In: 2019 IEEE International Conference on E-health Networking, Application Services (HealthCom). pp 1–6
30. Atallah R, Al-Mousa A (2019) Heart disease detection using machine learning majority voting ensemble method. In: 2019 2nd International Conference on new Trends in Computing Sciences (ICTCS). pp 1–6
31. Dimitriadou E, Weingessel A, Hornik K (2001) Voting-Merging: An ensemble method for clustering. pp 217–224
32. Pedregosa F, Varoquaux G, Gramfort A, Michel V, Thirion B, Grisel O, Blondel M, Prettenhofer P, Weiss R, Dubourg V, Vanderplas J (2011) Scikit-learn: machine learning in python. J Mach Learn Res:2825–2830
33. Shah D, Patel S, Bharti SK (2020) Heart disease prediction using machine learning techniques. SN Comput Sci 1:345. https://doi.org/10.1007/s42979-020-00365-y
34. Rajdhan A, Agarwal A, Sai M et al (2020) Heart disease prediction using machine learning. Int J Res Technol 9:659–662
35. Li JP, Haq AU, Din SU et al (2020) Heart Disease identification method using machine learning classification in e-healthcare. IEEE Access 8:107562–107582. https://doi.org/10.1109/ACCESS.2020.3001149
36. Urbanowicz RJ, Meeker M, La Cava W et al (2018) Relief-based feature selection: Introduction and review. J Biomed Inform 85:189–203. https://doi.org/10.1016/j.jbi.2018.07.014
37. De Jay N, Papillon-Cavanagh S, Olsen C et al (2013) mRMRe: an R package for parallelized mRMR ensemble feature selection. Bioinformatics 29:2365–2368. https://doi.org/10.1093/bioinformatics/btt383
38. Muthukrishnan R, Rohini R (2016) LASSO: A feature selection technique in predictive modeling for machine learning. In: 2016 IEEE International Conference on Advances in Computer Applications (ICACA). IEEE, pp 18–20
39. Sun Y, Todorovic S, Goodison S (2010) Local-learning-based feature selection for high-dimensional data analysis. IEEE Trans Pattern Anal Mach Intell 32:1610–1626. https://doi.org/10.1109/TPAMI.2009.190
40. Yadav SS, Jadhav SM, Nagrale S, Patil N (2020) Application of machine learning for the detection of heart disease. In: 2020 2nd International conference on innovative mechanisms for industry applications (ICIMIA). pp 165–172
41. Bharti R, Khamparia A, Shabaz M et al (2021) Prediction of heart disease using a combination of machine learning and deep learning. Comput Intell Neurosci 2021:1–11. https://doi.org/10.1155/2021/8387680
42. LeDell E, Poirier S (2020) H2O AutoML: Scalable automatic machine learning. In: 7th ICML Work Autom Mach Learn
43. Blagus R, Lusa L (2017) Gradient boosting for high-dimensional prediction of rare events. Comput Stat Data Anal 113:19–37. https://doi.org/10.1016/j.csda.2016.07.016
44. Van Rossum G, Drake FL (2009) Python 3 Reference manual. Createspace, Scotts valley, CA

45. Core Team R (2021) R: A language and environment for statistical computing
46. Bisong E (2019) Google AutoML: Cloud Vision. Building machine learning and deep learning models on google cloud platform. Apress, Berkeley, CA, pp 581–598

Subsurface Lithology Classification Using Well Log Data, an Application of Supervised Machine Learning

Atul Kumar Patidar, Sarthak Singh, and Shubham Anand

1 Introduction

Lithology delineation is an important aspect of geology, it involves study of the physical and chemical properties of rocks at the surface as well as in the subsurface. Delineation of lithology is crucial in the subsurface exploration of natural resources. Accurate prediction of subsurface stratigraphy is required to find economically viable prospects for oil and gas exploration [1]. Prediction of the petrophysical rock properties of subterranean formations is essential during the exploration stage, especially in the wild cat regions, and this requires systematic investigations of geological, geophysical, and geospatial [16, 18] data. Machine Learning (ML) [13, 19] and Artificial intelligence (AI)-based techniques are being utilized to precisely handle and analyze large datasets in order to reduce human error [2]. These models can also assist in the classification of subsurface lithologies as well as the mapping of the lateral distribution of reservoir properties [3]. Lithology classification has traditionally been done in the oil and gas business by analyzing the drill cuttings, core data, or well log interpretation. These processes are time-consuming and inconsistent due to the involvement of biases of a geologist and interpreters [4]. Due to reservoir heterogeneity and a large amount of well log data, the abilities of conventional methods to reliably classify subsurface lithologies becomes increasingly constrained once an oil and gas field reached to its development stage [5]. This necessitates the use of computer-aided models for lithology identification. Many industries, including geoscience and petroleum, have been transformed by these modern digital applications. The AI and ML-based algorithms are successfully used to interpret large, complex geological, geophysical, and petrophysical data sets that are primarily gathered from well logs and drilling operations in order to gain insights and generate

A. K. Patidar (✉) · S. Singh · S. Anand
Department of Petroleum Engineering and Earth Sciences, Energy Cluster, University of Petroleum and Energy Studies (UPES), 248007, Dehradun, Uttarakhand, India
e-mail: apatidar@ddn.upes.ac.in

A. Ramdane-Cherif et al. (eds.), *Machine Intelligence and Data Science Applications*, Algorithms for Intelligent Systems, https://doi.org/10.1007/978-981-99-1620-7_18

predictive subsurface models [6]. ML algorithms are especially well suited for tasks like subsurface lithology classification and hydrocarbon reservoir characterization because they can spot patterns and correlations in data that may not be immediately apparent to human interpreters [7, 8]. The use of such algorithm-based models can improve the predictability of any data-intensive organization.

The petroleum industry has long adopted computer-based techniques to discover patterns and trends in order to determine the hydrocarbon-bearing reservoirs. Initially, a core analysis method was used to determine the formation type and its physio-chemical properties [9]. Nowadays, machine learning models are being used in lithology classification, well log correlation, missing data prediction, porosity and permeability analysis and seismic data processing and interpretation [5] [10–12]. By using these digital modeling techniques, historical datasets are trained through tests and training sets to forecast the most desirable model, which reduces interpreter's time and enhance accuracy [20]. ML algorithms are classified into supervised and unsupervised learning categories based on the training labels [12, 14]. For lithology identification, supervised learning algorithms are frequently used because studies have demonstrated that supervised models are more accurate than unsupervised ones for subsurface modeling [15]. Recent evaluations of the effectiveness of several ML techniques have found that the Random forest machine learning algorithm has greater accuracy than other algorithms like Support Vector Machine (SVM) Algorithm, XGBoost, Artificial Neural Network (ANN), and Decision tree classifiers especially for lithology prediction [21].

The current study uses supervised machine learning (Random forest algorithm) to classify the subsurface litho-facies of the Kupe field in the Taranaki basin, New Zealand (https://www.nzpam.govt.nz/), using open-source well log data. The geophysical log data contains readings of the gamma ray, spontaneous potential, neutron porosity, and resistivity at the respective depths. Approximately 9000 data points (70% training and 30% testing) are used to develop and train the model and predict the subsurface lithology of the adjacent well. The current study demonstrates the relationship between the well log data and the precision of predicted lithology for the Kupe South-4 well using two different machine learning algorithms, decision tree, and random forest. The comparison between the accuracy of prediction from these two techniques is also discussed in the interpretation section. The predicted lithology is plotted in the form of litholog and compared with the actual lithology for a better understanding of the readers.

2 Overview of the Study Area and Data Used

The study area is the Kupe field, which is located in New Zealand's offshore Taranaki basin (Fig. 1). Kupe field is spread across the area of 256 km^2, and has an estimated reserve of more than 225 bcf of gas and 25 mmbbl of oil (https://data.gns.cri.nz/pbe/). The paleocene farewell formation, which primarily includes medium to coarse grain sandstones with good porosities and permeabilities, is the principal reservoir in the central field region [17]. The permeability of these sandstones exceeds 400-500mD at some places [17]. Upper stratigraphic intervals also contain claystone, siltstone, and limestone. Overall, sediments from the late Cretaceous to Recent constitute the overall stratigraphy for the Taranaki Basin. The open-source well log data is used in current research includes geophysical logs mainly the gamma rays, spontaneous potential, neutron porosity, and resistivity at the respective depths. An example of log motive plot is shown in Fig. 2.

Brief details of the log used are as follows. The caliper log provides the diameter of the wellbore and helps to identify the caving and swelling of the shale-rich formation. The sonic log records the travel time interval of the sound wave between a source and a receiver. Sonic log data provides a formation's interval transit time that helps in determining the formation porosity [5]. The gamma-ray log measures the natural gamma-ray value emitted by rocks due to their radioactive decay. High gamma ray values indicate the presence of shale-rich formation. The Laterolog measures the resistivity of the formation and gives an idea of the hydrocarbon-bearing zone due to its difference in resistivity compared to adjacent rocks. The neutron porosity log records the attenuation of emitted neutrons and helps in determining the porosity of the formation. Measures the bulk density of the formation and helps to identify oil and gas-bearing zones. The photoelectric factor log measures the photoelectric effect of the formation and helps in the prediction of the rock's petrophysical properties. The spontaneous potential log measures the potential difference between the surface and the wellbore. It helps to determine the permeability of formation and water

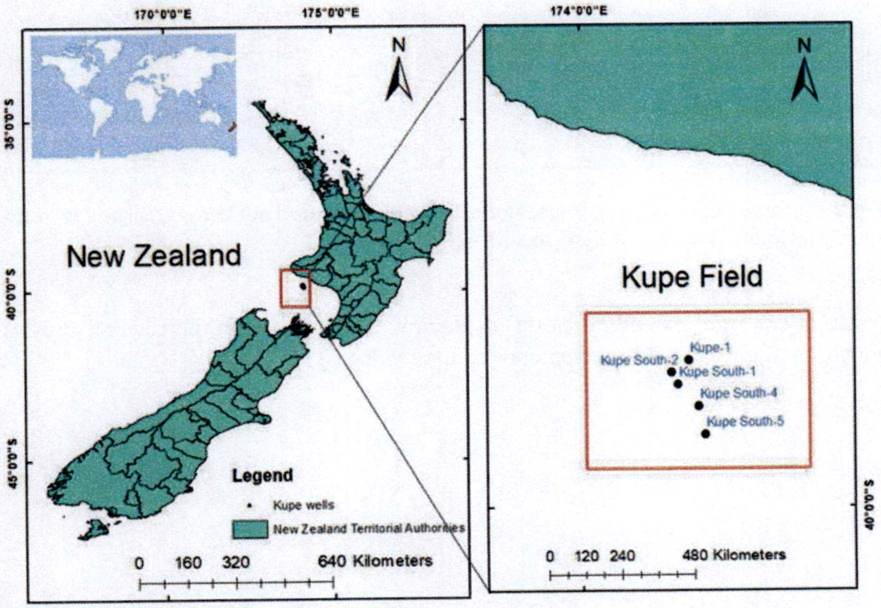

Fig. 1 Location map of the Kupe field (study area) in the Taranaki basin, New Zealand

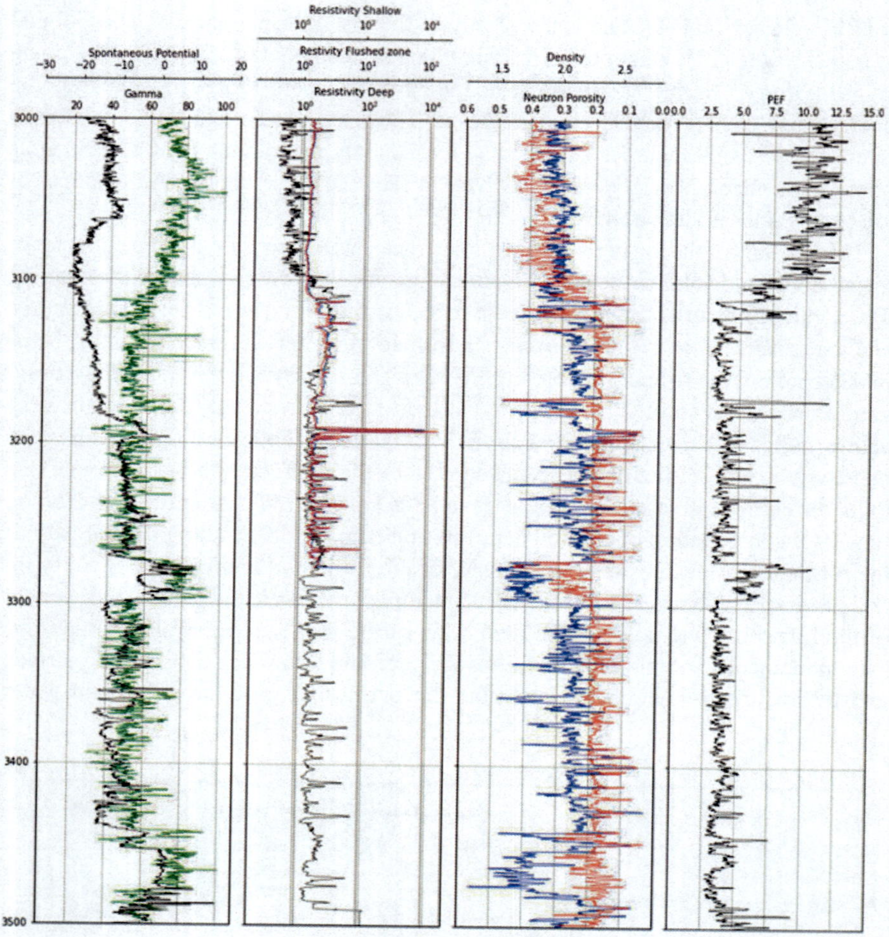

Fig. 2 Spontaneous potential, gamma ray, Shallow and deep resistivity, density, neutron porosity, and photoelectric factor log of Kupe south-1 well

resistivity. A brief example (subset) of the raw data set used in the current study is shown in Table 1 for the convenience of the readers.

Table 1 A subset of collected well log data of the well Kupe south-1 showing values of various geophysical logs with respect to Depth. It is only an example for the reference of the reader. The actual datasets are approximately 9000 points. The units and abbreviations used in well logs are as follows. CALI–Caliper log (Inches), DT–Sonic log (ms/ft), GR–Gamma Ray log (GAPI), LLD–Laterolog deep (Ohm m), LLS–Laterolog shallow (Ohm m), NPHI—Neutron Porosity (vol/vol), RHOB–Bulk density (g/cc), PEF–Photoelectric factor (b/e), SP–Spontaneous Potential (mV)

Depth	CALI	DT	GR	LLD	LLS	NPHI	RHOB	PEF	SP
3000.15	13.2425	87.9995	65.9963	1.8906	2.24494	0.326723	1.77929	11.398	−18.938
3000.3	12.6331	88.6881	69.8126	2.0539	2.45729	0.3057	1.79689	11.2806	−18.9383
3000.45	12.93	88.2509	71.0014	2.0561	2.47558	0.305705	1.8076	10.906	−19.4531
3000.6	13.172	90.3119	67.4997	1.98451	2.39532	0.300314	1.7988	10.9842	−19.954
3000.76	13.2111	89.4998	66.9379	1.9537	2.3321	0.317397	1.79	11.4524	−18.9377
3000.91	13.516	89.6885	67.1893	1.93562	2.32355	0.311502	1.78521	11.883	−18.4377
3001.06	13.7902	91.2495	70.5622	1.92062	2.28176	0.3184	1.7813	11.5309	−19.422
3001.21	12.633	94.5599	73.875	1.86409	2.22078	0.317398	1.7705	11.0001	−19.4377
3001.37	12.9529	91.9389	73.9356	1.82912	2.17891	0.298801	1.7666	11.117	−18.9385
3001.52	13.117	89.8752	77.3763	1.8698	2.20962	0.283699	1.7695	11.1018	−17.9191
3001.67	13.117	90.188	74.9988	1.9042	2.25888	0.280811	1.7695	11.3364	−15.9138

3 Methodology Adopted

Initially, the well log data from the open-source platform (https://www.nzpam.govt.nz/) is collected. The raw data had nonuniformity, hence the outlier values were removed during the data conditioning process and continuous data points between 3000 and 3500 m depth were used for modeling purposes. The correlation plots (scatterplot) were prepared to display the strength, direction, and various relationships between two geophysical properties (quantitative variables). A cross plot between neutron porosity and density values is generated to determine the lithology type in a particular well. Then the preparation of the ML model is done. Two models are used in current research work for comparison purposes and to get a more accurate correlation (i) Decision tree classifier and (ii) Random forest classifier. Finally, the accuracy check of both the models is done and predicted lithologies are plotted and compared with actual lithology. A detailed flowchart is shown in Fig. 3.

4 Data Interpretation and Modeling Results

In the current research work, two ML algorithms are used to predict the subsurface lithology using well logs of the above-mentioned basin. Typically, the ordinal, binary, and multiclass classification problems are solved using a decision tree algorithm [11]. But many times, it has been cited that this algorithm tends to overfit training datasets which is its biggest drawback. At the other end, random forest is used to fit

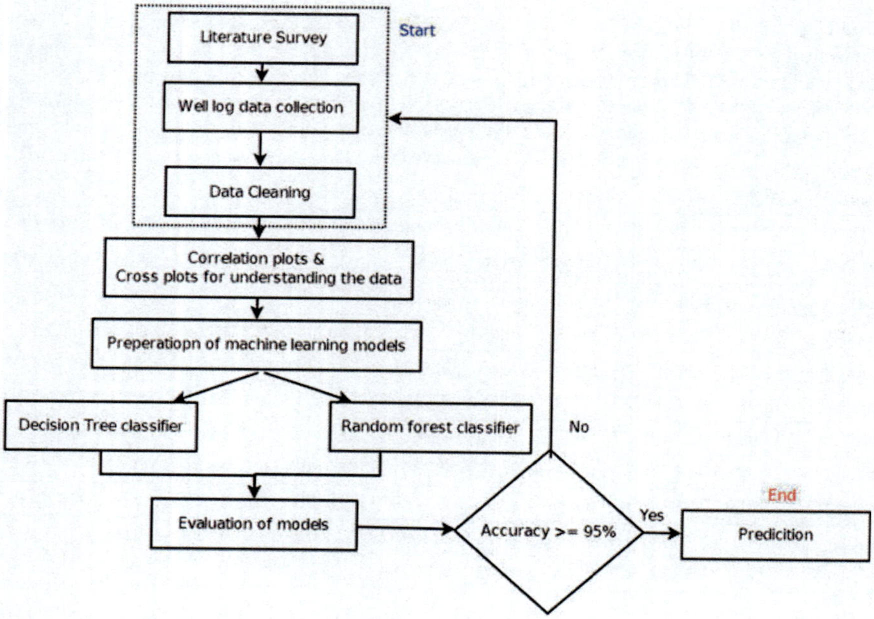

Fig. 3 Flowchart illustrating the adopted methodology for the current research work

several decision tree classifiers on different samples of the dataset. The final outcome (prediction) is done based on the majority. In a random forest classification, averaging is used which helps in controlling the overfitting problem and it also enhances the prediction accuracy. A random forest classifier can also be called a random decision forest. The model used an ensemble learning approach (use of many decision trees and takes a majority vote to enhance predictive accuracy). Based on the size and complexity of the dataset, there can be multiple trees in a single random forest model.

4.1 Pre-model Preparation Data Overview

The correlation coefficient denoted by ρ(X, Y), is defined as ρ(X, Y) = Cov(X, Y)/(SD(X)SD(Y)), where ρ(X, Y) is used to summarize the trend between random variables. The value for correlation ranges from -1 to 1. Where if the value of the correlation coefficient is close to -1, then it is indicative of a very strong negative correlation, and if the value of the correlation coefficient is close to $+1$ then it is an indication of a strong positive correlation. The effectiveness of a machine learning model in predicting the target variable depends upon the quality of the training dataset [3]. The correlation matrix heatmaps are useful for the identification of any outliers in the training or testing dataset and it is also useful for selecting the variables

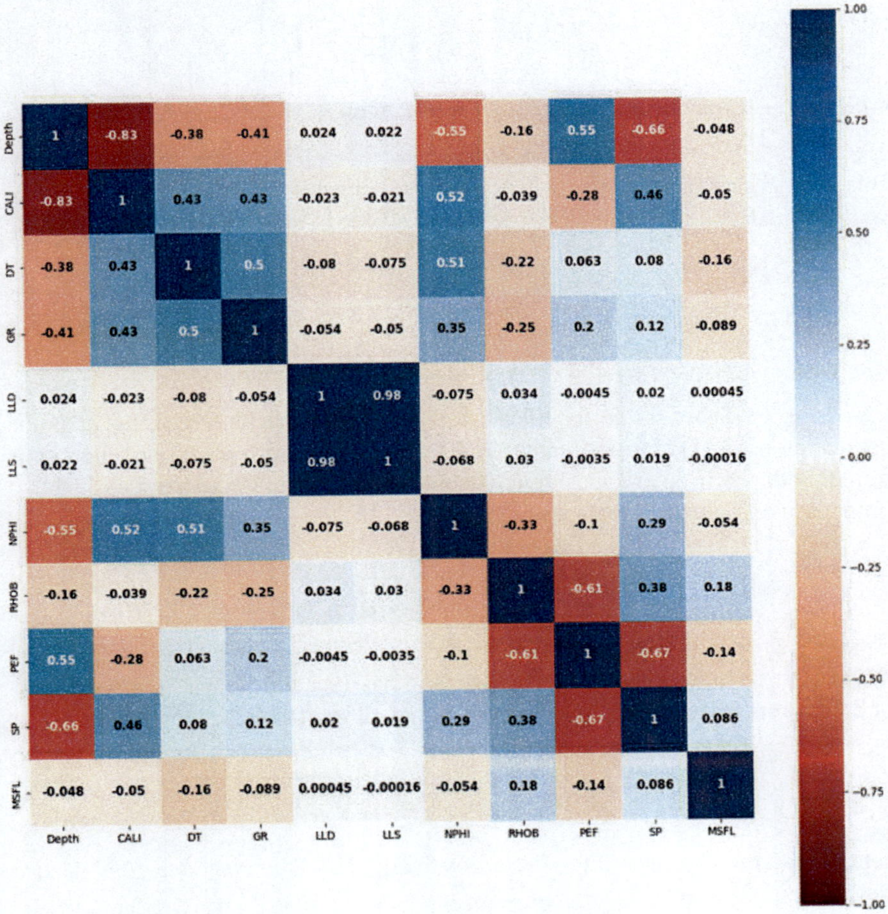

Fig. 4 Correlation matrix for well log data of Kupe south-1

for training the machine learning model. In Fig. 4, it can be observed that certain variables like LLD and LLS have a very strong positive correlation.

4.2 *Evaluating ML Model*

Precision: Precision is the index of a machine learning model's performance. It is defined as the ratio of correctly classified positive samples to the total classified positive samples. It helps us to imagine the credibility of the model in categorizing the model as positive. The precision of a model is 1.0 if it produces no false positives.

$$\text{Precision} = \frac{TP}{TP + FP} \tag{1}$$

where TP—True Positives and FP—False Positives

Recall: Recall is the fraction of the retrieved instances. It is also known as sensitivity. It is the measure of the model which is correctly identifying true positives. It gives us an estimate of how precisely the model is capable of identifying the relevant data.

$$\text{Recall} = \frac{TP}{TP + FN} \tag{2}$$

where TP—True Positives and FN—False Negatives

F1-Score: F1-Score is the main evaluation metric in machine learning. It is the estimation of the model's accuracy on a dataset. The main goal of the F1-score is to join both precision and recall metrics in a single metric. F1-Score is designed to properly work with an imbalanced dataset.

$$F = \frac{2}{precision^{-1} + recall^{-1}} \tag{3}$$

Accuracy: Accuracy is the metric for assessing classification models. It is the number of data points that are predicted correctly out of all available data points.

$$\text{Accuracy} = \frac{TP + TN}{TP + FP + FN + TN} \tag{4}$$

here, TP—True Positives, FP—False Positives, TN—True Negatives, FN—False Negatives

Macro Average: Macro Average is the mean of scores of all the classes. It is calculated using the arithmetic mean of the scores. It calculates the metric of each class independently and after that, it takes the average. Whereas Support is the number of samples of the true responses which are there in each class of the target values.

The Neutron porosity (NPHI) and Bulk density (RHOB) cross plot is often used to delineate the lithology of the targeted section. It can be observed that the majority of data points have Gamma Ray values ranging from (50 to 80 API) which is indicative of dominantly Shale free (Sandstone) formation with thin streaks of Claystone (Fig. 5). Few data points with high GR values are also obvious in the plot.

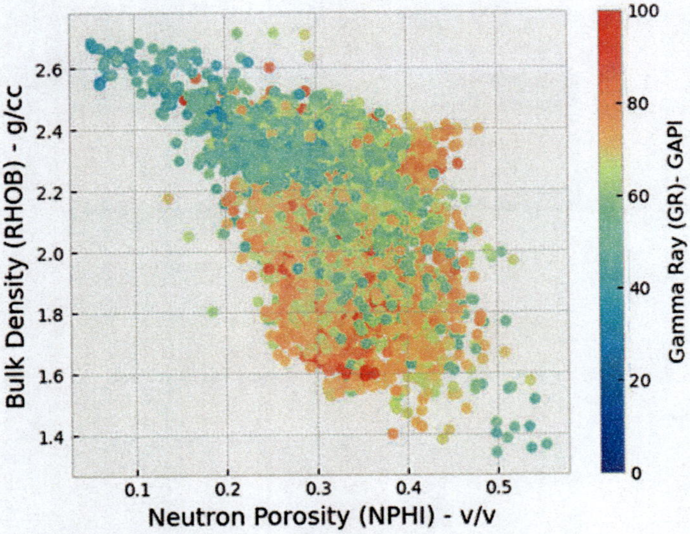

Fig. 5 A cross plot/Scatter plot between the Neutron porosity and Density. GAPI Gamma values are plotted with different colors

Figure 6 shows a confusion matrix that illustrates the random forest classifier that has predicted 1132 values of claystone correctly and only 37 values are labeled incorrectly (only 3%), similarly for claystone/thin-bedded sand 1108 values are predicted correctly and only 45 are labeled incorrectly (4%) and in case of sandstone/minor clay or sandstone/siltstone interbedded almost all the values are predicted correctly.

The accuracy of the models was compared where the decision tree classifier has an overall accuracy of 94% (Table 2) and the values of precision, recall, and F1-score are also significant.

The classification report for the random forest classifier shows that the random forest classifier has an accuracy of 97% (Table 3) which is relatively higher than the above-mentioned decision tree classifier report in Table 2. The random forest classifier also has a high precision and F1-score as compared to the decision tree classifier. This indicates that model 2 is better than model 1. A higher F1-score indicates the effectiveness of the model on the dataset.

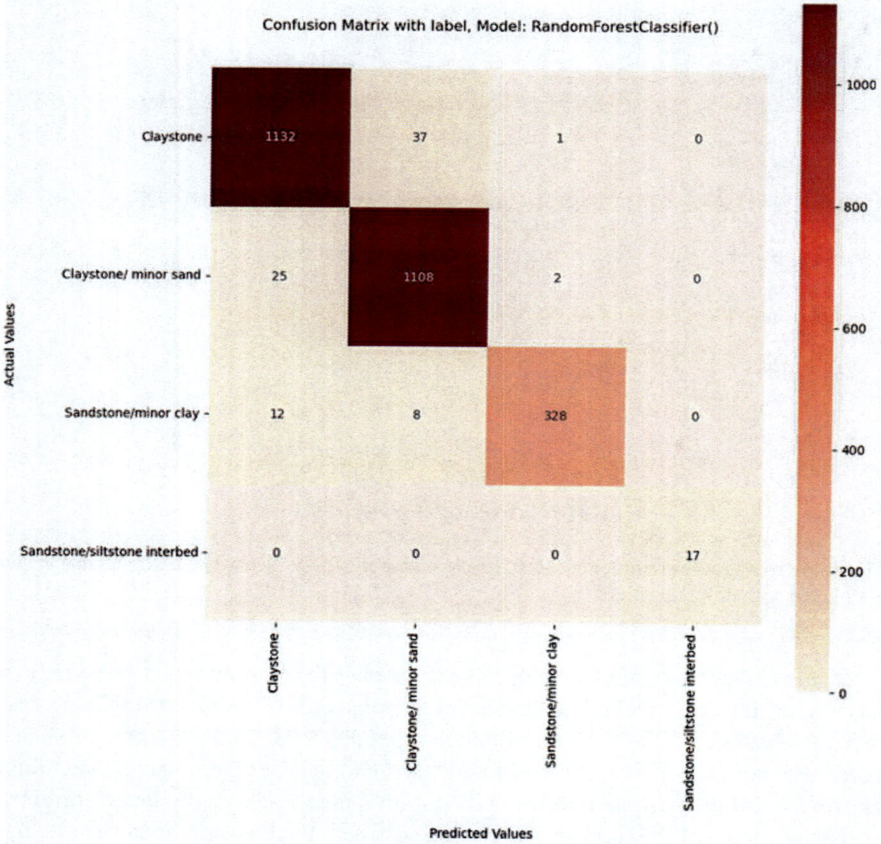

Fig. 6 Confusion matrix (Kupe south 1) showing correctly/ incorrectly predicted values by the random forest classifier

Table 2 Classification report (Decision Tree Classifier)

Lithology	Precision	Recall	F1-score	Support
Claystone	0.95	0.94	0.94	1170
Claystone/minor sand	0.94	0.95	0.94	1135
Sandstone/minor clay	0.93	0.95	0.94	348
Sandstone/siltstone interbedded	1	0.82	0.9	17
Accuracy			*0.94*	*2670*
Macro avg	0.95	0.91	0.93	2670
Weighted avg	0.94	0.94	0.94	2670

Table 3 Classification report (Random forest classifier)

Lithology	Precision	Recall	F1-score	Support
Claystone	0.97	0.97	0.97	1170
Claystone/minor sand	0.96	0.98	0.97	1135
Sandstone/minor clay	0.99	0.94	0.97	348
Sandstone/siltstone interbedded	1	1	1	17
Accuracy			*0.97*	*2670*
Macro avg	0.95	0.91	0.93	2670
Weighted avg	0.97	0.97	0.97	2670

Finally, two lithologs are prepared, the actual litholog of Kupe south-4 (based on drilled lithology in well) and a predicted litholog of a similar interval using the random forest classifier (Fig. 7). Comparison between actual vs predicted lithologs (Fig. 7) illustrates almost similar lithology in both the wells. The random forest classifier successfully predicted the subsurface lithology of the adjacent well with an accuracy of 97% in comparatively less time than traditional methods (Table 3).

The identification of sedimentary facies can also be done using this type of lithology classification model [1]. Hence, we support the statement that says the discipline of earth sciences greatly benefits from supervised machine learning models. For the exploration and production sector, these models offer highly precise, scalable, and affordable solutions. The well logs from nearby regions can be used to effectively predict the lithology during the exploration or appraisal drilling. This useful information can be applied to field development plans, drilling mud selection, drilling rate of penetration selection, and many other crucial variables.

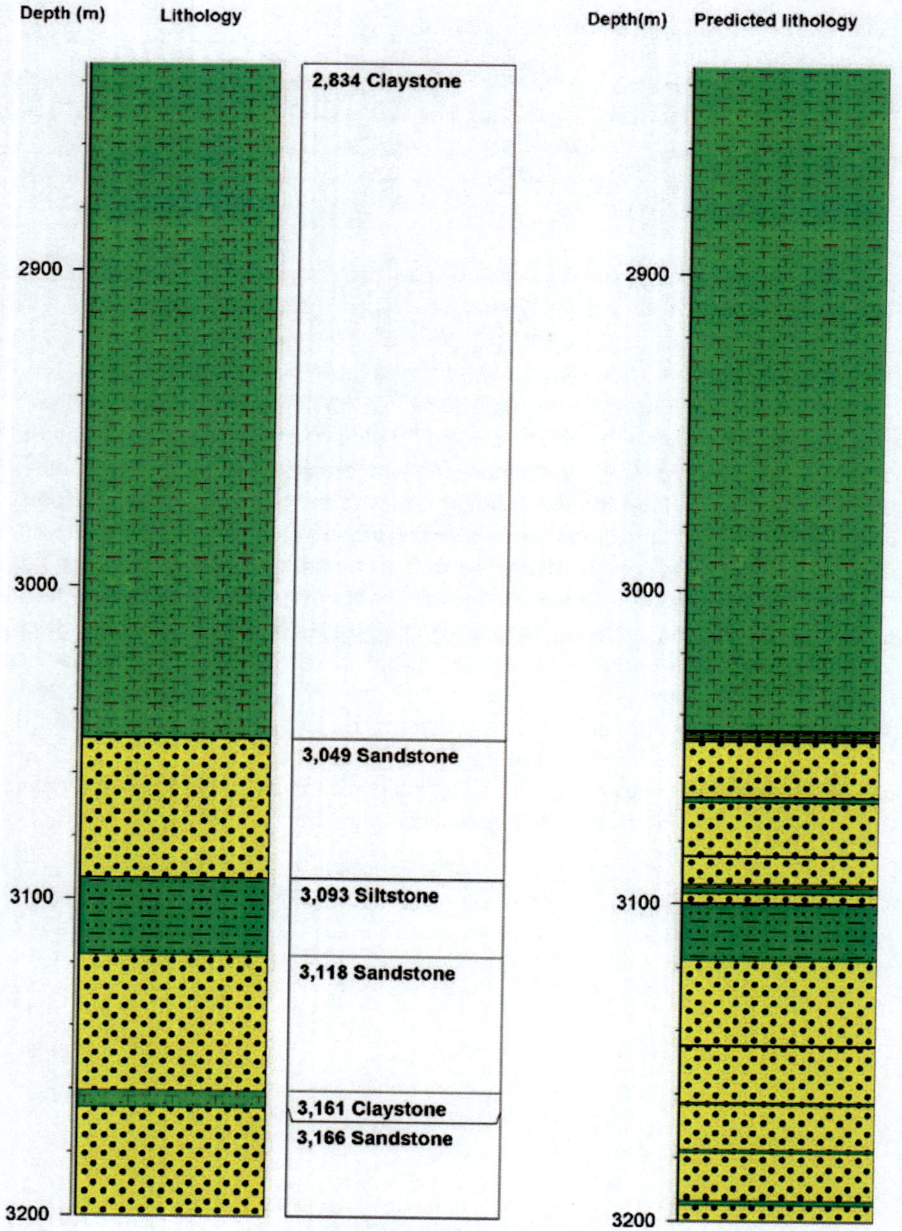

Fig. 7 Litholog showing actual lithology drilled in the well Kupe south-4 (left) and predicted lithology (right) using Random Forest classifier

5 Conclusion

The conclusion of the current study is listed below

a. Supervised Machine learning model can be used with confidence while interpreting well log data to identify/classify subsurface lithology.
b. The current work state that the random forest classifiers are more accurate in predicting subsurface lithology compared to decision tree classifiers.
c. The random forest classifier was able to predict the subsurface lithology of adjacent well with an accuracy of 97% while the decision tree classifier shows 94% accuracy in our dataset.
d. The authors also state that the output of the ML model for lithology classification is not 100% accurate and highly dependent on the expertise of a modeler and used algorithm.

Acknowledgements The authors sincerely acknowledge the UPES, Dehradun. No specific grant from funding organizations in the public, private, or non-profit sectors has been used to support this research.

Conflict of Interest Authors declare that the publishing of this paper is devoid of any conflicts of interest.

References

1. Zhang J, He Y, Zhang Y, Li W, Zhang J (2022) Well-logging-based lithology classification using machine learning methods for high-quality reservoir identification: a case study of baikouquan formation in Mahu Area of Junggar Basin. NW China. Energies 15(10):3675. https://doi.org/10.3390/en15103675
2. Li W, Hu W, Abubakar A (2020) Machine learning and data analytics for geoscience applications-Introduction. Geophysics 85(4). https://doi.org/10.1190/geo2020-0518-spsintro.1
3. Mahmoud, A.A., Elkatatny, S., Al-AbdulJabbar, A. (2021) Application of machine learning models for real-time prediction of the formation lithology and tops from the drilling parameters. Journal of Petroleum Science and Engineering, 203,108574, DOI: https://doi.org/10.1016/j.petrol.2021.108574
4. Liu H, Wu Y, Cao Y, Lv W, Han H, Li Z, Chang J (2020) Well Logging Based Lithology Identification Model Establishment Under Data Drift: A Transfer Learning Method. Sensors 20(13):3643. https://doi.org/10.3390/s20133643
5. Joshi D, Patidar AK, Mishra A et al (2021) Prediction of sonic log and correlation of lithology by comparing geophysical well log data using machine learning principles. GeoJournal. https://doi.org/10.1007/s10708-021-10502-6
6. Xie Y, Zhu C, Zhou W, Li Z, Liu X, Tu M (2018) Evaluation of machine learning methods for formation lithology identification: A comparison of tuning processes and model performances. J Pet Sci Eng 160: 182–193. https:// doi.org/https://doi.org/10.1016/j.petrol.2017.10.028
7. Sfidari E, Kadkhodaie-Ilkhchi A, Rahimpour-Bbonab H, Soltani B (2014) A hybrid approach for litho-facies characterization in the framework of sequence stratigraphy: A case study from

the South Pars gas field, the Persian Gulf basin. J Petrol Sci Eng 121:87–102. https://doi.org/10.1016/j.petrol.2014.06.013
8. Ren X, Hou J, Song S, Liu Y, Chen D, Wang X, Dou L (2019) Lithology identification using well logs: A method by integrating artificial neural networks and sedimentary patterns. J Petrol Sci Eng 182:106336. https://doi.org/10.1016/j.petrol.2019.106336
9. Salehi SM, Honarvar B (2014) Automatic identification of formation lithology from well log data: a machine learning approach. J Pet Sci Res 3(2):73. https://doi.org/10.14355/jpsr.2014.0302.04
10. Khilrani N, Prajapati P, Patidar AK (2021) Contrasting machine learning regression algorithms used for the estimation of permeability from well log data. Arab J Geosci 14:2070. https://doi.org/10.1007/s12517-021-08390-8
11. Mishra A, Sharma A, Patidar AK (2022) Evaluation and development of a predictive model for geophysical well log data analysis and reservoir characterization: machine learning applications to lithology prediction. Nat Resour Res 31:3195–3222. https://doi.org/10.1007/s11053-022-10121-z
12. Okoroafor ER, Smith CM, Ochie KI, Nwosu CJ, Gudmundsdottir H, Aljubran M (2022) Machine learning in subsurface geothermal energy: Two decades in review. Geothermics 102:102401. https://doi.org/10.1016/j.geothermics.2022.102401
13. Bhatia K, Arora S, Tomar R (2016) Diagnosis of diabetic retinopathy using machine learning classi-fication algorithm. In: 2016 2nd International Conference on Next Generation Computing Technolo-gies (NGCT). Dehradun, India, 2016, pp 347–351. https://doi.org/10.1109/NGCT.2016.7877439
14. Bhattacharya S, Timothy C, Mahesh P (2016) Comparison of supervised and unsupervised approaches for mudstone lithofacies classification: Case studies from the Bakken and Mahantango-Marcellus Shale, USA. J Nat Gas Sci Eng 33(2):1119–1133. https://doi.org/10.1016/j.jngse.2016.04.055
15. Mohamed IM, Mohamed S, Mazher I, Chester P (2019) Formation Lithology Classification: Insights into Machine Learning Methods. SPE Annual Technical Conference and Exhibition, Calgary, Alberta, Canada,. https://doi.org/10.2118/196096-MS
16. Nath A, Koley B, Saraswati S et al (2022) Geospatial analysis of short term shoreline change behavior between Subarnarekha and Rasulpur estuary, east coast of India using intelligent techniques (DSAS). GeoJournal. https://doi.org/10.1007/s10708-022-10683-8
17. Jumat, N., Shalaby, M.R. & Aminul Islam, M. (2018) Integrated reservoir characterization of the Paleocene Farewell Formation, Taranaki Basin, New Zealand, using petrophysical and petrographical analyses. Journal of Petroleum Exploration and Production Technology 8, 685–701, DOI: https://doi.org/10.1007/s13202-017-0420-5
18. Choudhury T, Kandwal A (2021) Machine learning based analytical approach for geographical analy-sis and prediction of Boston City crime using geospatial dataset. GeoJournal. https://doi.org/10.1007/s10708-021-10485-4
19. Hayat, A.; Morgado-Dias, F.; Bhuyan, B.P.; Tomar, R. Human Activity Recognition for Elderly People Using Machine and Deep Learning Approaches. Information 2022, 13, 275. https://doi.org/10.3390/info13060275.
20. Bergen KJ, Johnson PA, De Hoop MV, Beroza GC (2019) Machine learning for data-driven discovery in solid Earth geoscience. Science 363 (6433). https://doi.org/10.1126/science.aau0323.
21. Merembayev T, Kurmangaliyev D, Bekbauov B, Amanbek YA (2021) Comparison of machine learning algorithms in predicting lithofacies: case studies from Norway and Kazakhstan. Energies 14(7):1896. https://doi.org/10.3390/en14071896

Using Blockchain in Pharmaceutical Industry to Prevent Vaccine Counterfeiting

Sarishma Dangi, Utkarsh Garg, Diksha Bisht, and Ravi Tomar

1 Introduction

Healthcare is currently the most discussed topic in the world, and pharmacy plays an important role in accomplishing the objectives. In order to give an efficient solution for patient treatment, technologies are now being utilized in the healthcare industry. The majority of machinery used in healthcare centers are fully or partially automated and operate under the control of a certain authority, due to technologies such as artificial intelligence, machine learning, the internet of things, and others. The pharmaceutical business is confronted with a number of issues that must be addressed in the near future, since the global pandemic has demonstrated how disruptive these issues are to the smooth operation of the organizations involved in the process. When humanity is suffering from a big pandemic ailment like COVID-19, the rise in drug adulteration such as counterfeit medicines, misleading sub-branding, and internet scams puts people's lives in jeopardy. When the entire country is suffering and fighting a severe condition, vaccination becomes the only hope for them to protect themselves from a deadly infection [1].

Global outbreak of diseases has always interfered with and influenced the growth of human history. Thousands of people die, civilizations fall apart, and economies collapse as a result of such diseases, which have proven as a humongous challenge to the world's leading regimes. The COVID-19 pandemic in India is part of the 2019 worldwide coronavirus disease pandemic, caused by coronavirus (SARS-CoV-2) which causes severe acute respiratory syndrome. Until December 2019, when an outbreak occurred in Wuhan, China; the virus was unknown. The first reported incidence of presence of COVID-19 in India was reported on January 30, 2020.

S. Dangi (✉) · U. Garg · D. Bisht
Graphic Era Deemed to be University, Dehradun, Uttarakhand, India
e-mail: sarishmasingh@gmail.com

R. Tomar
Persistent Systems, Pune, India

A. Ramdane-Cherif et al. (eds.), *Machine Intelligence and Data Science Applications*, Algorithms for Intelligent Systems, https://doi.org/10.1007/978-981-99-1620-7_19

At present, India has the most confirmed cases in Asia with 18.7 crores confirmed cases of COVID-19 infection and 40.3 lakh deaths as of July 12, 2021. India has the second-highest number of confirmed cases with 3.08 crores and the third-highest number of COVID-19 deaths with 4.08 lakh [2].

In this scenario, the rapid rollout of vaccine is an important aspect and the implementation of vaccination campaign is crucial for India with a population of 139 crores. Former health minister of India indicated in September 2020 that the country hoped to approve and distribute vaccines by the first quarter of 2021. With 3006 vaccination centers, India began its vaccination program on January 16, 2021. Health staff and frontline workers that deal directly with COVID patients were to be the first recipients. The Drug Controller General of India (DCGI) licensed the Oxford—AstraZeneca vaccine "Covishield" for emergency use on January 1, 2021. DCGI also approved BBV152, commonly known as Covaxin, a local vaccine produced by Bharat Biotech in collaboration with the Indian Council of Medical Research and the National Institute of Virology for interim emergency use authorization on January 2. DCGI approved Russia's Sputnik V vaccine for emergency use in India on April 12, 2021. Dr. Reddy's Laboratories, the vaccine's local distributor, claimed that the vaccine would be ready for administration in India by late May 2021 [3].

In terms of utilizing technologies for goods, manufacturing, distribution, and supply chain management, the pharmaceutical business has seen a significant growth, adoption, and development in the last decade. These sectors are responsible for delivery of medicine or pharmaceuticals, which are a high-demand commodity in the health sector as of today. On the other hand, major adulteration concerns such as inventory risk, counterfeit risk, financial risk, regulatory risk, supply failure risk, falsely labeled pharmaceuticals, online scams exist in the pharmaceutical supply chain. These risks can lead to disruptions in supply chain management, resource waste, and most importantly counterfeiting of medicine, which poses a severe danger to patient's life. In this era of digitalization, when all items are available online, the threat to this business has grown. As a result, the need for an effective solution to reduce such threats and provide a safe mechanism for administration of vaccine is required which must ensure accountability [4]. Table 1 represents the available and authorized vaccines in India, their vaccine design, storage temperature, and the manufacturing company.

Table 1 COVID-19 Vaccines in India

Vaccine name	Vaccine design	Storage temperature	Manufacturing company
Covishield	Non-replicating chimpanzee adenovirus vaccine vector (ChAdOx1)	2–8 °C	Serum Institute of India, Pune, India
Covaxin	Inactivated-virus vaccine	2–8 °C	Bharat Biotech Ltd. Hyderabad, India
Sputnik V	Inactivated human adenovirus Ad5 and Ad26 with Spike proteins inserts	− 18 °C*	Dr. Reddy's Laboratories, Hyderabad, India

Vaccination not only improves a patient's recovery rate, but it also immunizes the human body, which is vulnerable to the attack. Due to the adulteration in the supply chain, counterfeiting, tampering, false labeling, fishing scam attacks, and black marketing, people's lives are in danger. In order to deal with such problems, a model is needed that can maintain the supply chain while also protecting it from manipulation and unnecessary interference. Blockchain can be used to develop a method that can provide a firm grasp on the monitoring of commodities. Blockchain is defined as a distributed public ledger comprising of continuous growing chain of blocks into a peer-to-peer network which offers key features like immutability, accountability, transparency of data, autonomy, and security. Using blockchain provides such features by using distributed architecture and consensus protocol mechanism where each block is capable for doing operations with data independently [5].

In this work, we aim to use blockchain and solve the problem of vaccine counterfeiting by proposing a model which can be used for vaccine management throughout the country. In particular, the contributions of this work are outlined as follows:

- To propose a blockchain-based model to prevent vaccine counterfeiting and provide a platform for open access to vaccine-related information and management.
- To discuss the challenges faced in real life toward vaccine distribution and administration in Indian landscape.
- To provide an implementation for the proposed blockchain-based vaccine supply management system using object-oriented programming language, i.e., Java.

The rest of the paper is organized as follows: Sect. 2 provides the review of existing literature in the area. Section 3 discusses the key challenges faced during vaccine distribution and administration. Section 4 provides the methodology used for implementation of the proposed work. Section 5 represents how well the proposed implementation of blockchain scales with respect to number of transactions. Section 6 summarizes the work and points toward future research directions.

2 Related Work

The technology infrastructure for constructing decentralized applications is provided by blockchain. It consists of a linked list of blocks connected together by hash pointers where each block comprises a set of transactions related to a digital asset. Blockchain is based on hash-based data structures, which provide features such as collision avoidance, data binding, and data concealment. Claudia Antal et al. [6] proposed a system that employs blockchain to ensure data integrity as well as immutability of vaccination beneficiary registrations thereby preventing identity theft and impersonation.

Smart contracts were designed for monitoring and tracking appropriate vaccine distribution environment as per the vaccine producer's safe handling requirements, and thus allowing all the network peers to be aware of that information. With the

COVID-19 vaccine distribution circumstances in mind, a prototype is built which utilizes the Ethereum test network and is named "Ropsten." In 2020, Jun Shen et al. [20] developed a vaccine blockchain system that uses blockchain and machine learning techniques. This system was intended to facilitate vaccination traceability along with smart contract functions and it may also be used to combat vaccine expiration and record fraudulent activities. Machine learning algorithms may also offer suggestions to vaccination practitioners and corresponding recipients thereby helping them to opt for better immunization procedures, policies, and vaccines.

In the scenario of a pandemic crisis just like the present one, blockchain applied to the health sector can provide novel and efficient solutions to improve numerous activities linked with the prevention and control of diseases leading to improved clinical risk management [8]. An incentive-based approach for COVID-19 using blockchain was proposed by Manoj et al. [22] to keep track for every individual person's COVID-19 related information while also keeping a specialized personal color scheme that can indicate whether an individual is tested positive or is suspected of having the virus.

Smart contracts are used as software agents which can be deployed within the blockchain network using a second-generation blockchain platform like Ethereum. Without the intervention of third parties, smart contracts can automatically carry out the specified terms within the agreement and thus verify the trustworthy transactions. Dounia Marbouh et al. [10] proposed a blockchain-based system that uses Ethereum-based smart contracts along with oracles in order to track the reported information from trusted sources about the total number of detected new cases, deaths and the number of recovered cases. It also presented algorithms that can easily capture the interactions between different stakeholders which are involved in the network.

Hence, blockchain can assist track vaccinations and ensure they haven't been tampered with, as well as keep a proper track of patient's immunization records and provide the proof for vaccination especially in the wake of the fact that COVID-19 vaccination required two doses separated by a time gap. While reviewing literature, it was observed that blockchain has huge potential to solve the challenges related to vaccine administration. A lot of significant work can be done in the area to push the corresponding adoption of blockchain and related technologies to resolve the existing issues [11]. The challenges related to vaccine distribution and administration are discussed at length in the next section.

3 Challenges Toward Vaccine Distribution and Administration

India is renowned as the "pharmacy of the world" since it is the world's largest supplier of medications and produces 60% of the world's vaccinations. The irony is that, despite being one of the world's largest vaccine producers, India faces one of the world's greatest hurdles in immunizing its population of nearly 1.39 billion humans. The pandemic has pushed the already overloaded healthcare system to a breaking point in many sections within the country. While urban areas are facing the heat, it is the rural area of India which is hit hard especially due to lack of vaccines and means of administration as well as distribution.

3.1 *Counterfeiting of Vaccines*

Consumers may be harmed by inadequate and fraudulent medical supplies, which fail to treat the diseases for which they were prescribed. They cause people to lose faith in drugs, healthcare practitioners, and healthcare institutions. Antimalarial drugs and antibiotics are two of the most regularly reported substandard and counterfeited pharmaceuticals. Both generic and innovator drugs, ranging from highly expensive cancer treatments to very affordable pain treatments, can be faked. They're sold illegally on the streets, on unregulated websites, in pharmacies, clinics, and even hospitals. In low- and middle-income nations, about one out of every ten medical items is substandard or counterfeit.

COVID-19 vaccinations, like other medications, have made their way into the illegal market in India. A medical officer from a primary health centre (PHC) in Satyanarayanapuram, Andhra Pradesh was arrested on May 15, 2021 for alleged participation in the case. Dr. NS Raju of G Konduru PHC was allegedly caught red-handed in possession of five doses of Covaxin and six doses of Covishield [12]. Similarly, the bogus COVID vaccination shots fraud has affected nearly 2,000 people in Mumbai and 500 people in Bengal [21].

3.2 *Temperature-Controlled Storage*

COVID vaccines are highly sensitive and require frequent quality checks, so they must be distributed under appropriate conditions such as temperature controlled cold-chain distribution mechanisms. In countries with warmer climates, such as India, cold-chain logistics are especially challenging. According to the International Air Transport Association (IATA), Centre of Excellence for Independent Validators in Pharmaceutical Logistics, 25% of vaccines are deteriorated by the time they arrive at their destination and temperature errors cost nearly $34.1 billion per year. Some vaccines in development may also need to be stored at –70 °C temperature. Despite the fact that India has a 28,000-unit cold storage network designed for use by the government's universal immunization program, experts as well as logistics corporations have pointed out that existing companies do not have the potential to carry vaccines at a temperature range which is below –25 °C.

An estimate of the vaccine cold-chain space required was made based on the assumption that 18% of the population would be vaccinated within six months. To arrange the facility for cold-chain space necessary for the vaccine, the Union Health

Ministry of India has already initiated a cold-chain augmentation strategy. However, in order to make the COVID vaccine program successful, the government will need to enlist the aid of the business sector to ensure that protection for all. Even if the government retains responsibility for immunizing vulnerable groups, it lacks the resources to reach people in rural areas of the country.

3.3 *Quality Assurance*

To ensure and validate the transcript of any type of vaccine's lifecycle and voyage, a properly documented chain of custody must be maintained. Given the long and proven history of counterfeiting within the pharmaceutical supply chain, traceability thus becomes a key prerequisite. As the demand for Covid vaccines is through the roof, every dose has to be tracked and validated with proper information so as to guarantee that counterfeits are not distributed to vulnerable groups. The answer could be found in blockchain as it provides security, transparency, accountability, traceability, and immutability. Blockchain provides a mechanism to record transactions on a distributed ledger. The tamper-proof feature of blockchain allows all participants in a supply chain network to record transactions at each stage of the product's journey, and it's a great tool for preventing counterfeit doses from entering the pharmaceutical supply chain in the first place.

3.4 *Administration of the Vaccine*

The last stretch for the vaccine's voyage may be the most difficult one. Determining the sites for vaccination delivery is a big barrier, similar to the issues arisen with COVID testing. In this scenario, India will need to greatly expand its existing infrastructure in order to provide 400–500 million doses of a COVID vaccine by the first two quarters of 2021.

Strengthening the vaccine cold-chain network, expanding the stock of ancillary products such as syringes/glass vials/needles, and training healthcare professionals in an appropriate manner are all a part of this capacity-building effort [19]. If this does not happen, the COVID vaccination may not be available to those who need it the most. The databases of healthcare and frontline personnel, as well as the procurement of ancillary products, are in advanced stages of preparation in order to deliver the vaccination.

3.5 *Post-Launch Challenges*

Distributors throughout the country must also consider the risk of recalls, which will place more strain on an already stressed vaccine supply chain. Logistically, it is impossible to vaccinate everyone within a quarter or two. Distributors will have to therefore devise separate techniques for distribution of vaccines in a timely manner and consider it carefully for each wave. Due to the fact that many companies are still using decades-old technologies, the vaccine's distribution will be more complicated than ever before.

Many logistics organizations' out-of-date back-office systems are incapable of handling the scale as well as the complexity of this humungous task at hand. This is why many businesses are relying on flexible and scalable SaaS apps for supply chain management [7] and re-engineering their processes to take use of cutting-edge technologies like artificial intelligence [15], IoT [9, 14, 16], and blockchain [17, 18] to boost efficiency.

4 Methodology and Implementation

The rollout of the COVID-19 vaccine has been the most fervently anticipated product introduction in the history of vaccination. Thus, any kind of failure within the supply chain is not acceptable. As billions of people globally wait for immunization in the following months and years; it will undoubtedly become the world's most complex yet comprehensive large-scale logistical endeavor. In this work, we propose a blockchain-based methodology to digitize as well as decentralize the conventional scheme thereby allowing for transparent COVID-19 vaccination distribution monitoring and administration throughout the country.

In order to prevent counterfeiting of vaccines, a simple blockchain-assisted implementation is provided which can serve as a cross checking point for end users to check out information related to their vaccinations. This blockchain-assisted model stores vaccination data on a distributed ledger, ensuring information immutability and transparency. It gives truthful information on the delivery of vaccinations to the recipients. The end user can validate the quality of vaccine by checking for it before getting the shot administered. Once administered, the information will be pushed onto the ledger in the form of a transaction with all the necessary details bundled together to avoid tampering.

The entire blockchain-based system is separated into four segments: blockchain or the ledger for information storage, vaccine distribution, vaccine administration, and vaccine supply regulated by blockchain. In the following sections, we'll go through the system's major segments and how they're implemented utilizing self-enforcing smart contracts [13].

4.1 Blockchain

A blockchain is simply a collection of blocks which are linked together to form a connected chain which can be traced. Each block in the blockchain will have its own unique digital fingerprint called hash, including the preceding block's digital fingerprint. There are unique enforcing mechanisms to preserve the integrity of the blockchain and the information being stored inside it. The functions and modules utilized in the proposed blockchain-assisted implementation are discussed below.

Block. The Block class is used to construct blocks by assembling transactions. The block's hash will be stored in hash, while the previous block's hash will be saved in previous Hash. The contents of the block are saved as data, while the timestamp of the block is recorded as timeStamp. calculateHash() is a hash-generating function which preserves the integrity of these transactions and thus of the entire block as well. In a blockchain network, a Merkle root is the hash of all the hashes of all the transactions that constitutes a block. Figure 1 represents the structure of blocks and how genesis block serves as a root for other blocks to derive information. Figure 2 represents the smart contract for this part of the system.

Transaction. Transactions are recorded using unique hashes which are an immutable cryptographic signature. Each transaction will contain a specific amount of data, such as the sender's public key, the receiver's public key, the value or inputs which are to be stored such as references to prior transactions proving the sender has funds to send, and so on. Outputs demonstrate the number of relevant addresses received in the transaction and a cryptographic signature proving that the address owner is the one initiating the transaction and that the data hasn't been tampered with. Figure 3 represents the smart contract for execution of a transaction by taking in the necessary inputs from the vaccine administrator and the end user combined.

StringUtil. StringUtil is a utility class that uses the SHA-512 algorithm to generate a signature from a string and returns it as a string. SHA-512 is a hashing algorithm with 512-bit hash values that performs a hashing function on some data given to it. It is a member of the SHA-2 family of cryptographic hash functions and MessageDigest Class is used to calculate cryptographic hashing value in Java under

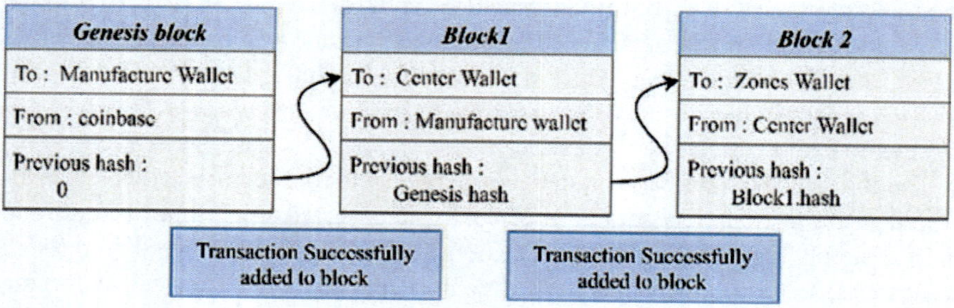

Fig. 1 Blockchain and Genesis block

```
Smart Contract: Block
State:
    {hash, previousHash, merkleRoot, data, timestamp, nonce}
    ArrayList<Transaction> transactions = new ArrayList<Transaction>()
Function calculateHash
    Input:
    Output: hashCalculated
    Begin:
        hashCalculated = StringUtil.applySha512(previousHash + data +
        Long.toString (timeStamp) + Integer.toString (nonce) + merkleRoot)
        return hashCalculated
    End
Function mineBlock
    Input: difficulty
    Output:
    Begin:
        merkleRoot = StringUtil.getMerkleRoot (transactions)
        String target = StringUtil.getDificultyString (difficulty)
        while !hash.substring ( 0, difficulty).equals (target)
            nonce++
            hash = calculateHash()
        end while
        System.out.println ("Block Mined!!! : " + hash)
    End
```

Fig. 2 Smart contract for creating and mining block

the package java.security. getInstance() is a static method that initializes the algorithm. applyECDSASig() method is used to apply ECDSA Signature and returns the result as bytes whereas verifyECDSASig() method verifies the signature and returns the result as boolean. Elliptical Curve Digital Signature Algorithm (ECDSA) is a digital signature method that uses ECC to generate the key pairs required in the digital signature signing and verification procedure. The smart contract demonstrating the binding conditions for generating hashes is provided in Figs. 4 and 5.

Wallet. A blockchain wallet is a digital wallet that stores and manages bitcoin and ether for customers in the traditional cryptocurrency-based implementation of blockchain. Coin ownership is transferred and stored on the blockchain as transactions in crypto-currencies, and participants have a unique address to which funds can be sent and received. Wallets can simply store these addresses in their most basic form; however, most wallets also include software that allows users to create new transactions and then store them onto the blockchain. Similarly, in the case of vaccine supply chain funds, the number of vaccines delivered to and from is taken into consideration. The *getBalance()* method is used to retrieve a wallet's balance whereas *sendFunds()* method is used to send funds from one wallet to another wallet.

Smart Contract: Transaction

```
State:
    {transactionId, PublicKey sender, PublicKey recipient , value , byte[] signature, sequence = 0}
    ArrayList <TransactionInput> inputs = new ArrayList <TransactionInput>()
    ArrayList <TransactionOutput> outputs = new ArrayList <TransactionOutput>()
Function processTransaction
    Input:
    Output: True or False
    Begin:
        if signatureVerification() == false then
            System.out.println ("Transaction Signature failed to verify")
            return false
        end if
        for TransactionInput i:inputs
            i.UTXO = Vaccine_supply_blockchain.UTXOs.get (i.transactionOutputId)
        end for
        if getInputsValue() < Vaccine_supply_blockchain.minimumTransaction then
            System.out.println ("Transaction Inputs too small: " + getInputsValue())
            System.out.println ("Please enter the amount greater than "
+Vaccine_supply_blockchain.minimumTransaction)
            return false
        end if
        leftOver = getInputsValue() - value
        transactionId = calulateHash()
        outputs.add (new TransactionOutput (this.recipient, value, transactionId))
        outputs.add (new TransactionOutput (this.sender, leftOver, transactionId))
        for TransactionOutput o:outputs
            Vaccine_supply_blockchain.UTXOs.put (o.id , o)
        end for
        for TransactionInput i:inputs
            if i.UTXO == null then continue end if
            Vaccine_supply_blockchain.UTXOs.remove(i.UTXO.id)
        end for
        return true
    End
```

Fig. 3 Smart contract for transaction

4.2 Vaccine Distribution

Vaccine distribution is a critical part of the vaccine supply chain, and getting COVID-19 shots to 1.38 billion people is a challenging task that will push supply systems to their breaking point. In our proposed system, the hierarchy is carefully separated into levels where the vaccine distribution chain begins with the manufacturer, moves through the central government, then to the zones, and finally to the vaccination centers. *Manufacturer*, *Center*, and *Zones* are the three primary sub-sections belonging to this module. Figure 7 represents this hierarchy and the relationship between them. The sub-sections discussed are elaborated in detail as follows.

Manufacturer. Vaccine manufacturers play a critical role in the dissemination of vaccines. According to the Indian scenario, DCGI has approved the use of three vaccines: Covishield, Covaxin, and Sputnik. Covishield is manufactured by Serum

```
Smart Contract: StringUtil
State:
Function applySha512
    Input: input
    Output: String
    Begin:
        try
            MessageDigest digest = MessageDigest.getInstance ("SHA-512")
            byte [] hash = digest.digest (input.getBytes ("UTF-8"))
            StringBuffer hexString = new StringBuffer()
            for j=0 to j less than hash.length
                hex = Integer.toHexString (0xff & hash[j])
                if hex.length()==1 then hexString.append ('0') end if
                hexString.append (hex)
            end for
            return hexString.toString()
        end try
        catch (Exception e)
            throw new RuntimeException (e)
        end catch
    End
```

Fig. 4 Smart contract for string utility

```
Smart Contract: Wallet
State:
    {PrivateKey private_Key, PublicKey public_Key}
    HashMap <String,TransactionOutput> UTXOs = new HashMap <String,TransactionOutput>()
Function sendFunds
    Input: PublicKey recipient_val, value
    Output: Transaction
    Begin:
        ArrayList<TransactionInput> inputs = new ArrayList<TransactionInput>()
        totalValue = 0
        for Map.Entry<String, TransactionOutput> item : UTXOs.entrySet()
            TransactionOutput UTXO = item.getValue()
            totalValue = totalValue + UTXO.value
            inputs.add (new TransactionInput (UTXO.id))
            if totalValue > value then break end if
        end for
        Transaction newTransaction = new Transaction (public_Key, recipient_val , (int)value, inputs)
        newTransaction.generateSignature (private_Key)
        for TransactionInput input : inputs
            UTXOs.remove (input.transactionOutputId)
        end for
        return newTransaction
    End
```

Fig. 5 Smart contract for wallet

```
Smart Contract: Manufacturer

State:
    {sputnik, covaxin, covishield, manufacture_vaccine, center_demand, initial_vaccine}
Function Vaccine_availability
    Input: sputnik, covishield, covaxin
    Output: manufacture_vaccine
    Begin:
        initial_vaccine= sputnik + covishield + covaxin
        manufacture_vaccine = sputnik + covishield + covaxin
    End
Function Check_availability
    Input: manufacture_vaccine, center_demand
    Output: center_demand
    Begin:
        if center_demand > manufacture_vaccine then
                center_demand = manufacture_vaccine
                manufacture_vaccine = 0
        else
                manufacture_vaccine =  manufacture_vaccine - center_demand
        end if
    End
```

Fig. 6 Smart contract for manufacturer

Institute, covaxin is manufactured by Bharat Biotech, and sputnik is manufactured by Dr. Reddy's Laboratory, as previously indicated.

In our blockchain system, manufacturers have direct contact with the central government, and they are only responsible to the center for vaccine availability and shortages. This necessary bottleneck ensures that no vaccine enters the supply chain apart from the institutions which are authorized and approved by the Government of India. The smart contract for manufacturer is provided in Fig. 6.

Central. In India, central government is responsible for distributing vaccines all over the country. Firstly, the center will inquire about the needs of each zone, and then it will compile a list of all the zones' demands and transmit it to the manufacturer. The manufacturer checks the demand of the center and then sends the available vaccine to it, which the center then distributes to each zone as per the requirement. The smart contract for central is given in Fig. 8.

Zones. We propose the division of India into six basic zones, i.e., north, south, east, west, north-east, and central. These six basic zones have different vaccine dose requirements. They send their requirements one by one, and it compiles a list of all zones demand before sending it to the center. After a cycle, we take zones demand and combine them with previously unmet demand as a result, there is a new demand for comprehensive requisition. When the vaccines are sent from the center to the zones, the zones split the doses equitably. Zones are in charge of distributing vaccines to vaccine centers in their respective regions. Vaccines reach the residents of the country

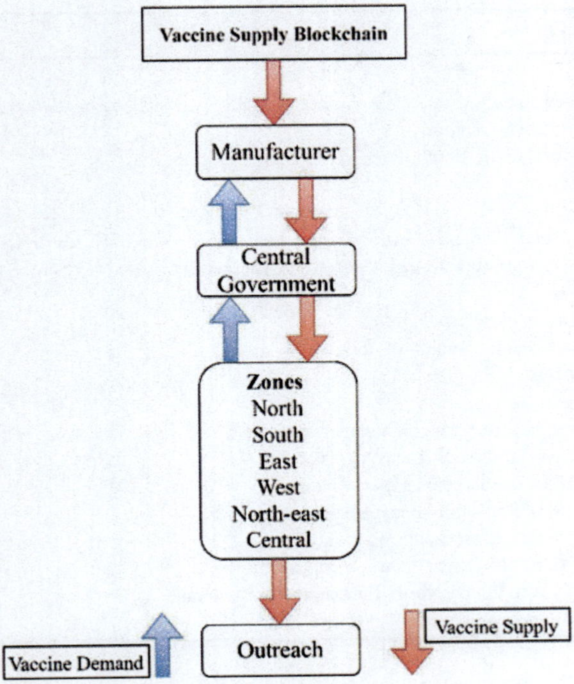

Fig. 7 Vaccine supply chain permissioned hierarchy

```
Smart Contract: Center
State:
     {zone_demand, center_vaccine}
Constructor
     Input: zone_demand
     Output:
     Begin:
          this.zone_demand=zone_demand
          Manufacturer object_Manufacturer=new Manufacturer(zone_demand)
          center_vaccine= object_Manufacturer.Check_Availability()
          Zones object_Zones=new Zones()
          object_Zones.Update_demand(center_vaccine)
     End
```

Fig. 8 Smart contract for Central

Smart Contract: Zones

```
State:
    {North, South, East, West, Central, North_East, total_demand, zones_per_vaccine}
Function ZonesDemand
    Input: North, South, East, West, Central, North_East
    Output:
    Begin:
        total_demand=North+ South+ East+ West+ Central+ North_East
        Center object =new Center(total_demand)
    End
Function VaccineDistribution
    Input: center_vaccine
    Output:
    Begin:
        zones_per_vaccine= center_vaccine/6
        North =North- zones_per_vaccine
        East = East- zones_per_vaccine
        South = South-zones_per_vaccine
        West =West- zones_per_vaccine
        Central= Central-zones_per_vaccine
        North_East = North_East-zones_per_vaccine
    End
```

Fig. 9 Smart contract for zones

in this manner, and the process runs smoothly. The smart contract for functioning of zones is provided in Fig. 9.

4.3 *Vaccine Administration and Verification*

The system will input all of the information, such as the Aadhar card, OTP, vaccine id, and zone id, and then all this information will be combined and will generate a unique key. This unique key will be then combined with time stamp and nonce for increasing security and then the combination of all these will be converted into hash and the transaction will be added to the Blockchain. In the verification process, the user can verify all the information by providing their Aadhar id. The smart contract for vaccine administration is provided in Fig. 10.

4.4 *Vaccine Supply Blockchain*

Vaccine supply blockchain is working as a bridge between the blockchain functions, vaccine distribution, and vaccine administration. It will serve as a connecting class

Smart Contract: Vaccine_Administration

```
State:
    {aadhar_card, OTP, vaccine_id, zone_id, unique_key}
Function Personal_Details
    Input: aadhar_card, OTP, vaccine_id, zone_id
    Output:
    Begin:
        unique_key=zone_id+vaccine_id+aadhar_card+OTP
    End
```

Fig. 10 Smart contract for vaccine administration

for all the other classes. It is in-charge of declaring wallets, transferring doses to the manufacturer's wallet, as well as the center and zones. It will also check if the blockchain is legitimate after adding blocks, using the isChainValid() method, which iterates over each block in the blockchain. It will compare the hashes of the previous and current blocks and return a result in the form of a Boolean. Smart contract for supply of vaccine is provided in Fig. 11.

5 Results

The proposed blockchain-assisted system was implemented in Object-Oriented programming language Java in Eclipse IDE. The system was designed from scratch with the help of fundamentals of OOPS concepts. The system is permission less toward the end user and permissioned when moving backwards. In India, central government is responsible for vaccine distribution so we have assumed that the central government has central authority in the system. The sequence diagram to demonstrate the working of vaccine supply is provided in Fig. 12, whereas the sequence diagram demonstrating the workflow from distribution till administration is represented in Fig. 13.

Figure 14 represents the graph between the number of transactions and the transaction time taken by the blockchain system. We can observe from the graph that as the number of transactions increases the time taken for executing all the transactions will also increase. However, the latency is not an issue as these transactions are being pushed every second. Therefore, the presented system scales well with respect to sudden increase in pushing of transaction without breaking. On the other hand, the latency for the number of transactions less than 1,00,000 is significantly less and can therefore be considered negligible.

Smart Contract: Vaccine_supply_blockchain

```
State:
    ArrayList<Block> blockchain = new ArrayList<Block>()
    HashMap<String,TransactionOutput> UTXOs = new HashMap <String,TransactionOutput>()
    {difficulty = 3, minimumTransaction = 0, Wallet walletM, Wallet wallet}
    {Wallet wallet1, Wallet wallet2, Wallet wallet3, Wallet wallet4, Wallet wallet5,
     Wallet wallet6, Transaction genesisTransaction}
Function main
    Input:
    Output:
    Begin:
        Security.addProvider ( new org.bouncycastle.jce.provider.BouncyCastleProvider() )
        genesisTransaction = new Transaction ( coinbase.public_Key, walletM.public_Key,
Manufacturer.intial_vaccine , null)
        genesisTransaction.generateSignature (coinbase.private_Key)
        genesisTransaction.transactionId = "0"
        genesisTransaction.outputs.add (new TransactionOutput (genesisTransaction.reciepient,
genesisTransaction.value, genesisTransaction.transactionId))
        UTXOs.put (genesisTransaction.outputs.get (0).id, genesisTransaction.outputs.get (0))
        System.out.println (" Creating and Mining Genesis block! ")
        Block genesis = new Block ("0")
        genesis.addTransaction (genesisTransaction)
        addBlock (genesis)
        System.out.println("Manufacturer wallet " + walletM.getBalance())
        Block block1 = new Block(genesis.hash)
    block1.addTransaction(walletM.sendFunds(walletC.public_Key,Manufacturer.center_demand ))
        block1.addTransaction(walletC.sendFunds(wallet1.public_Key, Zones.zones_per_vaccine))
        block1.addTransaction(walletC.sendFunds(wallet2.public_Key, Zones.zones_per_vaccine))
        block1.addTransaction(walletC.sendFunds(wallet3.public_Key, Zones.zones_per_vaccine))
        block1.addTransaction(walletC.sendFunds(wallet4.public_Key, Zones.zones_per_vaccine))
        block1.addTransaction(walletC.sendFunds(wallet5.public_Key, Zones.zones_per_vaccine))
        block1.addTransaction(walletC.sendFunds(wallet6.public_Key, Zones.zones_per_vaccine))
        addBlock(block1)
        Block block2 = new Block(block1.hash)
        addBlock(block2);
        isChainValid();
    End
```

Fig. 11 Smart contract for vaccine supply blockchain

6 Conclusion and Future Work

The Cowin site is assisting India in conducting immunization efforts across the country. According to current scenario, each person who wishes to be vaccinated must register on the website and provide their Aadhar card as well as an OTP that will be sent to their registered mobile number. The user can access its vaccination certificate from the portal but our blockchain system is transparent to public. We are working toward the counterfeiting of vaccines hence the blockchain system should be transparent, trustworthy, and efficient. UIDAI (Aadhar Id), OTP, vaccine id, zone id, and time stamp all together contribute to the formation of a unique key. All these information are not a confidential one and anyone should be able to validate and verify it at any point of time using its Aadhar ID.

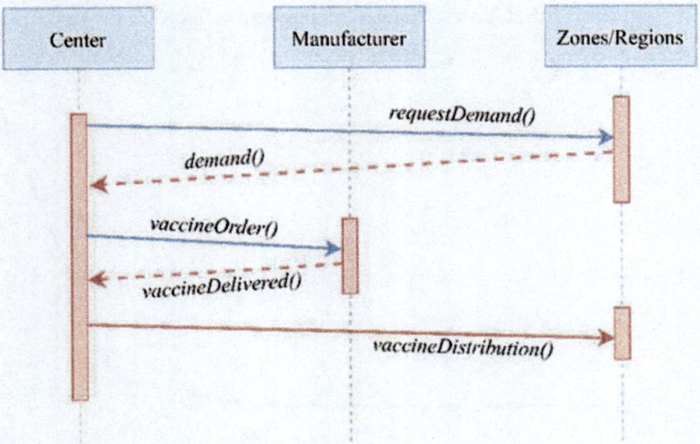

Fig. 12 Sequence diagram for vaccine supply chain

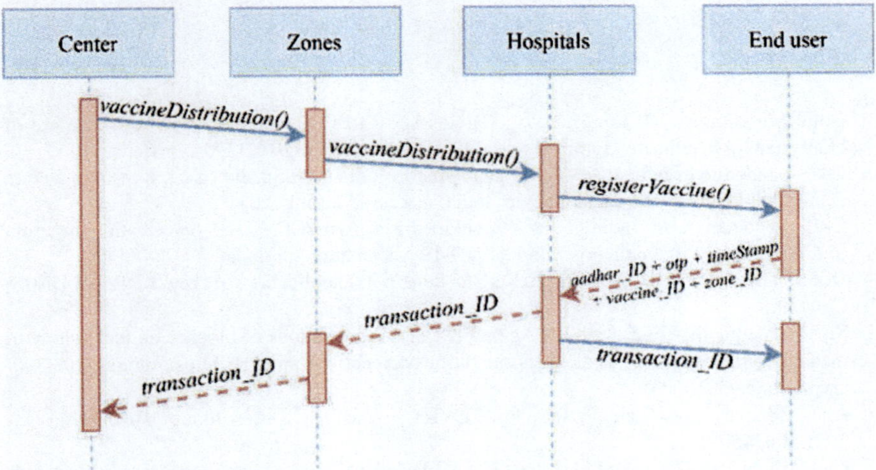

Fig. 13 Sequence diagram from distribution to administration

The study and implementation proposed in this research work is modest and has used the basic concepts of Object-Oriented Programming. Using advanced concepts of OOPs in Java, we can make the blockchain system more secure and reliable in the future. In the proposed blockchain system, the country is divided into zones, with the assumption that vaccines are given equitably. We can work toward further dividing zones into states and states into regions, as well as dividing vaccines based on demand and population of the various states in the country. Our research is a small step toward utilizing blockchain technology using Object-Oriented Programming Language Java to address the critical vaccination supply chain issue.

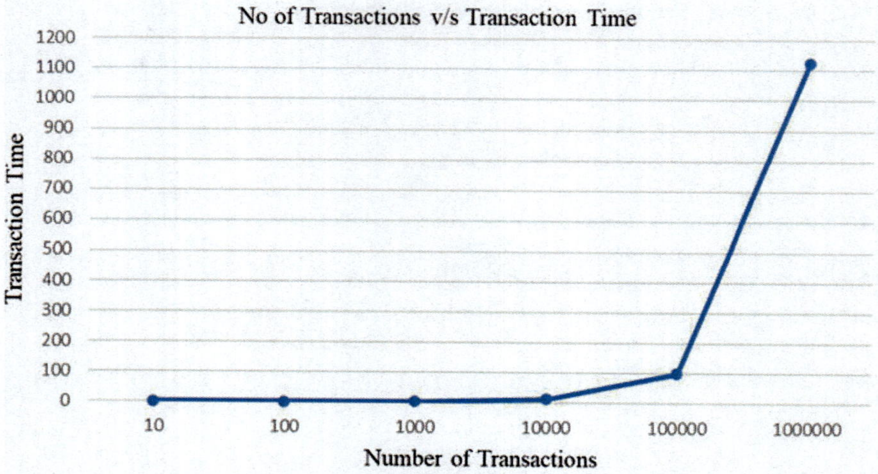

Fig. 14 Performance in terms of transaction time and scale

References

1. Hölbl M, Kompara M, Kamišalić A, Zlatolas LN (2018) A systematic review of the use of blockchain in healthcare. Symmetry (Basel) 10. https://doi.org/10.3390/sym10100470
2. C.-19 pandemic in India, COVID-19 pandemic in India—Wikipedia (n.d.). https://en.wikipedia.org/wiki/COVID-19_pandemic_in_India. Accessed 12 July 2021
3. C–19 vaccination in India. https://en.wikipedia.org/wiki/COVID-19_vaccination_in_India (n.d.). https://en.wikipedia.org/wiki/COVID-19_vaccination_in_India
4. Di Francesco P, Maesa D, Mori (2020) Blockchain 3.0 applications survey. J Parallel Distrib Comput
5. Siyal AA, Junejo AZ, Zawish M, Ahmed K (n.d.) Applications of blockchain technology in medicine and healthcare : Challenges and future perspectives. pp 1–16. https://doi.org/10.3390/cryptography3010003
6. Antal C (2021) Blockchain platform For COVID-19 Vaccine, 2. https://doi.org/10.1109/OJCS.2021.3067450
7. Guo J, Cengiz K, Tomar R (2021) An IOT and blockchain approach for food traceability system in agriculture. Scalable Comput Pract Exp 22:127–137
8. Fusco A, Dicuonzo G, Atti VD, Tatullo M (2020) Blockchain in Healthcare : Insights on COVID-19
9. Choudhury T, Khanna A, Toe TT, Khurana M, Nhu NG (2021) Blockchain applications in IoT ecosystem. Springer, Cham, Switzerland
10. Marbouh D, Abbasi T, Maasmi F, Omar IA, Debe MS, Salah K, Jayaraman R, Ellahham S (2020) Blockchain for COVID-19: review, opportunities, and a trusted tracking system. Arab J Sci Eng 45:9895–9911. https://doi.org/10.1007/s13369-020-04950-4
11. AA., Sarishma Dangi PR (2022) Integrating blockchain with education: proposed model, prospects and challenges. In: Transform. through blockchain technol. pp 87–111. https://doi.org/10.1007/978-3-030-93344-9_4
12. C. vaccines find way to black Market, Andhra Pradesh: Covid vaccines find way to black market|Vijayawada News–Times of India (n.d.). https://timesofindia.indiatimes.com/city/vijayawada/covid-vaccines-find-way-to-black-market/articleshow/82952946.cms. Accessed 12 July 2021

13. Negi D, Sah A, Rawat S, Choudhury T, Khanna A (2021) Block chain platforms and smart contracts. In: Choudhury T, Khanna A, Toe TT, Khurana M, Gia Nhu N (eds) Blockchain applications in ıot ecosystem. eaı/springer ınnovations in communication and computing. Springer, Cham. https://doi.org/10.1007/978-3-030-65691-1_5
14. Sille R et al. (2022) Review study on blockchain frameworks for security ıssues in IoT devices. In: 2022 International symposium on multidisciplinary studies and ınnovative technologies (ISMSIT). Ankara, Turkey, pp 876–881. https://doi.org/10.1109/ISMSIT56059.2022.9932744
15. Liu L, Han M, Zhou Y, Parizi RM, Korayem M (2020) Blockchain-based certification for education, employment, and skill with incentive mechanism. https://doi.org/10.1007/978-3-030-38181-3_14
16. Panarello A, Tapas N, Merlino G, Longo F, Puliafito A (2018). Blockchain and iot integration: A systematic survey. https://doi.org/10.3390/s18082575
17. Tomar R (2019) Sarishma. Maintaining trust in VANETs using blockchain, Ada User J. 40:236–241. https://doi.org/10.1145/3431235.3431244
18. Sharma A, Sarishma R, Tomar N, Chilamkurti BG (2020) Kim, Blockchain based smart contracts for internet of medical things in e-healthcare. Electron 9:1–14. https://doi.org/10.3390/electronics9101609
19. Sarishma A, Gupta P (2021) Mishra, Blockchain based framework to maintain chain of custody (coc) in a forensic ınvestigation. Commun Comput Inf Sci: 37–46. https://doi.org/10.1007/978-3-030-81462-5_4
20. Yong B, Chen H (2020) A blockchain based system for safe vaccine supply and supervision
21. T. get fake C. vaccine shots in alleged scam in India, Thousands get fake COVID vaccine shots in alleged scam in India–CBS News (n.d.). https://www.cbsnews.com/news/fake-covid-vaccine-shots-india/. Accessed 12 July 2021
22. Mk M, Srivastava G, Rama S, Somayaji K (n.d.) An ıncentive based approach for COVID-19 using blockchain technology

Reviewing Metaverse: Applications, Challenges, and Integration with Other Technologies

Aastha Verma and Sarishma Dangi

1 Introduction

The term "metaverse" first appeared in the science fiction novel Snow Crash in 1992, written by Neal Stephenson [1]. Metaverse is a combination of two words, "meta," meaning beyond, and the universe, so metaverse can be translated into a universe that is beyond ours. In Snow Crash, the metaverse was defined as the virtual environment that is parallel to ours and where people interact with their virtual avatars. With technologies like artificial intelligence, augmented reality, virtual reality, etc., we are moving toward turning this sci-fi world into reality. From movies like The Matrix, Ready Player One, and Avatar, people have already gotten a glimpse of what the metaverse can be in the near future.

Now that we've moved on to cutting edge technologies like Web 3.0, Edge computing, Robotics, etc., we realize the concept of the metaverse is not that far away. It has moved on from being a theoretical concept toward an engineering problem, from being a fiction to non-fiction, and since Facebook was renamed to "meta," "metaverse" has become a new buzzword. People are excited to learn and research about metaverse to be a part of this technological revolution. Companies are also moving their research and investment to secure their spots in the metaverse.

In a presentation at the annual Connect conference 2021, Facebook CEO Mark Zuckerberg announced that the company is rebranding as "meta" and shared his vision of building a more immersive and embodied internet in the form of a metaverse. He shared his vision of building a creative economy out of the metaverse where developers and contributors can develop and contribute to building this metaverse technology that is built around people, connecting them with a feeling of presence. With the development of the metaverse, the way we connect will change completely. Think about attending the meeting through holograms with augmented reality, or

A. Verma · S. Dangi (✉)
Graphic Era University, Dehradun, India
e-mail: sarishmasingh@gmail.com

A. Ramdane-Cherif et al. (eds.), *Machine Intelligence and Data Science Applications*, Algorithms for Intelligent Systems, https://doi.org/10.1007/978-981-99-1620-7_20

teleporting in time and space to experience the ancient city or explore the deep sea. All of this can be possible in the metaverse. "*In order to unlock the potential of the metaverse, there needs to be interoperability,*" Mark Zuckerberg said. Presence, Avatars, Home Space, Teleporting, Interoperability, Privacy, Safety, Virtual Goods, and Natural Interfaces are stated as some key features to build the metaverse. With this vision, Meta will be investing $150 million to train the next generation of creators to build immersive learning content. For monetizing the metaverse, developers can build virtual products and create a marketplace regulated with NFTs and cryptocurrencies.

Though it may appear unreal to some, components of metaverse technology have already been developed within online video games such as World of Warcraft, Minecraft, Fortnite, Second life, Roblox, The Sandbox, VRChat, and others, as well as some early projects such as Active Worlds and The Palace. Meta is also working on projects like Horizon, Cambria, etc., to make Metaverse more accessible. Google also developed its AR glasses in 2013, though they were not commercially successful but it surely was a brand-new experience for the people.

In the past few years, like any other developing technology, the meaning and concept of the "metaverse" have also evolved, and new features keep being added. By some researchers, it can be defined as the "post-reality universe," a perpetual and persistent multiuser environment merging physical reality with digital reality [2] with access points like VR (Virtual Reality), AR (Augmented Reality), and MR (Mixed Reality). While some refer to it as a fully immersive three-dimensional digital environment [3]. We define the metaverse as the immersive, multisensory, embodied, and interactive universe that can be integrated with the real world to create a form of the internet where we can communicate with the feeling of presence and gestures.

The vast possibilities this new emerging domain of technology has to offer in the near future are revolutionary.

As IT companies are focusing on the development of technologies like AR, VR, and MR and the IT giants like Meta are focusing on the metaverse, this field of technology shows great future potential. In the next 5–10 years, we will be able to experience this more significantly. As we move from desktops to the web to smartphones, technology is growing rapidly. Imagine holding meetings and conferences with people interactive with holograms, medical students practicing surgery in the metaverse, or you are attending virtual concerts of your favorite artist; all of this can be achieved, and all is possible in the metaverse. In this paper we will be

- Reviewing the research that is happening in the metaverse.
- Revisiting the history to outline the development of the present-day metaverse.
- Challenges that are being faced in the metaverse.
- To state the relationship of the metaverse with other technologies like blockchain, AI (Artificial Intelligence), and IoT (Internet of Things) (Fig. 1).

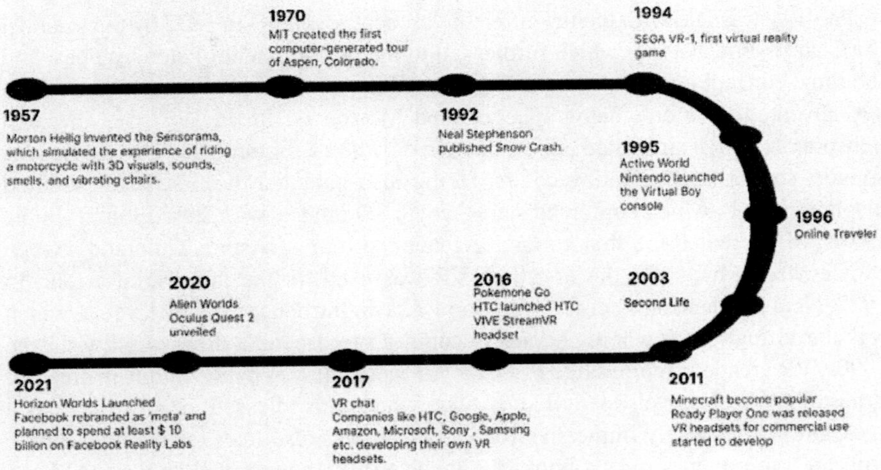

Fig. 1 A brief history of metaverse

2 Related Work

Every emerging or existing technology has a large number of researchers working on it, defining it, enhancing it, and portraying it in their own way, from different points of view. The same goes for metaverse. Every existing work has a different description of the concept, adding more dimensions and depth to it. So, it is important to get to know them properly. When it comes to defining the metaverse, some define it as "the interconnected web of social, networked, immersive environments" [2]. Comparing metaverse with the "heterotopia" [4] and referring it to as a paradox conjures up an image of an upsetting matrix. That can be true if we think about the security and privacy challenges of metaverse or the "*metaverse addiction.*" The solution of the security issues may be found in the blockchain technology [5], as it is believed that the blockchain will provide the fundamental infrastructure of the metaverse and AI can help make that metaverse more intelligent [6]. Metaverse characteristics include multi-technology, sociality, and hyper-spatiotemporality [7]. With the development of 3D internet or web 3.0 people are moving from 2D internet toward a more dimensional internet that can be more interactive and robust. Metaverse is a crucial concept that will make that happen. Powered with cutting edge technologies like blockchain, AI, ML, edge computing, etc., the fiction will soon be a reality or we can say that immersive virtual reality will soon be the reality.

From its debut in Snow Crash till now, the term "metaverse" has evolved in terms of technology. Before that, in 1935, fiction writer Stanley Weinbaum published the book Pygmalion's Spectacles, in which the main protagonist explores a virtual world using a pair of multisensory goggles. To experience virtual reality, we are highly dependent on technologies like extended reality, mixed reality, virtual reality, and augmented reality. If we look at history, the foundation of the modern-day Web 3.0

technology was laid with the invention of "binocular vision" in 1838 by the scientist Sir Charles Wheatstone, which further led to the development of stereoscopes, and the same concept is used in the modern-day VR headset. Then the first VR machine was invented by a cinematographer named Morton Heilig in 1957. He invented Sensorama, which simulated the experience of riding a motorcycle with 3D visuals, sounds, smells, and vibrating chairs. Heilig also patented the first head-mounted display in 1960, which combined stereoscopic 3D images with stereo sound. In the 1970s, MIT created the first computer-generated tour of Aspen, Colorado, Aspen Movie Map, which was the first time VR was used for the purpose of a tour. In 1982, Neal Stephenson's sci-fi novel, Snow Crash, introduced the metaverse, which was the virtual place where characters could go to escape a dreary reality. In the 1990s, VR arcade machines like SEGA VR-1 were enjoyed by the public in arcades. Ernest Cline's hit novel Ready Player One, which was released in 2011, gave us a peek into a completely immersive reality, which was also later made into a movie with the same name. In this book, the metaverse was portrayed as the OASIS, a place where the players are represented by their avatars. Since 2010, companies like Google, Microsoft, Sony, Samsung, and Meta have been creating more immersive reality tools like HoloLens headsets, VR headsets, virtual conference rooms, etc.

Many studies have been conducted to determine how we can apply the concept of the metaverse in various domains such as medicine, education, gaming, food supply chain management [8], social networking, and e-commerce, in conjunction with technologies such as Edge Intelligence [9], AI, and blockchain [6]. But the access points that join our physical world with the metaverse: augmented reality, mixed reality, and virtual reality are necessary to experience the metaverse. But the line between these access points is a bit blurry. The difference between the terms extended reality (XR), VR, AR, and MR should be clearly understood.

By connecting the actual environment to a digital one, the next-generation computing platform known as "extended reality" (also known as "XR") has transformed how we work, study, connect, and play. XR is the umbrella term used to refer to AR, MR, and VR. The term has been popularized in the past few years as technology has been improving rapidly [10]. In this research work [11], it has been clearly defined what you mean by the "virtual world" and how it is related to our physical world. VR is defined as a simulated experience that can be similar or different from the real world. Apart from Virtual reality terms like Virtual World, Visual environment and immersive virtual reality are also used to refer to the technology. Since the development of sensorama and the advent of current VR headgear, virtual reality applications and usage have expanded beyond simple entertainment.

Whereas, AR is a form of experience in which the real world is integrated with digital information in real time [12]. Some popular examples of augmented reality are Snapchat, Pokemon GO, Google Glass, etc. The main difference between VR and AR is that, whereas AR uses the real-world environment and integrates it with digital information, VR requires tools such as a VR headset to create a virtual environment. A virtual reality can be completely imaginary, whereas augmented reality is more interactive with the real world. We can say that augmented reality is mounting the virtual objects in the real-world environment. The technology that can implement both AR and VR on the same device or application is called mixed reality. Mixed reality is also defined as the "strong AR" by a few experts, and the term sometimes stands for a little blurry [13].

3 Applications

3.1 *Metaverse in Healthcare*

XR technology has been actively used in healthcare [14] technology, and it is also expected that the metaverse will be really helpful in this field. From the training and education of healthcare professionals to finding assistance in real-life surgeries, the metaverse can really revolutionize the industry. John Hopkins University successfully carried out the world's very first surgery using Xvision, an AR-based spinal surgery support system. This method is approved by the FDA and is being used in surgeries from 2020. Other than that, Metaverse is also used in the rehabilitation of mental illness [15] and psychotherapy in patient treatment [16]. The application of AR and VR technologies for the advancement in the field of medical science can also be seen in past works [17].

3.2 *Metaverse in Gaming*

Metaverse has been popular in the gaming industry for a very long time. With the development of games like Minecraft, The Sandbox, Chain of Alliance, and Fortnite, we can already have a peek into the metaverse. With the development of Meta's Horizon Worlds, the vision of the modern metaverse is much clearer. Horizon Worlds is an ever-expanding social universe where you can hang out with friends, interact with new people, play games, attend events, and enjoy many more fun activities. It comprises over 10,000 worlds and experiences to explore. The online gaming mogul, Roblox Corp., has also developed the metaverse, which garners almost 47 million users on a daily basis with a diverse experience.

3.3 *Metaverse in Education*

With virtual classrooms, education [18] can become more about practical experience than the theoretical basis of it. Students can study astronomy by virtually experiencing outer space in the metaverse. History students can learn about the Mughal dynasty by teleporting into that time with the metaverse. Students studying earth science can study the Amazon rainforest by not physically going there. Even in the case of school closure, like we experienced during the COVID-19 pandemic, students can still socially interact [19]. A student can attend any school or college around the globe without physically moving there. A smart testing system can also be adopted using the metaverse [16]. Also, children who are suffering from diseases like autism can get special training needed for their skill development and education [20]. As for the credibility of this metaverse-based education system can be governed by the blockchain itself [21].

3.4 Metaverse and Metaeconomics

Since we are talking so much about various applications of the metaverse, monetization of the metaverse and making it commercially successful is also important. So, we need to think about various ways in which we can make the metaverse profitable for people. Metaverse can be seen as an integrated design space for developers and designers to develop and design [20]. The metaverse can also be used in the transportation industry with the help of an intelligent transportation system [22]. To encourage the purchase of goods and services in virtual commerce, the alignment of application design paradigms with the elements influencing customer behavior is essential. Virtual commerce transforms online shopping from 2D product catalogues to 3D immersive virtual places by integrating immersive technologies like augmented reality and virtual reality. To encourage the purchase of goods and services in virtual commerce, the alignment of application design paradigms with the elements influencing customer behavior is essential [23]. Their relationship must be clarified, along with any potential application design enhancements [24]. Apart from that, developing a meta-society comprising various meta-enterprises can also help in building a strong meta-economics [25].

Big fashion brands like Adidas, Nike, Puma, Gucci, etc. are also working on virtual fashion for your avatars, inventing the concept of cyber-space fashion. For example, you have seen the styling options from various brands on Snapchat. Developers can also develop metaverse products and sell them in the metaverse to do the business. As for the money transactions in the metaverse, they can be done with the help of cryptocurrency and NFTs [26].

3.5 Metaverse in Tourism

We can also diversify tourism with the metaverse. With XR, anything is possible, from online virtual tours of one of the world's highest peaks to experiencing the Colosseum in the medieval era through augmented reality. Even in the midst of the lockdown, you can still travel to any corner of the world or have a virtual tour of the Natural History Museum. The metaverse of tourism would aim to improve the travel experience of consumers by leveraging the benefits of virtual tourism. People can also use the metaverse to attend various events like live concerts, conventions, after-parties, award functions, etc. (Table 1).

4 Relationship of Metaverse with Other Technologies

Metaverse cannot function on its own without the assistance of other technologies. For it to be intelligent, secure, scalable, accessible, and dynamic, it is dependent on technologies that were developed and are constantly being improved upon by many organizations. To be commercially successful, it needs a state-of-the-art system that incorporates various technologies in perfect harmony. Metaverse also needs a user-friendly interface that would be simple for people to learn yet would provide them with deep functionality.

4.1 Metaverse Empowered with Blockchain

We know that a blockchain is a decentralized, distributed ledger that consists of blocks of records securely linked together via cryptography, making it immutable and secure. These blocks are then chained together cryptographically to form a linear, time-stamped series of verifiable transactions [5]. It is also believed that blockchain will be the fundamental infrastructure of the metaverse. Also, we know that cryptocurrencies like Bitcoin, Ethereum, etc. are based on the same technology, and the transections in the metaverse will take place via cryptocurrency. Other than that, the immutability and security that are provided by the blockchain add on to its advantage of being used [6]. Thus, creating decentralized blockchain-based virtual spaces for mixed reality that can be shared, mapped, and reused among different applications and users [27].

4.2 Metaverse and Artificial Intelligence

Artificial intelligence is the new normal in the field of technology. Almost all applications use AI models in one way or another, whether it is an OTT platform or Social Networking sites, every application incorporates an AI model to make intelligent and more user friendly. In metaverse, there will be various bots, speech recognition features, NLC in games [6] and many more features that will be powered with AI creating a more immersive experience [25].

Table 1 XR Gadgets with cutting edge technology

Gadgets	Images	Manufacturers	Description
Magic Leap 2 AR Headset	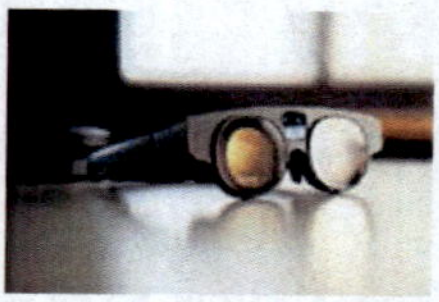	Magic Leap	Lightweight immersive futuristic XR headset
SenseGlove Nova AR glove	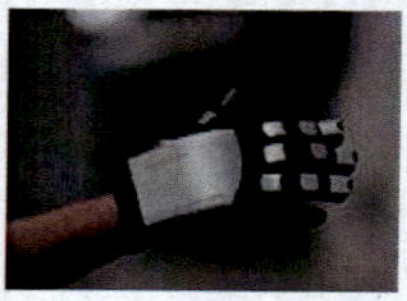	Senseglove	Haptic gloves that make digital object feels like they are really in your hand
HTC VIVE Wrist Tracker	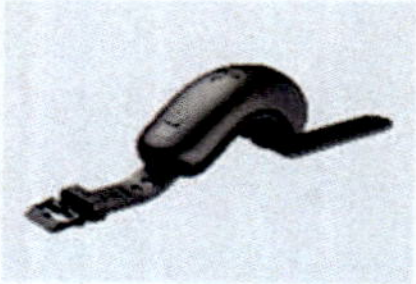	HTC	Lightweight gadget to track objects and movements in 3D space
Google AR Glasses		Google	Translates languages in real time right on the lenses
VITURE One wearable gaming glasses		VITURE	Provides the user with portable VR screen and can be control with their voice
HTC VIVE Pro 2 VR headset		HTC	Gives immersive VR experience in 5k resolution
Varjo XR-3 mixed reality headset		Varjo	Delivers powerful, photorealistic mixed reality experience

4.3 *Metaverse and Internet of Things (IoT)*

IoT or Internet of things is referred to the collective network of connected devices and technology that facilitates the communication between various devices, technologies as well as cloud. The dependencies of metaverse on IoT in current scenario are massive. As you can't get any immersive experience without using the devices like camera, motion sensors, VR headsets, etc. The proliferation of intelligent IoT devices [28] in our daily contexts creates possibilities for brand-new services and apps that can enhance our quality of life [29]. Since it is hard to carry a bundle of controllers for various IoT devices, users can do without physical controllers. Virtual environments (AR/MR/XR) demonstrate prominent aspects of visualizing unseen instances and their activities, including Wi-Fi and user personal information. Additionally, AR may show people how smart speakers and cameras transmit IoT data, alerting them to potential risks in user-IoT interactions. As a result, consumers can leverage platforms for AR visualization to handle their IoT data.

5 Challenges

The underlying AR and VR technologies provide a variety of difficulties for the Metaverse. Both technologies are persuasive and have the power to change users' thoughts, feelings, and actions. The high cost of the equipment is a deterrent to widespread use, although it is anticipated that it will be lessened over time. Four groups of risks associated with AR can be distinguished: those that pertain to data privacy, psychology, ethics and morality, physical well-being, health, and safety. On a physical level, harmful mishaps have occurred as a result of users' attention being diverted by location-based AR applications. A psychological problem that needs to be avoided is information overload. Unauthorized enhancement and the fabrication of facts to support biased viewpoints are moral dilemmas. The greatest danger is associated with data gathering and sharing with third parties.

5.1 *Privacy and Security*

With the expansion of metaverse the data online will increase exponentially. Also, the user information has to be secured. Though the basic infrastructure of the metaverse will of blockchains still the question of privacy and security stands firm [30]. Suppose if a company having a meeting in a virtual conference room in metaverse, what is the assurance that the conversation taking place there is not heard by any rival company. Also, the XR system may render users' control over their own personal data ineffective. Due to the fact that many AR/VR systems rely on integrations or third-party apps with questionable security, it's feasible that hackers will find a way to steal critical data, including personal information. Later, they might hijack and take over XR remotely to pretend to be someone in a social collaboration at work [10]. The extra layer of data may turn out to be a cybersecurity risk [31]. Spatial doxing and volumetric capturing can violate people's privacy. More crucially, Metaverse actors may be persuaded to create user psychographic profiles based on emotional user input. These profiles may be used to draw unintentional conclusions about behavior that support algorithmic prejudice.

5.2 *Scalability*

The virtual world may have scalability concerns just like many other technologies. Since the metaverse will be vast, having millions of users worldwide, having their own virtual space, comprising virtual as well as real features. To mimic the characteristics of the real world that is governed by the laws of physics is itself a challenge because we can never absolutely replicate it on computers. Currently, the metaverse works on the traditional centralized model which is not scalable and has many limitations. But hypothetically the metaverse will be built on blockchains which follows a decentralized approach. Hence there are many peer-to-peer approaches that has been proposed to divide the computational and communication load in ways to address the scalability needs of the metaverse [3].

5.3 *Social Issues*

An excessive use of the Internet can result in a medical, social, or mental disease known as a cyber-syndrome. Electronic devices have shrunk and gotten lighter as interactive ways have continued to emerge. People are using the Internet more and more as a result of equipment simplification. The metaverse and the physical world are interconnected at the same time. The problem of cyber-syndrome is made even more problematic by the blurring of the actual and virtual worlds and the Metaverse's intense immersion. Standards must be established for the Metaverse because it is a multidimensional virtual world that is intimately related to reality [7]. Aside from that, we also need to raise awareness of metaverse ethics, as many people take advantage of their anonymity and the fact that they are in a virtual environment to engage in illegal activity digitally, which has a negative psychological impact on both them and other online users. A ten-year old can access a mature virtual realm by pretending to be older than he actually is, hence there is a problem with parental control or the age element in the metaverse.

5.4 *Health Concerns*

Motion sickness, nausea, and vertigo are some of the most often mentioned health issues associated with VR. The weight of VR headsets prevents extended use due to head and neck strain. Extended VR use has the potential to cause addiction, social isolation, and withdrawal from physical reality, frequently together with body neglect. Open social worlds are also known to be plagued with toxic and antisocial conduct, such as cyberbullying, harassment, and griefing. Virtual reality settings and graphic depictions. Related to data ethics, artificial intelligence algorithms and deep learning techniques can be utilized to create VR deep fake avatars [2] and identity theft [10].

5.5 *Interoperablity*

There are problems with interoperability between Metaverses developed by various companies. Compatibility of Metaverses with the Real World (including currency compatibility and circulation issues, as well as legal dispute resolution). Presently, interoperability in virtual worlds refers to a loosely coupled set of data, format, and information standards, the majority of which are concerned with the transfer of 3D models and objects between different virtual world settings. Despite the fact that several standards have been put out, none of them have found widespread acceptance. Though real interoperability also incorporates communication protocol, location, identification, and currency standards [9], it is not just restricted to the transmission of 3D objects in virtual worlds [27]. The general acceptance of standards in any field of computing typically results from a lot of work, collaboration, and lobbying, where the most challenging hurdles may not really be technological or computational but rather social or political.

6 Conclusion

The underlying AR and VR technologies provide a variety of difficulties for the Metaverse. Both technologies are highly priced and need a lot of improvement to make the metaverse more immersive. Since metaverse is the future, all the tech companies are extensively working on providing their customers with the best virtual reality experience. From making VR games to conference rooms and virtual concerts, we can say that the metaverse is booming rapidly. Google has invested US$ 39.5 million in private equity funds for all metaverse projects. The company is trying to incorporate the metaverse technology in Maps, YouTube, etc. to make them immersive. Microsoft acquired Activision Blizzards investing US$ 70 billion securing a major spot in metaverse. Meta has also put forward its motives to monetize and commercialize the metaverse with investing about US$ 10 billion in its meta platforms. NVIDIA has also launched "Omniverse," its metaverse platform. Tech Mahindra launched one of its kind "Meta Village" a digital twin of Paragaon in Maharashtra to gamify learning on the Roblox platform. In this paper, we have gone through a brief history of the metaverse and how researchers and engineers are working toward its advancement. We have identified some challenges that are being faced in its development and also in its applications in various domains like healthcare, education, etc. Also, we can conclude that this decade is the decade of the metaverse.

References

1. Stephenson N (1992) Snow crash
2. Mystakidis S (2022) Metaverse. Encyclopedia 2(1):486–497. https://doi.org/10.3390/encyclopedia2010031
3. Dionisio JDN, Burns WG, Gilbert R (2013) 3D virtual worlds and the metaverse: current status and future possibilities. ACM Comput Surv 45(3). https://doi.org/10.1145/2480741.2480751
4. van der Merwe D The metaverse as virtual heterotopia
5. Nakamoto S Bitcoin: a peer-to-peer electronic cash system. www.bitcoin.org
6. Yang Q, Zhao Y, Huang H, Xiong Z, Kang J, Zheng Z (2022) Fusing blockchain and AI with metaverse: a survey. IEEE Open J Comput Soc 3:122–136. https://doi.org/10.1109/ojcs.2022.3188249
7. Wang HN et al A survey on metaverse: the state-of-the-art, technologies, applications, and challenges
8. Guo J, Cengiz K, Tomar R (2021) An IOT and blockchain approach for food traceability system in agriculture. Scalable Comput Pract Exp 22:127–137
9. Lim WYB et al (2022) Realizing the metaverse with edge intelligence: a match made in heaven. http://arxiv.org/abs/2201.01634
10. Hui S, Chuah W Why and who will adopt extended reality technology? Literature review, synthesis, and future research agenda
11. Girvan C (2018) What is a virtual world? Definition and classification. Education Tech Research Dev 66(5):1087–1100. https://doi.org/10.1007/s11423-018-9577-y
12. Arena F, Collotta M, Pau G, Termine F (2022) An overview of augmented reality. Computers 11(2). https://doi.org/10.3390/computers11020028
13. Speicher M, Hall BD, Nebeling M (2019) What is mixed reality? In:Conference on human factors in computing systems - proceedings, May. https://doi.org/10.1145/3290605.3300767
14. Sharma A, Sarishma, Tomar R, Chilamkurti N, Kim B-G (2020) Blockchain based smart contracts for Internet of Medical Things in e-healthcare. Electronics 9:1609. https://doi.org/10.3390/electronics9101609
15. Lee CW (2022) Application of metaverse service to healthcare industry: a strategic perspective. Int J Environ Res Public Health 19(20):13038. https://doi.org/10.3390/ijerph192013038
16. Huh S (2022) Application of the computer-based testing in Korean medical licensing examination, the emergence of a metaverse in medical education, journal metrics and statistics, and appreciation to reviewers and volunteers. J Educ Eval Health Prof 19. https://doi.org/10.3352/jeehp.2022.19.2
17. Gupta I, Dangi S, Sharma S (2022) Augmented reality based human-machine interfaces in healthcare environment: benefits, challenges, and future trends. In:2022 international conference on wireless communications signal processing and networking (WiSPNET), pp 251–257

18. Kumar M, Singh TP, Choudhury T, Gupta SC (2019) ICT—The Smart Education System in India. In: 2019 international conference on contemporary computing and informatics (IC3I), Singapore, pp 279–282. https://doi.org/10.1109/IC3I46837.2019.9055562
19. Kye B, Han N, Kim E, Park Y, Jo S (2021) Educational applications of metaverse: possibilities and limitations. J Educ Eval Health Prof 18. https://doi.org/10.3352/jeehp.2021.18.32
20. Lee JH et al (2022) Development and application of a metaverse-based social skills training program for children with autism spectrum disorder to improve social interaction: protocol for a randomized controlled trial. JMIR Res Protoc 11(6). https://doi.org/10.2196/35960
21. Dangi S, Aggarwal A, Rastogi P (2022) Integrating blockchain with education: proposed model, prospects and challenges. In: Transformations through blockchain technology, pp 87–111
22. Njoku JN, Nwakanma CI, Amaizu GC, Kim DS (2022) Prospects and challenges of metaverse application in data-driven intelligent transportation systems. IET Intell Transp Syst. https://doi.org/10.1049/itr2.12252
23. Shen B, Tan W, Guo J, Zhao L, Qin P (2021) How to promote user purchase in metaverse? A systematic literature review on consumer behavior research and virtual commerce application design. Appl Sci (Switz) 11(23). https://doi.org/10.3390/app112311087
24. Jeong H, Yi Y, Kim D (2022) An innovative e-commerce platform incorporating metaverse to live commerce. Int J Innov Comput, Inf Control 18(1):221–229. https://doi.org/10.24507/ijicic.18.01.221
25. Wang FY, Qin R, Wang X, Hu B (2022) Metasocieties in metaverse: metaeconomics and metamanagement for metaenterprises and metacities. IEEE Trans Comput Social Syst 9(1):2–7. https://doi.org/10.1109/TCSS.2022.3145165
26. Rawat HVS, Bisht D, Kumar S, Dangi S (2022) Rise of blockchain-based non-fungible tokens (NFTs): overview, trends, and future prospects. In: Machine intelligence and data science applications: proceedings of MIDAS 2021, vol 132, p 1
27. Ryskeldiev B, Ochiai Y, Cohen M, Herder J (2018) Distributed metaverse: creating decentralized blockchain-based model for peer-to-peer sharing of virtual spaces for mixed reality applications. In:ACM international conference proceeding series, Feb. https://doi.org/10.1145/3174910.3174952
28. Choudhury T, Khanna A, Toe TT, Khurana M, Nhu NG (2021) Blockchain applications in IoT ecosystem. Springer, Cham, Switzerland
29. Lee L-H et al (2021) All one needs to know about metaverse: a complete survey on technological singularity, virtual ecosystem, and research agenda. http://arxiv.org/abs/2110.05352
30. Tomar R (2020) Maintaining trust in VANETs using blockchain. ACM SIGAda Ada Lett 40:91–96
31. Sille R et al (2022) Review study on blockchain frameworks for security issues in IoT devices. In: 2022 international symposium on multidisciplinary studies and innovative technologies (ISMSIT), Ankara, Turkey, pp 876–881. https://doi.org/10.1109/ISMSIT56059.2022.9932744
32. Slater M, Sanchez-Vives MV (2016) Enhancing our lives with immersive virtual reality. Front Robot AI 3. https://doi.org/10.3389/frobt.2016.00074

Smart D: An IoT-Based Smart Dustbin for Better Waste Management

Sharad Singhania, Avita Katal, Rishabh Sharma, Sahil Agarwal, and Vishakha Sharma

1 Introduction

Recent advancements in digital communication, particularly the use of wireless sensor devices, have offered numerous prospects for developers and researchers of many sophisticated smart systems developed for socially significant purposes [1]. Everyone is focusing on smartphones, intelligent systems, smart home automation, and smart water systems, among other things. The Internet of Things enables all people and things to become smarter and more connected to the Internet [2]. Smart bin is a trash collecting bin that monitors the amount of waste in the dustbin and sends warning signals to city authorities, enabling them to clean the dustbin. This form of trashcan will be extremely beneficial in areas where the frequency of individuals using the garbage changes, as timely checks will not suffice [3]. Other capabilities include the automatic closing of the gates utilizing motor and an Ultrasonic Sensor if the trashcan is full, as well as the recognition of anything around the trash to use an IR Sensor, that can avoid rubbish from stacking up from around dustbin.

The tremendous growth in population in recent years needs more garbage transport. As a result, a legal waste management system is critical to preventing the spread of some harmful diseases [4]. There are several dustbins located across the city, including hospitals, businesses, and other organizations. Ultrasonic sensors connect these bins to a microcontroller-based design. An ultrasonic sensor uses to quantify the number of wastes in the garbage can, and it is constantly sent to the microcontroller. The information was sent to the cloud through Wi-Module before being displayed in the web service as the status of the waste level in the bin [5]. With the growing number of keen urban areas, a large number of responsibilities must also be met. A sensible way of life begins with cleanliness, and cleanliness begins with

S. Singhania · A. Katal (✉) · R. Sharma · S. Agarwal · V. Sharma
School of Computer Science, University of Petroleum and Energy Studies (UPES), Dehradun 248007, India
e-mail: avita207@gmail.com

A. Ramdane-Cherif et al. (eds.), *Machine Intelligence and Data Science Applications*, Algorithms for Intelligent Systems, https://doi.org/10.1007/978-981-99-1620-7_21

the trash can. The general people will have their garbage properly disposed of if the dustbins are properly placed and gathered. The underlying issue in the current waste administration framework in many Indian cities is the unfavorable situation of dustbins [6].

In this paper, Smart Dustbin is proposed which is integrated with hardware components like NODEMCU, Servo Motor, and Ultrasonic Sensor. These components help in detecting the status of the dustbin. If the dustbin is empty the mobile application will show "Empty" status and the dustbin will automatically open the lid on detection of the human hand. If the dustbin is full the mobile application will show "Full" status and the dustbin lid will not open and the dustbin will generate a "Dustbin is Full" message from the application.

The subsections of the paper are as follows: Sect. 2 describes the background, Sect. 3 discusses the components used, Sects. 4 and 5 show the methodology and the circuit diagram, respectively. Furthermore, Sect. 6 shows the algorithm and Sect. 7 discusses the applications in this domain.

2 Background

This section covers the work done in the domain of smart dustbin or smart waste management.

The authors in [7] proposed IoT-enabled garbage cans. These dumpsters use RFID tags to monitor trash and are connected to a web-based online system; the web host computes credits and refreshes the virtual wallet database based on the weight of garbage added. It also examines the number of filled garbage cans and announces the state of each trashcan on the municipality server. It alerts them when the dumpster is full and recommends the shortest method for emptying all of the trashcans based on the capacity of the municipal rubbish loading vehicles. When a vehicle completes a route, its capacity is estimated and updated based on the number of garbage cans provided.

The authors in [8] created an Internet of Things-based garbage collecting device based on an Arduino or Raspberry Pi board and an open IoT platform. An Arduino microcontroller, a Wi-Fi module, an ultrasonic sensor, and a heap battery are all part of the suggested design. The Arduino microcontroller is used to collect data from the ultrasonic sensor and the load cell. An ultrasonic sensor detects the elevation of the garbage in the segment, as well as the gravity of the trashcan from the pile cell. The data appears on the LCD display. To follow the garbage framework, Thingspeak and an open IoT conversation are used. The executive may track and organize garbage disposal in this framework.

Smart dustbin was created by the authors of [9]. The proposed smart trashcan is built on an Arduino Uno board and is linked to GSM, GPRS, and sensors. The sensors are utilized to check the dustbin's minimum threshold. The result in different levels has already been determined. If the waste level exceeds the prescribed level, a continuous alert is sent to the proper authorities until the rubbish is recovered, and the externally placed LED becomes red. The LED glows green when the trash bin is empty. This alarm system is activated by sensors on the GSM modem. The specific authority is given a time limit (say, 24 h), and if she or he fails to perform the role, an alarm is sent to the proper authorities.

The authors in [10] proposed smart dustbin utilizing an esp8266 module with two primary pins attached to the sensor, the trigger pin and the echo pin. It aids in the preservation of the environment and the reduction of human work. When the smart bin senses an object on top of it, the lid/cover opens. This is due to the ultrasonic sensor, which is located on both the top and bottom of the lid. When the trash surpasses the specified threshold, the data is sent to another node, the raspberry-pi, which utilizes socket programming to output the appropriate house number into the screen.

The authors in [11] presented a new waste management method for Smart Cities. The suggested trash disposal or monitoring system is seen as vital for modern smart cities (MSC). The Internet of Things (IoT) may be employed in both IS and MSC, culminating in a well-organized operational strategy for the future. Special methods can be utilized to develop the technology used in our trash treatment process in order to give a high level of service (QoS).

The authors in [12] designed smart bin for the management of waste. The message is transmitted to the control room via the smart dustbin's sensors. The bin is equipped with an ultrasonic sensor, an infrared sensor for sensing the amount of trash, and an unnamed gas sensor, all of which are linked to a Raspberry Pi microcontroller, that is programmed to deliver a message to the command center if the garbage is full or if it has not been disposed of for an extended period of time.

Garbage bins or dustbins located in public spaces are frequently overwhelmed in cities. It produces unsanitary circumstances for humans as well as attractiveness in that location, leaving a terrible odor. Smart D, an IoT-based smart dustbin is a solution used to avoid any such scenarios. The main aim of the proposed technique is not to overflow the trash in the bin. An IoT-based smart dustbin is designed that automatically closes its lid if full. This prevents people from overfilling the dustbin and also slipping the garbage out of it. The above-mentioned papers are either in complete or cannot be used in the present scenario as they have not developed any dashboard for the users for real-time monitoring. An android application is developed for the users for easy management of the dustbin.

3 Problem Statement

Many times, in a city the garbage bins or dustbins placed at public places are overloaded. It creates unhygienic conditions for people as well as ugliness in that place leaving a bad smell. To avoid all such situations a prototype called Smart D: IoT-Based Smart Dustbin is proposed. The main goal behind this prototype is to avoid people from littering around the dustbin when the dustbin is full. An IoT-based dustbin that will automatically close its lid if it gets filled is developed. This will prevent people from overfilling the dustbin and also garbage from slipping out of it.

4 Components Used

The prototype includes the various components that are listed below

A. NodeMCU—NodeMCU is open-source Lua-based firmware for Espressif's ESP8266 Wi-Fi System-on-aChip (SoC), which utilizes the SPI Flash File System (SPIFFS) file system on flash. The Espressif NON-OS SDK is used to build NodeMCU, which is written in C. The firmware was originally intended to be a partner item for the renowned ESP8266-based NodeMCU programming kits, but it is now open source and may be run on any ESP module. NodeMCU is an open-source system that pairs devices and transfers data over Wi-Fi using the ESP8266. Figure 1 shows the NodeMCU.
B. Ultrasonic Sensor—Ultrasonic sensors are in charge of identifying obstructions in the way. It is a four-pin gadget with the pin designations Vcc, Trig, Echo, and GND. It features two eyeballs that serve as both transmitter and receiver. The transmitter is in charge of transmitting ultrasonic waves that travel through the air and impact the target. Figure 2 shows the Ultrasonic sensor.
C. Infrared Sensor—The infrared sensor is in charge of sensing the line and level on the ground. An infrared sensor is made up of an emitter and a receiver. This is referred to as a photo-coupler or an optocoupler. Figure 3 shows the IR sensor.
D. Servo Motor—A servo motor is a form of rotary actuator that can precisely manipulate angular position. It has a motor and a feedback signal sensor. A servo drive is necessary to complete the device. The feedback sensor is employed by the drive to precisely control the motor's spinning position. Figure 4 shows the Servo Motor.

Fig. 1 NsodeMCU

Fig. 2 Ultrasonic sensor

Fig. 3 Infrared sensor

Fig. 4 Servo motor

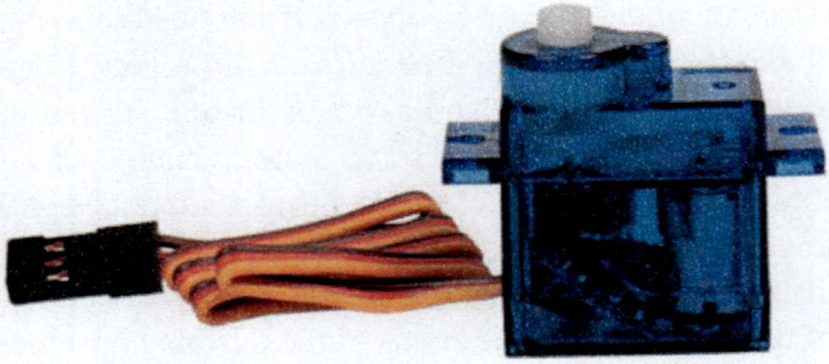

E. Neo-6M—The NEO-6M GPS module is a super high performance complete GPS receiver with a 25 × 25 × 4 mm ceramic antenna built in for efficient satellite search. The power and signal indications can be used to monitor the module's condition. The data backup battery enables the module to save data even when main power has been turned off inadvertently. Figure 5 shows the Neo-6M sensor.

Fig. 5 Neo-6M

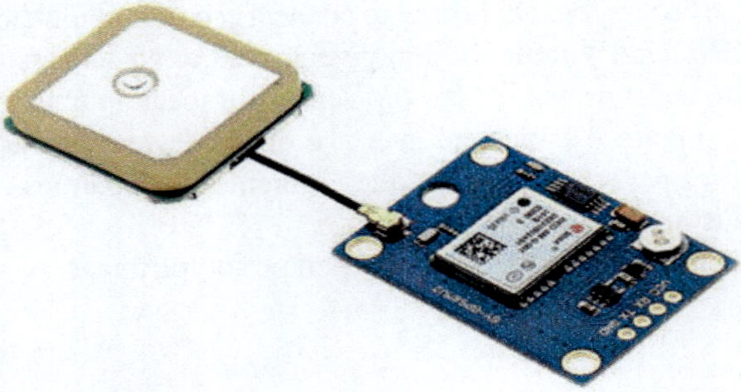

5 Methodology

To personalize the garbage collection status, a prototype is designed which can determine whether dustbin is full or not by using ultrasonic sensors and automatically opens the lid of the dustbin when a person wants to throw trash. A mobile application is developed to notify the authorities when it gets full, saving both money and time. The level of the dustbin is monitored in real-time. C programming language and React native are used to build the prototype. The ultrasonic sensor is utilized to estimate the waste level in the dustbin and infrared sensor is used to detect the presence of the foot or an object near the dustbin, when it detects, the servo motor is activated and the lid is opened if the dustbin is empty and green led is on. The NodeMCU is utilized to link it to a Wi-Fi server, which may upload the information to a specific database and assist us in sending the dustbin details to the receiver. The IR sensor triggers the servo motor, which opens the bin lid with the help of the rope linked to the lid. Neo-6M sensor is interfaced with microcontroller for location tracking. A cross platform mobile application is developed using React Native which displays the garbage level in the dustbin. This application can be used by the users or the staff to check the status of the dustbin conveniently.

6 Circuit Diagram

Figure 6 shows the circuit diagram of the proposed prototype. Pins D0 and D1 of the microcontroller are connected to the servo motor for the input instructions, Pins D2 and D3 are used to give input instructions to ultrasonic sensors and infrared sensors, respectively. Moreover, pin D4 is connected with buzzer for the input instruction to the buzzer. Pin D5 is used to connect eco port of ultrasonic sensor. Furthermore, pins GND and Vin are used to give ground to all sensors and voltage, respectively. Pin D6 is connected to Neo-6M sensor for location tracking.

Figure 7 shows the interface of the developed mobile application. It shows the three pages which depict the different functionalities of the android application like dustbin in use and full warning.

Figure 8 shows the developed prototype image.

7 Algorithm

Figure 9 shows the steps involved in the prototyping of the entire architecture. First the IR sensor and ultrasonic sensor are initialized. Once the initialization is done, the IR and the ultrasonic sensor starts collecting the desired information. The IR sensor is used to activate the servo motor on the basis of the desired output whereas the ultrasonic sensor is used to check whether the dustbin is full or not. Once all the

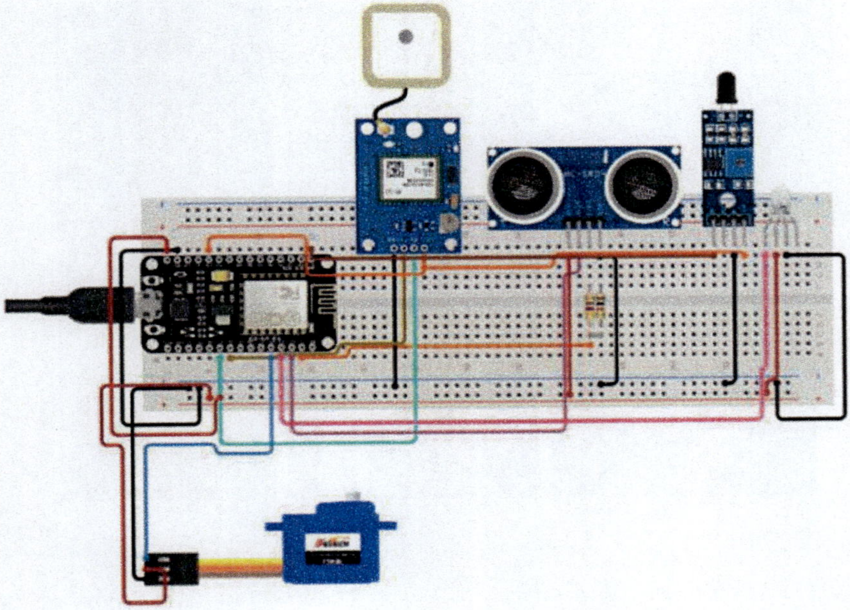

Fig. 6 Circuit diagram

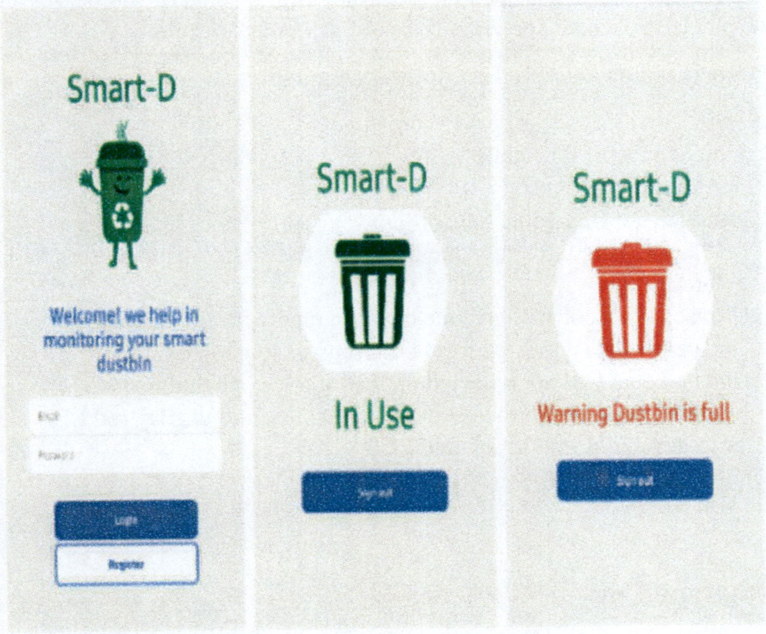

Fig. 7 Interface of mobile application

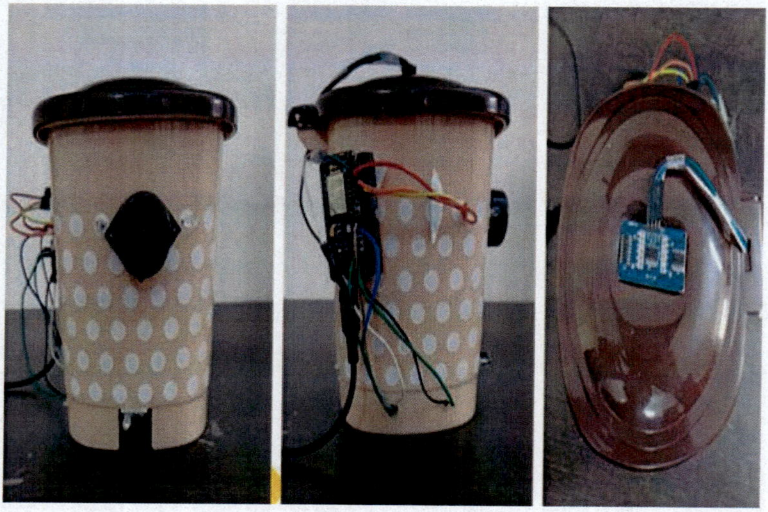

Fig. 8 Prototype image

values are collected from both sensors, the data is updated on the firebase which can be retrieved through mobile application. The ultrasonic sensor calculates the distance of the trash to the upper level and if the trash is full, it activates the buzzer and red light which alerts the user about the level of trash in the dustbin.

Algorithm 1: Smart Dustbin functionality algorithm

Step 1: Start
Step 2: Initialize the ESP Module
Step 3: Initialize the IP Sensor
Step 4: Initialize the Ultrasonic Sensor
Step 5: If IP Sensor digitalRead value goes high then don't display green LED else display green LED
Step 6: Ultrasonic Sensor will give the distance reading from the bin
Step 7: Update the data collected from both sensors to google firebase
Step 8: Updated data will get reflected on the mobile application
Step 9: If the distance is <10 then activate the buzzer and display red LED else set the buzzer to low and don't display red LED
Step 10: Stop

8 Prototype Cost

Table 1 shows the cost of each hardware used for making the smart dustbin.

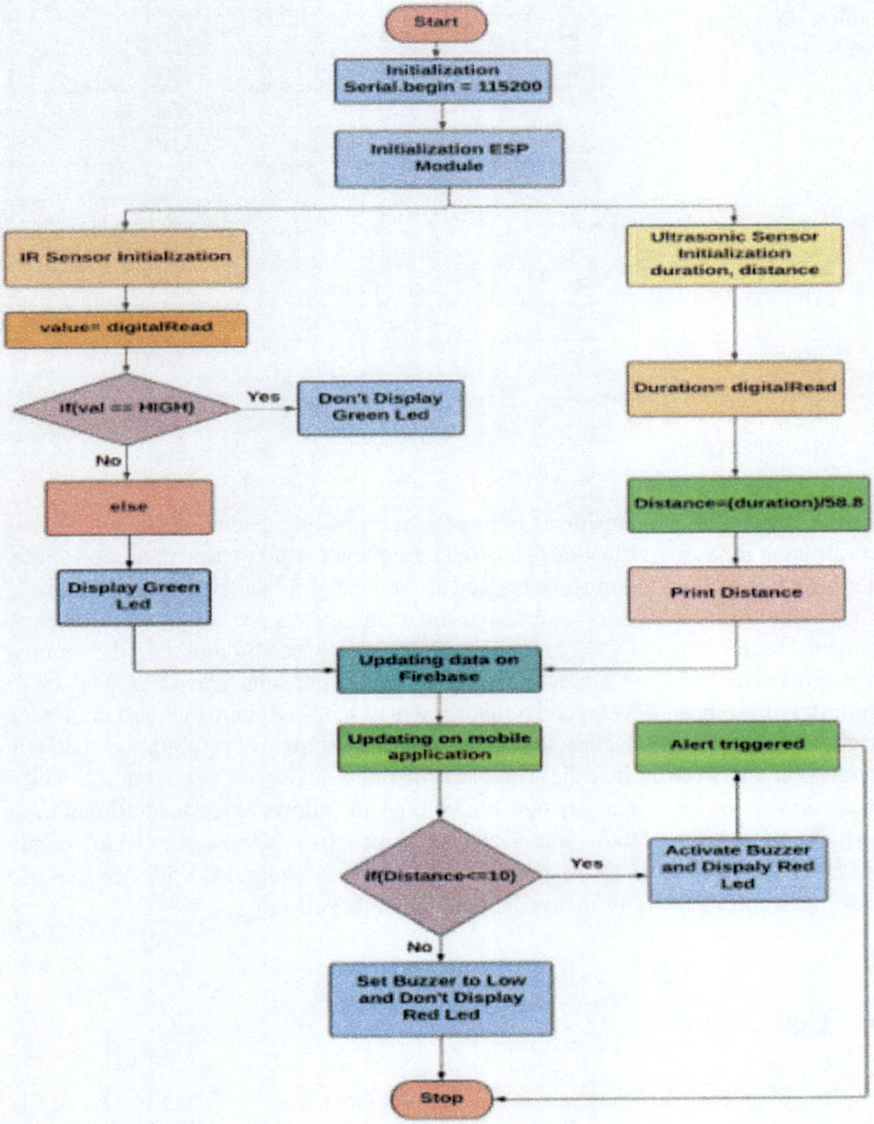

Fig. 9 Flowchart

Table 1 Cost of each hardware used

S.No	Hardware	Cost in Rs
1	Servo motor	500
2	Node MCU	760
3	Ultrasonic sensor	310
4	Infrared sensor	210
5	Dustbin model	780
6	Breadboard	240
7	Jumper wires	450
8	Buzzer and LED	220
Total		Rs 3470

9 Applications

A smart city requires its residents to live a smart lifestyle, and cleanliness is a basic requirement for a smart lifestyle [4]. As a consequence, smart waste reduction strategies based on technology must be created around the globe in effort to keep communities clean, healthier, better, and faster. Smart dumpsters are a benefit to the environment. Smart bins cut CO2 and other greenhouse gases by reducing the number of pickup visits [13]. As a result, streets are getting cleaner and safer. The smart dumpsters have been developed so that they may be filled with standard collection gear. Smart containers occupy less space than traditional recycling bins, reducing congestion and making it straightforward for individuals to move around the city. Mild compaction technique expands bin volume and allows to stash additional trash [14]. The time required to fill and clean smart bins will be decreased, making empty and clean bins available to the general public as it is integrated with firebase and real-time status of the bins are available to the authorities.

10 Conclusion

People's lives have been impacted by technological improvements in practically every way. It has made a significant contribution to improving people's living situations. With each single day, more emphasis is being focused on global activities related to the creation of green infrastructure. A smart city necessitates a smart life for its citizens, and sanitation is a basic prerequisite for a smart living. It is past time for electronics to be used in this field to produce waste disposal systems worthy of a smart city that are also effective enough yet to manage the ever-increasing amounts of garbage. The development of smart systems, particularly the creation of smart trash cans, is on the rise, owing to the availability of intelligent technology. In this paper, a smart dustbin is designed coupled with the various sensors to detect whether the dustbin is full or not. The NodeMCU is used to send data to firebase so that it can be accessible from anywhere. An android application is also built in to check the status of dustbin at ease. The future work includes the autonomous segregation of waste with the help of different sensors.

References

1. Singhvi RK et al (2019) IoT based smart waste management system: India prospective. In: Proceedings - 2019 4th international conference on Internet of Things: smart innovation and usages, IoT-SIU 2019. https://doi.org/10.1109/IOT-SIU.2019.8777698
2. Katal A, Singhania S, Sethi V (2020) Obstacle avoidance vehicle integrated with cloud. In: 2020 IEEE international conference for innovation in technology, INOCON 2020. https://doi.org/10.1109/INOCON50539.2020.9298430
3. Anagnostopoulos T et al (2017) Challenges and opportunities of waste management in IoT-enabled smart cities: a survey. IEEE Trans Sustain Comput 2:275–289
4. Folianto F, Low YS, Yeow WL (2015) Smartbin: smart waste management system. In: 2015 IEEE 10th international conference on intelligent sensors, sensor networks and information processing, ISSNIP 2015. https://doi.org/10.1109/ISSNIP.2015.7106974
5. Rohit GS, Chandra MB, Saha S, Das D (2018) Smart dual dustbin model for waste management in smart cities. In: 2018 3rd international conference for convergence in technology, I2CT 2018. https://doi.org/10.1109/I2CT.2018.8529600
6. Shyam GK, Manvi SS, Bharti P (2017) Smart waste management using Internet-of-Things (IoT). In: Proceedings of the 2017 2nd international conference on computing and communications technologies, ICCCT 2017, pp 199–203. https://doi.org/10.1109/ICCCT2.2017.7972276
7. Varude PS, Ranpise RR, Kedari RB, Darekar KB, Gupta A (2008) Top-k query based dynamic scheduling for IoT-enabled smart city waste collection. Int Res J Eng Technol 1376
8. Kanade P, Alva P, Prasad JP, Kanade S (2021) Smart garbage monitoring system using Internet of Things (IoT). In: Proceedings - 5th international conference on computing methodologies and communication, ICCMC 2021, pp 330–335. https://doi.org/10.1109/ICCMC51019.2021.9418359
9. Lincy FA, Sasikala T (2021) Smart dustbin management using IOT and Blynk application. In: Proceedings of the 5th international conference on trends in electronics and informatics, ICOEI 2021, pp 429–434. https://doi.org/10.1109/ICOEI51242.2021.9452988
10. Praveen Aet al (2020) IoT based smart bin: a swachh-bharat initiative. In: Proceedings of the international conference on electronics and sustainable communication systems, ICESC 2020, pp 783–786. https://doi.org/10.1109/ICESC48915.2020.9155626
11. Murugaanandam S, Ganapathy V, Balaji R (2018) Efficient IOT based smart bin for clean environment. In: Proceedings of the 2018 IEEE international conference on communication and signal processing, ICCSP 2018, pp 715–720. https://doi.org/10.1109/ICCSP.2018.8524230
12. Aravindaraman BA, Ranjana P (2019) Design of a monitoring system for waste management using IoT. In: Proceedings of 1st international conference on innovations in information and communication technology, ICIICT 2019. https://doi.org/10.1109/ICIICT1.2019.8741499
13. Abdallah M, Adghim M, Maraqa M, Aldahab E (2019) Simulation and optimization of dynamic waste collection routes. Waste Manage Res 37:793–802
14. Santos AA, da Silva AF, Gouveia A, Caetano N, Felgueiras C (2022) Recyclable waste collection—increasing ecopoint filling capacity to reduce energy for transportation. Energy Rep 8:430–436

Blockchain-Based Framework for Indian Retail Market in SMEs

Sarishma Dangi, Riya Sharma, Ravi Tomar, and Ambica Prakash Mani

1 Introduction

E-commerce has witnessed an exponential growth over the last few decades. It is affecting the lifestyle and behavior patterns of user. It has also provided impetus to the retail industry by becoming one of the most important contributors of retail sector. It has bought fresh opportunities as well as novel challenges to the field especially for small to medium-scale retailers who are competing with large scale enterprises.

E-commerce is offering smooth means to retailers for establishing their business. While the other side of the coin reflects the constraints on investment in the distribution channel as pointed by [1] in 2012.

With the onset of Covid-19 pandemic, the situation has become worse. A considerable portion of the market was forced to integrate their needs with online platforms for conducting business. A retailer in an Indian context is a person who single handedly or jointly supports a small-scale business. This business needs constant presence for conducting transactions. Under such scenario, it becomes difficult for the retailer to strike profitable deals with respect to the market conditions. Moreover, even if the retailer is able to find out the correct deals at the right time; then also the large-scale enterprises stand united to interfere in myriad ways.

The key contributions of this work are outlined as follows:

- To discuss the applicability of blockchain in resolving the issues faced by small-scale Indian retailers.

S. Dangi (✉) · R. Sharma · A. P. Mani
Graphic Era Deemed to be University, Dehradun, India
e-mail: sarishmasingh@gmail.com

R. Sharma
e-mail: riyasharma.comm@geu.ac.in

R. Tomar
Persistent Systems, Pune, India

A. Ramdane-Cherif et al. (eds.), *Machine Intelligence and Data Science Applications*, Algorithms for Intelligent Systems, https://doi.org/10.1007/978-981-99-1620-7_22

- To provide a framework based on blockchain which can unitedly share the fair market prices among retailers.
- To provide a discussion on finding the balance between centralized and decentralized aspects of using blockchain as compared to the traditional model.
- To provide smart contracts for retailers and large corporations which can be used to support legal dimension.

The rest of the paper is organized as Sect. 2 provides a review of the related work done in the area of retail industry and blockchain. Section 3 provides the necessary technical background including concepts related to blockchain and retail market scenario in India. Section 4 discusses the prevalent challenges faced by retailers in India with a special focus on Covid-19 as an influencing factor. In Sect. 5, we provide the proposed conceptual framework which is layered in nature. It provides the answer to the question of degree of centralization or decentralization in relation to E-commerce and retail. Lastly, the work is concluded.

2 Related Work

In this section, we briefly outline the existing works done in the area of retail market integration with the concept of blockchain.

S.No	Year	Author(s)	Purpose	Findings	Scope of work
1	2022	Bhardwaj, Garg, Gajpal	Examine the factors influencing the intention of small and medium enterprises (SMEs) in India to adopt blockchain technology in their supply chains	Proposed a novel integrated technology adoption framework	Future research can further engage in longitudinal studies to explore the factors that impact blockchain technology adoption in supply chains across various types of organizations
2	2020	Ragh Satya Sai Medid (1)	To highlight the scope of blockchain in retail industry	Identified the factors that holds significance in the blockchain-based retail industry	Application of blockchain technology missing
3	2021	Lanzini, Ubacht, Greeff (2)	To explore which factors influence SMEs' intention to adopt blockchain technology	Study primarily influenced by organizational rather than by technological and environmental factors	The study is based on a literature review, empirical study can be undertaken

(continued)

(continued)

S.No	Year	Author(s)	Purpose	Findings	Scope of work
4	2022	Jain, Shrivastava, Paul et al.	To explore the role of blockchain technology in building a trustworthy yet collaborative environment in SME clusters through the principles of community self-governance based on the work of Nobel Laureate Elinor Ostrom	Study developed and present a blockchain commons governance framework for the three main dimensions, i.e., interaction, autonomy, and control	Future studies can critically examine the current theories present in the blockchain discourse
5	2020	Dutta, Choi, Somani, Butala	Identifies possible opportunities and challenges in societal impacts, current state-of-the-art technologies the blockchain, and major trends	This article provides a thorough review of 178 research articles and covers several aspects of blockchain technology in SC and logistics management	Research needs to be done to understand the implementation and adoption challenges for blockchain in e-commerce so that it can eliminate the issues related to product counterfeiting and tampering of data by intruders
6	2020	Choudhury, Goswami, and Gurung	To develop mechanisms are required that will assist people in following the social and physical distancing norms in public places	Propose a privacy-preserving blockchain-based framework that allows people, through use of their smartphones and other communication devices, to protect themselves from infections as they go about their day-to-day activities	The scope of study is limited and needs empirical testing of the proposed framework

(continued)

(continued)

S.No	Year	Author(s)	Purpose	Findings	Scope of work
7	2020	Miraz, Ghozali, Sharif	Critically analyze the variables of the current research framework of the blockchain implementation in retail market	Study implemented blockchain in retail market for increasing customer benefits in order to improve the retail industrial supply chain activities	Needs more factors analysis to adapt this study more effectively in retail industry
8	2018	Gupta, Bedi	Propose a novel approach for EWM using blockchain-based smart contracts	Proposed EWM using smart contracts to bring coordination among producers, importers, retailers, and recyclers of EEEs	The proposed model can be empirically tested in the future research

3 Technical Background

In this section, we briefly describe the blockchain and its working. We also cover the peer-to-peer network, smart contracts, and consensus mechanisms used by blockchain. Lastly, we cover the working of India SMEs and retail market.

In 2008, the concept of blockchain was conceptually introduced by Satoshi Nakamoto in a white paper [2]. Originally it was intended as a means to achieve complete decentralization while maintaining trust in the network. Blockchain when integrated with the monetary field led to the boom in cryptocurrencies especially Bitcoin [3]. Slowly, we started to believe in a shift of working models from a centralized approach to a decentralized approach.

Blockchain is primarily a chain of blocks where each block stores the hash of its previous block. Every block comprises transactions or exchanges which occur between participating entities. These very participating entities are embodied with the responsibility of maintaining and sustaining the blockchain. Every participant has equal rights with regard to access of blockchain and every participant is free to check the transactions for fraud or any malpractices. Use of blockchain leads to dilution in centralized structures which have existed and operated for decades such as banks, governments, electricity grids, auditors, and large-scale business enterprises [4]. Blockchain is paving the way to a decentralized future.

Although it is still in its development phase, blockchain still finds application in areas like smart grid management, smart cities [5], Internet of Things [6, 7], healthcare [8], sustainable systems, vehicular ad-hoc networks [9], financial market, food supply chain management [10], security [11, 12], etc.

Consensus refers to the process of arriving or agreeing on an agreement. There are multiple entities who participate in achieving consensus. Theoretically, entire network must arrive at the exact same consensus. Practically, it is not possible at all times. There are multiple mechanisms to achieve consensus in the network [13].

From now on, we will be using user and retailer term interchangeably as they imply toward the same entity in the network.

4 Challenges Faced by Retail Sector

The challenges faced by the traditional supply chain are as follows:

- **Opaqueness in upstream and downstream side**: Due to lack of visibility of end party on both the side creates the chances of fraud, violation of legal framework and code of conduct. With the help of blockchain technology every change made will be easily visible and authenticity of each smart contract [14] is maintained [15].
- **Lack of means to fulfill sudden increase in demand**: The online presence of business provides access to global customers which can lead to sudden increase in demand. This sudden increase in demand is difficult to cater without strong and transparent supply chain management system.
- **Minimal usage of advance technologies**: Traditional supply chain management does not use the advance technologies that poses a threat to business in the dynamic global world [16].
- **The maintenance of records is done manually**: The traditional businesses keep the record manually which is extremely tedious and may lead to loss of information with the increase in the level of operations.
- **Presence of multiple middlemen**: The retailers bear the burden of multiple intermediaries as it increases the cost of product and reduces the profit margins. If these intermediaries are removed and the retailers are directly able to trade with manufacturers through blockchain technology, the retailers can compete in the global market with the giants.
- **Slow settlement mechanism**: As the warehouses receive the stock it takes days, weeks, or months to settle the transactions which acts as hindrance in the complete trade cycle. A supply chain built with blockchain using smart contracts will enable immediate payments once the contract criteria are met.

5 Proposed Conceptual Framework

Blockchain offers a real-time connection between various distributed entities across a network. This can be achieved by using a custom-made application which allows the user/retailer to access the blockchain [17]. Once the user generates an account and connects with the API, then it is ready to do transactions across the network.

We propose the use of smart contracts which are legally backed by the government regulating authorities. Smart contracts make the process smooth, trustworthy, and easily verifiable.

The proposed conceptual framework is represented in Fig. 1. The functionality of each layer is specific and carefully orchestrated for smooth functioning of workflow. Layer-wise description along with primary responsibilities is described as follows.

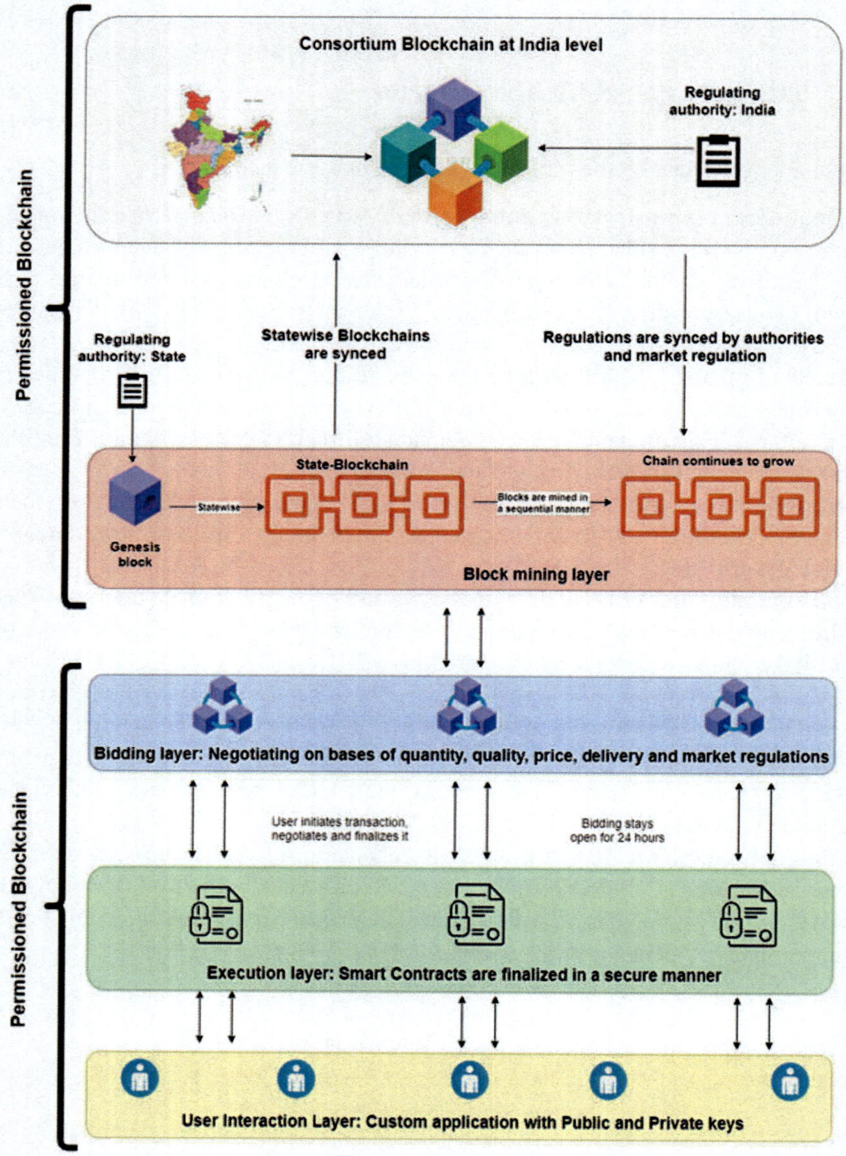

Fig. 1 Proposed blockchain-based conceptual framework for SME retailers

5.1 User Interaction Layer

When a retailer wishes to become a part of the blockchain to conduct business, she/he has to register with their GST number, business name, UIDAI, and keywords. Former pre-requisites are for generating a unique ID for the user including a set of public and private keys denoted as {k, K}. The latter, i.e., keywords have a very specific role to play in the blockchain. These keywords rotate around the type of business or resource being dealt with. Specifying keywords leads to easy segregation of market dealers, suppliers, and other entities. It enables easier finding of business prospects for the retailer. These keywords can be modified and changed via the user profile. Whenever a retailer wishes to sell or buy, she/he can go to application and make a bid.

5.2 Execution Layer

Execution layer particularly deals with the generation, negotiation, and finalizing of the contract. It is a secure layer and has to be maintained in a trustworthy manner since smart contracts are initialized as well as finalized by this layer. Execution layer has the authority to generate three types of smart contracts, described as follows:

- *Prospect Smart Contract:*

    Purpose: To **initiate** a business transaction
    Initialized by: Buyer or Seller

  Input parameters:

    Digital Signature,
    BC_State_ID,
    BC_Center_ID,
    Keyword(s),
    Quantity,
    Quality,
    Price,
    Location,
    Hash of user input.

  Execution layer: Creates a prospective smart contract and embeds a time counter in the contract itself which will automatically time out after 24 h.

- *Accepted Smart Contract:*

    Purpose: To **accept** a business transaction

Initialized by: Buyer or Seller

Input parameters:

Digital Signature,
BC_State_ID,
BC_Center_ID,
Keyword(s),
Quantity,
Quality,
Price,
Location,
Hash of user input.

Execution layer: Creates an accepted smart contract and sends it to the parties concerned.

- *Final contract:*

Purpose: To **legally bind** a business transaction
Initialized by: Buyer or Seller

Input parameters:

Digital Signature,
BC_State_ID,
BC_Center_ID,
Keyword(s),
Quantity,
Quality,
Price,
Location,
Digital Signature,
Timestamp,
Hash of user input.

Execution layer: Creates a final smart contract and embeds legal implications within the contract. The contract must be signed and verified by the parties involved and the execution layer, respectively.

Once the user bids, execution layer will take in the entered parameters. Making a bid will lead to generation of a *prospect smart contract* which is tentative and negotiable in nature. This smart contract is then circulated in the sub-network generated in an ad-hoc manner around the keyword(s) specified. Maximum lifetime of a prospect smart contract is 24 h after which it will expire and a new bid has to be done.

Whenever a prospective buyer wishes to agree on the prospect smart contract, s/he can digitally sign the contract and pass it back to the execution layer. Once both the parties are decided and *prospect smart contract* has the buyer's signature, only then the execution layer is going to generate another smart contract, i.e., *final smart contract*; which is legally binding in nature. Once the contract is generated, it is sent to both the parties who encrypt as well as sign the document using their private key and signature in a timestamped manner.

5.3 Bidding Layer

This layer is responsible for handling and segregating the multiple smart contracts initiated by the entities within the blockchain. The segregation is proposed on the basis of keywords which depict the name of the commodity or resource in most cases. Using specific keywords helps in generating ad-hoc subnetworks within the combined market.

5.4 Block Mining Layer

A time counter has to be decided by the state-level authorities for mining of a block. The purpose of regulating authority is to monitor and make the choices which a centralized system must make. Decentralization in true form is not practical considering our present reality as discussed. We have proposed a balanced form of centralized control with the decentralized entities.

5.5 Consortium Layer

A time counter has to be decided by the state-level authorities for mining of a block. The purpose of regulating authority is to monitor and make the choices which a centralized system must make. We have proposed a balanced form of centralized control with the decentralized entities.

6 Conclusion

Blockchain has found applications in numerous areas over the last decade. The Indian Retail Market is a complex mix of small and medium-level enterprises who works in an integrated manner. In this work, we present a blockchain-based framework that can be used by both small and medium-level enterprises for smooth conduct of business across a wide range of spectrum.

References

1. Jing B, Chen X, Cai G (2012) Equilibrium financing in a distribution channel with capital constraint. Prod Oper Manag 21(6):1090–1101
2. Satoshi N, Nakamoto S (2008) Bitcoin: a peer-to-peer electronic cash system. Bitcoin, p 9. https://doi.org/10.1007/s10838-008-9062-0
3. World B, Beck R (2018) Beyond Bitcoin: the rise of
4. E-textiles C (2018) Towards the Internet of smart clothing: a review on IoT wearables and garments for creating intelligent. https://doi.org/10.3390/electronics7120405
5. Xie J et al (2019) A survey of blockchain technology applied to smart cities: research issues and challenges. IEEE Commun Surv Tutor 21(3):2794–2830. https://doi.org/10.1109/COMST.2019.2899617
6. Choudhury T, Khanna A, Toe TT, Khurana M, Nhu NG (2021) Blockchain applications in IoT ecosystem. Springer, Cham, Switzerland
7. Christidis K, Devetsikiotis M (2016) Blockchains and smart contracts for the Internet of Things. IEEE Access 4:2292–2303. https://doi.org/10.1109/ACCESS.2016.2566339
8. Sharma A, Tomar R, Sarishma, Chilamkurti N, Kim B-G (2020) Blockchain based smart contracts for internet of medical things in e-healthcare. Electronics 9(10):1609
9. Tomar RS (2020) Maintaining trust in VANETs using blockchain. ACM SIGAda Ada Lett 40(1):91–96
10. Guo J, Cengiz K, Tomar R (2021) An IOT and blockchain approach for food traceability system in agriculture. Scalable Comput Pract Exp 22:127–137
11. Sille R et al (2022) Review study on blockchain frameworks for security issues in IoT devices. In: 2022 international symposium on multidisciplinary studies and innovative technologies (ISMSIT), Ankara, Turkey, pp 876–881. https://doi.org/10.1109/ISMSIT56059.2022.9932744
12. Sengupta J, Ruj S, Das Bit S (2020) A comprehensive survey on attacks, security issues and blockchain solutions for IoT and IIoT. J Netw Comput Appl 149:102481. https://doi.org/10.1016/j.jnca.2019.102481
13. Wan S, Li M, Liu G, Wang C (2020) Recent advances in consensus protocols for blockchain: a survey. Wirel Netw 26(8):5579–5593. https://doi.org/10.1007/s11276-019-02195-0
14. Negi D, Sah A, Rawat S, Choudhury T, Khanna A (2021) Block chain platforms and smart contracts. In: Choudhury T, Khanna A, Toe TT, Khurana M, Gia Nhu N (eds) Blockchain applications in IoT ecosystem. EAI/Springer innovations in communication and computing. Springer, Cham. https://doi.org/10.1007/978-3-030-65691-1_5
15. Medida RSS (2020) Scope of blockchain technology in retail sector. IAEME. 0976-6375
16. Lanzini F (2021) Blockchain adoption factors for SMEs in supply chain management. J Supply Chain Manag Sci 2
17. Vardhan Singh Rawat H, Bisht D, Kumar S, Dangi S (2022) Rise of blockchain-based non-fungible tokens (NFTs): overview, trends, and future prospects. In: Machine intelligence and data science applications, pp 1–10
18. Abdella J (2018) Peer to peer distributed energy trading in smart grids: a survey. https://doi.org/10.3390/en11061560
19. Liu J, Liu Z (2019) A survey on security verification of blockchain smart contracts. IEEE Access 7:77894–77904. https://doi.org/10.1109/ACCESS.2019.2921624
20. Jain G, Shrivastava A, Paul J et al (2022) Blockchain for SME clusters: an ideation using the framework of Ostrom commons governance. Inf Syst Front 24:1125–1143

Comparative Analysis of Heart Disease Prediction Using Machine Learning Algorithms

Richa Choudhary, Sona Tyagi, Harsh Goyal, Yashraj Prasad, Ansh Tyagi, and Tanupriya Choudhury

1 Introduction

A heart condition is a potentially fatal one. Heart disease accounts for far too many fatalities each year. Cardiac muscle aging may lead to heart disease [1, 2]. Heart failure has a precise term that refers to the heart's inability to pump blood. Blood flow to the arteries may be inadequate, which can lead to CAD. Following a projection of the prevalence of heart failure patients, the World Health Organization (WHO) advises preventive measures to lower death rates. As the rates are increasing every year, preventive measures must be taken. The heart, the largest muscle in our body, pumps blood all over the body. Through some veins and arteries, the body's unpurified blood is sent to the heart, where it is subsequently transferred to the lungs to obtain oxygen. With each beat, the heart sends blood to the body. Sometimes, heart

Richa Choudhary and Tanupriya Choudhury are the corresponding authors and all authors contributed equally and are the first author.

R. Choudhary (✉) · S. Tyagi · H. Goyal · Y. Prasad · A. Tyagi
School of Computer Science, University of Petroleum and Energy Studies, Dehradun 248007, Uttarakhand, India
e-mail: richachoudhary.86@gmail.com

T. Choudhury (✉)
CSE Department, Symbiosis Institute of Technology, Symbiosis International University, Pune, Maharashtra 412115, India
e-mail: tanupriya.choudhury@sitpune.edu.in; tanupriya@ddn.upes.ac.in; tanupriya1986@gmail.com

Informatics Cluster, School of Computer Science, University of Petroleum and Energy Studies (UPES), Dehradun, Uttarakhand 248007, India

CSE Department, Daffodil International University, Daffodil Smart City, Birulia 1216, Bangladesh

Director Research (Honorary), The AI University, Cutbank Montana 59427, US

A. Ramdane-Cherif et al. (eds.), *Machine Intelligence and Data Science Applications*, Algorithms for Intelligent Systems, https://doi.org/10.1007/978-981-99-1620-7_23

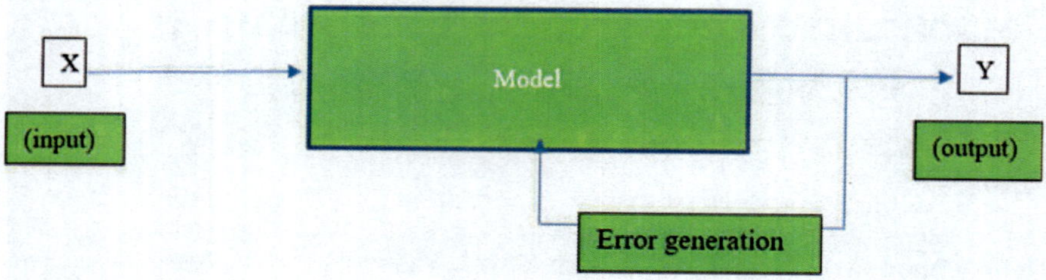

Fig. 1 Supervised learning

abnormalities can lead to heart attacks or strokes. Abnormalities include a spike in blood pressure, an obstruction in an artery or vein, and a number of other disorders. However, as our country and the rest of the world have developed to this point, a number of machine learning approaches are now accessible to identify heart attacks, which lowers the risk by a specific percentage and is still being developed today.

Today, there is a wealth of data that may be used to generate insights using variety of machine learning algorithms for heart attack prediction [3]. Machine learning [4] is used in a wide range of healthcare use cases. By processing vast amounts of data, machine learning technology can help medical practitioners create precise drug treatments that are matched to individual features. Different types of analysis techniques [5, 6] help to identify the best algorithm to detect such kinds of diseases pretty well in advance (Fig. 1).

2 Literature Review

This study uses machine learning techniques to predict heart attacks. In the research paper "Prediction of Heart Disease Using Machine Learning, Aditi Gavhane, Gouthami Kokkula, Isha Pandya, and Prof. Kailas Devadkar (Ph.D.)" multi-layer perceptron (MLP) is a neural network algorithm which is used to train the model for the given dataset. MLP contains many levels of input, output, and hidden layers. Hidden layers can be one layer or more layers between the input and output layers. In the final result, the system predicts whether the person will have CAD or not.

In the pertinent work "Prediction of Heart Diseases using Support Vector Machine, Yamala Sandhya" used SVM to predict heart attack detection. By taking into account dangerous factors connected with heart disease, the approach aids in using machine learning algorithms for predicting heart disease. Out of all the machine learning algorithms applied, SVM should have the highest accuracy for heart attack prediction [7].

In this study, Heart Disease Prediction System using k Nearest Neighbor Algorithm with Simplified Patient Health Parameters, authors predicted heart disease using data mining techniques. They have used simplified parameters to predict heart disease so it can be used in remote monitoring of patients. To simplify the parameters, the parameter weighting method has been used in this study along with the KNN algorithm. This has improved the accuracy with very few number of parameters required to train the model. A total of 8 parameters were used out of 13, as these 8 can be easily measured at home as well. Through the KNN method, it has been proved that these 8 parameters are accurate enough as compared to 13 using KNN, Naïve Bayes, and decision Tree. KNN was the most accurate out of all the algorithms at predicting heart attacks [8].

"A metaheuristic-enabled training system for ensemble classification technique for heart disease prediction, Paul T Sheeba, Deepjoyti Roy, Mohmmad Haider Syed" uses ensemble classification technique for the prediction of heart disease.

"Heart disease prediction, Md Mamun Ali, Bikash Kumar Paul, Kawsar Ahmed, Francis M. Bui, Julian M. W. Qunin, Mohmmad Ali Moni," uses a supervised machine learning approach and performed a comparative analysis for each approach. There MN-PCA model showed an accuracy of 91.7% for heart attack prediction [9].

3 Methodology

In order to detect the presence of or estimate the probability of having heart attacks from a huge dataset, an effective machine learning method was selected for this study among the alternatives in an open-access data mining platform. The body's severe pressure is what causes this sickness. Predictions of heart disease can be made using the dataset's wide variety of parameters. To examine the patients' performance, we used a dataset that contained 1024 unique data values across 7 parameters. The workflow and thorough design processes for the proposed system are outlined below:

1. Import the dataset
2. Exploration of data to figure out what they look like
3. Preprocessing of data
4. Splitting of data into attributes and labels
5. Applying the SVM algorithm
6. Applying the KNN algorithm
7. Applying the Random Forest algorithm
8. Applying the Logistic Regression algorithm
9. Calculating the accuracy and comparing them
10. Make predictions
11. Evaluation based on the results (Fig. 2).

3.1 Dataset

The Kaggle dataset which was used to forecast cardiac attacks was obtained. The dataset contains only properties with numerical values. A total of 13 attributes make

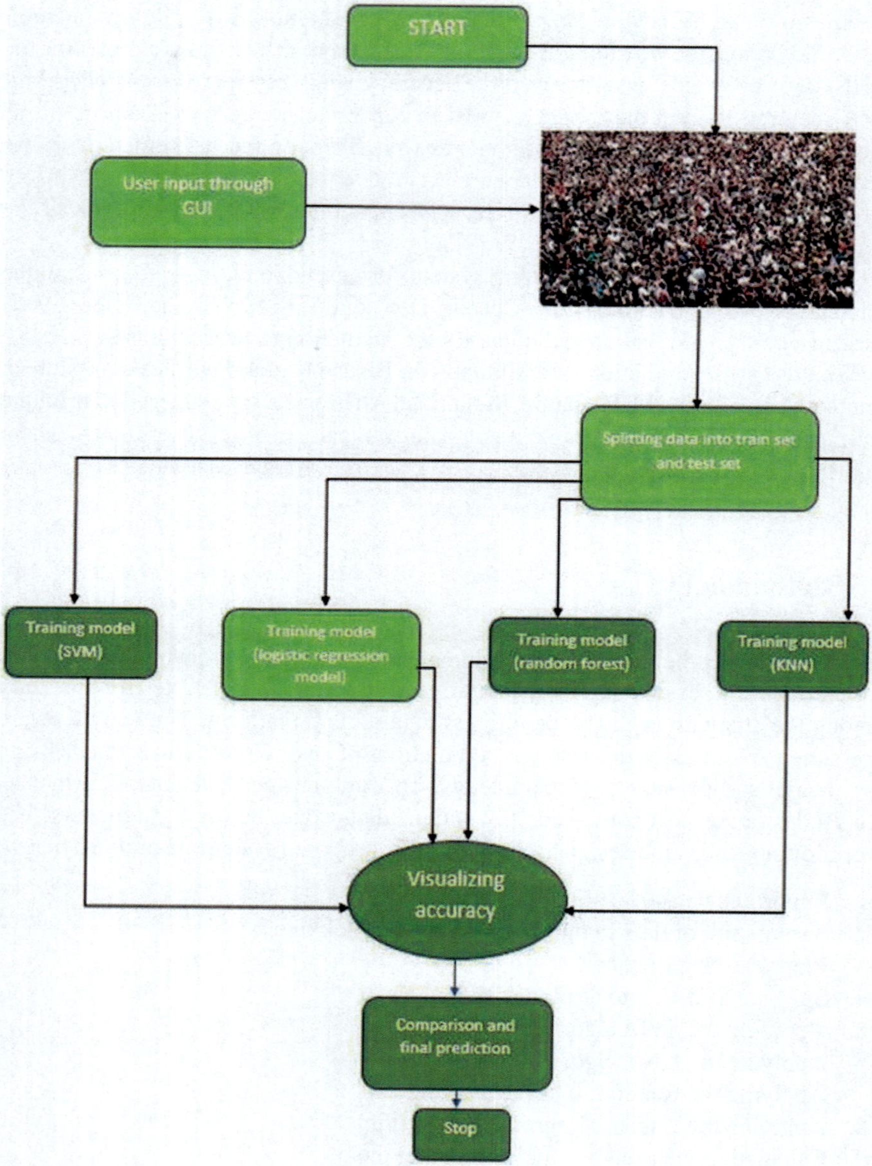

Fig. 2 Flowchart of heart disease prediction

up the dataset, and each one is used for exploration. There are 1024 records in our dataset. The dataset has the following attributes: age, sex, cholesterol, chest pain (0—No pain, 1—low pain, 2—moderate pain, 3—high pain), thalach, thal (hemoglobin level), Fasting blood sugar (1–120 mg/dl, 0–<120 mg/dl), Trestbps, restecg (0—normal; 1—having ST-T; 2—hypertrophy), exang (1—yes; 0—no), oldpeak, CA

(Number of major vessels (0–3) colored by fluoroscopy), and num (diagnosis of heart disease).

3.2 *Data Preprocessing*

The process of changing the dataset's columns is known as data processing. As our dataset does not have any outliers or null values, there is no need to do preprocessing. But since we have to take input from the users, in that case, we have to check whether the user has entered any null values.

3.3 *Train and Split Data*

The process of dividing the whole dataset into training data and test data. In this work, the training data account for 80% and testing data for 20% respectively.

3.4 *Machine Learning Algorithm*

To obtain high accuracy of heart attack patients, various machine learning algorithms were tested. The machine learning algorithms used are:

Support Vector Machine

Support vector machines (SVMs) are effective yet responsive supervised machine learning methods that acclimated both classification as well as regression. They are, however, frequently applied to categorization problems. SVMs are implemented in a different way than other machine learning algorithms. They have recently gained a lot of popularity thanks to their capacity to handle numerous continuous and categorical variables [8].

K Nearest Neighbor

K Nearest Neighbor is an elementary innovation that classifies the latest particulars or occurrences based on the resemblance between them and all above-stored illustrations. A data end is often categorized using the classification of its neighbors. KNN methods are falls under the category of supervised machine learning (ML) that can be applied to classification and regression predicting issues [10, 11]. However, it is primarily employed in the industry for classification and forecasting problems.

Random Forest

Random Forest is a segment of the supervised learning approach in machine learning algorithms. One might utilize ML arguments involving a couple of classifications and regressions. It is based on the concept of group learning, it involves combining different classifiers to address complex problems and improve model efficiency. Random Forest, as the tag suggested, is a classifier that operates the number of decision trees on the non-identical subdivision of the provided data file and means them to rise the dataset's predictive correctness. Rather than only using one decision tree, the random forest users anticipate the outcome based on the poll of the majority of projections using the forecasts from each tree.

Logistic Regression

In logistic regression, the output of a categorical dependent variable is predicted via logistic regression. The outcome must, therefore, be a distinct or rather categorical value. It can be either True or False, Yes or No, 0 or 1, etc. Rather than providing the exact values of 0 and 1, it offers the probabilities with standards ranging from 0 to 1. Except for the manner in which they are used, logistic regression and linear regression are very analogous. While logistic regression is previously owned to decode classification difficulties, linear regression is used to resolve the regression issue.

4 Results and Discussion

The result obtained from the Machine learning models is analyzed here. After applying all the algorithms on the dataset, we created a confusion matrix, where we calculated the recall, F1 score, precision, and accuracy of all algorithms. The accuracy of SVM turned out to be 84.39%, KNN 86.83%, Logistic Regression 82.93%, and Random Forest 84.88% (Fig. 3).

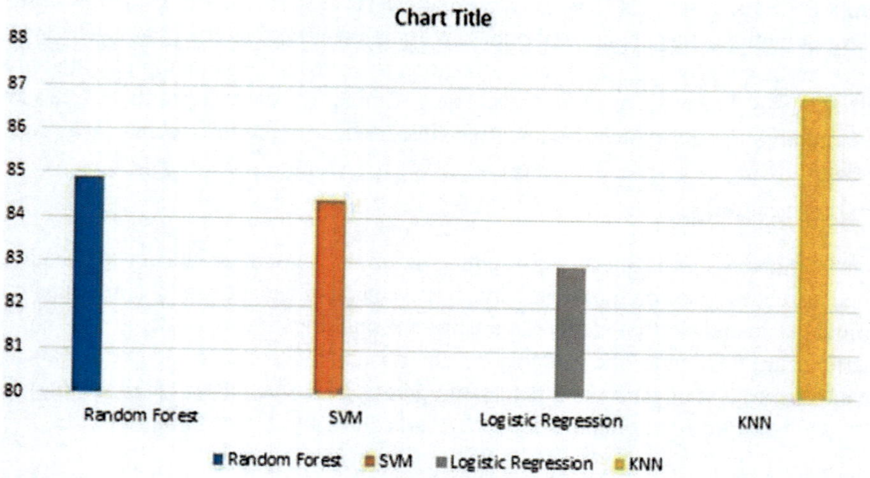

Fig. 3 Accuracy chart

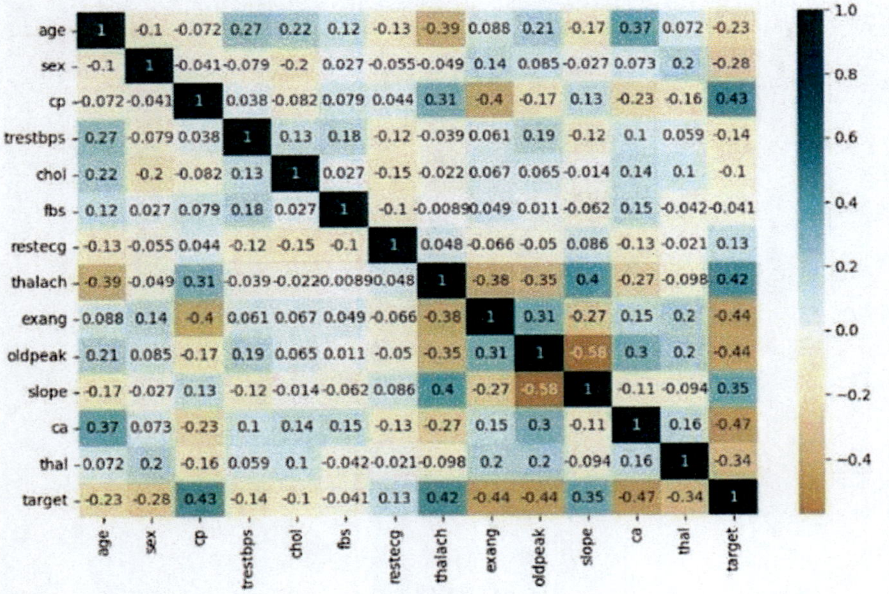

Fig. 4 Heat map

Heat maps are 2D representations of values in a data matrix. The correlation between the columns is determined using a heat map [12]. Heat maps help people focus on the most important parts of data visualizations by making it easier to see how many places and events are in a dataset (Fig. 4).

4.1 Confusion Matrix

A confusion matrix is an arrangement that resembles a table and contains true negative as well as true positive values and divine values. It is defined in the first segment, the initial of which is a true positive (TP), where values are classified as correct even though they weren't in fact true. The next one is a false positive (FP) through which the values that are determined, however, are mistakenly thought to be true. The tertiary is a false negative (FN), where the result was correct but was misclassified as negative. The last is true negative (TN), where the value was actually identified as being negative [5, 13] (Figs. 5, 6, 7, and 8).

Precision

Precision, usually referred to as "positive predictive value," provides the proportion of real positives. It is the proportion of actual positive results to all expected positive results. Out of all samples that were positively predicted, it only counts the positive samples that were successfully identified. The equation is

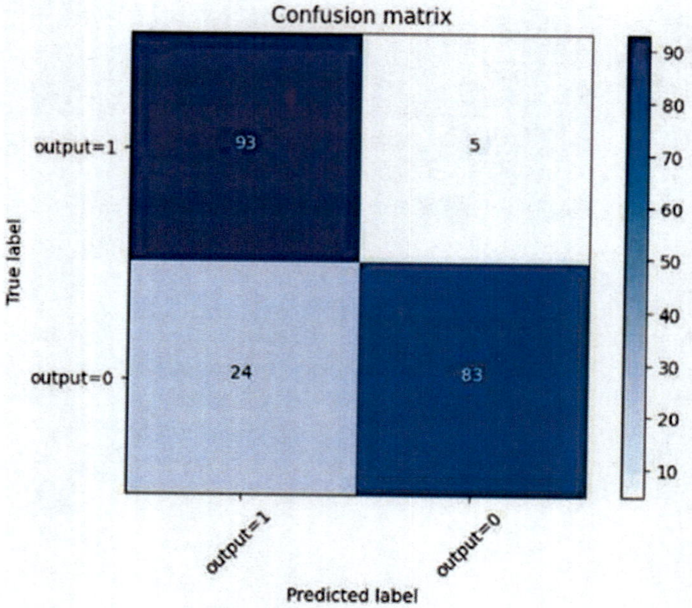

Fig. 5 SVM confusion matrix

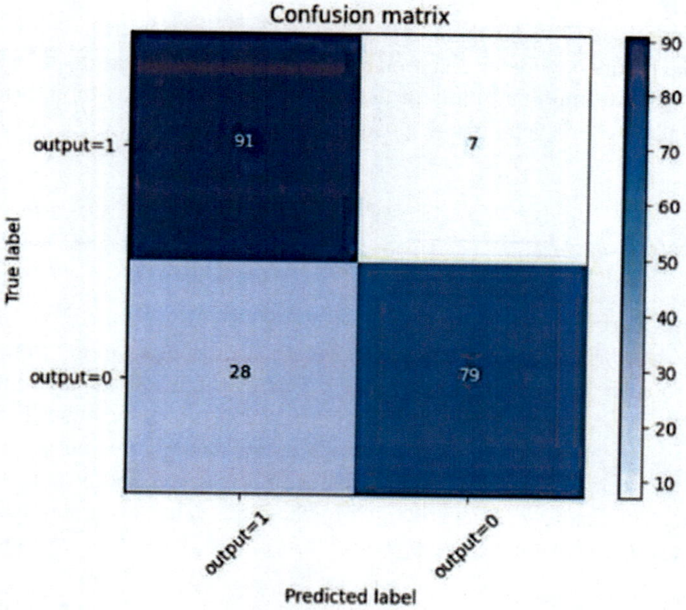

Fig. 6 KNN confusion matrix

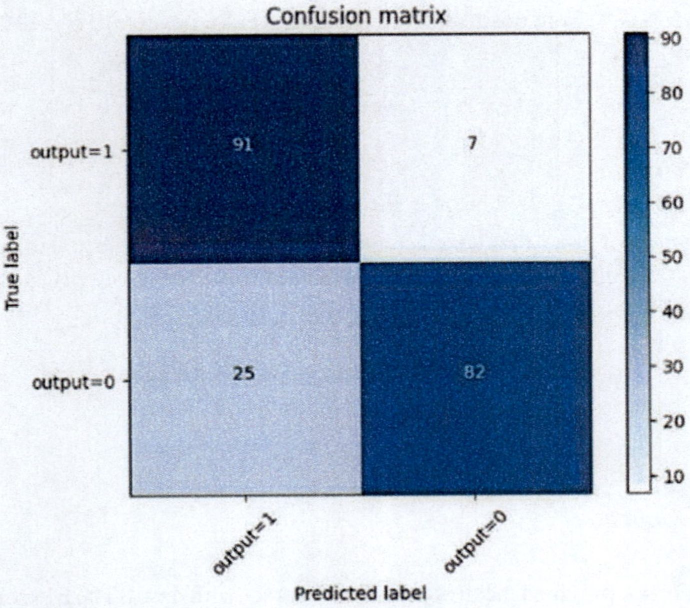

Fig. 7 Random forest confusion matrix

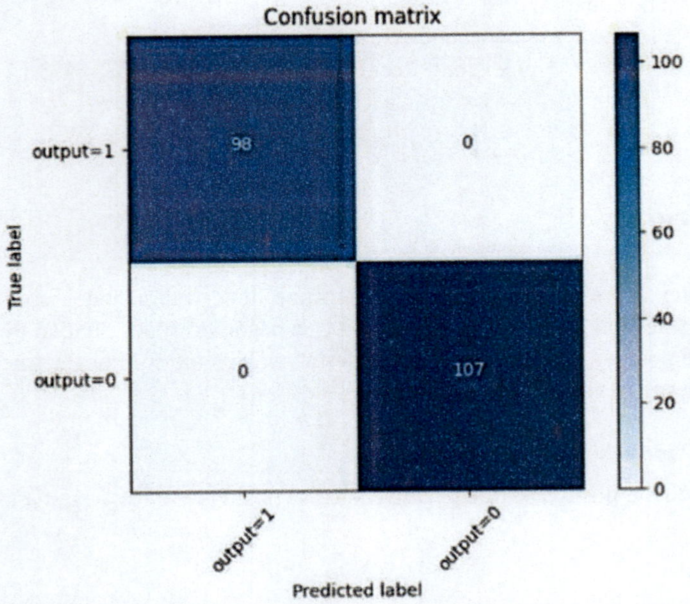

Fig. 8 Logistic regression confusion matrix

True positive/(True positive + False positive)

4.2 Recall

The proportion of cases that were actually positive but were projected to be positive is known as sensitivity (or true positive). Another name for sensitivity is recall. The equation is

True positive/(True positive + false negative)

4.3 F1 Score

The F1 score is a weighted mean of precision together with recall. The F1 score unifies both precision as well as recall into a single statistic by calculating the harmonic mean of a classifier's Evaluation. It is mainly used to differentiate the efficiency of two classifiers. The equation of F1 score is.

(2* precision* Recall)/(Precision + Recall)

4.4 Accuracy

An accuracy score is then applied to determine how well a model is doing. It is defined as the sum of the true positive and true negative values divided by the sum of the true positive and true negative values as well as the sum of the true positive and true negative values. The equation is

True positive + true negative
/(True positive + true negative + False positive + False negative)

5 Conclusion

In order to compare the best way of prediction, the researchers tested the use of different classification algorithms to forecast heart attacks. When utilizing several classification methods in machine learning, the research findings do not provide a noticeably different forecast. This experiment can be a useful tool for doctors to identify high-risk cases and provide appropriate advice. The classification model will be able to respond to more intricate questions regarding the disorders that cause heart attacks. The KNN, SVM, Logistic Regression, and Random Forest Classifier algorithms' findings on predictive accuracy indicate that the parameters being employed are reliable predictors of the presence of cardiac disorders.

References

1. Choudhury T, Kumar V, Nigam D, Mandal B (2016) Intelligent classification of lung & oral cancer through diverse data mining algorithms. In: 2016 international conference on microelectronics and telecommunication engineering (ICMETE), Ghaziabad, India, pp 133–138. https://doi.org/10.1109/ICMETE.2016.24
2. Choudhury T, Kumar V, Nigam D (2015) Intelligent classification & clustering of lung & oral cancer through decision tree & genetic algorithm. Int J Adv Res Comput Sci Softw Eng 5(12):501–510
3. Arshad I, Dutta C, Choudhury T, Thakral A (2018) Liver disease detection due to excessive alcoholism using data mining techniques. In: 2018 international conference on advances in computing and communication engineering (ICACCE), Paris, France, pp 163–168. https://doi.org/10.1109/ICACCE.2018.8441721
4. Sille R, Choudhury T, Sharma A, Chauhan P, Tomar R, Sharma D (2023) A novel generative adversarial network-based approach for automated brain tumour segmentation. Medicina 59:119. https://doi.org/10.3390/medicina59010119
5. Srivastava R, Tomar R, Sharma A, Dhiman G, Chilamkurti N, Kim B-G (2021) Real-time multimodal biometric authentication of human using face feature analysis. Comput, Mater Contin 49(1):1–19. https://doi.org/10.32604/cmc.2021.015466
6. Tomar R, Sastry H, Prateek M (2020) Establishing parameters for comparative analysis of V2V communication in VANET. J Sci Ind Res (JSIR) 79(01):26–29
7. Son Y-J, Kim H-G, Kim E-H, Choi S, Lee S-K (2010, Dec 30) Application of support vector machine for prediction of medication adherence in heart failure patients. Healthc Inform Res. Accessed 20 Sept 2022
8. Enriko IKA, Gunawan DA, Suryanegara M (2016, January) Heart disease prediction system using K-nearest neighbor algorithm with heart disease prediction system using k-nearest neighbor algorithm with simplified patient's health parameters. Accessed 9 Nov 2022
9. Almustafa KM (2020, July 2) Prediction of heart disease and classifiers' sensitivity analysis. BMC Bioinform. BioMed Central. Accessed 9 Nov 2022
10. Angorro DA (2020, May) Comparison of accuracy level of support vector machine (SVM) and K-nearest neighbors (KNN) algorithms in predicting heart disease. Accessed 9 Nov 2022
11. Reddy KSK, Kanimozhi KV (2022) Novel intelligent model for heart disease prediction using dynamic KNN (DKNN) with improved accuracy over SVM. In: 2022 international conference on business analytics for technology and security (ICBATS)

12. Sheebaa PT, Roy D, Syed MH, Mirjalili S, Patro SP, Vivekanandan T, Mohan S, Fitriyani NL, Wang H, Joo G, Khan MA, Guo C, Wang B, Ali L, … Li JP (2022, Oct 10) A metaheuristic-enabled training system for ensemble classification technique for heart disease prediction. Adv Eng Softw. Accessed 30 Nov 2022
13. Patro SP, Nayak GS, Padhy N (2021, Aug 11) Heart disease prediction by using novel optimization algorithm: a supervised learning prospective. Inform Med Unlocked. Accessed 9 Nov 2022

Extensive Review on Automatic Plant Leaf Disease Detection Using Image Analytics

Sanika Singh, Saurabh Mukherjee, and T. Choudhury

1 Introduction

India is known for agriculture. It is an agricultural country. The majority of the population depends on agriculture. Farming in India is world-famous. Indian farmers have wide variety of crop selection. Cultivation can become more comfortable and easier with technological support. In any environmental condition, some disease always affects crops and cultivated plants. Which causes lots of harm to the overall production. In most of the cases, the disease is affecting the plant's leaf, stem, and fruits. Hence, the detection of plant and plant-based diseases is very important and

Sanika Singh and Tanupriya Choudhury are corresponding authors and all authors contributed equally and are the first author.

S. Singh (✉) · S. Mukherjee
Department of Computer Science and Engineering, Banasthali Vidyapith, Jaipur, India
e-mail: sanikasingh39@gmail.com

S. Mukherjee
e-mail: msaurabh@banasthali.in

T. Choudhury (✉)
CSE Department, Symbiosis Institute of Technology, Symbiosis International University, Pune, Maharashtra 412115, India
e-mail: tanupriya.choudhury@sitpune.edu.in; tanupriya@ddn.upes.ac.in; tanupriya1986@gmail.com

Informatics Cluster, School of Computer Science, University of Petroleum and Energy Studies (UPES), Dehradun, Uttarakhand 248007, India

CSE Department, Daffodil International University, Daffodil Smart City, Birulia 1216, Bangladesh

Director Research (Honorary), The AI University, Cutbank Montana 59427, US

A. Ramdane-Cherif et al. (eds.), *Machine Intelligence and Data Science Applications*, Algorithms for Intelligent Systems, https://doi.org/10.1007/978-981-99-1620-7_24

plays an important role in planting. Most commonly, the reason for such diseases are viruses, fungi, and bacteria. Apart from this, sometimes environmental conditions, soil, water, and fertilizers become the reason for such diseases in plants.

If some techniques can be used to encounter these illnesses in the early stage so that corrective measures may be taken to protect the entire crop, then that will be very beneficial for farmers.

There are many image processing techniques available that can detect diseases and can classify them into categories based on their symptoms. Farmers use the conventional method to detect the disease, which is the use of the naked eye which can be limited to some area of cultivation but it cannot cover large crops. Also, it may be time-consuming and would require more effort.

Using image processing techniques, this can be done easily with the least effort, time, and with high accuracy. This can be cost-effective as well as it will save investment in pesticides. Lots of research has been done and many algorithms are developed to solve this issue.

Different researchers and authors have proposed different techniques for plant leaf disease detection accurately. This paper examines and surveys various methods for identifying and categorizing plant leaf diseases. In the first part, this paper includes an introduction to plant leaf disease and detection. The second part of the paper includes a literature survey that explains techniques used by different authors for plant leaf disease detection and classification, a similar approach has been used in human disease detection as well [1]. In the third part of the paper, a review table is included that gives quick information about all the techniques used by different authors. The conclusion is included toward the end of the paper.

2 Literature Review

The authors of paper [2] describe an image processing method for spotting sugarcane leaf disease. Sugarcane Mosaic, Red Stripe, Red Rot, Downy Mildew, and Brown Spot are the six disease types that the authors chose for their experiment. For image analysis [3], images are acquired in higher quality resolutions using formats like TIF, PNG, JPEG, and BMP.

Preprocessing involves converting RGB photos to grayscale and removing any unnecessary data. Segmentation finds the areas of a given image that are healthy and contain green pixels as well as any areas that may be sick. For feature extraction and illness identification, three algorithms such as Linear SVM, Non-linear SVM, and Multiclass SVM are used.

The authors of paper [4] describe an image processing method for identifying groundnut plant diseases [5].

Early leaf spot and Late leaf spot are the two main diseases that affect groundnut plants. After acquiring leaf photos in RGB, they are transformed into images in HSV color. It has been discovered that green pixels in an image speed up processing. The co-occurrence matrices technique is employed in the analysis of color and extraction of textural features.

A total of two methods are used for analyzing texture images for extracting texture features. The first approach is a structured one, whereas the second is a statistical one.

In this study, the author employed a statistical approach. The backpropagation method is used to classify and identify peanut disease.

Propagation and weight update are the two types of phases in backpropagation. The authors categorized four distinct illnesses.

The authors of paper [6] offer an image processing method for identifying illnesses in tomato leaves. Digital photos of tomato leaves with early blight, Powdery mildew, and other tomato diseases are gathered during the image capture phase. Techniques for picture improvement, smoothness, noise removal, image scaling, image isolation, and background removal are used during the preprocessing phase. To identify and categorize tomato infections, the author introduced Gabor wavelet processing and Support vector machines. Gabor wavelet transform is used in the feature extraction phase to produce feature vectors that will be used in the classification phase that follows.

Support vector machines (SVMs) are trained to recognize the category of tomato diseases during the classification phase. Feature vectors and associated classes are the inputs and outputs of SVM, respectively.

The authors of study [7] describe image processing methods for identifying bacterial infection in plants. Bacterial leaf scorch is a common illness that affects plants, and early diagnosis of this infection enhances plant growth [8]. Image acquisition is the initial step in the image processing process, which includes fundamental actions like taking a picture and transferring information to a format that a computer can understand. Using the K-means clustering approach for image segmentation, clustering is carried out to separate the images of the foreground and backdrop.

The leaf area is highlighted by subtracting clustered leaf images from the base images. Clustering is based on intensity mapping. In comparison to fuzzy logic, the K-means clustering technique reduces the need for manual cluster selection while still being simple and effective at detecting diseased areas. By utilizing ADSP target boards, additional implementation is feasible using FPGA tools.

In paper [9], the author put up a procedure for developing a system that automatically detects plant diseases. Through the k-means clustering algorithm and Otsu's classifier, research has predicted the area of the leaf that was infected.

In the suggested study, characteristics for both shape and texture were retrieved. Area, color axis length, eccentricity, solidity, and perimeter were among the shape-oriented features that were extracted in this study, whereas contrast, correlation, energy, and homogeneity were the texture-oriented features [9]. Finally, a neural network [5] classifier was used in this study which classifies data [10].

Genetic algorithms were used as an image segmentation technique in a paper [11] in order to automate the recognition and classification of affected plants. A small set of images were used as training and test sets of leaves for his four plants: banana, bean, rose, and lemon.

We used the color co-occurrence method to extract features that consider both texture and color information. Diseases were classified with accuracies of 86.54% and 95.71%, respectively, using the minimum distance criterion with k-means clustering and the SVM [11]. Combining the genetic algorithm with the minimum distance criterion classifier improves the performance by up to 93.63%.

The authors of paper [1] describe an image processing technique for the detecting of diseased citrus leaf regions. Four different citrus diseases are as follows: Citrus canker, anthracnose, overwatering, and citrus greening are just a few examples.

For the purpose of creating a database, the author presented a system in which picture acquisition comes first when using a digital camera to capture high-resolution images. In image preprocessing, color space conversion and picture enhancement are done. The domain of the discrete cosine transform is used to improve color images [12].

For color space conversion, the YCbCr and l*a*b* scheme is used. The author uses Gray-Level Co-Occurrence Matrix (GLCM) to see statistics like homogeneity, entropy, contrast, and energy utilizing the graycoprops function in feature extraction. SVMRBF and SVMPOLY are two varieties of support vector machine (SVM) classifiers employed to distinguish between citrus leaf diseases.

A study in the author's paper [13] used an E color difference algorithm to isolate affected regions, and used colors histogram and texture data for disease diagnosis, achieving an overall accuracy of 99.9%.

Various classifiers were used, including fine ANN, cubic SVM, boosted, and bagging tree classifiers. Compared to other classifiers, the bagging tree classifier outperforms with 99.5%, 100%, and 100% accuracy for RGB, HSV, and LBP functions, respectively. The accuracy rates of KNN, Cubic SVM, and Boosted tree classifiers were 88.9%, 90.1%, and 50.90%, respectively [14].

3 Review Table

This table basically includes the summary of the research papers studied while the research, the methods, and techniques used by different authors. In the first paper, the author used MATLAB as a tool and different SVM techniques to detect plant leaf disease. The author of the second paper has used the backpropagation method for this purpose. The dataset is made from groundnut leaves. Other authors and researchers have used methods like HSV histogram, RGB histogram, LBP method, K-mean clustering and intensity mapping, etc., for the purpose of digital leaf image disease identification (Table 1).

Table 2 contains a list of the pros and cons of each method used by different researchers in their research works. Each method have pros as well as cons. The clustering method used in paper [13, 15] is bad for the worst-case scenario. But are faster in terms of computation. The threshold method used by the author in paper [16] is the best fit for real-life scenarios. The edge detection method used by the

Table 1 Summary of techniques used for plant leaf disease detection

Paper	Techniques used
1. Matlab image processing for detection and prevention of leaf disease	Non-linear SVM, multiclass SVM, and linear SVM
2. Detection and classification of peanut leaf disease using back propagation algorithm	CCM, back propagation algorithm
3. Tomato leaves diseases detection approach using support vector machine	SVM, Gabor wavelet transform
4. Novel image processing algorithm for detecting bacterial leaf scorch (BLS)	K-means clustering algorithm, intensity mapping
5. Plant disease detection and their solution by image classification	Otsu's algorithm, K-mean clustering
6. Using image segmentation and soft computing techniques to find leaf diseases	CCM methodology, SVM, K-mean clustering
7. Citrus leaf disease detection using image processing methods	GLCM, SF-CES, SVMRBF and SVMPOLY classifier
8. Automated citrus disease identification using color histograms and textural descriptions based on symptoms	LBP (local binary patterns), the Delta E color difference-based technique, the RGB histogram, and the HSV histogram

author in paper [14] is best in the case when images are of high contrast around its regions.

4 Conclusions

This survey provides an overview and analysis of methods for employing image processing for identifying and classifying plant leaf diseases. For precise recognition and classification of illnesses, several authors have employed various algorithms and methodologies. The benefit of applying image processing techniques is that leaf diseases can be detected at the correct time and remedial action can be taken. The majority of researches employed classifiers like Artificial neural network [5], Support vector machine, and others for an increase in recognition rates. All the strategies in this study are time-saving and yield effective results.

Table 2 A review of segmentation methods used in the detection process

Segmentation techniques	Pros	Cons
Clustering method [13]	• Getting a homogeneous area is simple • Faster calculations • K-means performs better if the value of K is lower	• Worst case scenario conduct is bad • Requires clusters of similar size
Edge detection methods [14]	• Suitable for images with high contrast between regions	• Did not work well with images with many edges • Difficulty finding the correct edge of an object
Threshold method [16]	• Image prior knowledge not needed • Fast and low-cost computation • Easy to use and suitable for real-life scenario	• Threshold selection is important • Very sensitive to noise • The resulting image may lose spatial information
Clustering method [15]	• Homogeneous regions easily obtained • Faster computation • Lower K value, better K-mean operation	• Worst case scenario conduct is bad • Requires clusters of similar size

References

1. Gavhale KR, Gawande U, Hajari KO (2014) Unhealthy region of citrus leaf detection using image processing techniques. In: 2014 IEEE international conference for convergence of technology. 978-1-4799-3759-2/1
2. Mitkal P, Pawar P, Nagane M, Bhosale P, Padwal M, Nagane P (2016) Leaf disease detection and prevention using image processing using Matlab. Int J Recent Trends Eng Res (IJRTER) 02(02). ISSN: 2455-1457
3. Sarkar T, Salauddin M, Choudhury T et al (2021) Spatial optimisation of mango leather production and colour estimation through conventional and novel digital image analysis technique. Spat Inf Res 29:439–453. https://doi.org/10.1007/s41324-020-00377-z
4. Ramakrishnan M, Sahaya Anselin Nisha A (2015) Groundnut leaf disease detection and classification by using back propagation algorithm. In: IEEE ICCSP 2015 conference. 978-1-4799-8081-9/15
5. Sunil CK, Jaidhar CD, Patil N (2021) Cardamom plant disease detection approach using EfficientNetV2. IEEE Access 10:789–804. https://doi.org/10.1109/ACCESS.2021.3138920
6. Mokhtar U, Alit MAS, Hassenian AE, Hefny H Tomato leaves diseases detection approach based on support vector machines. In: 2015 IEEE. 978-1-5090-0275-7/15
7. Krishnan M, Sumithra MG (2013) A novel algorithm for detecting bacterial leaf scorch (BLS) of shade trees using image processing. In: 2013 IEEE 11th Malaysia international conference on communications, Kuala Lumpur, Malaysia, 26–28 Nov 2013. 978-1-4799-1532-3/13
8. Shahane SK, Tawale PV (2016) Prediction on crop cultivation. Int J Adv Res Comput Sci Electron Eng (IJARCSEE) 5(10)
9. Saradhambal G, Dhivya R, Latha S, Rajesh R (2018) Plant disease detection and its solution using image classification. Int J Pure Appl Math 119(14):879–884
10. Kumar P, Choudhury T, Rawat S, Jayaraman S (2016) Analysis of various machine learning algorithms for enhanced opinion mining using Twitter data streams. In: 2016 international

conference on micro-electronics and telecommunication engineering (ICMETE), Ghaziabad, India, pp 265–270. https://doi.org/10.1109/ICMETE.2016.19
11. Singh V, Misra AK (2016) Detection of plant leaf diseases using image segmentation and soft computing techniques. Inf Process Agric 8:252–277
12. Vinod K, Reddy N, Ajai SR (2019) Analysis of classification algorithms for plant leaf disease detection. In: Proceedings of the 2019 3rd IEEE international conference on electrical, computer and communication technologies (ICECCT 2019), pp 1–6
13. Zhang S, You Z, Wu X (2019) Plant disease leaf image segmentation based on super pixel clustering and EM algorithm. Neural Comput Appl 31:1225–1232
14. Badage A (2018) Crop disease detection using machine learning: Indian agriculture. Int Res J Eng Technol 5
15. Champaneri M, Chachpara D, Chandvidkar C, Rathod M (2020) Crop yield prediction using machine learning. Int J Sci Res 9
16. Choudhury T, Kumar V, Nigam D, Mandal B (2016) Intelligent classification of lung & oral cancer through diverse data mining algorithms. In: 2016 international conference on micro-electronics and telecommunication engineering (ICMETE), Ghaziabad, India, pp 133–138. https://doi.org/10.1109/ICMETE.2016.24

A Hash Based Feature Extraction Strategy for Vector Size Reduction in Text Sentiment Analysis

Mousumi Bhattacharyya, Anirban Mitra, Sadip Midya, Asmita Roy, and Anupam Ghosh

1 Introduction

Sentiment analysis is a technique for automatically classifying a large amount of content into positive or negative attitudes. Due to social media's phenomenal growth, organisations and businesses have started to use vast amounts of internet data for product innovation and proactive decision-making. The way people express their ideas has evolved substantially in recent years because to social media, forums, blogs, and other online communication platforms. The vast amount of unstructured data must be mined for pertinent information, such as consumer views of company branding, by the majority of enterprises and organisations. The application of sentiment analysis is not just limited to product or movie reviews. Researchers in disciplines including media, politics, sport, etc. have also employed it. Sentiment analysis, for instance, can be used to find out how people feel in online political conversations about a particular candidate or political party. For the Persian language, however, very little study has been done in order to develop models. However, sentiment analysis has

M. Bhattacharyya (✉) · A. Mitra · S. Midya · A. Roy
Sister Nivedita University, DG Block (Newtown), Action Area I, 1/2, Newtown, New Town, Kolkata 700156, West Bengal, India
e-mail: mousumi.b@snuniv.ac.in

A. Mitra
e-mail: anirban.m@snuniv.ac.in

S. Midya
e-mail: sadip.m@snuniv.ac.in

A. Roy
e-mail: asmita.r@snuniv.ac.in

A. Ghosh
Netaji Subhash Engineering College, Techno City, Panchpota, Garia, Kolkata 700152, West Bengal, India
e-mail: anupam.g@snuniv.ac.in

A. Ramdane-Cherif et al. (eds.), *Machine Intelligence and Data Science Applications*, Algorithms for Intelligent Systems, https://doi.org/10.1007/978-981-99-1620-7_25

been used extensively for the English language using both traditional and modern machine learning techniques [1]. Simple text categorization methods fall short of revealing hidden elements, making the creation of a tool or system that can precisely categorise the text's emotions necessary. The demand for sentiment analysis has grown as a result of its use in a variety of industries, including market research, corporate intelligence, e-government, web search, and email filtering. Deep learning and machine learning algorithms are frequently employed to solve business problems in today's competitive markets [2]. The main step in sentiment analysis is to determine the polarity of a text, image, audio, or video. Positive, negative, or neutral polarity are all possible. Emotional intelligence or opinion mining are some names for it. In essence, it is a method of gathering various structured and unstructured data from social networking sites or micro-blogging, then analysing that data in accordance with specified rules. Sentiment research can provide us with a clear understanding of the public's perceptions of various marketable goods, political parties, the stock market, and other topics. At addition to this, it can also give users advice on how to shop in various stores [3]. Two fundamental principles govern sentiment analysis: Rule-based strategy: this method makes use of a lexicon with polarised words. Rule-based approaches classify content into ordered groupings using a set of manually created linguistic rules. These rules instruct the system to use semantically relevant textual elements to identify appropriate categories depending on the content of a document. Each rule consists of an antecedent or pattern and a projected category [4]. Rule-based systems are comprehensible to humans and can be improved over time. There are certain disadvantages to this tactic, though. These systems, to begin with, call for in-depth knowledge of the industry. They need a lot of time because developing rules for a complex system can be challenging and frequently calls for in-depth research and testing. Rule-based systems are difficult to manage and do not scale well since adding new rules can change the results of the pre-existing rules. The second approach is one based on machine learning (ML): Using a training set, an ML model is trained to discern the sentiment of the words. Although this classic ML-based strategy outperforms the rule-based approach in terms of performance, it has drawbacks such manual feature extraction and word order, among others. Instead of manually setting criteria, machine learning text classification learns to create categories based on prior observations. By employing pre-labelled examples as training data, machine learning algorithms may be able to comprehend the numerous relationships between text fragments and that a particular output (i.e., tags) is predicted for a particular input (i.e., text). A "tag" is a predetermined classification or grouping that each provided text may fall into. The first step in training an NLP classifier using machine learning is feature extraction, which entails converting each text into a numerical representation in the form of a vector. One of the most popular techniques is the "bag of words," in which a vector represents the frequency of a word within a specified lexicon. Another popular method is TF-IDF method [5].

2 Related Study

People frequently use the social media site Twitter to share their thoughts and emotions on various occasions. However, the complex format of tweets makes processing harder. The very small size of the tweet format creates a whole new set of issues, such as the use of slang and acronyms [6]. It sometimes suggests that various classes may not always be mutually exclusive. Tweets may not always be categorically distinct, meaning that it may partially belong to one class and partially to another. This creates a problem for a discriminatory algorithm [7, 15]. With its wide social reach, Twitter has developed into a rich source of user attitudes and emotions, with far-reaching effects on how future products, services, or organizations will be run. Due to constraints imposed on the maximum number of characters that can be used in a single tweet, users regularly invent fresh and innovative ways to communicate often violating the traditional language syntax [8]. For conventional Natural language processing algorithms and methods that use Part of speech tags and dependency tags, the produced text may therefore be rather "noisy." For these challenges, a number of deep learning models have recently been put forth. Deep Learning is a machine learning technology that uses neurons from the neural network seen in the human brain as its design inspiration [9, 10]. Neurons are the fundamental building elements of the nervous system, and each neuron processes inputs to produce output. It enables the system to categorize and find patterns in the raw data without explicit programming. By enabling the machine to learn from and get better at similar experiences, machine learning evolves. Deep learning algorithms gradually learn from the provided data's simple to complex properties [11–14].

3 Motivation

According to the aforementioned survey, virtually every company produces enormous amounts of data in today's digital age. Machine learning algorithms are being used to extract useful patterns from this data that may be used to guide management and commercial decisions. Dimensionality reduction strategies may lessen the computational burden of machine learning algorithms by dramatically decreasing the time complexity of the training phase. This paper's main objective is to examine how dimensionality reduction techniques impact the performance of ML algorithms.

4 Proposed Methodology

A modified text mining engine can be proposed to extract feature. Steps of feature extraction are described below.

4.1 Data Acquirement

Using social media API offered by many social networking sites, the social media data, specifically data from Twitter, are harvested on a big scale. When extracting data, COVID-related hash tags and keywords are found. Based on the length of time and volume of data extracted, the API establishes limits on the data access rate. A generic data interchange format, such as CSV, will be used to extract and store all pertinent data based on the query in the centralized storage location. A broad range of cleaning factors must be applied to the data obtained from various social media platforms in order to obtain a well-defined set of data.

4.2 Data Pre-processing, Cleaning and Amalgamation

The gathered data from social media are in large volume and contain lots of noise and irrelevant words. A proper data cleaning and pre-processing are required. Hyperlinks, punctuation, digits, stop words, and emoji are among the things that are removed during the noise removal process.

4.3 Tokenization of the Data

Tokenization is the next stage. This method breaks down the entire tweet into words by removing all punctuation, including commas and blank spaces. Tokens are created by slicing up the input data into relevant words. After that, the tokens are utilized as an input for additional processing.

4.4 Lemmatization

Lemmatization creates a lemma by replacing a word's suffix with another one or by completely removing it. As an illustration, take the word "laugh", applying lemmatization retains "laughing, laughed, laughs" and for pluralisation purposes only "laugh" is selected.

4.5 Corpus Formation

Once the entire tweets are lemmatized tweet-wise vector can be formed comprising lemmas. Each vector is called document, and entire vector set is called corpus. The corpus is then used for feature extraction.

4.6 Feature Extraction

Text feature extraction is required to make a matrix from a text content, which will be later used in Machine Learning model. A traditional Bag of Word (BOW) model can be used to extract features from text document. But BOG suffers from some drawbacks; one of the major drawback is the size of vector. It depends upon total number of unique words. As tweets from different persons are considered as corpus, therefore, total number of unique words can be huge and it leads to a large vector. Finally we will get a sparse matrix. Apart from this, if a word frequently occurs in a single document but not present in the other documents will have high frequency count. But that word is not actually very important with respect to the entire corpus. To avoid this TF-IDF (Term frequency-Inverse document frequency) technique may be used. In this technique high frequency word can't dominant the classification. But, the major drawback of this technique is the size of the vector. Some less important words will also be mapped; whereas those will not take much role in classification.

To overcome the above mentioned problems, a hash function can be used to determine a fixed size vector. Instead of building a word wise vector, a hashing trick can build a vector of a pre-defined length by applying a hash function h to the features. A predefined list of positive and negative word will help to define the length of the vector. Along with this unimportant words can be eliminated from the vector. For all positive and negative unique words, an index will be generated and stored in dictionary. This can be further used in Machine Learning algorithm.

A modified dictionary based TF-IDF algorithm for feature extraction can be used to make final matrix from the text. In traditional TF-IDF algorithm the frequency of each unique word is measured. Neutral words doesn't play any role in polarity, but those also are treated as unique word. Therefore, the presence of more neutral words increases the matrix size. Not only that, more frequent neutral words may dominant the actual polarity of the corpus. However, the size of the matrix will be smaller if we simply take positive and negative terms into account. Additionally, a hashing function can be used to determine a word's frequency in order to calculate inverse document frequency and term frequency. That can also save the searching time. The details of the modified dictionary based TF-IDF algorithm is explained below:

Let D_1 and D_2 be two dictionaries based on unigram positive and negative words and x and y be the number of words of D_1 and D_2 respectively. Let N be the total number of tweets in corpus T. Therefore, T can be written as in Eq. (1)

$$T = T_1 \cup T_2 \cup T_3 \cup \cdots\cdots \cup T_N \quad (1)$$

Each word of D_1 and D_2 will be considered as term t. Therefore, total number of terms will be $(x + y)$. Term frequency (TF) of each term ti in given tweet Tj can be considered as in Eq. (2)

$$tf_{ti} = P(t_i \mid T_j) \qquad \text{[probability of term } t_i \text{ in a particular tweet } Tj\text{, where } i \text{ varies from 1 to } (x+y) \text{ and } j \text{ varies from 1 to } N.] \quad (2)$$

Document Frequency (DF) of each t_i in entire corpus T can be expressed as in Eq. (3)

$$df_{ti} = P(T_{t_i} \mid T) \qquad \text{[probability of tweet } T_j \text{ comprising term } t_i\text{, in entire corpus } T] \quad (3)$$

Therefore, Inverse Document Frequency (IDF) of each t_i in entire corpus T can be calculated as in Eqs. (4) and (5)

$$idf_{ti} = -log(df_{ti}) = -log(P(T_{ti}|T)) \quad (4)$$

$$(tf - idf)_{ti} = tf_{ti} * idf_{ti} = P(T_{ti}|T) * \log \frac{1}{P(T_{ti}|T)} \quad (5)$$

Let α or β be the weight to be applied with each $(tf - idf)_{ti}$ depending upon corresponding positive and negative dictionary respectively. Finally a feature vector TV_j for each tweet T_j can be calculated as in Eq. (6)

$$\begin{aligned} TV_ij &= \alpha * P(T_{ti}|T) * log1/(P(T_{ti}|T)) \text{ if } t_i \in D1 \\ & \qquad Or \\ &= \beta * P(T_{ti}|T) * log1/(P(T_{ti}|T)) \text{ if } t_i \in D2 \end{aligned} \quad (6)$$

To calculate the probability of each term a hash function will be more efficient. It can be designed to map every unique term of $D1 \cup D2$ in vector TV_j. The detailed algorithm is shown in Fig. 1.

However, to extract the term from each documents we considered unigram positive and negative words. But more accuracy can be achieved by using n-grams. N-grams are continuous word, symbol, or token sequences in a document. They can be described technically as the adjacent groups of items in a document. For example, 'happy' is a positive term but 'not happy' is negative. In this case, if unigram is used, then it can be mapped as positive emotion, whereas it should be mapped as negative emotion. To achieve more accuracy positive and negative bigrams and trigrams have been considered. Details of term extraction method has been described in Fig. 2. The

```
/*importing dictionary*/
D1 = positive_words /*Dictionary comprising all unigram positive words*/
D2 = negative_words /*Dictionary comprising all unigram negative words*/
N = total number of tweets /*downloaded from social media i.e. twitter*/
column = x + y = number of words in D1 + number of words in D2
tf = vector of size N*(x + y) /*will store the term frequency for each dictionary word*/
idf = vector of size 1*(x + y) /*will store the inverse document frequency for each
dictionary word*/
tfidf = vector of size N*(x + y) /*will store the term frequency- inverse document
frequency for each dictionary word*/
/*calculating idf*/
for j := 1 to N do
        T[j] ← choose jth tweet from corpus T
        for i := 1 to column do
                t[i] ← choose ith term from D1∪D2
                idf[h(t[i])] ← idf[h(t[i])] + 1 /*counting number of documents comprising
                each terms t[i], and the hash function h(t[i]) will map each t[i] in vector idf*/
        End for
End for
for i := 1 to column do
        if idf[i] <> 0 then
        idf[i] ← log(N/idf[i]) /*counting IDF for each terms t[i]*/
End for
/*calculating TF*/
for j := 1 to N do
        T[j] ← choose jth tweet from corpus T
        for i := 1 to length(Tj) do
                t[i] ← choose ith term from T[j]
                tf[j][h(t[i])] ← tf[j][h(t[i])]+ 1
        End for
        /*calculating no. of frequency of each terms ti from D1∪D2 in each tweet T[j]
and
        the hash function h(t[i]) will map each t[i] in vector TV[j].*/
        for i := 1 to column do
                tf[j][i] ← tf[j][i]/length(T[j])) /*calculating tf*/
        End for
        /*calculating tf-idf*/
        for i in 1 to column do
                tfidf[j][i] ← (tf[j][i]*idfi[i])
End for
End for
```

Fig. 1 Proposed algorithm

algorithm in Fig. 2 will generate sequences of tokens starting from different items of the list of tokens. The parameter n in the extract_ ngrams() function controls how many sequences to generate. 3 would be replaced by 2 to extract bigrams. After extracting trigrams and bigrams, the remaining tokens are considered for unigrams. The image in Fig. 3 describes the entire process. The first dataset comprises a comma

```
Step1: Extract trigrams from each tweet;
    extract_ngrams(df, n); // df is dataframe object and n is an +ve integer
    for each tweet of df do:
        split the tokens and store in a temporary list (tmp);
        while all the tokens of tmp is not counted:
            start ← 0
            for i:= start to (start+n-1) do:
                make tuple by adding tmp[i]
                add the tuple to output list (sequences)
            start←start+1
Step2: for all extracted trigram (tri: each tuple from sequences) do:
        if tri exists positive or negative trigram dictionary
            calculate hash based TF-IDF tri to list of token
Step3: repeat step1 and step 2 for bigrams;
```

Fig. 2 Term extraction method

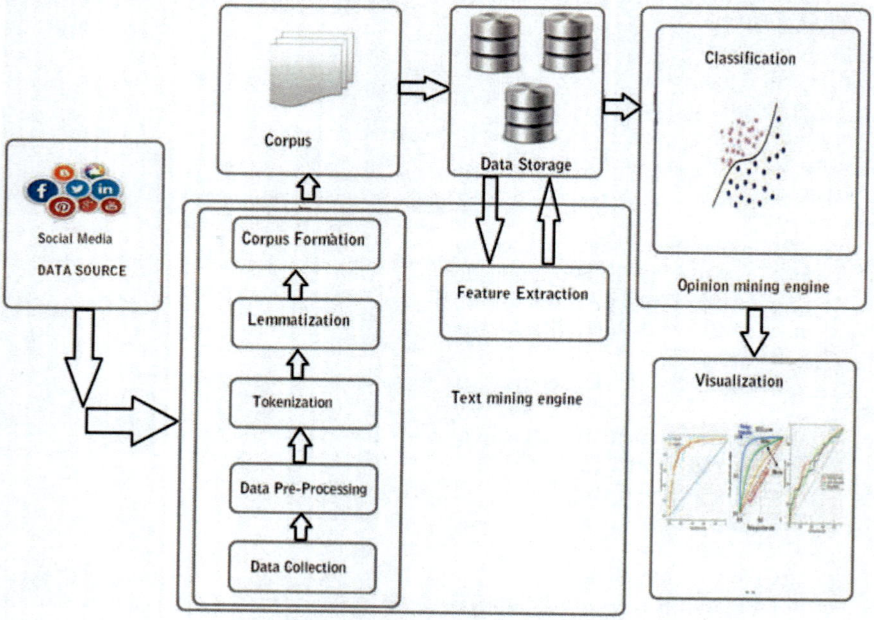

Fig. 3 Block diagram for the proposed feature extraction strategy

separated values with the following tabs: textID, sentiment, author, text, old_ text, aux_ id, new_ sentiment, selected_ text. Among these, text comprises tweets from different users and selected_ text column contains the valid tweets written in English. Sentiment comprises the basic sentiment labels such as happy, sad, anger, fun, etc.; whereas new_ sentiment contains only three category i.e. positive, negative and neutral.

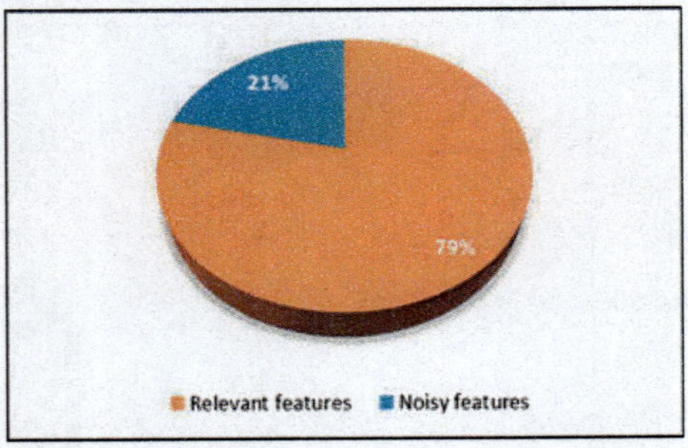

Fig. 4 Relevant features versus noisy features

We worked to train and validate the model's performance after the feature extraction and pre-processing processes have been completed. The gathered data set has been split into two sets: a training set and a testing set. The testing set is the one on which the experiment is run, whereas the training set is used to train the classifier (machine learnt model). Depending on the application, different training and testing dataset ratios may be used. We divided the dataset into 70% for training and the remaining 30% for testing. We validated the above mentioned feature extraction model by comparing the classified opinions of both the processes i.e. with traditional TF-IDF approach and the proposed modified TF-IDF approach while generating fixed length vector. The exact efficiency of the model has been measured. We used Naïve Bayes and SVM for classification and validation. Using Naïve Bayes we got 0.8565 accuracy in traditional TF-IDF and 0.8666 accuracy in modified TF-IDF. On the other side, we got accuracy 0.9086 with traditional TF-IDF approach, whereas we achieved 0.9091 accuracy with hash based modified approach. The graphical analysis of the entire process is shown in this section.

Figure 4 shows the ratio of the number of noisy tweets on the total number of relevant tweets. The values of the series are displayed as a percentage of each class Relevant variables, Noisy variables. Figure 5 describes the distribution of sentiments

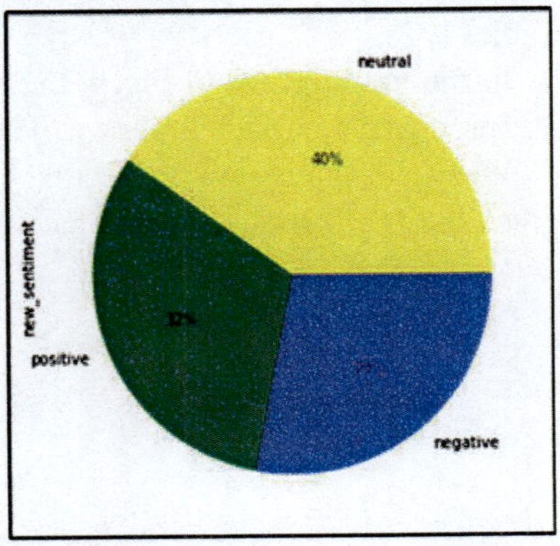

Fig. 5 Distribution of sentiments across all the tweets

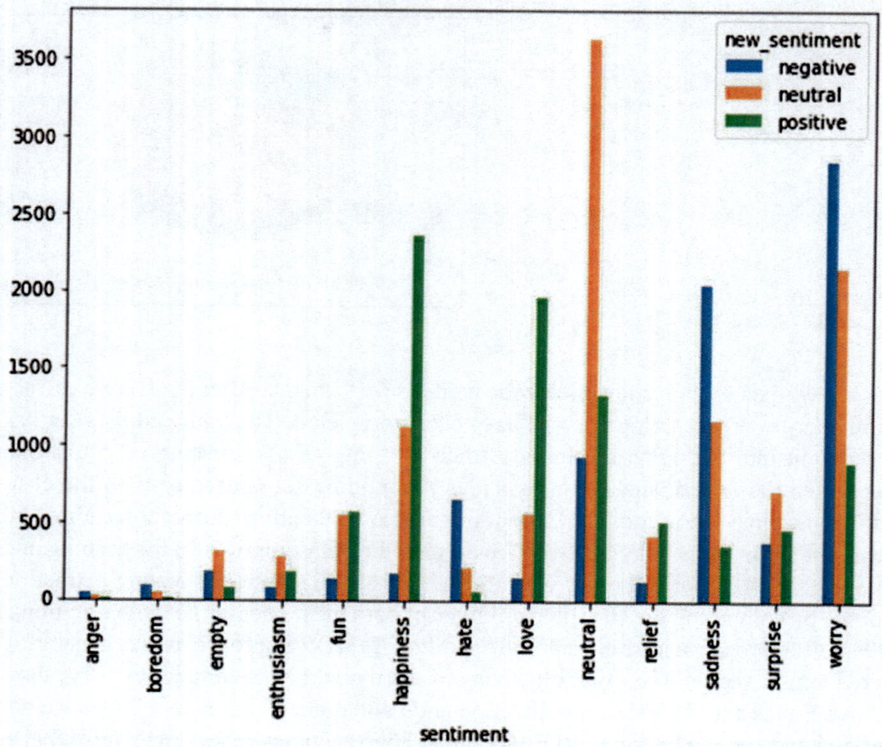

Fig. 6 Distribution of sentiment in each emotions

among the corpus. In the above image we can see the percentage of public tweets containing positive, negative and neutral sentiments.

In the graph shown in Fig. 6 we can see the distribution of sentiment in each emotions. For example, happiness, fun, etc. are considered as positive sentiment whereas hate, worry, etc. are considered as negative sentiment (Figs. 7, 8 and Tables 1, 2).

In the graph given in Fig. 9, the accuracy between the traditional approach and our proposed approach is shown. The result are shown in Table 3.

We can also measure the accuracy using unigram, bigram and trigrams. The chart given in Fig. 10 shows the comparison. The result is shown in Table 4.

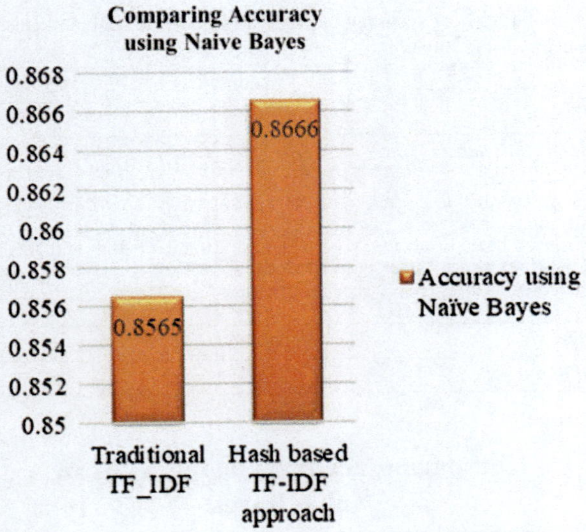

Fig. 7 Comparison on basis of accuracy between traditional TF-IDF and Hash based modified TF-IDF approach using Naïve Bayes

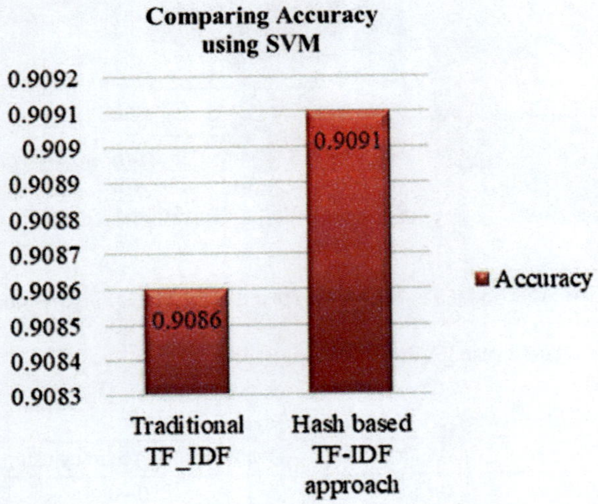

Fig. 8 Comparison on basis of accuracy between traditional TF-IDF and Hash based modified TF-IDF approach using SVM

We validated the result by using AUC-ROC. There are numerous distinct evaluation metrics for classification. The most well-liked metric is accuracy, which counts how frequently the model is right. Since it is simple to understand and having the most accurate estimations is frequently required, this metric is excellent. The chart given in Fig. 11 shows the AUC of our model.

Table 1 Comparison on basis of accuracy between traditional TF-IDF and Hash based modified TF-IDF approach using Naïve Bayes

Approach	Accuracy
Traditional TF-IDF	0.8565
Hash based TF-IDF approach	0.8666

Table 2 Comparison on basis of accuracy between traditional TF-IDF and Hash based modified TF-IDF approach using SVM

Approach	Accuracy
Traditional TF-IDF	0.9086
Hash based TF-IDF approach	0.9091

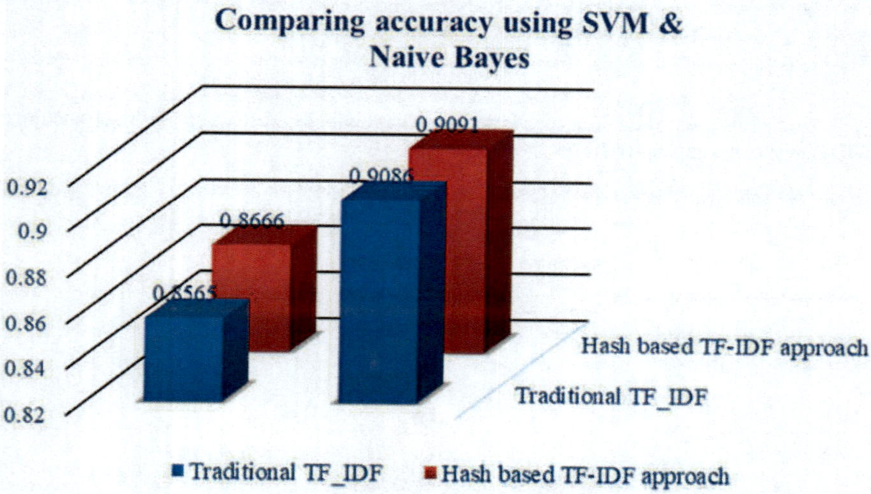

Fig. 9 Comparing accuracy using SVM and Naive Bayes

Table 3 Accuracy table between SVM and Naive Bayes

Approach	Accuracy using Naïve Bayes	Accuracy using SVM
Traditional TF_IDF	0.8565	0.9086
Hash based TF-IDF approach	0.8666	0.9091

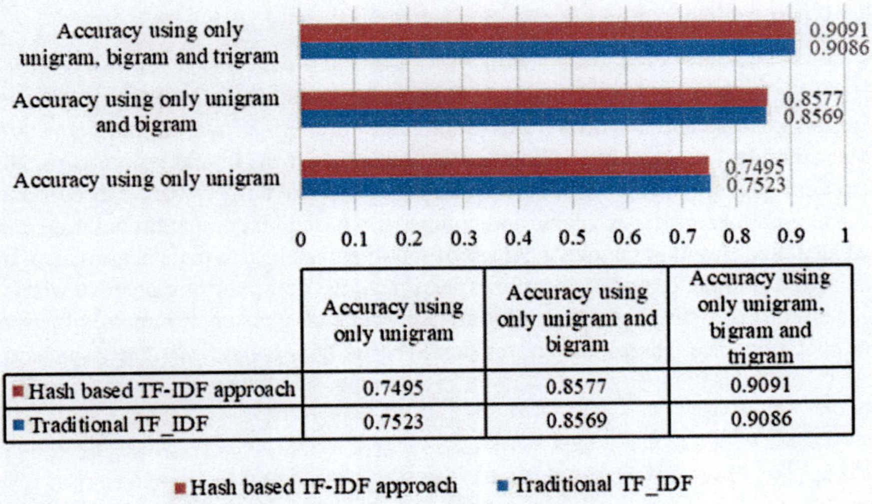

Fig. 10 Comparison of accuracy using unigram, bigram and trigram

Table 4 Accuracy using unigram, bigram and trigram

Approach	Accuracy using only unigram	Accuracy using only unigram and bigram	Accuracy using only unigram, bigram and trigram
Traditional TF_IDF	0.7579	0.8569	0.9086
Hash based TF-IDF approach	0.7495	0.8572	0.9091

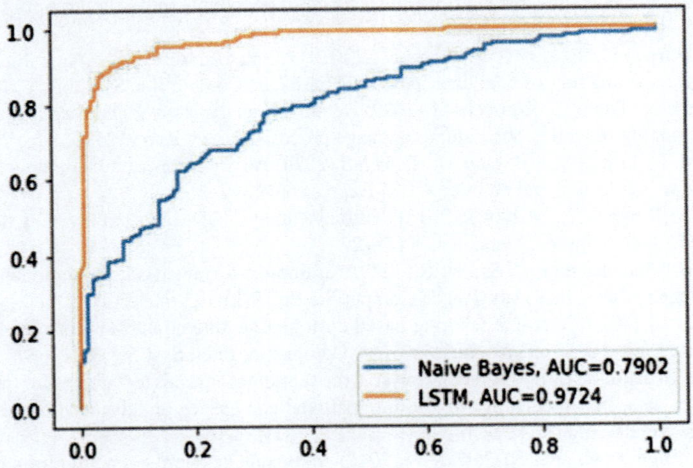

Fig. 11 AUC for validation

5 Conclusion

We used a hash based feature extraction technique; we also considered trigrams and bigrams positive and negative words along with unigrams. However, our algorithm is based on text data analysis. Whereas text with emoticons or text with images are also very popular way to express human emotions. So further work can be done focusing on text with emoticons, and comparison based on computational time can be calculated. Another major challenges of sentiment analysis is to detect sarcasm. In sarcastic comment or review, people express negative opinion using positive words. This is really difficult to analyze and completely depends upon environment, situation and specific topics. Therefore, ngrams should be analysis perfectly to detect sarcasm.

References

1. Dashtipour K, Gogate M, Adeel A, Larijani H, Hussain A (2021) Sentiment analysis of persian movie reviews using deep learning. Entropy 23(5):596
2. Ghorbani M, Bahaghighat M, Xin Q, Özen F (2020) ConvLSTMConv network: a deep learning approach for sentiment analysis in cloud computing. Journal of Cloud Computing 9(1):1–12
3. Yi S, Liu X (2020) Machine learning based customer sentiment analysis for recommending shoppers, shops based on customers' review. Complex & Intell Syst 6(3):621–634
4. Zhao W, Du S (2016) Spectral-spatial feature extraction for hyperspectral image classification: a dimension reduction and deep learning approach. IEEE Trans Geosci Remote Sens 54(8):4544–4554
5. Liu R, Shi Y, Ji C, Jia M (2019) A survey of sentiment analysis based on transfer learning. IEEE Access 7:85401–85412
6. Gupta B, Negi M, Vishwakarma K, Rawat G, Badhani P, Tech B (2017) Study of Twitter sentiment analysis using machine learning algorithms on Python. Int J Comput Appl 165(9):29–34
7. Liu H, Burnap P, Alorainy W, Williams ML (2019) A fuzzy approach to text classification with two-stage training for ambiguous instances. IEEE Trans Comput Soc Syst 6(2):227–240
8. Omuya EO, Okeyo G, Kimwele M (2022) Sentiment analysis on social media tweets using dimensionality reduction and natural language processing. Eng Rep e12579
9. Ullah MA, Marium SM, Begum SA, Dipa NS (2020) An algorithm and method for sentiment analysis using the text and emoticon. ICT Express 6(4):357–360
10. Heikal M, Torki M, El-Makky N (2018) Sentiment analysis of Arabic tweets using deep learning. Procedia Computer Science 142:114–122
11. Devi TS, Dhanalakshmi R, Sankar S (2020) An improved framework for sentiment analysis for college reviews. Int J Adv Trends Comput Sci Eng 9(2):1959–1963
12. Yi S, Liu X (2020) Machine learning based customer sentiment analysis for recommending shoppers, shops based on customers' review. Complex & Intell Syst 6(3):621–634
13. Tyagi P, Tripathi RC (2019) A review towards the sentiment analysis techniques for the analysis of twitter data. In: Proceedings of 2nd international conference on advanced computing and software engineering (ICACSE)
14. Rose S, Engel D, Cramer N, Cowley W (2010) Automatic keyword extraction from individual documents. Text Mining: Appl Theory 1–20
15. Liu R, Shi Y, Ji C, Jia M (2019) A survey of sentiment analysis based on transfer learning. IEEE Access 7:85401–85412

ARM-FT Autonomic Randomized Cloudlet Management Through Fault Tolerance

Bhupesh Kumar Dewangan, Tanupriya Choudhury, Madhu Khurana, and Teoh Teik Toe

1 Introduction

One of the most popular business models in the IT industry is cloud computing. However, cloud infrastructure establishment for provisioning is quite expensive, but for the long term, it gives more profit than investments. From the infrastructure perspective, server, network, and storage virtualization plays a vital role in the elastic scale of computational resources subscribed by end-users. The objective of cloud computing is to achieve low-cost management of infrastructure, performance

Bhupesh Kumar Dewangan and Tanupriya Choudhury are the corresponding authors and contributed equally and are the first authors.

B. K. Dewangan (✉)
Department of Computer Science and Engineering, OP Jindal University, Raigarh, India
e-mail: bhupesh.dewangan@gmail.com

T. Choudhury (✉)
CSE Department, Symbiosis Institute of Technology, Symbiosis International University, Pune, Maharashtra 412115, India
e-mail: tanupriya.choudhury@sitpune.edu.in; tanupriya@ddn.upes.ac.in; tanupriya1986@gmail.com

Informatics Cluster, School of Computer Science, University of Petroleum and Energy Studies (UPES), Dehradun, Uttarakhand 248007, India

CSE Department, Daffodil International University, Daffodil Smart City, Birulia, Bangladesh 1216, India

Director Research (Honorary), The AI University, Cutbank Montana 59427, US

M. Khurana
University of Gloucestershire, The Park, Cheltenham GL50 2RH, UK

T. T. Toe
Nanyang Technological University (NTU), Singapore, Singapore
e-mail: 4ttteoh@ntu.edu.sg

A. Ramdane-Cherif et al. (eds.), *Machine Intelligence and Data Science Applications*, Algorithms for Intelligent Systems, https://doi.org/10.1007/978-981-99-1620-7_26

monitoring, avoid under/over provisioning, and satisfy the agreements signed with consumers. In 2019, to achieve these objectives more efficiently, Self-management Aware Autonomic Resource Management Technique (SMART) was proposed by Dewangan et al. [1]. SMART is responsible for workload and Cloudlets (CLD's) management based on self-optimization methods.

The structure of this paper is introduced followed by recent developments in relevant areas and analysis of the recent works with different parameters, and then identifications of issues and challenges followed by methodology and results with the conclusion of the research conducted.

Through fault tolerance, Autonomic Randomized Cloudlet Management is a vital concept for cloud computing [2, 3]. This expertise allows us to do efficient management of resources within a cloud environment, thus it helps to ensure that the services are available even in the event of system failure or other faults. In this research paper, it has been discussed how Autonomic Randomized Cloudlet Management operates, its advantages, and the potential challenges involved in it. Autonomous randomized cloudlet management operates by dividing the tasks randomly across multiple nodes so as to increase the utilization of resources while providing fault tolerance against node failures or other faults [4]. Different factors such as task size, energy consumption levels, and latency requirements are taken into consideration while scheduling the tasks on various nodes such that can be done more efficiently and effectively without even sacrificing their performance or reliability. Since there are redundant copies running on various nodes which can take over if necessary, this guarantees that any single point of failure is not the reason for the disruption of service availability. In addition to this, we have reduced response times and reduced overall cost since very few resources need to be allocated for the better utilization rates achieved from randomization techniques applied by the autonomous systems. However, there are some disadvantages leading to challenges of this autonomous randomized cloudlet management like ensuring sufficient security measures are taken into consideration while dealing with the sensitive data which are dumped into remote servers, monitoring network traffic flows between the distributed segments, and managing the unexpected spikes in demand caused by sudden changes in user behavior. Also, another key factor here that might be taken into consideration is scalability because ample amounts of requests are sent at a time then the performance may suffer significantly due to overloaded networks or insufficient processing power available at each node. Therefore, contingency planning needs to be done beforehand such that appropriate countermeasures can be taken to reduce any relevant issues arising during the runtime operations. In a nutshell, the autonomic randomized cloudlets provide us important improvements over the traditional methods used to manage the cluster environments; proper implementation combined with robust security protocols will allow them to reach their full potential of distributing reliable high-quality services that the users expect in today's digital age [5–7].

2 Recent Development in Cloudlet Management

Author Saxena et al. introduces a system by executing probes three genuine responsibility datasets specifically, Google Bunch Dataset, with the best-fit approach [8]. Author Rajabzadeh et al. proposed a model by checking the situation with assets and dissecting the acquired information that has prompted the appropriate arrangement and solidification of virtual machines through designated relocations brilliantly. When the cloudlets are created and available for execution, it depends upon the different parameters which is actually affected the service-level-agreement (SLA) [9].

Masdari and Khezri present a broad study and scientific classification of the prescient CLD movement approaches adjusted for the cloud DCs. For this reason, it initially gives the central points of interest in regard to the CLD movement and afterward characterizes them in light of their applied expectation calculation. It outlines the fundamental commitments of each plan and depicts how forecast techniques are coordinated into the CLD movement process, to make it more successful. Besides, an examination of the prescient movement plans is given. At long last, the finishing up comments and future exploration regions are determined [10]. Aldossary presents a thorough writing survey regarding the matter of dynamic asset the board (i.e., CLDs union and asset provisioning) in distributed computing conditions, alongside a general conversation of the firmly related works. The results of this research can be use to improve the utilization, minimizing the operational cost and energy consumption of available resources in the resource pool [11]. Gholipour et al. presented an energy-based cloudlet management through QoS parameters and achieved significant outcomes [12].

Sivagami and Easwarakumar propose another CLD Relocation for the authorization of cloud server farm foundation dependability through cutting-edge recuperation system of VN (Virtual Organization) requests. The whole number Direct Programming model is carried out for dissecting the course traffic which thinks about undeniably related mathematical elements to pick the ideal CLD in the ideal way. Then, at that point, the Elective Switch Distinguishing Proof Calculation decides the new CLD movement and steering is achieved. We utilized Cloudsim to carry out and gauge the viability of the proposed DFTM calculation. Then, we contrasted our outcomes and existing approaches as far as by and large execution time, normal asset use time, circle use, and other execution measurements. Trial results represent that the proposed issue of recuperation calculation improves the survivability of the CLD-based cloud server farm [13].

Ali et al. sets up a scientific categorization of tremendous energy utilization issues and their connected arrangements. The creators cover all parts of energy utilization by Cloud Datacenters and break down a lot more examination papers to figure out the improved answer for proficient energy utilization [14]. Khan et al. conducted AI-based resource management for different cloudlets [15]. Arshad et al. presented another calculation called the energy effectiveness heuristic utilizing virtual machine union to limit the high energy utilization in the cloud. By setting two limits, has

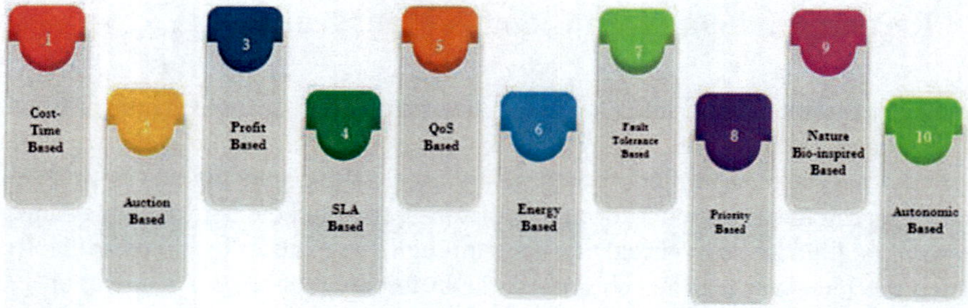

Fig. 1 Performance metrics in cloudlets management

are ordered into three fundamental classes. The planned model redistributes virtual machines starting with one actual host and then onto the next to limit energy utilization. The consequences of the proposed calculation have been gotten as far as virtual machine relocations, execution debasement brought about by movement, administration level understanding infringement, and execution time, showing a huge improvement over cutting-edge methods [16]. Agbaje et al. conducted a study on different assets on the board procedures embraced in writing. We focused significantly on researching game-hypothetical accommodation for the administration of required assets, as a likely arrangement in displaying the asset portion, booking, provisioning, and load adjusting issues in distributed computing. This paper presents a study of a few game-hypothetical procedures carried out in distributed computing assets on the board. In light of this overview, we introduced a rule to help the reception and use of game-hypothetical assets in the executive's methodology [17]. Liang et al. studied the general design of CDCs. Thusly, movement rules are laid out for the one-layered and multi-layered confided in CLDs. A high-pertinence heterogeneous CDC asset the board calculation in view of confided in CLD relocation (HTCLD2) is then proposed. The proposed further develops load adjusting while at the same time guaranteeing CLD execution. As obeserved, the proposed method is performing better as compared to other three researches conducted in this field, as shown by trial results [18].

Most of the above studies in cloudlet management are focused on improving the efficiency of resource management over cloud servers. The outcome of their research is mainly analyzed on the 10 parameters shown in Fig. 1.

The above studies are also compared based on the different parameters as shown in Table 1.

3 Working of ARM-FT

The working architecture of the proposed method ARM-FT is shown in Fig. 2.

Table 1 Comparative analysis of recent developments in cloudlets management based on different performance metrics

Heading level	Real-time developments	Simulation	Standalone system	Case studies (on paper only)
Cost-time	*	*		*
Auction	*			*
Profit	*		*	
Service level agreement	*	*	*	
QoS	*	*		*
Energy/power		*		
Fault tolerance	*		*	*
Priority	*	*	*	
Nature/bio-inspired algorithms		*	*	*
Autonomic computing	*	*		

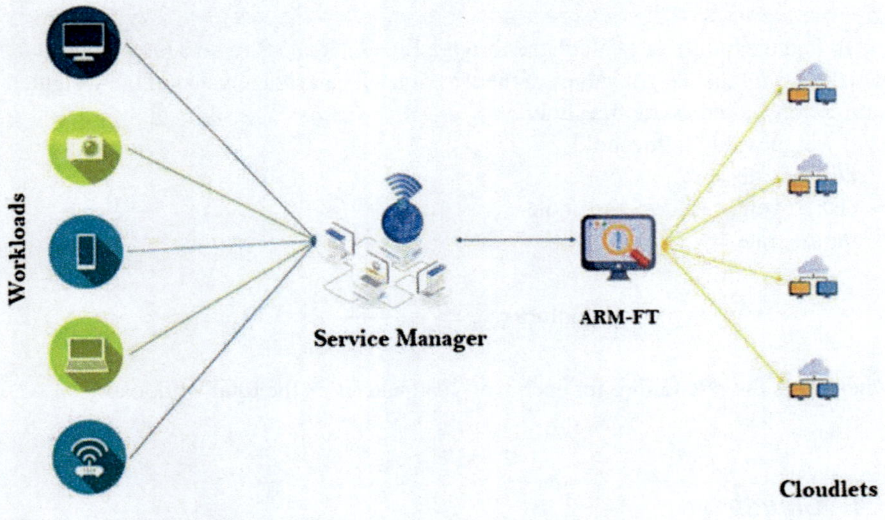

Fig. 2 ARM-FT architecture

3.1 Workloads

The service request sent by different clients to the service providers, basically known as service managers. The size of workloads at a time cannot be predicted, it is totally dependent upon the service requirements and computation size of the service. The workloads can be calculated through Eq. 1.

$$x = \frac{w_h}{w_t} * 100 \tag{1}$$

where x is the size of the workload, W_h is the workload per hour, and W_t is the total workload.

3.2 Service Manager

The platform through which user can submit the request for the service. The service manager is responsible for diverting the request to the right cloudlets, which should not be under or over-utilized.

3.3 ARM-FT

The main objective of the proposed ARM-FT is to find whether the available cloudlets are under or over-utilized. This can be obtained by calculating the failure rate as follows:

To find the failure rate calculated through Eq. (2), it needs to set a threshold value, which uses all the weight values of the workload. The average value of the weighted sum is considered as the threshold.

if weight value < threshold
failure_rate = 2;
else if weight value > threshold
failure_rate = 1;

$$\text{Failure rate} = \frac{\text{failure}(R_i)}{n} \tag{2}$$

where R_i is the rate failure for each workload and 'n' is the total workload.

3.4 Cloudlets

The set of computable machines which are responsible to compute the service requests. While executing the service request, same time the execution time, resource cost, and utilization has been calculated.

Execution time

$$\text{Execution time} = \sum_{i=1}^{n} \left(\frac{\text{RC}_i - \text{RS}_i}{n} \right) \tag{3}$$

where RC_i is the request that is workload completion time and RS_i is the workload submission time.

4 Experimental Results and Discussion

The proposed framework ARM-FT is first simulated in Cloudsim 3.0. In this experimental research, 20 cloudlets configure with random values of memory, bandwidth, and cpu utilization. The main objective of ARM-FT is to find the fault in cloudlets and separate it from the service pool. Table 2 presents the experimental outcomes of cloudlets utilization based on memory, bandwidth, and cpu uses.

The best CLD identified as presented in Table 2 is to allocate the workloads submitted by the cloud user.

Table 2 ARM-FT best cloudlets

CLDs	Resource cost	Energy consumption	Cld utilization	Best Cld values
CLD16	308.3492	0.5593	0.1185	309.0270
CLD6	301.1154	0.5283	0.0565	301.7002
CLD11	285.7385	0.5552	0.1104	286.4041
CLD15	271.8542	0.5380	0.0759	272.4680
CLD20	256.1547	0.5702	0.1404	256.8653
CLD2	241.3947	0.5364	0.0729	242.0040
CLD4	225.9429	0.5644	0.1288	226.6361
CLD10	209.8294	0.5544	0.1087	210.4924
CLD1	194.2727	0.5654	0.1308	194.9689
CLD12	182.0642	0.5337	0.0673	182.6652
CLD13	166.0541	0.5291	0.0583	166.6415
CLD5	149.6783	0.5880	0.1759	150.4422
CLD3	133.2470	0.5535	0.1070	133.9076
CLD19	117.4434	0.5396	0.0793	118.0623
CLD7	101.5918	0.5330	0.0661	102.1909
CLD14	**84.8098**	**0.5785**	**0.1570**	**85.5453**
CLD18	**68.4232**	**0.5380**	**0.0760**	**69.0373**
CLD8	**51.6316**	**0.5517**	**0.1034**	**52.2868**
CLD9	**34.7114**	**0.5139**	**0.0278**	**35.2532**
CLD17	**17.6389**	**0.5737**	**0.1474**	**18.3600**

5 Conclusion

The ARM-FT is implemented and simulated in the Cloudsim simulator. The objective of ARM-FT is to find the best Cloudlets among available cloudlets to compute the service request submitted by the users. The initial set-up was established by 20 cloudlets in a simulation environment, and for these cloudlets, memory, bandwidth, and cpu utilization have been evaluated. Based on this utilization, the cloudlets utilization has been calculated. The threshold value has been calculated based on the utilization value. In this research, the threshold value $\delta = 100$ is obtained, and based on this, below δ cloudlets are identified as fault cloudlets that are under-utilized, and cannot be computed for the service request as desired. In the future, the cloudlets will be provisioned to the service request, and based on the machine learning approach, the over-utilized cloudlet will be identified.

References

1. Dewangan BK, Agarwal A, Venkatadri M, Pasricha A (2019) Design of self-management aware autonomic resource scheduling scheme in cloud. Int J Comput Inf Syst Ind Manag Appl 11:170–177
2. Tomar R, Khanna A, Bansal A, Fore V (2018) An architectural view towards autonomic cloud computing. In: Satapathy S, Bhateja V, Raju K, Janakiramaiah B (eds) Data engineering and intelligent computing. Advances in intelligent systems and computing, vol 542. Springer, Singapore. https://doi.org/10.1007/978-981-10-3223-3_55
3. Choudhury T, Dewangan BK, Tomar R, Singh BK, Toe TT, Nhu NG (2021) Autonomic computing in cloud resource management in Industry 4.0. Springer
4. Singhal A, Sarishma, Tomar R (2016) Intelligent accident management system using IoT and cloud computing. In: 2016 2nd international conference on next generation computing technologies (NGCT), Dehradun, India, pp 89–92. https://doi.org/10.1109/NGCT.2016.7877395
5. Dewangan BK, Agarwal A, Choudhury T, Pasricha A, Chandra Satapathy S (2020) Extensive review of cloud resource management techniques in Industry 4.0: issue and challenges. Softw: Pract Exp 1–20
6. Dewangan BK, Jain A, Choudhury T (2020) AP: hybrid task scheduling algorithm for cloud. Rev d'Intelligence Artif 34:479–485
7. Singh BK, Danish M, Choudhury T, Sharma DP (2021) Autonomic resource management in a cloud-based infrastructure environment. In: Choudhury T, Dewangan BK, Tomar R, Singh BK, Toe TT, Nhu NG (eds) Autonomic computing in cloud resource management in Industry 4.0. EAI/Springer innovations in communication and computing. Springer, Cham. https://doi.org/10.1007/978-3-030-71756-8_18
8. Saxena D, Singh AK, Buyya R (2021) OP-MLB: an online CLD prediction based multi-objective load balancing framework for resource management at cloud datacenter. IEEE Trans Cloud Comput
9. Rajabzadeh M, Toroghi Haghighat A, Rahmani AM (2020) New comprehensive model based on virtual clusters and absorbing Markov chains for energy-efficient virtual machine management in cloud computing. J Supercomput 76(9):7438–7457
10. Masdari M, Khezri H (2020) Efficient CLD migrations using forecasting techniques in cloud computing: a comprehensive review. Clust Comput 23(4):2629–2658

11. Aldossary M (2021) A review of dynamic resource management in cloud computing environments. Comput Syst Sci Eng 36(3):461–476
12. Gholipour N, Arianyan E, Buyya R (2020) A novel energy-aware resource management technique using joint CLD and container consolidation approach for green computing in cloud data centers. Simul Model Pract Theory 104:102127
13. Sivagami VM, Easwarakumar KS (2019) An improved dynamic fault tolerant management algorithm during CLD migration in cloud data center. Futur Gener Comput Syst 98:35–43
14. Ali SA, Affan M, Alam M (2019, January) A study of efficient energy management techniques for cloud computing environment. In: 2019 9th international conference on cloud computing, data science & engineering (confluence). IEEE, pp 13–18
15. Khan T, Tian W, Zhou G, Ilager S, Gong M, Buyya R (2022) Machine learning (ML)–centric resource management in cloud computing: a review and future directions. J Netw Comput Appl 103405
16. Arshad U, Aleem M, Srivastava G, Lin JCW (2022) Utilizing power consumption and SLA violations using dynamic CLD consolidation in cloud data centers. Renew Sustain Energy Rev 167:112782
17. Agbaje MO, Ohwo OB, Ayanwola TG, Olufunmilola O (2022) A survey of game-theoretic approach for resource management in cloud computing. J Comput Netw Commun
18. Liang B, Dong X, Wang Y, Zhang X (2022) A high-applicability heterogeneous cloud data centers resource management algorithm based on trusted virtual machine migration. Expert Syst Appl 197:116762

A Bibliometric Analysis of Autonomic Cloud Computing

Bhupesh Kumar Dewangan, T. Choudhury, Hussain Falih Mahdi, and Ved Prakash Mishra

1 Introduction

Autonomic Computing [1] is a way to deal with advanced issues in programming frameworks. It is a self-management processing model named after, and designed on, the autonomic computing system. An autonomic processing framework [2] would control the working of computer applications and frameworks without input from the client, and service managers, that the autonomic computing system controls the frameworks without cognizant contribution from the human. The objective of autonomic processing is to make frameworks that run themselves prepared to do a significant level of work, while at the same time keeping the framework's complexity imperceptible to the client. It alludes to the self-managing qualities of conveyed resources,

B. K. Dewangan
Department of Computer Science and Engineering, OP Jindal University, Raigarh, India

T. Choudhury (✉)
CSE Department, Symbiosis Institute of Technology, Symbiosis International University, Pune, Maharashtra 412115, India
e-mail: tanupriya.choudhury@sitpune.edu.in; tanupriya@ddn.upes.ac.in; tanupriya1986@gmail.com

Informatics Cluster, School of Computer Science, University of Petroleum and Energy Studies (UPES), Dehradun, Uttarakhand 248007, India

CSE Department, Daffodil International University, Daffodil Smart City, Birulia 1216, Bangladesh

Director Research (Honorary), The AI University, Cutbank Montana 59427, US

H. F. Mahdi
Department of Computer and Software Engineering, University of Diyala, Baquba, Iraq
e-mail: hussain.mahdi@ieee.org

V. P. Mishra
Amity University Dubai, Dubai International Academic City, Dubai, UAE
e-mail: vmishra@amityuniversity.ae

A. Ramdane-Cherif et al. (eds.), *Machine Intelligence and Data Science Applications*, Algorithms for Intelligent Systems, https://doi.org/10.1007/978-981-99-1620-7_27

adjusting to flighty changes while concealing natural complexity to administrators and clients. It Started by IBM in 2001, this drive eventually intended to computer frameworks equipped for self-administration, to conquer the quickly developing complexity of processing framework the executives, and to diminish the obstruction that complexity stances to additional development.

The paper is organized by introducing autonomic computing followed by the literature review and then the methodology has been discussed. The bibliometric analysis has been presented and then concluded the work. recent developments in relevant areas, and analyzing the recent works with different research.

One of the fastest-growing technologies having the potential to revolutionize the way businesses and individuals use cloud services is Autonomic cloud computing [3, 4]. A bibliometric analysis of autonomic cloud computing was done where it examines how this new technology is applicable and hence discussed in scholarly literature. The researchers can attain insights into trends, patterns, and areas of research that are most popular among the scholars in this field by analyzing the publications related to autonomic cloud computing. The very initial step in carrying out a bibliometric analysis consists of identifying the important sources of the collection of data like journals or conference proceedings with reference to autonomic cloud computing. The researchers can gather the information from them once these sources have been identified regarding the topics covered by each and every publication as well as the authorship details like affiliation or the number of citations received per article/paper published on the topic. This type of data is good for an assessment of the topics which are receiving more attention, and also for the topics where the person is leading the research efforts within academics when it comes to autonomous cloud systems design implementation strategies and so on. The following steps involve the usage of citation indices (for example Google Scholar) and other metrics tools (for example Scopus) to analyze citation networks, and thus recognize the influential papers/authors depending on their impact factor scores; calculate h-indexes; track collaboration patterns across countries institutions; determine co-occurrence frequencies between keywords; and so on. Whatever is happening within the academic circle based on Autonomic Cloud Computing technologies, these analyses will help to build up a clear understanding and provide useful insights into recent trends, future directions, and so on. Lastly, when all the important and relevant data has been collected, analyzed, and interpreted, then we have the conclusion and recommendation part. These might have suggestions on the methods to do better for the existing practices and utilize resources effectively to develop new applications. Thus, the mission is to create a comprehensive picture of the state of Autonomic Cloud Computing today, tomorrow, and beyond. In this way, industry professionals might make structured decisions when it comes to investing in implementing solutions to their organizations [5, 6].

2 Review of Literature

Parashar, M., and Hariri, S., present a prologue to autonomic registering, its difficulties, and valuable open doors. The authors first give an outline of the design of the sensory system and use it to spur the autonomic registering worldview. At the same point, frame the vital difficulties of autonomic figuring and present an outline of existing autonomic processing frameworks and applications [7]. Figure 1 shows the autonomic computing reference architecture by IBM in 2006.

In 2005, Sterritt, R. writes about the most recent autonomic frameworks examination and advances to impact the business; it looks behind AC, summing up what it is, the present status of-the-craftsmanship research, related work and drives, features exploration and innovation move issues and finishes up with further and suggested accordingly [9].

The author presented a system and simulated it through MATLAB considering wavered resource requests and portability of end client gadgets. The proposed autonomic profound Q-learning-based technique fundamentally works on the presentation of the calculation offloading by limiting the inertness of administration processing. The absolute power utilization because of various offloading choices is likewise read up for similar review reasons which show the proposed approach as energy proficient regarding the cutting-edge calculation offloading arrangements [10].

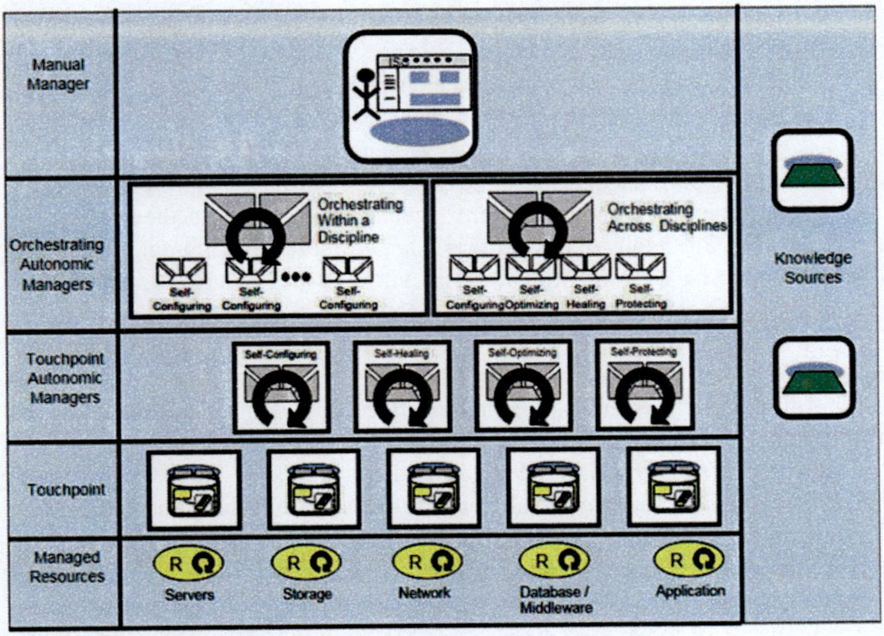

Fig. 1 Autonomic computing reference architecture [8]

The author introduces a system worked over the CloudSim toolbox with an autonomic asset provisioning structure enlivened by the system [11].

In 2018, Ghobaei-Arani, M., presented a half breed asset provisioning approach for cloud services that depends on a blend of the idea of autonomic figuring and support learning (SL). Likewise, we present a structure for autonomic resource provisioning which is roused by the cloud layer model. At long last, we assess the adequacy of our methodology under two certifiable responsibility follows. The implementations outcomes show that the proposed approach diminishes the absolute expense by up to half, and expands the asset usage by up to 12% contrasted and the other systems [12].

Author Elgendy, I. A., introduces a resource provisioning and calculation offloading model with information security to address the limits of such gadgets. Moreover, a coordinated model, which mutually considers security, calculation offloading, and asset distribution, is planned to limit investment utilization of the whole framework. At last, an offloading calculation is created with point-by-point cycles to decide the ideal calculation offloading choice for MUs. Reproduction results demonstrate the way that our model and calculation can altogether work on the exhibition of the whole framework contrasted and neighborhood execution and full offloading plans [13].

The author applied autonomic comping in BCIs on independence. The author then thinks about the effect of BCI innovation on every one of these capacities. Albeit on first look, BCIs exclusively upgrade self-assurance since they reestablish or further develop capacities, they will show that there are other positive, yet additionally adverse consequences on client independence, which require further philosophical and moral conversations [14].

The author developed oneself advanced energy-productive asset the executive's procedure has been thinking about, which gives the ideal answer for augmenting the asset usage and recognizing the flawed assets to abstain from deluding of planning. This technique apportions the assets to the jobs presented by the cloud client by the satisfaction of the QoS with less SLA infringement rate and boosts the asset use for minimum price. The oddity of this paper is to apply Antlion Streamlining calculation to track down the ideal asset. The exploratory outcomes are proof of the most extreme execution of the proposed work [15].

The author simulated a technique that is reproduced in the cloudsim toolbox, assesses the exhibition measurements with an alternate arrangement of responsibilities, and the perception of this exploration and its trial results and relative examination with existing systems are proof of most extreme execution [16].

The author proposed a SMART algorithm, which is simulated in cloudsim3.0 and also executed in Amazon's real cloud platform. In this, the author proposed a SLA-oriented cloudlet management system, which they have achieved and are able to make the system efficient in terms of cost and execution time [17].

Author presented energy-oriented asset planning for the cloud is proposing. Absolute asset usage of every asset has been working out and energy is streamlining through antlion improvement calculation to stay away from intense usage of force. The asset is related to its best usage esteem and dole out to submit jobs according to

Table 1 Comparative analysis of cloud models based on autonomic computing through different parameters

Research	Research parameters				
	Autonomic computing	Execution time	Cost	Energy	SLA
[9]	✓	✗	✓	✗	✗
[10]	✗	✓	✗	✓	✗
[11]	✗	✓	✓	✗	✓
[12]	✓	✗	✗	✗	✗
[13]	✗	✓	✗	✓	✗
[14]	✗	✗	✓	✗	✓
[15]	✗	✗	✗	✗	✓
[2]	✓	✓	✓	✗	✗
[17]	✓	✗	✗	✓	✗
[18]	✓	✓	✓	✓	✗
[19]	✓	✓	✓	✓	✗
[20]	✓	✓	✗	✗	✗
[21]	✓	✓	✓	✓	✗

need premise. The trial aftereffects of the proposed work are investigated with the existing autonomic structures, and it is noticed that the proposed work is performing the most extreme [18].

Before opting cloud, trust plays a vital role to opt it. In this research, the authors actually focused on the trust parameter. To continue this research, they have presented how the service providers can increase trust between the user and service providers [19].

Cloud storage plays a vital role in the cloud business, to mind this, the author presented some security models in which they have secured the cloud storage through comparative analysis done between MD5 and RSA algorithms [20].

The energy efficient model was proposed by Priti Sirohi in 2021, in this, the author researched about service selection model for the smart city and they enhanced the energy efficiency through the proposed system [21].

The above study is also compared based on different parameters, and presented in Table 1.

3 Methodology

Figure 2 is presenting the scheme of bibliometric analysis for autonomic computing. The steps of the analysis are discussed separately.

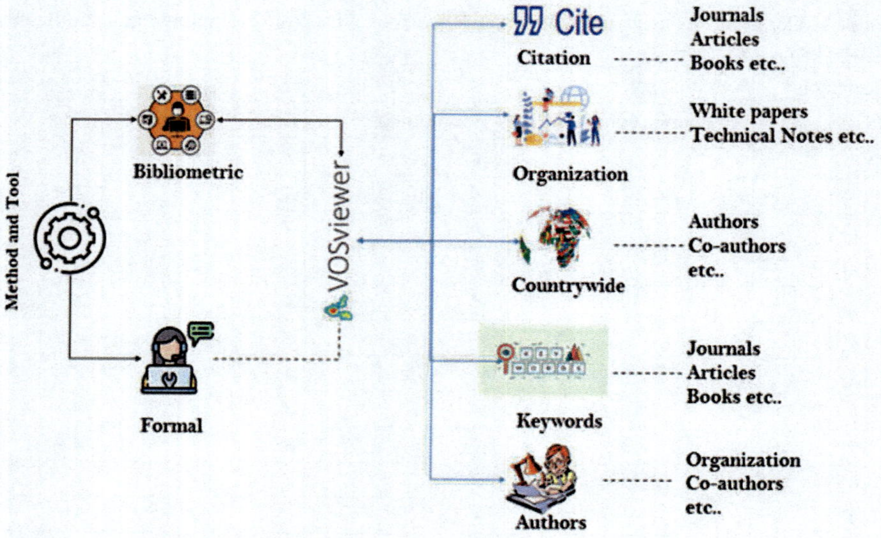

Fig. 2 Bibliometric analysis through VoS viewer

The dataset for this bibliometric analysis has been downloaded from https://www.scopus.com/ in which articles from 2019 to 2022 are considered. Then the dataset has been processed through VoS viewer https://www.vosviewer.com/.

4 Results and Analysis

The dataset is processed through VoS viewer, and the results are obtained as follows.

4.1 Analysis Based on Citations

The complete dataset with autonomic computing keywords was analyzed and it is observed that most of the publications are cited from alam m.g.r. (2019). Based on the results, we observed that most of the publications were done in 2019. Figure 3a, b shows the bibliometric outcomes.

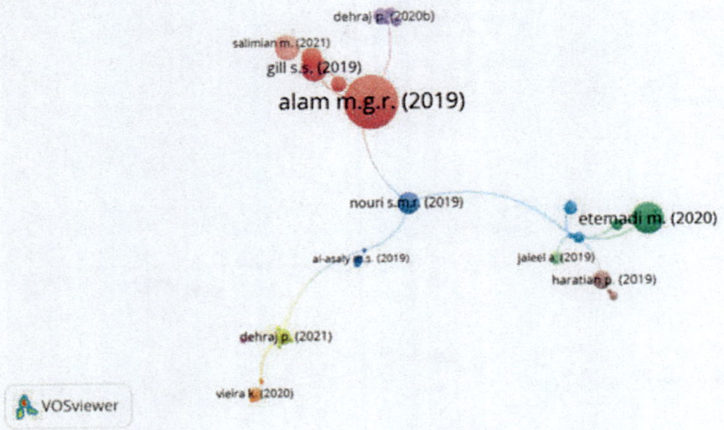

(a)

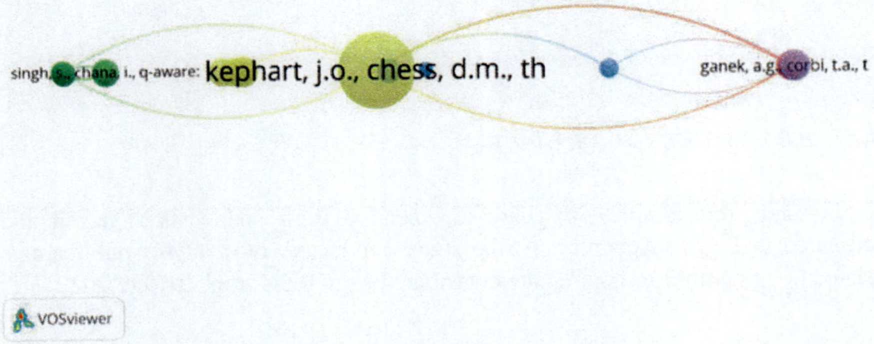

(b)

Fig. 3 Analysis of dataset: **a** based on citation, **b** based on cited articles

4.2 Analysis Based on Co-authorship of Authors

The co-authorship analysis has been done and it is presented in Fig. 4. As we obtained the results through VoS viewer, author agullar j. is having the Maximum co-authorship with autonomic computing publications.

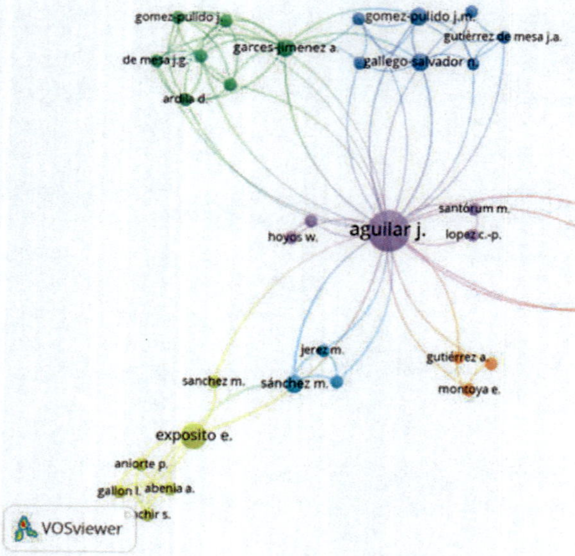

Fig. 4 Bibliometric analysis based on co-authorship

4.3 Analysis Based on Country

In this step, the complete database downloaded from Scopus is analyzed and presented in Fig. 5. According to the study conducted, most of the publications related to autonomic computing are contributed from India from 2019 to 2022.

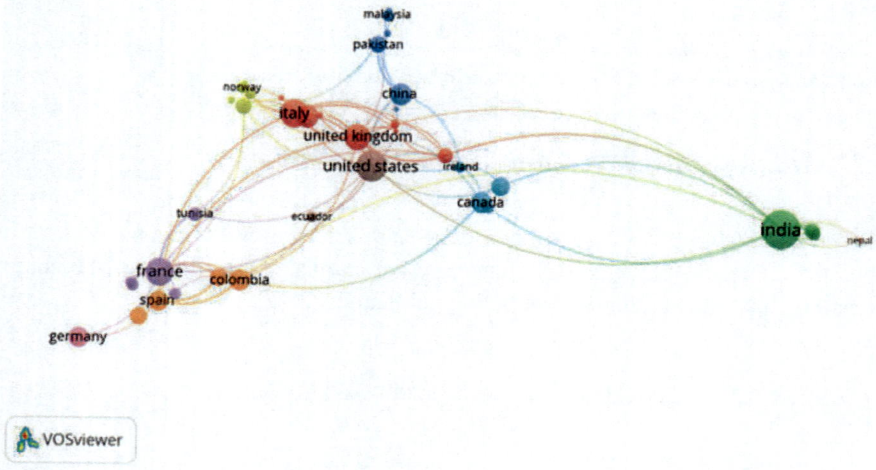

Fig. 5 Bibliometric analysis based on country

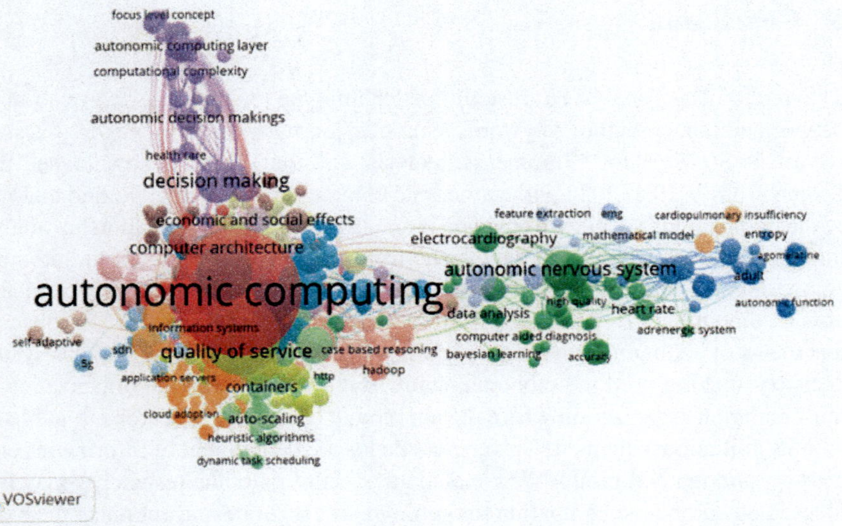

Fig. 6 Bibliometric analysis based on keywords

4.4 Analysis Based on Keywords

As per the analysis, we have observed that the most common keywords are quality of services, autonomic nervous system, decision-making used in autonomic computing research, and it is presented in Fig. 6.

4.5 Analysis Based on Organization

Based on the analysis, we found the following organization is the most common organization where autonomic computing research is published with its affiliations—cemisid, Universidad de los an, department of informatics engineering, and tepuy d group, and it is presented in Fig. 7.

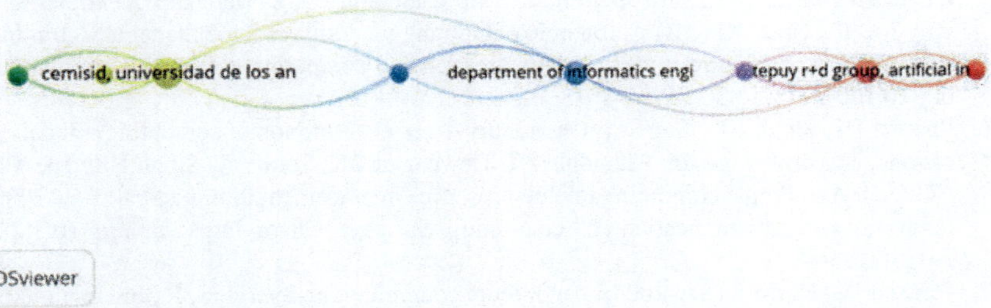

Fig. 7 Bibliometric analysis based on organizations

5 Conclusion

The bibliometric analysis is useful for finding the publications co-relationship between author, co-author, keywords, counties, journals, citations, etc. In this study, we use VoS viewer for bibliometric analysis, and the dataset is downloaded from Scopus from 2019 to 2022 with autonomic computing keywords. We find that most of the publications are cited from alam m.g.r. (2019). Where, co-authorship analysis finds that author agullar j. is having the maximum co-authorship with autonomic computing publications, in addition to this, most of the publications related to autonomic computing are contributed from India from 2019 to 2022. We also observed that the most common keywords are quality of services, autonomic nervous system, decision- making used in autonomic computing research, and the organization are most common organization where the autonomic computing research are published with its affiliations- cemisid, Universidad de los an, department of informatics engineering, and tepuy d group. We conclude that, as of date, the researcher may have focused on autonomic computing for getting more accurate and automate their software operations as the results says that very few researchers have been contributed in this research area.

References

1. Choudhury T, Dewangan BK, Tomar R, Singh BK, Toe TT, Nhu NG (2021) Autonomic computing in cloud resource management in Industry 4.0. Springer
2. Tomar R, Khanna A, Bansal A, Fore V (2018) An architectural view towards autonomic cloud computing. In: Satapathy S, Bhateja V, Raju K, Janakiramaiah B (eds) Data engineering and intelligent computing. advances in intelligent systems and computing, vol 542. Springer, Singapore. https://doi.org/10.1007/978-981-10-3223-3_55
3. Singhal A, Sarishma, Tomar R (2016) Intelligent accident management system using IoT and cloud computing. In: 2016 2nd international conference on next generation computing technologies (NGCT), Dehradun, India, pp 89–92. https://doi.org/10.1109/NGCT.2016.7877395
4. Dewangan BK, Agarwal A, Choudhury T, Pasricha A, Chandra Satapathy S (2020) Extensive review of cloud resource management techniques in industry 4.0: issue and challenges. Softw Pract Exp (October 2019), 1–20
5. Singh BK, Danish M, Choudhury T, Sharma DP (2021) Autonomic resource management in a cloud-based infrastructure environment. In: Choudhury T, Dewangan BK, Tomar R, Singh BK, Toe TT, Nhu NG (eds) Autonomic computing in cloud resource management in Industry 4.0. EAI/Springer innovations in communication and computing. Springer, Cham. https://doi.org/10.1007/978-3-030-71756-8_18
6. Sharma DP, Singh BK, Gure AT, Choudhury T (2021) Autonomic computing: models, applications, and brokerage. In: Choudhury T, Dewangan BK, Tomar R, Singh BK, Toe TT, Nhu NG (eds) Autonomic computing in cloud resource management in Industry 4.0. EAI/Springer innovations in communication and computing. Springer, Cham. https://doi.org/10.1007/978-3-030-71756-8_4
7. Parashar M, Hariri S (Sep 2004). Autonomic computing: an overview. In: International workshop on unconventional programming paradigms. Springer, Berlin, Heidelberg, pp. 257–269

8. Computing A (2006) An architectural blueprint for autonomic computing. IBM White Paper 31(2006):1–6
9. Sterritt R (2005) Autonomic computing. Innov Syst Softw Eng 1(1):79–88
10. Alam MGR, Hassan MM, Uddin MZ, Almogren A, Fortino G (2019) Autonomic computation offloading in mobile edge for IoT applications. Futur Gener Comput Syst 90:149–157
11. Bhardwaj T, Sharma SC (2019) An autonomic resource provisioning framework for efficient data collection in cloudlet-enabled wireless body area networks: a fuzzy-based proactive approach. Soft Comput 23(20):10361–10383
12. Ghobaei-Arani M, Jabbehdari S, Pourmina MA (2018) An autonomic resource provisioning approach for service-based cloud applications: a hybrid approach. Futur Gener Comput Syst 78:191–210
13. Elgendy IA, Zhang W, Tian YC, Li K (2019) Resource allocation and computation offloading with data security for mobile edge computing. Futur Gener Comput Syst 100:531–541
14. Friedrich O, Racine E, Steinert S, Pömsl J, Jox RJ (2021) An analysis of the impact of brain-computer interfaces on autonomy. Neuroethics 14(1):17–29
15. Agarwal A, Venkatadri M (2019) Self-characteristics based energy-efficient resource scheduling for cloud. Procedia Comput Sci 152:204–211
16. Dewangan BK, Agarwal A, Venkatadri M (Dec 2018) Autonomic cloud re-source management. In 2018 fifth international conference on parallel, distributed and grid computing (PDGC). IEEE, pp 138–143
17. Pasricha A (2019) Sla-based autonomic cloud resource management framework by antlion optimization algorithm. Int J Innov Technol Explor Eng (IJITEE) 8:119–123
18. Dewangan BK (2019) A self-optimization based virtual machine scheduling to workloads in cloud computing environment. Int J Eng Adv Technol (IJEAT) 8(4):91–96
19. Shende P (2013) The sliding window method: an environment to evaluate user behavior trust in cloud technology. Int J Adv Res Comput Commun Eng 2(2):1158–1162
20. Baghel SK (2012) Defense in depth for data storage in cloud computing. Int J Technol 2:58–61
21. Sirohi P, Al-Wesabi FN, Alshahrani HM, Maheshwari P (2021). Energy-efficient cloud service selection and recommendation based on QoS for sustainable smart cities. Appl Sci 11(20):9394

Analyzing the Online Customer Post-purchasing Behavior: Sentimental Analysis Approach

T. Praveen Kumar, A. Mansurali, and Tanupriya Choudhury

1 Introduction

The consumer decision-making process is the method taken by consumers to identify their needs, to search for information, to evaluate alternatives, and then to make their purchasing decisions. Economic and psychological factors affect consumer behavior and are also affected by external factors, such as societal values, and norms. Customer decision-making activity is a comprehensive and complex process and involves all facets, from problem identification to attitudes and behavior after purchase. Each customer's needs are different, and these are the needs that impact the

T. Praveen Kumar (✉)
School of Business and Management, Christ (Deemed to be University), Bengaluru, Karnataka 560076, India
e-mail: praveenkumarreach@gmail.com

A. Mansurali
Department of Management, Central University of Tamil Nadu, Thiruvarur, Tamil Nadu 610001, India
e-mail: mansurali.a@christuniversity.in

T. Choudhury
CSE Department, Symbiosis Institute of Technology, Symbiosis International University, Pune, Maharashtra 412115, India

Informatics Cluster, School of Computer Science, University of Petroleum and Energy Studies (UPES), Dehradun, Uttarakhand 248007, India

CSE Department, Daffodil International University, Daffodil Smart City, Birulia 1216, Bangladesh

Director Research (Honorary), The AI University, Cutbank Montana 59427, US

T. Choudhury
e-mail: tanupriya.choudhury@sitpune.edu.in; tanupriya@ddn.upes.ac.in; tanupriya1986@gmail.com

A. Ramdane-Cherif et al. (eds.), *Machine Intelligence and Data Science Applications*, Algorithms for Intelligent Systems, https://doi.org/10.1007/978-981-99-1620-7_28

decision-making process. Decisions may be ordered, unstructured, compared, evaluated, selected, and purchased on the basis of the consumer's view of a particular product from the point of view of a variety of products.

2 Review of Literature

Puschmann and Powell [1] in this article argue that the assumptions coded in sentiment analysis as a method, constraints that are sometimes de-emphasized in public discourse, are accompanied by a range of constraints, both with regard to their technical limitations. Tomar et al. [2] propose a parameterized approach for comparative study in an intercommunication-based application. Kumar et al. [3] in their paper have compared machine learning algorithms for Sentiment analysis on Twitter data. Sánchez-Núñez et al. [4] in his report performed a science mapping analysis of OMSAMC research in order to provide an overview of scientific work over the past two decades in this interdisciplinary area and to show trends that may be the basis for future innovations in this field. Mehra and Choudhury [5] proposed sentiment analysis of text using SVM. Kuznetsov et al. [6] Businesses are using sentiment analysis to analyze the competitive environment and monitor the customers' opinion. In this paper, authors had presented Media Placement news analysis tool targeted at measurement of brand recognition and media coverage in online media and [7] also aims to implement sentiment analysis for marketing research. Reference [8] also covers the how sentiment analysis can be used with News media. Poecze et al. [9] proposed that YouTube gamers were included among the participants. They had focused on the consistency of their interaction on Facebook to identify significant differences in terms of their user-generated Facebook metrics and commentary sentiments. Similarly, [10] also proposes to use sentiment analysis on social media. So Methodologically, ANOVA and sentiment analysis were applied. Wang et al. [11] described that in the midst of post-purchase variables and consumer loyalty, this conceptual paper explores relationships. Based on post-purchase variables in e-commerce, ideas have also been established on how consumer loyalty and perceived risk can differ [12]. Asthana et al. [13] in their paper have proposed an architecture to do sentiment analysis using speech recognition using K-means clustering, Naïve Bayes, and decision tree in medical field and similarly [14] has also indicated the usefulness of sentiment analysis in a medical emergency (Covid-19). Bhuvaneswaran et al. [15] proposed that the usefulness of this research is to analyze the satisfaction of customers in the city of Chennai with Flipkart.com users. They also tried to find out about the online shopping characteristics of Flipkart users in Chennai City. Based on the above literature the secondary data was collected to analyze the post-purchase behavior of the consumers using the sentiment analysis. Two major e-commerce giants Flipkart and Amazon are the source of the data. Customer feedback for three different categories (Electronics, Apparel, and Home & Kitchen Appliances) has been considered to determine the sentiment score.

3 Statement of the Problem

Digitization and continuous improvement of innovative technology completely transform the way consumers purchase items across all segments in India. After Walmart bought majority stocks of Flipkart, giving it a foothold in the industry, which is expected to grow to USD 200 billion by 2026. As per 2020, Filpkart is having 31.9% of the market share in India, while Amazon is almost catching them by having 31.2% of market share. It is important to understand the customers' sentiment to uplift their services and minimize the errors. The main goal of the study is to understand the customer sentiments for the same type of product categories among the two e-commerce giants and to understand the sentiments for different categories of products in the same e-commerce platform.

4 Objective of the Study

To understand the post-purchase customer sentiments toward different product categories in the same e-commerce portal.

5 Methodology

See Table 1.

Table 1 Methodology

S. no.	Research design	Descriptions
1	Source of data	Flipkart and Amazon portal
2	Statistical tool	Sentiment analysis
3	Sample size	3 product categories from each portal—electronics, apparel, and home and kitchen appliances For each category—50 reviews
4	Total sample	300 reviews

6 Extraction and Selection of Data

The study's data was derived from online customer reviews on the Flipkart and Amazon e-commerce platforms. The sentimental analysis was conducted using approximately 300 written inputs for the study. The reviews of Indian consumers are taken into account (geographically). Each portal includes three product categories: electronics, home and kitchen appliances, and apparel. Each product category has 50 reviews. This study's analysis is divided into three steps:

Step1: Text mining

Step2: Sentiment score measurement

Step3: Student sentiment classification.

R Studio was used to conduct the research for this report.

7 Analysis

7.1 Word Frequency

Table 1 includes word frequency extracted from the Amazon and flip kart online customer reviews/feedback (input data). The word frequency is used to identify the common and frequently occurring words in the online customer reviews (input data). And the next step is to create the document matrix after pre-processing the data. The resulting Table 2 displays the word frequency (customer feedback) of the input text document, and for the top 10 most common terms to visualize word frequency, the bar chart was graphed and illustrated as Fig. 1.

Table 2 Word frequency—Amazon versus Flipkart

S. no.	Word	Frequency (Amazon)	Word	Frequency (Flipkart)
1	Product	141	Good	151
2	Good	78	Product	126
3	Use	75	Use	72
4	Amazon	71	Qualiti	64
5	Qualiti	70	Look	54
6	Will	54	Flipkart	53
7	Buy	52	Will	50
8	Time	51	Can	49
9	Day	48	Install	47
10	Look	48	Price	46

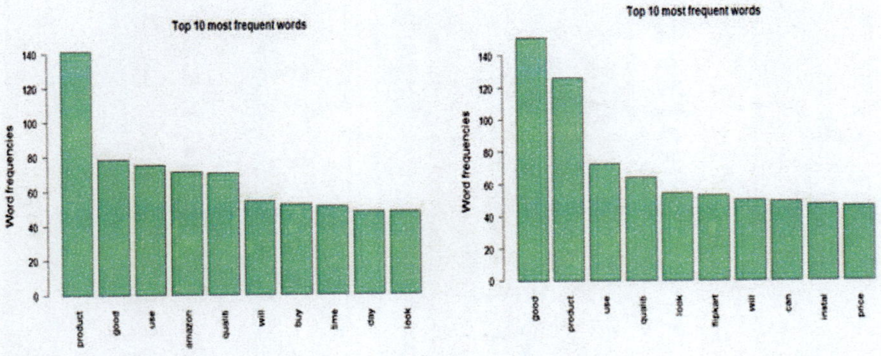

Fig. 1 Word frequency Amazon versus Flipkart

For Amazon, the maximum word that sometimes happens is "product" and for Flipkart it is "good". There are also no negative terms in the bar map, such as "not" demonstrates that most feedback does not remark on negative phrases such as "not good") for both the portals.

7.2 *Word Cloud Generation*

Word Cloud is an image that consists of keywords contained inside a matrix of text documents. The magnitude of each word point in a picture depends on the frequency in the text frame of each word. By using the word frequency data frame (Table 1), the word cloud was generated and is shown below.

Word cloud provides some potential in more research by displaying additional phrases, words such as "even," "time," and "can," which can provide more insight into the most commonly encountered words and support increase in a deeper understanding of the emotions of the user.

7.3 *Word Association*

This method was used to examine the relationship in the input of the user between the most frequent words and is used to view the point of view of the customer on these phrases.

In Fig. 2, the words most commonly associated with the three chosen terms (Lower Corlimit = 0.25 is set) are shown. For Amazon, the production specifies that "good" has the highest retention association of 35% and the lowest association of 28% with must. So, the most frequently occurring word ("good") is optimistic. Similarly, for

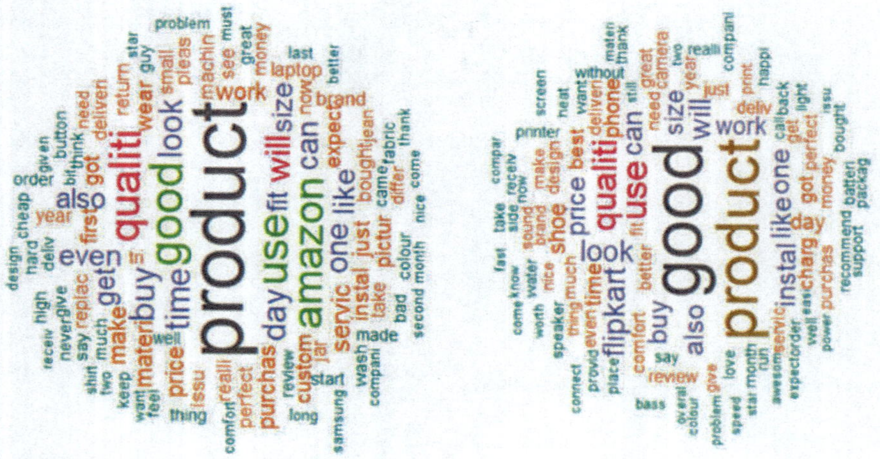

Fig. 2 Word clouds—Amazon versus Flipkart

Flipkart the performance states that "good" has the highest 38% frontal association and the lowest 26% wise association (see Fig. 3).

7.4 Sentiment Scores

Using sentiment ratings, the sentiment of clients toward Amazon was measured. There are optimistic, negative, and neutral feelings classified here. It is also represented by a numerical scale. In assessing the positive or negative strength unit of the feeling enclosed in a customer input, the numeric scale provides better information. We generated sentiment scores using the Syuzhet kit, which includes four sentiment dictionaries and a sentiment recovery technique. The "get" sentiment function takes two arguments: a vector of characters (of phrases or words) and a method, which specifies which of the four available sentiment extraction techniques Syuzhet, Bing, Afinn, and NRC are the four techniques that will be used. A different scale is used for each strategy and the findings are therefore slightly different.

From Fig. 4, the Syuzhet vector summary results show that for Amazon, the first (maximum) element has a value of 11.7 and for Flipkart the value is 13.45. It implies that in the first answer (line) in the text file, the sum of the sentiment scores of all relevant words adds up to 11.7 and 13.45, respectively. The evaluation scale ranges from −1 (most negative) to +1 (most positive) for sentiment scores (indicating most positive). The summary results of Syuzhet vector table indicates the mean value is 3.90 and 4.483 for Amazon and Flipkart, which are greater than zero and its positive. This indicates the mindset of customers toward Amazon and Flipkart is positive (see Fig. 5).

Word	clumpsi	maintain	morn	regret	walk	must
Good	0.35	0.35	0.35	0.35	0.35	0.28

Word	test	plastic	thought	build	duplic	higher
Product	0.33	0.32	0.31	0.31	0.3	0.3

Word	season	dosa	batter	hot	certain	bracket
Use	0.43	0.43	0.39	0.36	0.36	0.35

Word	front	especi	look	though	expens	wise
Good	0.38	0.35	0.33	0.33	0.32	0.26

Word	bluf	wrong	wipe	happi	cancel	clean
Product	0.33	0.33	0.33	0.29	0.26	0.26

Word	write	thin	window	week	adopt	status
Use	0.46	0.34	0.33	0.31	0.27	0.26

Fig. 3 Word Association—Amazon versus Flipkart

Sl No.	1	2	3
1	11.7	0	0

Sl No.	Min.	1st Qu	Median	Mean	3rd Qu	Max.
1	0	0	0	3.9	5.85	11.7

Sl No.	1	2	3
1	13.45	0	0

Sl No.	Min.	1st Qu	Median	Mean	3rd Qu	Max.
1	0	0	0	4.483	6.725	13.45

Fig. 4 Summary results of Syuzhet vector—Amazon versus Flipkart

The Bing vector sentiment rating is measured by the binary scale in that −1 indicates negative sentiment and +1 indicating positive sentiment and afinn vector sentiment is measured through integer scale ranging from −5 to +5. Based on the Bing and Afinn vectors results of Customer feedback (Amazon and Flipkart). The median value of sentiment scores is equal to 0 and is considered to be a positive.

Sl No.	Vector	1	2	3
1	Bing	-21	0	0
2	Affin	-4	0	0

Sl No.	Vector	Min.	1st Qu	Median	Mean	3rd Qu	Max.
1	Bing	-21	-10.5	0	-7	0	0
2	Affin	-4	-2	0	-1.333	0	0

Sl No.	Vector	1	2	3
1	Bing	-10	0	0
2	Affin	13	0	0

Sl No.	Vector	Min.	1st Qu	Median	Mean	3rd Qu	Max.
1	Bing	-10	-5	0	-3.33	0	0
2	Affin	0	0	0	4.333	6.5	13.0

Fig. 5 Summary results of Bing, and Affin vector—Amazon versus Flipkart

7.5 Three Vectors—Comparisons

Meanwhile, the study employed three distinct methods and four distinct scales. Before comparing them, it is a good idea to adjust their performance to a standard scale. The simple scales are been transformed using R's sign function; the signal function converts all positive values to 1 and all negative values to −1, while all zeros remain 0 (see Fig. 6).

For both the portals, the first element of the first row (vector) is 1, which implies that Syuzhet measured a positive sentiment score for the first method, i.e., for the first answer (line) in the document. For Amazon, other techniques, on the other hand, show opposite outcomes, i.e., those with negative feelings, but for Flipkart only for Bing vector opposite outcomes were there.

7.6 Findings

Upon applying machine learning algorithms and calculating customer feedback sentiment score for both the distinct words of the business were recognized. Potential sentimental words or related words to the described online teaching were nice, product,

Fig. 6 Normalize scale and compare three vectors—Amazon versus Flipkart

	1	2	3
1	1	0	0
2	-1	0	0
3	-1	0	0

	1	2	3
1	1	0	0
2	-1	0	0
3	1	0	0

use, efficiency, Amazon, and Flipkart. The aim of detecting these terms is to explore which words have the most effect on consumer opinion, and which ones provide insight into the feedback. Good is a word commonly spoken, occurring in several inputs, in optimistic sentiment. This may mean that both portals are good for reviews or suggestions. According to "Syuzhet" vector mean value for both companies are positive (3.9 and 4.483), i.e., there is no significant difference of sentiment between customers of Amazon and Flipkart. According to "Bing" vector mean value for both companies are negative (−7 and −3.33), i.e., there is no significant difference of sentiment between customers of Amazon and Flipkart. According to "Affin" vector mean value for Amazon is negative (−1.33) but for Flipkart it is positive (4.333), i.e., there is significant difference of sentiment between customers of Amazon and Flipkart.

7.7 Conclusion

The findings demonstrated and acknowledged the different emotions (focused words) of customers toward Amazon and Flipkart. Using a lexicon sentimental analysis framework and the definitions of Affin, Bing, and Syuzhet vectors, the sentiment was graded as positive or negative. That also shows that the difficulty level of analyzing the qualitative data (text data or opinions) can be solved using these NLP tools. It has a huge potential in every domain to understand the feelings, emotions, and needs and then can be analytically addressed in every level. Though there are lots of difficulties in data collection, data cleaning, and codification, but the end results have the potential to give a significant competitive advantage for a company to understand its existing customer. Hence, they can offer better service and resolve issues more efficiently.

7.8 Managerial Implications

Customers have the similar kind of emotions toward both the companies, from choosing the words or showing positive or negative sentiments. Like these e-commerce retail companies, this study can be applied in other businesses also. Besides the findings, further study is required to understand the different classifications of emotions. More reviews and comparison with other e-commerce companies also can be done to understand the difference in sentiments in a larger aspect.

References

1. Puschmann C, Powell A (2018) Turning words into consumer preferences: how sentiment analysis is framed in research and the news media. Social Media+ Society 4(3):2056305118797724
2. Tomar R, Sastry H, Prateek M (2020) Establishing parameters for comparative analysis of V2V communication in VANET. J Sci Ind Res (JSIR) 79(01):26–29
3. Kumar P, Choudhury T, Rawat S, Jayaraman S (2016) Analysis of various machine learning algorithms for enhanced opinion mining using twitter data streams. In: 2016 international conference on micro-electronics and telecommunication engineering (ICMETE), Ghaziabad, India, pp 265–270. https://doi.org/10.1109/ICMETE.2016.19
4. Sánchez-Núñez P, Cobo MJ, De Las Heras-Pedrosa C, Peláez JI, Herrera-Viedma E (2020) Opinion mining, sentiment analysis and emotion understanding in advertising: a bibliometric analysis. IEEE Access 8:134563–134576
5. Mehra S, Choudhury T (2018) Sentiment analysis of user entered text. In: 2018 international conference on computational techniques, electronics and mechanical systems (CTEMS), Belgaum, India, pp 457–461.https://doi.org/10.1109/CTEMS.2018.8769136.
6. Kuznetsov D, Gavrilov I, Benkovich N, Charnetskaya T, Yavorskiy R (2019) Media placement: using sentiment analysis in brand reputation maintaining
7. Rambocas M, Gama J (2013). Marketing research: the role of sentiment analysis (No. 489). Universidade do Porto, Faculdade de Economia do Porto
8. Kechaou Z, Wali A, Ammar MB, Karray H, Alimi AM (2013) A novel system for video news' sentiment analysis. J Syst Inf Technol
9. Poecze F, Ebster C, Strauss C (2018) Social media metrics and sentiment analysis to evaluate the effectiveness of social media posts. Procedia Comput Sci 130:660–666
10. Neri F, Aliprandi C, Capeci F, Cuadros M, By T (2012) Sentiment analysis on social media. ASONAM 12:919–926
11. Wang S, Anitsal M, Anitsal I (2016) A review of e-commerce: the influence of post-purchase factors on relationships between customer loyalty and perceived risk
12. Hernández B, Jiménez J, Martín MJ (2010) Customer behavior in electronic commerce: the moderating effect of e-purchasing experience. J Bus Res 63(9–10):964–971
13. Asthana R, Dhanetwal S, Choudhury T, Tomar R (2020) Study of sentiment analysis in mental health using speech recognition and image to text conversion. In: Das A, Nayak J, Naik B, Dutta S, Pelusi D (eds) Computational intelligence in pattern recognition. Advances in intelligent systems and computing, vol 1120. Springer, Singapore. https://doi.org/10.1007/978-981-15-2449-3_9
14. De Las Heras-Pedrosa C, Sánchez-Núñez P, Peláez JI (2020) Sentiment analysis and emotion understanding during the COVID-19 pandemic in spain and its impact on digital ecosystems. Int J Environ Res Public Health 17(15):5542. https://doi.org/10.3390/ijerph17155542. PMID: 32751866; PMCID: PMC7432069
15. Bhuvaneswaran S, Prabhu A, Subramani AK (2015) A study on customer satifaction towards Flipkart, Chennai. ZENITH Int J Bus Econ Manag Res 5(6):62–68

Synergizing Open Geospatial Data from Cloud Platform and Machine Learning (ML) Algorithms in Open Source Environment for Crop Yield Estimation

Kamal Pandey, Alex E. Frederick, Sharon, Aditya Raj, Mukesh Kumar, Harish Karnatak, and Deepak Lal

1 Introduction

Agriculture is a business fraught with risk and it is the primary source of livelihood for about 58 per cent of India's population [1]. Agriculture is the provider of food security and is important to the economy of any country. As per 2018, the agriculture sector employed more than 50c/o of the Indian workforce and contributed 17–18% to the country's GDP [2]. Crop production primarily depends on geographical [3], biological, political, socio-economic factors [4], precipitation and temperature variables [5, 6]. Accurate information about the nature of historical crop yields is an important modelling input that aids in improving the performance of the model and in turn farmers, agribusinesses and governmental bodies get benefited in managing the risk and establishing the proper policies for thing like crop insurance [7]. Crop size and commodity production information has a direct influence on policies for international trade in grains and other commodities. Public agencies make policies dealing

K. Pandey (✉)
Indian Institute of Remote Sensing, 4 Kalidas Road, Dehradun, Uttarakhand 248001, India
e-mail: kamal@iirs.gov.in

K. Pandey · A. E. Frederick · Sharon · M. Kumar · D. Lal
Sam Higginbottom University of Agriculture Technology and Sciences, Prayagraj, Uttar Pradesh, India
e-mail: deepakl@shiats.edu.in

H. Karnatak
Indian Institute of Remote Sensing, Dehradun, Uttarakhand 248001, India
e-mail: harish@iirs.gov.in

A. Raj
School of Computer and Systems Sciences, Jawaharlal Nehru University, New Delhi 110067, India

A. Ramdane-Cherif et al. (eds.), *Machine Intelligence and Data Science Applications*, Algorithms for Intelligent Systems, https://doi.org/10.1007/978-981-99-1620-7_29

with agricultural subsidies, incentive and crop support programs, crop insurance, procurement, stock management and distribution of grains based on the production estimates of crop size. Traditionally ground survey was conducted to estimate crop yield, this method is subjective, time-consuming and costly. To address these problems empirical and crop growth [8] were developed using weather data, but these methods also have number of issues like complexity involved in the spatial distribution of weather stations. For such methods, data demand and data processing have inbuilt complexity and causes delay in the availability of information. Final estimates from these methods become available only after the crops are actually harvested. So, there is a need to reduce the cost and timeliness and also to increase the accuracy of yield prediction. The thrust of this study is to explore the machine learning-based method for predicting crop yield (wheat crop) and compare the performance of different machine learning algorithms.

2 Related Studies

In the modern era, machine learning is playing a vital role in model development in many thematic application domains like crop modelling, forest ecosystem modelling, disaster modelling, etc. Crop modelling in itself is a very broad area and includes different aspects of agricultural management practices, e.g. crop yield forecast [9], drought management [10], crop acreage estimation and management [11], horticulture acreage estimation and monitoring [12], cropping pattern analysis [13]. Different modelling approaches have been attempted by many researchers for the estimation of crop yield. Multiple linear regression, M5-Prime regression trees, perceptron multilayer neural networks, support vector regression and k-nearest neighbour were compared for their predictive accuracy for crop yield prediction [14]. Satellite imagery has been used to evaluate the output of the Artificial Neural Network (ANN) and the Convolution Neural Network (CNN) by defining different soil and atmospheric parameters to predict crop growth [15]. ANN has proven to be a very effective tool to improve modelling and prediction performance. Reference [16] used crop prediction methodology to analyse the potential of most common Artificial Neural Network (ANN), feed forward backpropagation neural network, to predict and approximate crop yield on the basis of various variables like type of soil, PH, Nitrogen, Phosphate, Potassium, Organic Carbon, Calcium, Magnesium, Sulphur, Manganese, Copper, Iron, depth, temperature, rainfall, humidity. Random Forest is one of the popular machine learning algorithms which have been extensively tested for crop yield prediction by many researchers and have produced accurate results [17–20]. Reference [21] has done a comprehensive review of applications of machine learning techniques in agriculture production systems. Integration of machine learning-based techniques with sensor data can evolve farm management systems into a real-time artificial intelligence-enabled system that provide rich recommendations and insights for farmer decision support and action. Liakos argues that the data generated in modern agricultural operations provided by a variety of different sensors that enable a better understanding of the operational environment (an interaction of dynamic crop, soil, and weather conditions) and the operation itself (machinery data) leads to more accurate and faster decision-making when used along with machine learning based methods. Satellite data-derived data products provide useful feature vector for machine learning methods. While experimenting with machine learning algorithms for crop yield estimation [22] used the features vectors such as Normalised Difference Vegetation Index (NDVI), Standard Precipitation Index (SPI) and Vegetation Condition Index (VCI) to increase the reliability of the work. The training was conducted in a way where all input features were used in pair with yield as their output value.

3 Material and Methods

This section contains the details of the site chosen for the experiment, the dataset used and the approach adopted to carry out the study. The first section describes the specific features of the experimental site and highlights the importance of agriculture in the site. The second section list the datasets used in the study, typical characteristics of the data and sources of the datasets. The third section contains the details and flow of the methodology adopted to carry out the study.

3.1 Experimental Site

The study site was Uttar Pradesh, which is one of the northern states of India and is the fifth largest State of India having an area of 240,928 km^2, which is 7.34% of India and it lies between latitude 24° to 31° and longitude 77° to 84° East. Uttar Pradesh economy being predominately agrarian, it is greatly reliant on the performance of activities such as agriculture, horticulture, animal husbandry, dairying and fisheries. Contribution of agriculture to the state income was 31.7% in 2011 (**Census 2011**). The state is divided into 9 agro climatic zones, viz., Tarai Region, Western Plain Region, Central Western Region, South Western Region, Central Plain Region, Bundelkhand Region, North Eastern Plain Region, Eastern Plain Region, Vindhyachal Region (http://upagripardarshi.gov.in/Staticpages/JayadAgroclaimetikZone.aspx). In 2018–19 Uttar Pradesh had 12.83 Million Ha of its area under cultivation, which produced a total yield of 71.95 tones/ha. About 65% of state population depends on agriculture in some form or the other, and climate is the primary factor that determines the production (see Fig. 1).

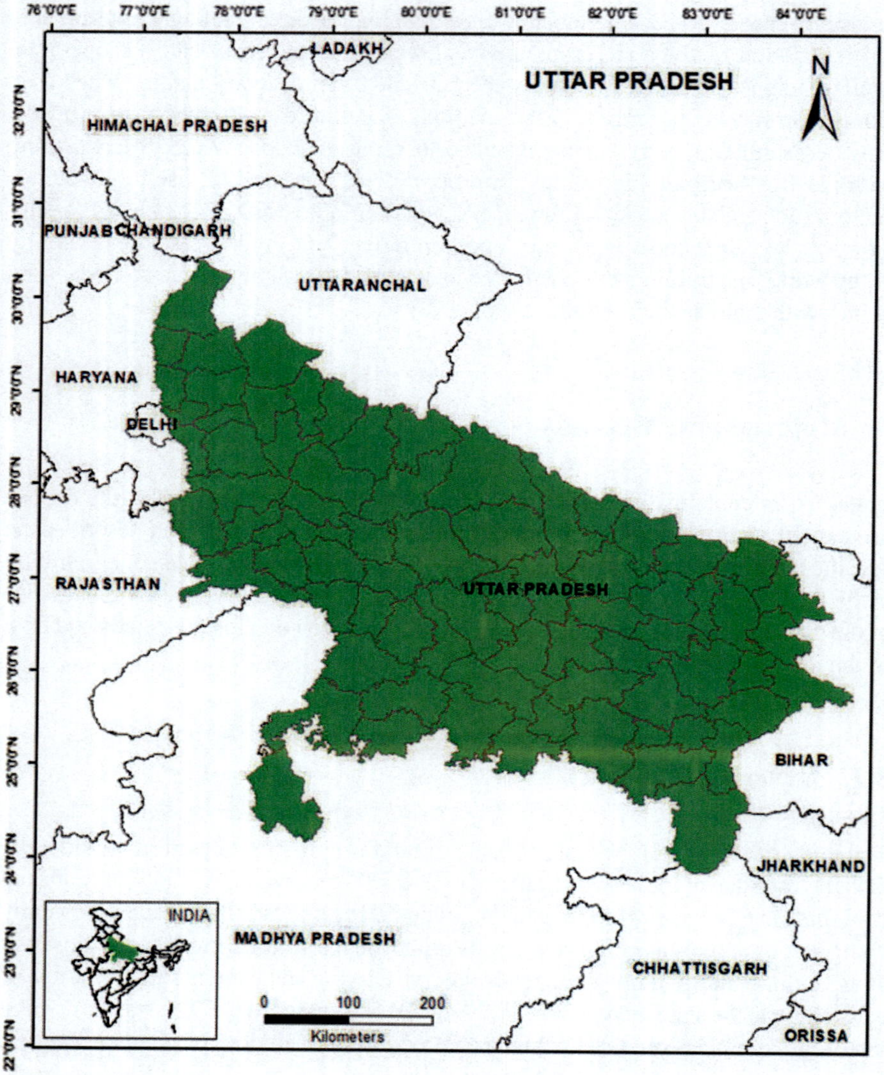

Fig. 1 Geolocation of the experimental site

3.2 Data Used

The data utilised for this paper are acquired for the years from 2001 to 2014 for Uttar Pradesh, India. The factors considered in this investigation are yearly district-wise 'Mean Temperature', 'Mean Relative Humidity', 'Mean Pressure', 'Mean Wind Speed', 'Mean Wind Direction', 'Mean Rainfall', 'Surface Net Solar Radiation', 'Evapotranspiration', 'NDVI'. Meteorological factors are taken from ERA5 which

Table 1 Broad description of the data used in the study

Name	Resolution	Source
ERA5	9 km	European centre for medium-range weather forecasts
NDVI	250 m	Google earth engine
Yield	District wise	data.gov.in

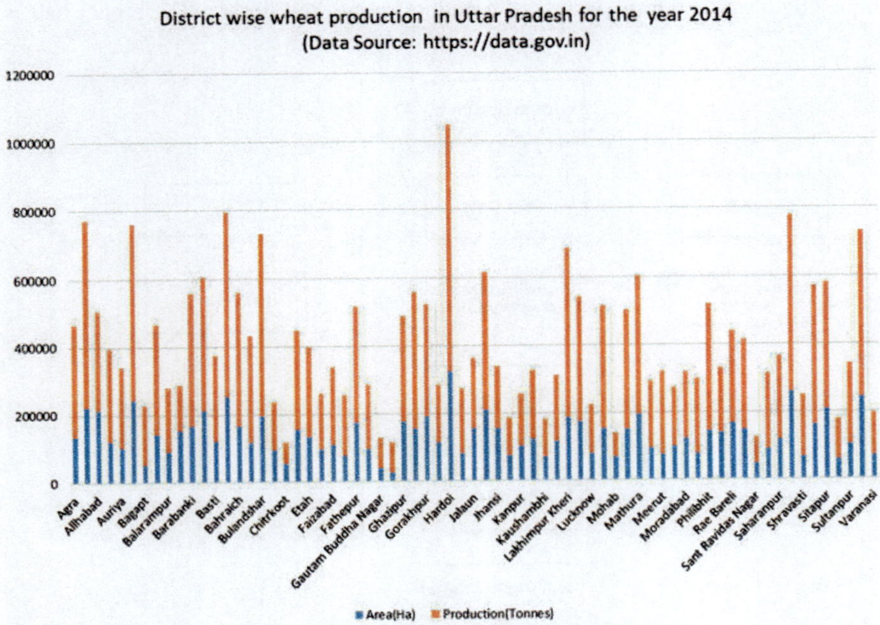

Fig. 2 Wheat production in the state of Uttar Pradesh, India in year 2014

is available as monthwise average from 2001 onwards, it was changed over to yearly average information, NDVI information is taken from MODIS satellite data products through Google Earth Engine and the yield data is collected from https://data.gov.in (Table 1). Wheat crop is taken for the study as it is one of the major food crops of Uttar Pradesh (http://sap.ipni.net/article/uttar-pradesh) is grown in large areas across all districts of the state (Fig. 2) and also contributes significantly to the economy of the state [23, 24].

3.3 Methodology

The methodology (Fig. 3) adopted in the study consists of (i) Noise removal, (ii) Data standardisation and normalisation, (iii) Data spitting for training and validation,

(iv) Model building, (v) Cross-validation, (vi) Hyperparameter tuning, (vii) Model validation, (viii) Performance analysis of models.

A real-world data generally contains noises, missing values, and attribute may have different scales which cannot straightforwardly be used for machine learning models. Dataset was scanned for null values by dataframe.isnull() function from pandas library and for attributes which don't contribute to the process were removed by dataframe.drop() function. The preprocessed dataframe is split into two dataframe in the 60:40 ratio by train_test_split function from Scikit library. The bigger half is

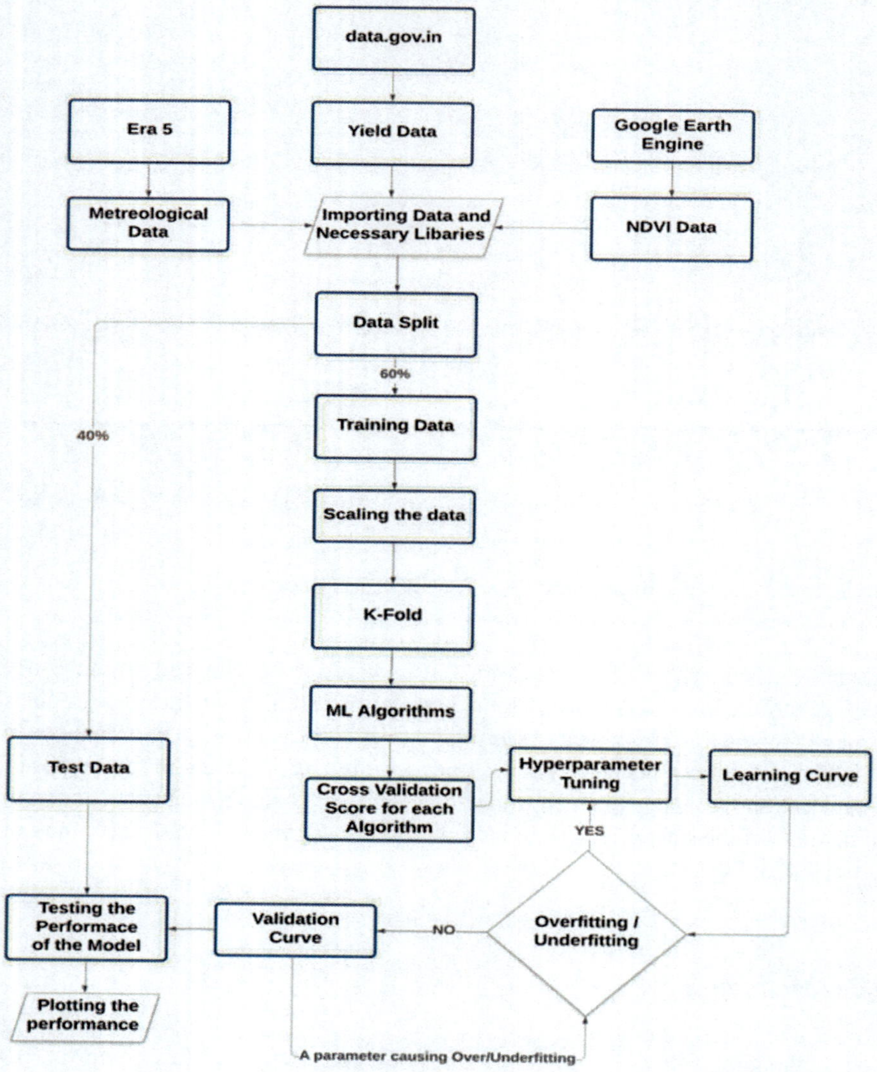

Fig. 3 Flowchart of methodology

used for training the machine learning algorithm and the other half is kept aside for testing the algorithm.

Attributes/Features in the dataframe consist of different scales and attributes/ features with different scales do not contribute equally to the process of training. Thus, to deal with this, standardisation is required, standardisation refers to shifting the distribution of each attribute to have a mean of zero and a standard deviation of one ($\mu = 0, \sigma = 1$). Cross-validation procedure was applied to evaluate the performance of machine learning algorithm on a limited sample, in order to understand how the model will perform on the unseen data which is not used in training. For this study k-Fold was used to randomly split the training data into 7 equal parts out of which, 6 parts were used for training the algorithm and 1 part was used for validation.

The error metric was computed using the .score() function. Each iteration of K-Fold CV provides an r^2 score. Each score was appended to a list and then the mean value was calculated in order to determine the overall accuracy of the model.

Architecture of a model is defined by parameter which governs the training process of that model and these parameters are known as hyperparameter. A hyperparameter controls the capacity of the model, i.e. how flexible the model is, how many degrees of freedom it has in fitting the data. Hyperparameter settings have a big impact on the prediction accuracy of the trained model. Optimal hyperparameter settings often differ for different datasets. Therefore, they should be tuned for each dataset. Model can have one or more hyperparameters and often times the optimal parameters are not known for designing the model and therefore to find that optimal combination of parameters, tuning of different parameter is required and this process of tuning is known as Hyperparameter tuning. Grid search is arguably the most basic hyperparameter tuning method. With this technique, we simply build a model for each possible combination of all of the hyperparameter values provided, evaluating each model and selecting the architecture which produces the best results. There are different optimization algorithms but, in this study, GridSearchCV was used. It evaluates models for a given hyperparameter vector using cross-validation. Cross-Validation scores after tuning the hyperparameters are shown in Table 2.

Each algorithm was fitted with best combination of parameters, returned by GridSearchCV and again their performance was calculated using the Cross Validation

Table 2 Cross-Validation Score of algorithms before and after tuning the algorithm and time taken to tune the algorithm

Algorithm name	Before tuning	After tuning	Time taken to tune (s)
	Accuracy (r^2)	Accuracy (r^2)	
SVR RBF-Kernel	0.90	0.92	2.92
Random forest regression	0.72	0.77	3527.61
Gradient boosted regression	−0.033	0.90	149.67

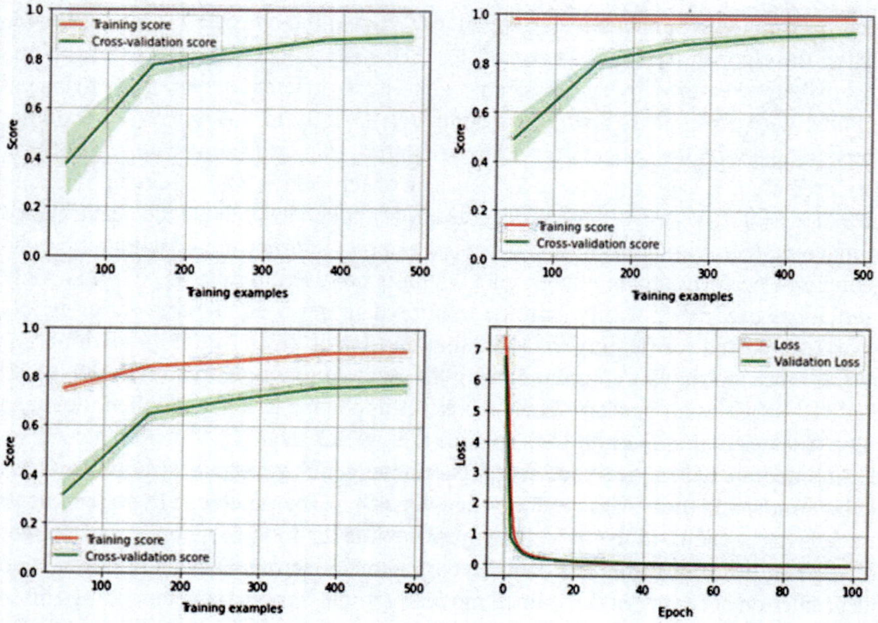

Fig. 4. **a** Learning curve of gradient booster tree, **b** Learning curve of support vector regression, **c** Learning curve of random forest, **d** Learning curve of artificial neural network

function, and learning curve was plotted to check whether the algorithm is under-fitting or over-fitting. Figure 4a–d shows the learning curve of each algorithm. If there is under-fitting or over-fitting, the hyperparameter tuning is performed again with specific parameters set. For RF, GBRT and SVR accuracy metric was used to plot the learning curve but for ANN loss metric was used because in ANN accuracy metric is reserved for classification task only. As shown in Table 2, r^2 value of RF is 0.77 even after tuning and this is also shown in its learning curve plot due to this reason RF forest is not considered further in the study.

A Validation Curve is an important diagnostic tool that shows the influence of a single hyperparameter influence on the training data and the dynamics of the plot gives an insight of whether a hyperparameter is causing an estimator to Underfit or Overfit. All the machine learning models were fitted on the training data using model.fit() function and after the models were trained, they were tested on the testing data by model.predict() function.

4 Results

The sevenfold cross-validation process, along with the split-test, was used to evaluate each algorithm, with the same random seed to ensure that the same splits were made to the training data and that each algorithm was evaluated in exactly the same way. Cross-validation gives an estimate on how the algorithm will perform on the unseen/ test data. Each estimator was provided with some parameter and cross-validation score was calculated in which it was observed that the score of Gradient Booster Trees was the lowest as shown in Table 2. Parameter tuning and optimization were performed on each algorithm with different combinations of parameters handled by GridSearchCV, which improved the prediction accuracy of all algorithms. Interestingly, Gradient Booster Trees value of coefficient of determination increased to 0.9 from –0.033. But the performance of the model is not estimated by coefficient of determination only; it is estimated by various other factors like time taken in tuning and making the prediction which is clearly observed case of Random Forest (Table 2). When parameter tuning was applied to increase the prediction accuracies, random forest took the longest to tune but still had the lowest accuracy. Low accuracy of model can be a cause of under-fitting and over-fitting also. Learning curves and validation curves were plotted to find out if the low accuracy of random forest is the cause of under-fitting and over-fitting. They were also plotted for the other model as well. Learning curve showed that the models were not either under-fitting or over-fitting. It was assumed then that the Random Forest is not suited for the dataset considered for the study and it was not considered for the study.

Gradient Booster Trees, Support Vector with RBF kernel were then compared to Artificial Neural Network. ANN was initially created with 3 hidden layers and 20 nodes in each layer and coefficient of determination value received from this configuration was 0.7, which is insignificant. In order to improve the value of the coefficient of determination 3 more hidden layers were added to the model with different number of nodes, but the coefficient of determination did not improve significantly. Then the number of epochs increased, which increased the number of times the model was allowed to pass through the training dataset, allowing model to reduce the losses and in turn increase the coefficient of determination.

Three errors metric were used to show the performance of each algorithm and from error statistic showed in Table 3 it is observed that all models performed very well on the datasets, but these statistics alone are not sufficient enough to understand the performance of the model.

Table 3 Performance evaluation of the algorithms via error metrices

Algorithm	R^2	MSE	RMSE
Gradient Booster Trees	0.92	0.028	0.169
Support Vector with RBF	0.93	0.025	0.160
Artificial Neural Network	0.92	0.028	0.167

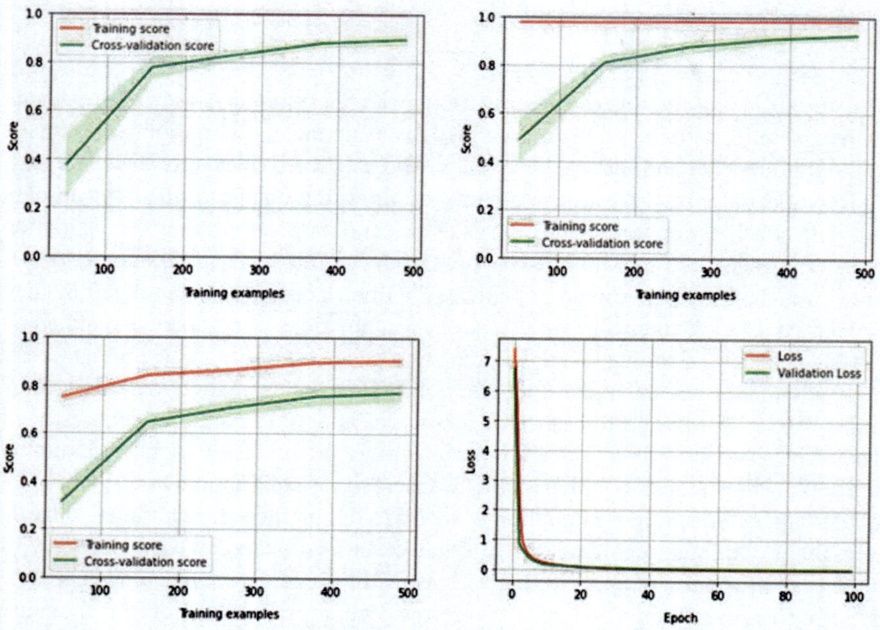

Fig. 5 Comparison plot of predicted value with actual data (left) and Scatter plot of Predicted yield against Observe Yield (right) for Support Vector Machine with RBF kernel

Comparison plot and Scatter plot for individual algorithms are presented in Figs. 5, 6 and 7. In scatter plot few outliers can be seen but most of the points are close to regressed diagonal line which means that the estimator prediction was correct. On comparing scatter plot of Artificial Neural Network with Gradient Booster Trees and Support Vector Machine with RBF kernel, it is observed that the points are more spread out which suggest that GBRT and SVR made prediction with more accuracy. This can also be seen from the comparison plot that ANN made predication with less accuracy when compared to the other method.

From Fig. 8, a general observation can be made that none of the algorithms produces prefect Gaussian look, but for all the algorithms most error occurred near zero this implies that most of the predicted values were correct. But all three algorithms have different frequencies of correct prediction and it can be observed from the plot that SVR has the highest frequency of correct prediction. GBRT had a higher frequency of correct prediction than ANN, but its peak in the graph is slightly on the negative side of 0 which means even though prediction were correct, there was a percentage of error involved with it. ANN had the lowest frequency of correct prediction.

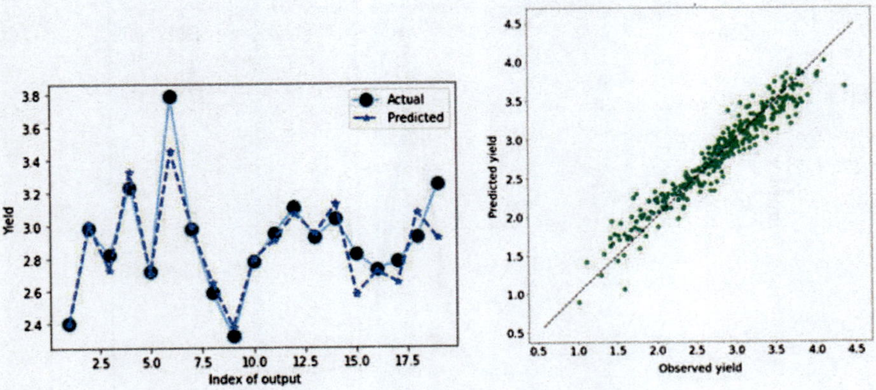

Fig. 6 Comparison plot of predicted value with actual data (left) and Scatter plot of Predicted yield against Observe Yield (right) for Gradient Booster Trees

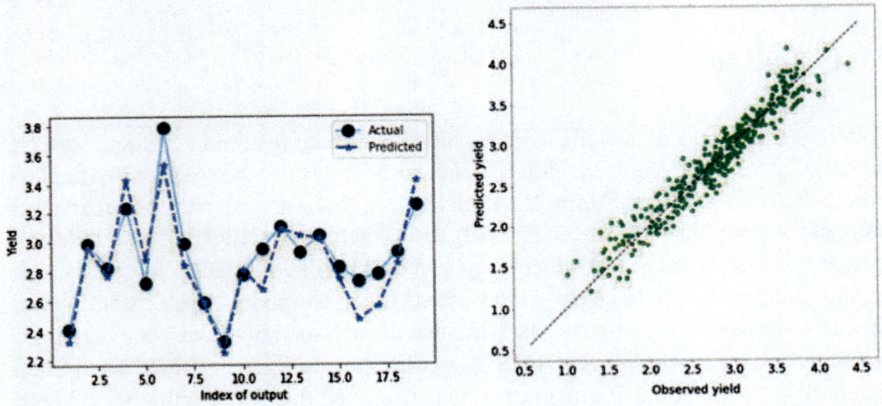

Fig. 7 Comparison plot of predicted value with actual data (left) and Scatter plot of Predicted yield against Observe Yield (right) for Artificial Neural Network

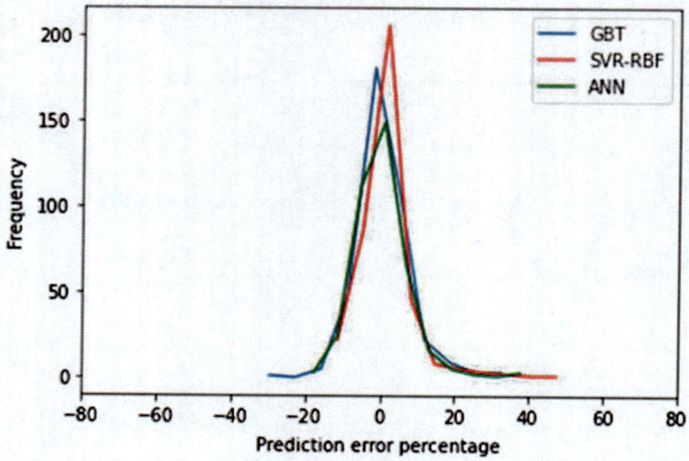

Fig. 8 Histogram of Model Prediction error percentage

5 Conclusion

Four regression-based machine learning models were compared in current research, namely, the Support Vector Machine (SVM), Gradient Booster Tree (GBT), Random Forest (RF) and Artificial Neural Network (ANN). Dataset used in this approach was composed of weather parameters, NDVI, Area sown, Production and Yield. Performance of each model was evaluated by sevenfold cross-validation method and to ensure that the best parameters were used to train the model, GridSearchCV was used to optimise the parameters. Performance of each model was inspected by three error metrics, mean squared error (MSE), root mean squared error (RMSE) and Coefficient of determination (R-squared). It was observed that SVM outperformed RF, GBT and ANN with R^2 of 0.93, RMSE of 0.160 and MSE of 0.025. The performance of the RF was not up to the mark. The study concludes that machine learning techniques are capable of producing accurate results and can contribute to effective decision-making process.

Acknowledgements The authors duely acknowledge the motivation and support extended by the Director, IIRS, Dean (Academics) and Group Head (PPEG), Indian Institute of Remote Sensing, Dehradun. The authors are thankful to Dean, Faculty of Engineering and Technology, SHUAT, Prayagraj, Uttar Pradesh for facilitating the research study. The moral support and guidance provided by the Group Head, GTOP Group, IIRS is also duely acknowledged. Last but not the least the authors are also thankful to Dr. Shakti S. Suryavanshi, Civil Engineering Department, SHUAT, Prayagraj, Uttar Pradesh for providing critical inputs during the research study.

References

1. "IBEF Report," Agriculture in India: Industry Overview, Market Size, Role in Development...| IBEF, Jan 2021. https://www.ibef.org/industry/agriculture-india.aspx. Accessed 06 May 2021
2. Sunder S (29 Jan 2018) India economic survey 2018: farmers gain as agriculture mechanisation speeds up, but more R&D needed. The Financial Express. https://www.financialexpress.com/budget/india-economic-survey-2018-for-farmers-agriculture-gdp-msp/1034266/. Accessed 06 May 2021
3. Ghodsi R, Yani RM, Jalali R, Ruzbahman M (2012) Predicting wheat production in Iran using an artificial neural networks approach
4. Fortin JG, Anctil F, Parent L-É, Bolinder MA (2011) Site-specific early season potato yield forecast by neural network in Eastern Canada. Precis Agric 12(6):905–923. https://doi.org/10.1007/s11119-011-9233-6
5. Lobell DB, Field CB (2007) Global scale climate–crop yield relationships and the impacts of recent warming. Environ Res Lett 2(1):014002. https://doi.org/10.1088/1748-9326/2/1/014002
6. Paswan R, Begum SA (2013) Regression and neural networks models for prediction of crop production. Undefined
7. Jørgensen SE (Sep 1994) Models as instruments for combination of ecological theory and environmental practice. Ecol Model 75–76©:5–20. https://doi.org/10.1016/0304-3800(94)90003-5
8. Murthy CS, Thiruvengadachari S, Raju PV, Jonna S (1996) Improved ground sampling and crop yield estimation using satellite data. Int J Remote Sens 17(5):945–956. https://doi.org/10.1080/01431169608949057
9. Panu US, Sharma TC (2002) Challenges in drought research: some perspectives and future directions. Hydrol Sci J 47(sup1):S19–S30. https://doi.org/10.1080/02626660209493019
10. Wu B, Li Q (2012) Crop planting and type proportion method for crop acreage estimation of complex agricultural landscapes. Int J Appl Earth Obs Geoinf 16:101–112. https://doi.org/10.1016/j.jag.2011.12.006
11. Deb D, Singh A, Kumar S (2018) Remote sensing techniques in horticulture, pp 39–48
12. Singh NJ, Kudrat M, Jain K, Pandey K (2011) Cropping pattern of Uttar Pradesh using IRS-P6 (AWiFS) data. Int J Remote Sens 32(16):4511–4526. https://doi.org/10.1080/01431161.2010.489061
13. Gonzalez-Sanchez A, Frausto-Solis J, Ojeda-Bustamante W (2014) Predictive ability of machine learning methods for massive crop yield prediction. Span J Agric Res 12(2):313–328. https://doi.org/10.5424/sjar/2014122-4439
14. Priyanka T, Soni P, Malathy C (2019) Agricultural crop yield prediction using artificial intelligence and satellite imagery Teresa. Eurasian J Anal Chem 13(SP):6–12
15. Dahikar SS, Rode SV (2014) Agricultural crop yield prediction using artificial neural network approach. Int J Innov Res Electr Electron Instrum Control Eng 2(1):683–686
16. Geetha V, Punitha A, Abarna M, Akshaya M, Illakiya S, Janani AP (Jul 2020) An effective crop prediction using random forest algorithm. In: 2020 international conference on system, computation, automation and networking (ICSCAN), pp 1–5. https://doi.org/10.1109/ICSCAN49426.2020.9262311
17. Jeong JH et al (2016) Random forests for global and regional crop yield predictions. PLoS ONE 11(6):e0156571. https://doi.org/10.1371/journal.pone.0156571
18. Khaki S, Wang L, Archontoulis SV (2020) A CNN-RNN framework for crop yield prediction. Front Plant Sci 10. https://doi.org/10.3389/fpls.2019.01750
19. Marques Ramos AP et al (2020) A random forest ranking approach to predict yield in maize with uav-based vegetation spectral indices. Comput Electron Agric 178:105791. https://doi.org/10.1016/j.compag.2020.105791
20. Liakos KG, Busato P, Moshou D, Pearson S, Bochtis D (2018) Machine learning in agriculture: a review. Sens Switz 18(8):1–29. https://doi.org/10.3390/s18082674

21. Tiwari P, Shukla P (2020) Artificial neural network-based crop yield prediction using NDVI, SPI, VCI feature vectors. In: Advances in intelligent systems and computing, vol 933, Springer Verlag, pp 585–594. https://doi.org/10.1007/978-981-13-7166-0_58
22. http://upagripardarshi.gov.in/staticpages/JayadAgroclaimetikZone.aspx. Accessed 17 May 2021
23. Dey A, Mangalasseri A, Gautam Y (2020) Wheat production in Uttar Pradesh–a study on the growth and instability over time. Int J Curr Microbiol Appl Sci 9:550–555. https://doi.org/10.20546/ijcmas.2020.903.064
24. Yang C, Peterson CL, Shropshire GJ, Otawa T (1998) Spatial variability of field topography and wheat yield in the Palouse region of the Pacific Northwest. Trans ASAE 41(1):17–27. https://doi.org/10.13031/2013.17147

Comparative Analysis of Music Genre Classification Framework Based on Deep Learning

Sai Siddhant Pattanaik, Prakhar Jain, Puneet Sharma, Siddharth Rathore, and Abhijit Kumar

1 Introduction

The inspiration behind this project stemmed from our desire to build a music recommendation tool and a music generation tool. Besides the real life application, this is a very prominent problem in Machine Learning due to the presence of inherent musical metadata. Pairing it along with concepts of signal processing and the underlying mathematics behind it ought to produce substantial results. Prediction of movie genres is an exciting subject with several applications, including building recommendation systems for viewers, assessing movie box office success, and comprehending the movie's concept, to name a few.

In the proposed work explores the different frontiers of training of the neural network to categorise music genres based on the features obtained during the feature extraction phase of the audio files. Simultaneously, the proposed method exploits the Frequency and Time Domain Audio processing of the audio data using the principles of Digital Signal Processing.

This paper consists of seven sections. The first section is Introduction. Introduction deals with the detailed overview of the work carried out followed by stature of the paper. The next section is literature review. In this section the explores the various work done in this area and also distinguish the gaps in research done so far. The next section is proposed methodology. This section describes a detailed approach followed along with the foundation of the approach followed. The next section deals with the implementation of the work. The next is result and discussion section, which establishes the efficiency of the proposed models along with the graphs, tables and figures. The second last section is conclusion. The last section is bibliography.

S. S. Pattanaik · P. Jain · P. Sharma · S. Rathore · A. Kumar (✉)
Informatics Cluster, School of Computer Science, University of Petroleum and Energy Studies (UPES), Dehradun 248007, Uttarakhand, India
e-mail: abhijitkaran@hotmail.com

A. Ramdane-Cherif et al. (eds.), *Machine Intelligence and Data Science Applications*, Algorithms for Intelligent Systems, https://doi.org/10.1007/978-981-99-1620-7_30

2 Literature Review

There were ample work is being done in the area of music genre classification. One of the base paper music genre classification of audio signal published in IEEE Transactions (2002) has established the GTZAN data-sets. The dataset description has been discussed in further section of the paper. In this paper [9] radial basis kernel function used with SVM and achieved accuracy of 80.80%. The paper [7] pioneer in automated feature extraction for musical pattern. It employed CNN to extract features using least anticipatory knowledge. The paper [11] employed K-NN classifier and obtained 77.77% precision across all genres applied with 200 features. The authors [13] incorporated the trial research with a group of Facebook users in order to validate the suggested recommendation system structure. The suggested framework's success is measured by the amount of user satisfaction attained. Based on the referred literature, we identified the gaps in the area and we explored the efficient feature training approaches on different models to categorise music genres of the audio files. Furthermore, the proposed method exploits the Frequency and Time Domain Audio processing of the audio data using the principles of Digital Signal Processing.

3 Methodology

In this section, we discuss our workflow and methods performed to meet the objectives of this project. We've divided the project into three major components. Although there's some overlap between the signal processing and feature extraction sections, they're still broad enough to be accounted for separately.

1. Digital Signal Processing of the audio
2. Pre-processing and Feature Extraction of the processed audio
3. Modelling Deep Learning Architectures (Fig. 1)

3.1 Audio Signal Processing

This section pertains to the analysis and processing of audio using principles of signal processing. To perform the necessary transformations on the audio in order to obtain

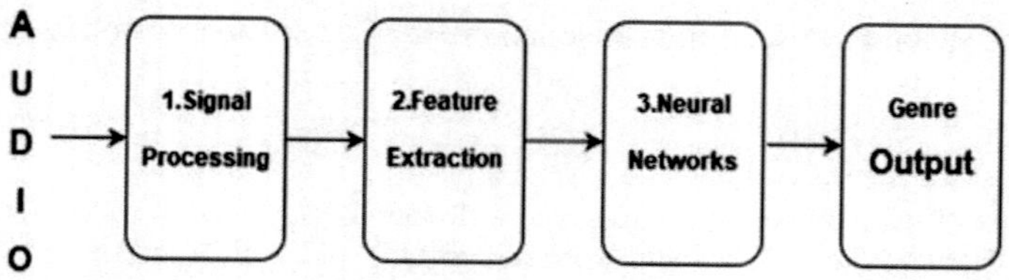

Fig. 1 Methodology overview

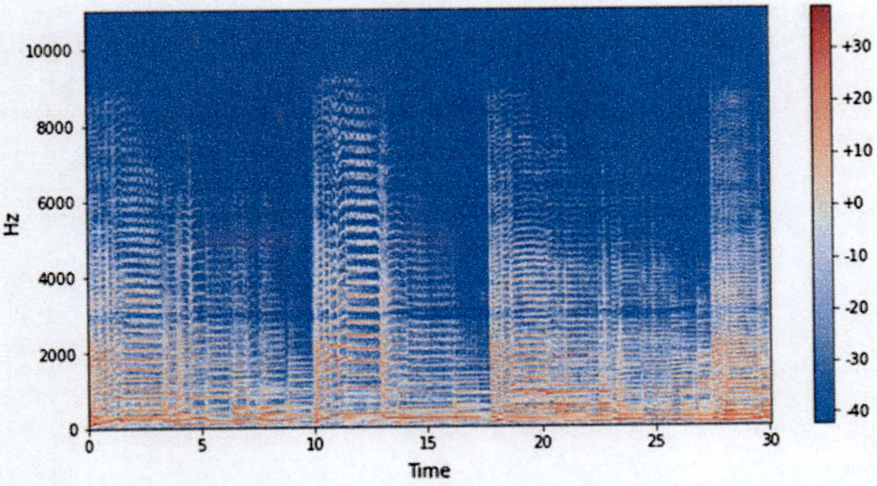

Fig. 2 Jazz00002 spectrogram

the useful features, we use the Librosa API. Librosa is used to transform the wav files into mel-spectrograms and chromagrams, in which the time and feature values are stored as a 2D array.

A spectrogram is a visual representation of the signal strength, or "loudness," of a signal across time at various frequencies contained in a waveform (because amplitude is represented on the z-axis). As a result, one may examine whether there is more or less energy at different frequencies, such 50 Hz versus 500 Hz, as well as how energy levels may alter over time (Fig. 2).

Short-time Fourier transformations can be used in conjunction with binning methods to convert an audio recording into its matching Chromagram. The twelve different pitch classes are closely related to the chromagram. The characteristics derived by Chromagrams assist us in analysing music whose pitches may be classified (often into twelve categories).

The mel-frequency cepstrum (MFC) is a visualisation of an audio signal's short-term power spectrum, created by the linear cosine transform of a log power spectrum on a nonlinear mel frequency scale. Mel-frequency cepstral coefficients (MFCCs) are used to create MFCs. These coefficients serve as a numerical representation of the audio, and as a result, they are useful features (Fig. 3).

To calculate the MFCCs, the following steps are followed.

3.1.1 Windowing

It must be analysed over a short period of time for stable audio features. As a result, speech analysis must be performed on brief segments of audio signal that are believed to be stationary. The audio signal is then split into segments, each with a 25 ms width and a 10 ms interval between them.

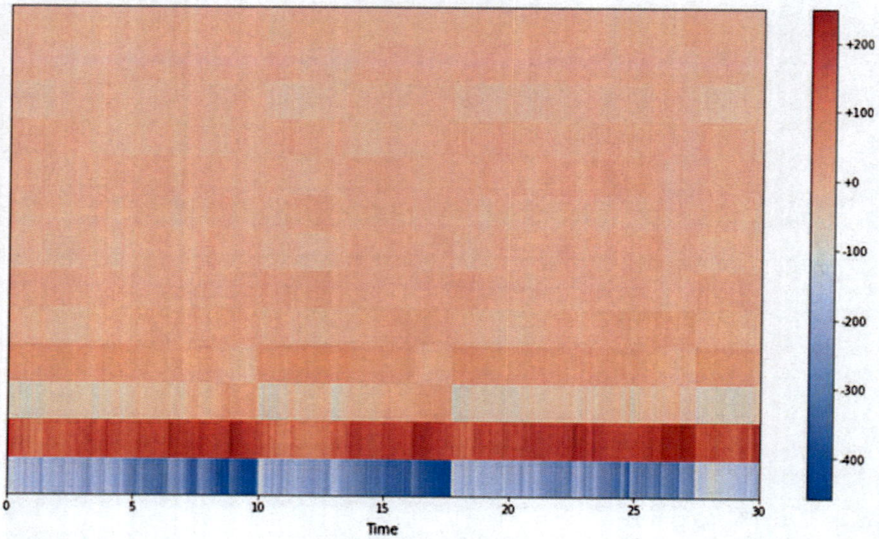

Fig. 3 Jazz00002 MFCC

3.1.2 Discrete Fourier Transform Spectrum

The windowed frames are converted into a magnitude spectrum with the help of DFT. The Discrete Fourier transform (DFT) is a mathematical tool used to convert temporal data into frequency domain data.

3.1.3 Applying Log Function

In comparison to lower energy, humans are less sensitive to changes in audio signal energy. The log function has a similar property: its x gradient is higher at low input values, but it is lower at high input values. To replicate the human hearing system, we apply log to the output of Mel-filter.

3.1.4 Mel Spectrum

A Mel is a frequency measurement metric perceived by the human ear. As, the human auditory mechanism does not appear to perceive pitch in a linear manner, it does not correspond linearly to the actual frequency of the tone. Below 1 kHz, the Mel scale has a linear frequency spacing, and above 1 kHz, it has a logarithmic frequency spacing (Fig. 4).

Mel can be approximated using physical frequency as follows:

$$m = 2595 \log_{10}\left(1 + \frac{f}{700}\right)$$

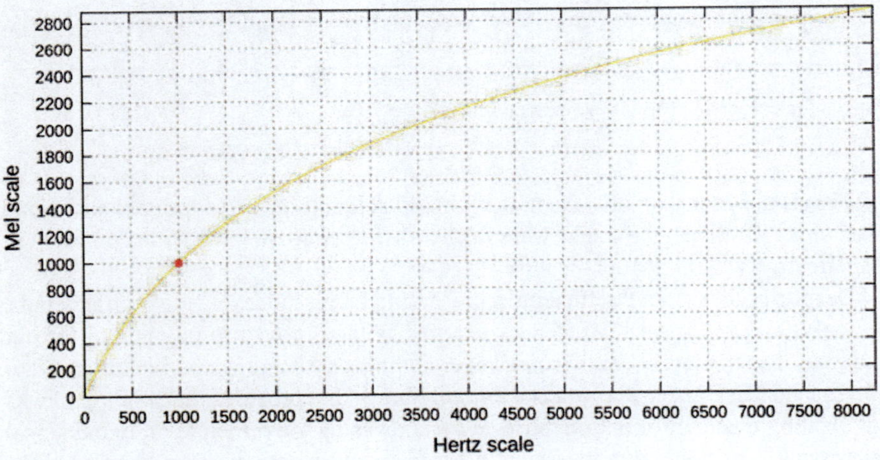

Fig. 4 Mel scale versus hertz scale Plot

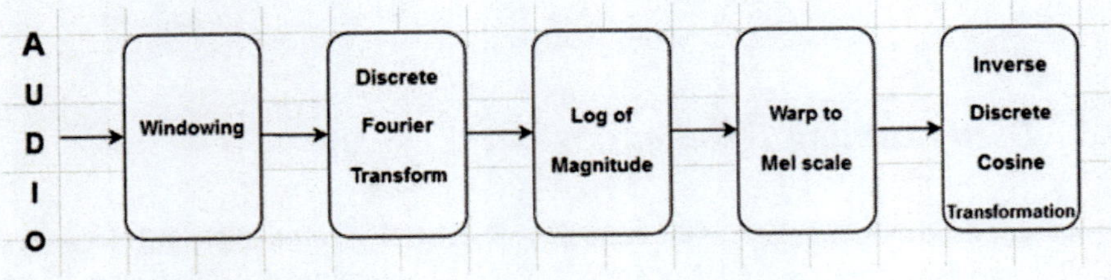

Fig. 5 MFCC calculation

3.1.5 Inverse Fast Fourier Transform (IDFT)

We're doing the inverse transform of the output from the previous step in this phase. A cepstrum is the inverse of the log of the magnitude of the signal. After executing the idft procedures, the MFCC model takes the first 13 coefficients of the signal (Fig. 5).

3.2 Pre-processing and Feature Extraction

The dataset that we used for our proposed model was GTZAN, in which the audio has a sample rate 22050 Hz. Due to variations in length in each track at around 30 s, we trim each track to 29 s to make sure we process files that are of equal duration. Therefore, the total amount of samples to be dealt with turns out to be 22050*29 = 639450.

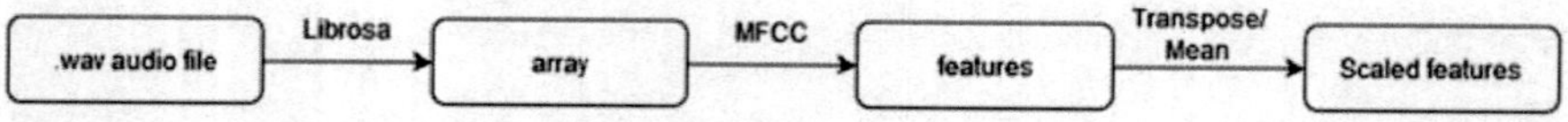

Fig. 6 Pre-processing and feature extraction workflow

The dataset consists of 10 genres with 100 songs each. Furthermore, each song lasts up to 30 s. Then the tracks were further divided into 10 parts with a duration of 3 s, effectively making 10000 songs.

Despite higher order coefficients representing increased levels of spectral details, we only take the first 13 MFCCs as features as they proved to be optimal for our use case. The models become increasingly complex when a high number of cepstral coefficients are chosen. Librosa is a Python library that provides the building blocks for creating music information retrieval systems using the principle of digital signal processing. The method to extract MFCCs from audio data discussed earlier in the paper is being done by this very library.

All MFCCs and associated labels are stored in a separate JSON file and are processed for further use (Fig. 6).

3.3 Deep Learning Techniques

In our proposed model, fusion of the concepts of both Convolutional Neural Networks (CNN) and Recurrent Neural Networks (RNN) are employed. CNNs tend to be excellent when working with images (spectrograms in our case). Long short-term Memory has feedback connections which standard feed-forward neural networks do not. Because of this property, such an RNN has the capability to process entire sequences of data (such as speech or video) which is very relevant in our use case. The next section goes into further discussion regarding the implementation.

4 Implementation of the Proposed Deep Learning Models

4.1 Convolutional Neural Networks

Ordinary neural networks are extremely similar to convolutional neural networks. They're made up primarily of neurons with learnable weights and biases. Each neuron's behaviour is determined by its weights.

Input → Convolution layer → Pooling layer → Fully connected layer → Dense layer → Output

Input layer contains the input data. The output of each layer is the input to the next layer (Fig. 7).

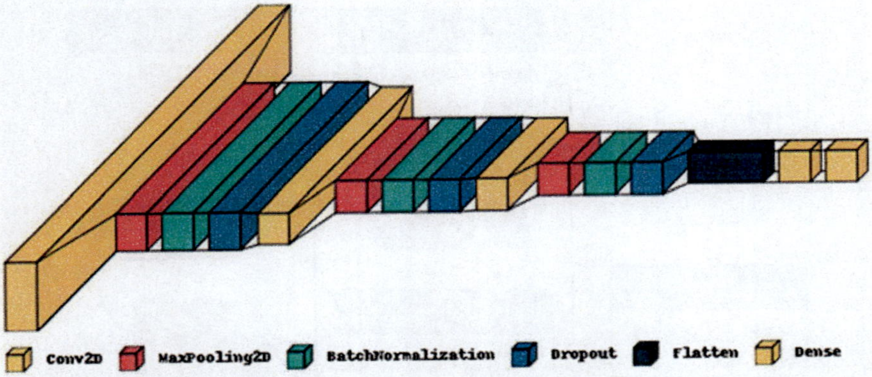

Fig. 7 Visualisation of our CNN model

- Convolution layer: It's also called feature extractor layer. It just recognises the most basic characteristics (horizontal and vertical edges). It also has ReLU activation, which reduces negative values to zero.
- Pooling layer: It is used to minimise the spatial volume of input following convolution between two convolution layers.
- FC layer: It involves neurons, weight, and biases. It inter-connects neurons of one layer to another layer neurons.
- Dense layer: Its the last layer of CNN residing afterwards of FC layer. Each Neurons in the dense layer provides input to the next layer, which makes the networks as a densely connected neural network layer. In models, the dense layer has been discovered to be the most frequently employed layer.
- Dropout Layer: In a neural network, dropout is done layer by layer. Sets the neuron values to 0 in a nutshell (Fig. 8).

4.2 Recurrent Neural Networks

Recurrent Neural Networks work on the premise of preserving a layer's output and feeding it back into the input to predict the layer's output. RNNs, unlike other Neural Network (NN) approaches, provide time-related context-based information to make judgments based on cycles of connections.

LSTM: Long-Short Term RNNs with memory are a specific type of RNN that can learn long-term dependencies. Their default behaviour is to remember information for a longer time.

RNNs have a memory that keeps all of the calculations' data. Because it achieves the same result by completing the same operation on all inputs or hidden layers, it uses the same parameters for each input. Unlike other neural networks, the complexity of the parameters is reduced (Figs. 9 and 10).

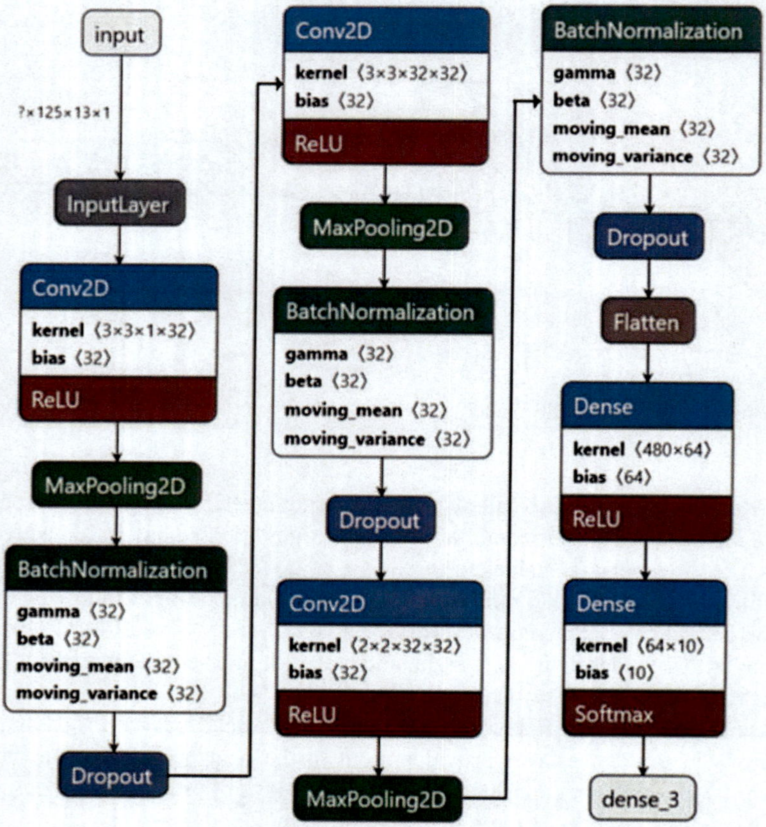

Fig. 8 CNN architecture

Fig. 9 RNN architecture

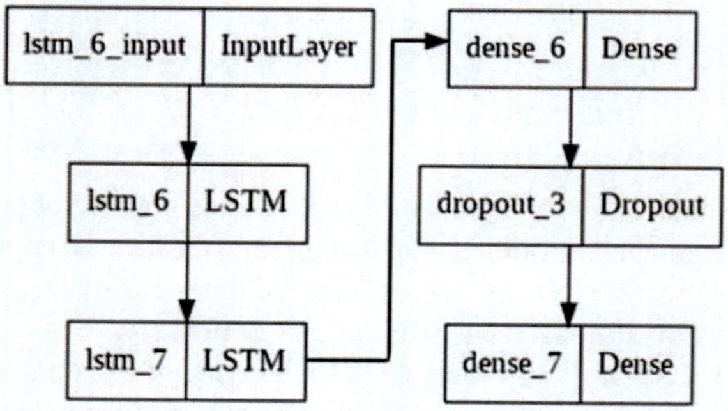

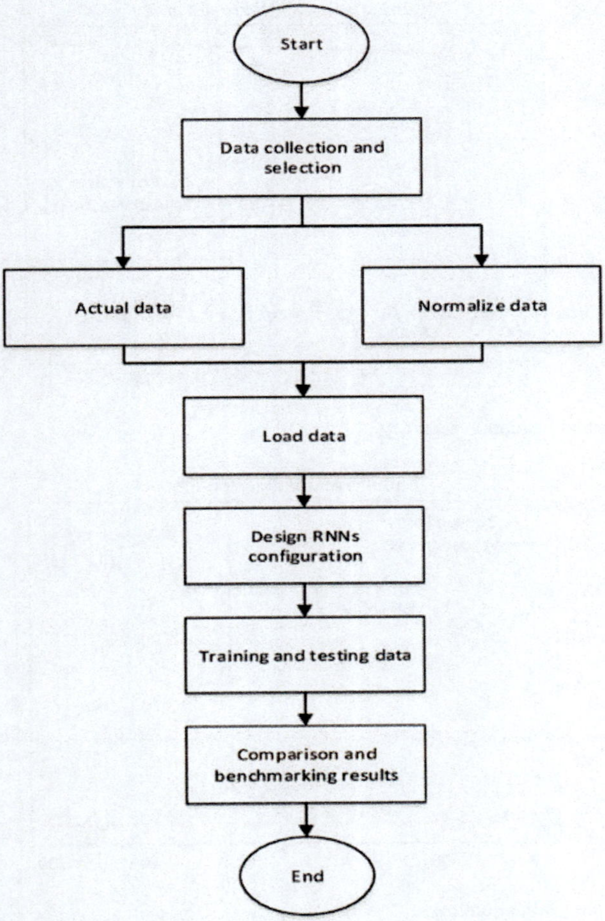

Fig. 10 RNN workflow

5 Results and Discussion

5.1 Datasets

For this project, we utilised the famous GTZAN Dataset, featured in the article, "Musical genre classification of audio signals" by G. Tzanetakis et al. published in IEEE Transactions of Audio and Speech Processing, 2002 [1] (Figs. 11, 12, 13, 14, 15, 16, 17 and 18).

The dataset consists of a total of 1000 audio tracks each lasting 30 s. There are 10 genres of 100 tracks of each in the dataset. The audio files is of .wav format, and each tracks are 22,050 Hz Mono 16-bit. The genres are as follows (Tables 1, 2 and 3):

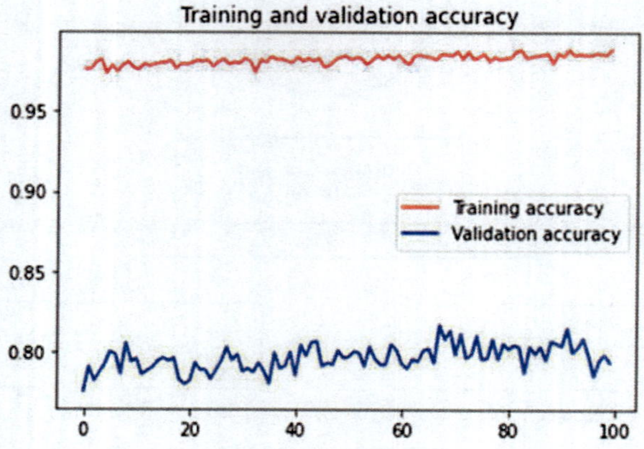

Fig. 11 RNN model validation accuracy

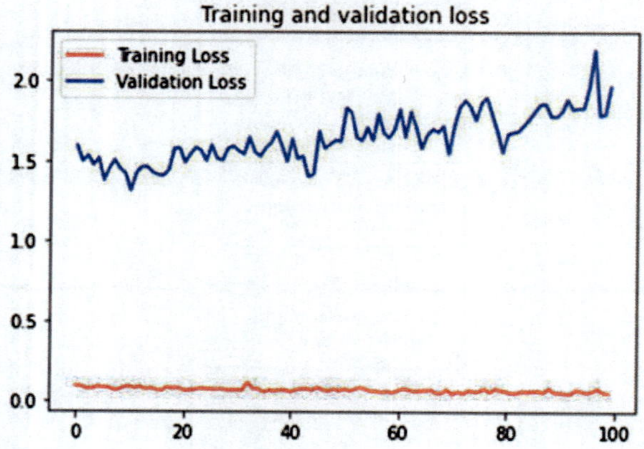

Fig. 12 RNN model validation loss

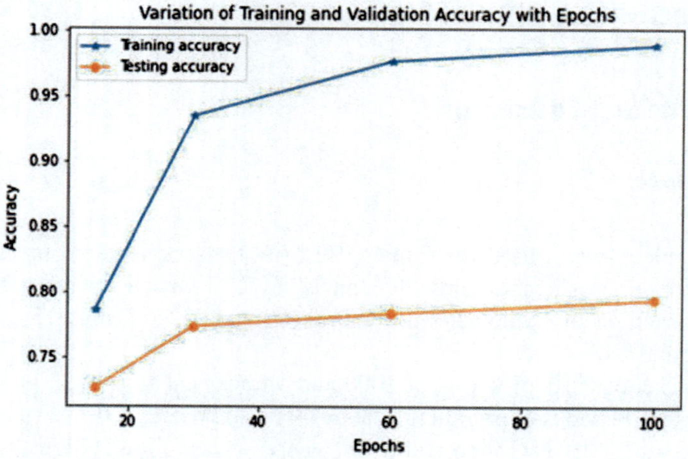

Fig. 13 RNN model validation accuracy with epochs

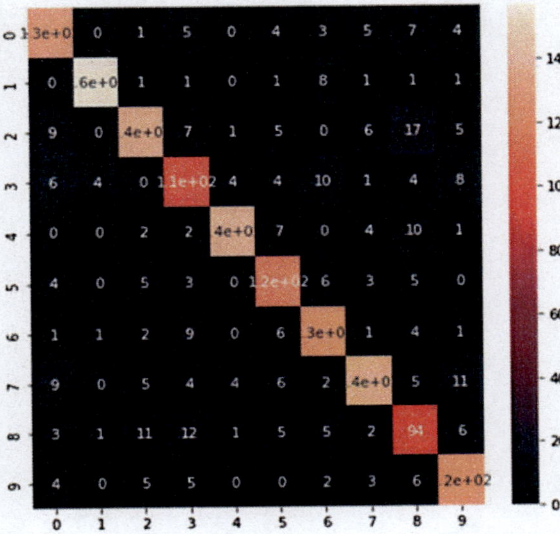

Fig. 14 RNN model confusion matrix

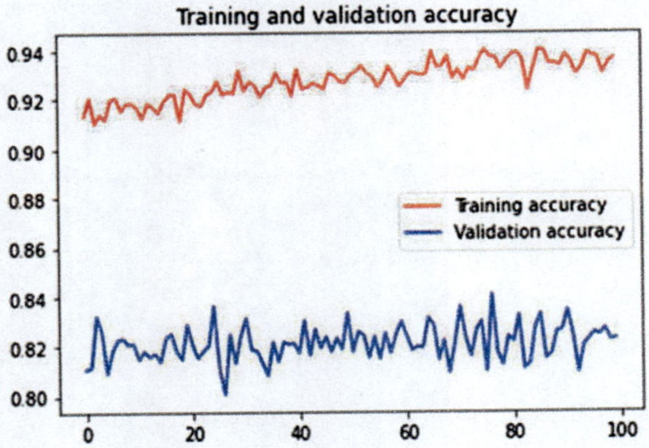

Fig. 15 CNN model accuracy

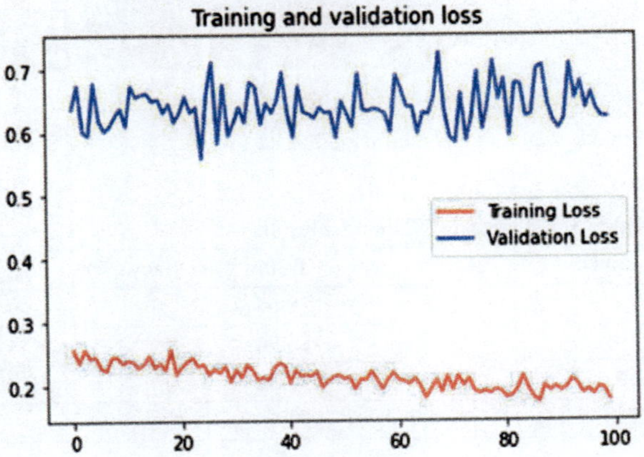

Fig. 16 CNN model validation loss

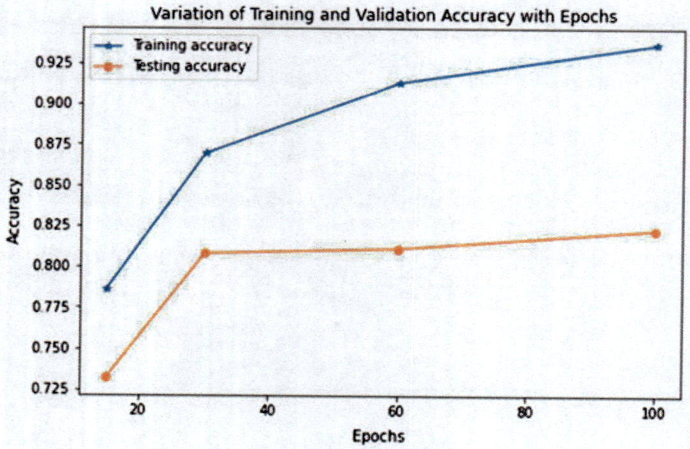

Fig. 17 CNN model accuracy with epochs

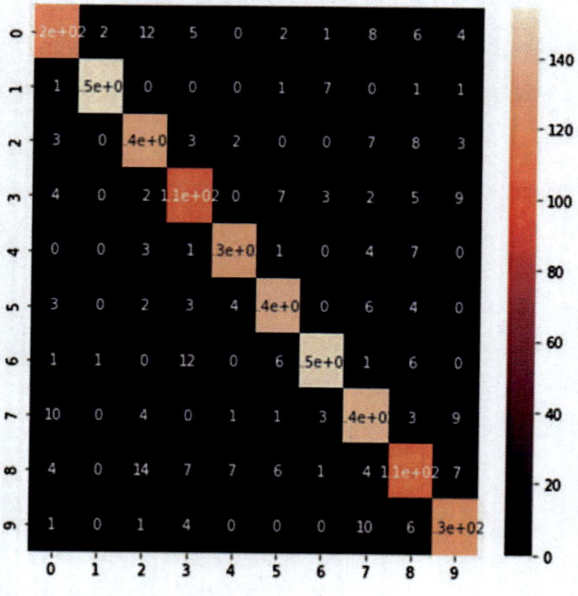

Fig. 18 CNN model confusion matrix

Table 1 Enhanced CNN model results

Epochs	Training accuracy (%)	Testing accuracy (%)
15	78.64	73.2
30	86.94	80.8
60	91.32	81.1
100	93.6	82.25

Table 2 Enhanced RNN model results

Epochs	Training accuracy (%)	Testing accuracy (%)
15	78.6	72.6
30	93.5	77.4
60	97.64	78.35
100	98.87	79.4

Table 3 Comparison with other reviewed articles

Author	Method	Accuracy (%)
Elbir and Aydin	CNN	81.8
Our Model	CNN	81.38
Tang, Chui, Yu, Zhiliang and Wong	LSTM	50 to 60
Proposed Model I	LSTM	**76.9**
Zhang, Lei , Xu and Xing	CNN (60 Epochs)	79.9
	CNN (100 Epochs)	81.9
Proposed Model II	CNN (60 Epochs)	**81.1**
	CNN (100 Epochs)	**82.25**

1. Blues
2. Classical
3. Country
4. Disco
5. Hiphop
6. Jazz
7. Metal
8. Pop
9. Reggae
10. Rock

6 Conclusion

The goal of this research is exploration and exploitation of deep learning and machine learning models to automatically classify music genres. A through examination of related literature has demonstrated. And based upon the capabilities of these models, a baseline model has been identified against which our research could be measured.

In the result section it is evident that the CNN model is already comparable to other works that have been previous work, the LSTM model on the other hand is worth consideration as it has produced significantly better results. This paper proposed two models based on LSTM and CNN. The proposed models achieved accuracy of 76.9% using LSTM, while using CNN the best accuracy achieved is 82.25%.

6.1 Future Works

As extracted features are important concern in classification approach, the features in audio signals like the spectrogram feature, which is essentially taken from the musical signal, is one of the best features that produced good results. Some more parameters could be introduced to further improve the results as well.

Upon further optimisation of the model, the eventual goal is to deploy the project and make it usable for the end-user. Currently, a FastAPI backend for image uploading has been created and after the initial few iterations, the web application takes a piece of music input via uploading or through a URL and puts out a JSON object containing the classification results.

References

1. Tzanetakis G, Cook P (2002) Musical genre classification of audio signals. IEEE Trans Speech Audio Proc 10(5):293–302
2. Pachet F, Cazaly D (2000) A classification of musical genre. In: Proceeding of the RIAO content-based multimedia information access Conference, Paris, France
3. Aucouturier J, Pachet F (2003) Representing musical genre: a state of the art. J New Music Res 32(February 2015):83-93
4. Serwach M, Stasiak B (2016) GA-based parameterization and feature selection for automatic music genre recognition. In: Proceedings of 2016 17th international conference computational problems of electrical engineering, CPEE 2016
5. Choi K, Fazekas G, Sandler M (2016) Automatic tagging using deep convolutional neural networks. ISMIR
6. Lidy T, Rauber A (2005) Evaluation of feature extractors and psycho-acoustic transformations for music genre classification. In: IZMIR, pp 34–41
7. Li TLH, Chan AB, Chun AHW (2010) Automatic musical pattern feature extraction using convolutional neural network. In: (IJACSA) of the international multiconference of engineers and computer scientists (IMECS 2010), I, pp 546–550
8. Sturm BL (2013) Alexander lerch: an introduction to audio content analysis: applications in signal processing and music informatics. Comput Music J 37(4):90–91
9. Nkambule T, Ajoodha R (2021) Classification of music by genre using probabilistic graphical models and deep learning models. In: Sixth international congress on information and communication technology (6th ICICT). Springer, pp 1–6
10. Choi K, Fazekas G, Sandler M (2016) Explaining deep convolutional neural networks on music classification
11. Rajput NK, Grover BA (2022) A multi-label movie genre classification scheme based on the movie's subtitles. Multimed Tools Appl. https://doi.org/10.1007/s11042-022-12961-6
12. Vinayak S, Sharma R, Singh R (2016) MOVBOK: a personalized social network based cross domain recommender system. Indian J Sci Technol 9(31):1–10
13. Singh RK, Singh P, Bathla G (2020) User-review oriented social recommender system for event planning. Ingénierie des Systèmes d Inf. 25(5):669–675

Image Quality Evaluation of Various Pan-Sharpening Techniques Using Landsat-8 Imagery

Greetta Pinheiro and Sonajharia Minz

1 Introduction

The processing of remotely sensed data efficiently and effectively is severely hampered by the neoteric and rapid growth in the availability of data from various sensors as well as by their very heterogeneous natures. However, this rise in RS and ancillary datasets creates the opportunity to combine multimodal datasets to enhance the effectiveness of processing strategies for the application at hand [1]. So, for a wide range of applications, multisource data fusion has drawn a lot of interest from researchers all around the world. The data generated from the remote sensors onboard the satellite platforms are observed to have a significant difference in spectral and spatial resolutions and with different types of properties of the data. This is valid for all sensors, including those that are onboard the same platform (eg: MS and PAN sensors).

The process of PS involves combining MS images with high spectral resolution and low spatial resolution with high spatial resolution PAN images [2]. Spectral resolutions from the MS image and spatial resolution from the PAN image are combined to create the final fused product with high spectral and spatial resolution [3]. PS algorithms are designed for visual enhancement of imagery, and in addition to that, they can also maximize spectral preservation [4] and is useful for quantitative analysis like identifying targets, classifying land cover, defining urban and rural areas, land cover mapping, and geological mapping, among others [5, 6].

Images from optical RS satellites are typically captured in either PAN or MS sensors. While a PAN image only has one broad spectral band and a spatial resolution which is higher than that of a MS image, there are several bands in a MS image with

G. Pinheiro (✉) · S. Minz
Jawaharlal Nehru University, New Delhi 110067, India
e-mail: greetz.pinheiro@gmail.com; greett17_scs@jnu.ac.in

S. Minz
e-mail: sona.minz@gmail.com

A. Ramdane-Cherif et al. (eds.), *Machine Intelligence and Data Science Applications*, Algorithms for Intelligent Systems, https://doi.org/10.1007/978-981-99-1620-7_31

narrow spectral bandwidths [7, 8]. IHS PS, the BT, the ESRI PS transformation, IHS, the SM technique, the spectral sharpening technique called GS technique, and the PCA technique are examples of common PS algorithms. In these techniques, the intensity of PAN is used in place of the intensity of MS, which is calculated by combining the digital numbers (DN) from the MS bands. The spectral content of the MS image is not retained by the pan-sharpened image as a result.

When using the BT, each band present in the MS image are multiplied with the PAN band and division is performed on the sum of the values of the MS bands. Utilizing this technique enhances the contrast between colors with dark and light tones. The value in the PAN band is divided among the color bands using a ratio generated from the proportions in the original color image.

The IHS technique is based on color space conversion and is frequently used in RS research. The RGB color space of the image is transformed into the IHS color space for IHS. Spatial information is extracted from the PAN image using a high pass filter is used to replace the MS image. Reverse IHS transformation creates a multi-spectral image with HR.

With different or the same satellites, optical-based PAN and MS images can both be obtained simultaneously over the same geographical locality on Earth. The high spectral resolution is offered by MS images while high spatial resolution is offered by PAN images. To generate MS images with high spatial resolution, PS is the image fusion established on the pixel level of MS images with PAN images. An RS optical image is created by each satellite at a different level of resolution.

High spatial resolution sensors are expensive and requires large amounts of storage. The images are sharpened using PS techniques, which also strengthens visual interpretation and increases the related images' classification accuracy. The PS technique may be conceptualized as adding the necessary sharpness detail information to the MS image after removing the pertinent information from the PAN image in the fundamental context.

In this study, PS was applied to Landsat 8 satellites' MS bands using the PAN band of the same satellite. As a result, MS images from Landsat-8 have higher spatial resolutions of that of the PAN band. The most widely used PS techniques, including HPFA, GS, IHS, HSV transform, SFIM, PCA, SM, and BT, were preferred despite the fact that numerous techniques have been developed. The ten image quality metrics for assessing the PS images used in the study are RMSE, ERGAS, bias, DIV, PCC, Q, CMC, RASE, CML, and MSE.

2 Study Area

The study area is San Francisco which is one of the largest cities in the United States and part of California State. It is famous for its tourist destinations, facilities and landmarks, beaches, pleasant weather, special cuisine, and multicultural community. San Francisco is the financial, business, cultural, and commercial hub of the United States and has a long history and tradition. The area used in this study is located on

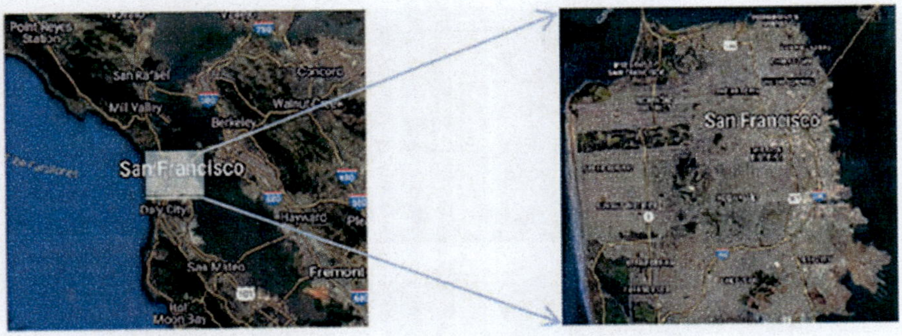

Fig. 1 Study area: San Francisco

the Pacific coast. Figure 1 shows the San Francisco city area. The coordinates of the study area are 37° 46′ 26.2992″ N and 122° 25′ 52.6692″ W.

The Landsat 8 OLI/TIRS combined sensor image with path 44 and row 34 of 18th March 2014 is used for the study. Blue, Green and Red bands i.e., Bands 2, 3 and 4 respectively of the Landsat 8 are used as the MS image with low-resolution and a radiometric resolution of 12 bits. The high-resolution PAN imagery having a radiometric resolution of 12 bits [9] is band 8 of Landsat 8. PAN imagery, Fig. 2a has a spatial resolution of 15 m, while MS imaging has a spatial resolution of 30 m, Fig. 2b.

3 Methodology

RS application like cartography, feature extraction or detection, and image-based GIS applications requires satellite data having high spectral and spatial resolution. Such data are created using PS techniques which fuses the PAN band having high spatial resolution and the multi-spectral bands with high spectral and low spatial resolution to obtain high spectral and spatial resolution multi-spectral images [10]. This section introduces the techniques which are used in the PS process.

For PS, many researchers have developed various techniques over the years. But neither of the previous comparative studies has considered the evaluation of the above-mentioned eight PS techniques using more image quality metrics. These PS techniques are studied and their respective image quality is evaluated using ten different image quality metrics such as RMSE, RASE, bias, ERGAS, PCC, Q, DIV, CMC, CML, and MSE in this research. This evaluative study throws light into the spectral and spatial image quality of the PS techniques and helps in deciding the apt PS technique for the particular application using the different evaluation image quality metrics. The methodology is primarily divided into two sections: the spectral quality evaluation of the pan-sharpened images and the PS of the low-resolution MS images using different PS techniques.

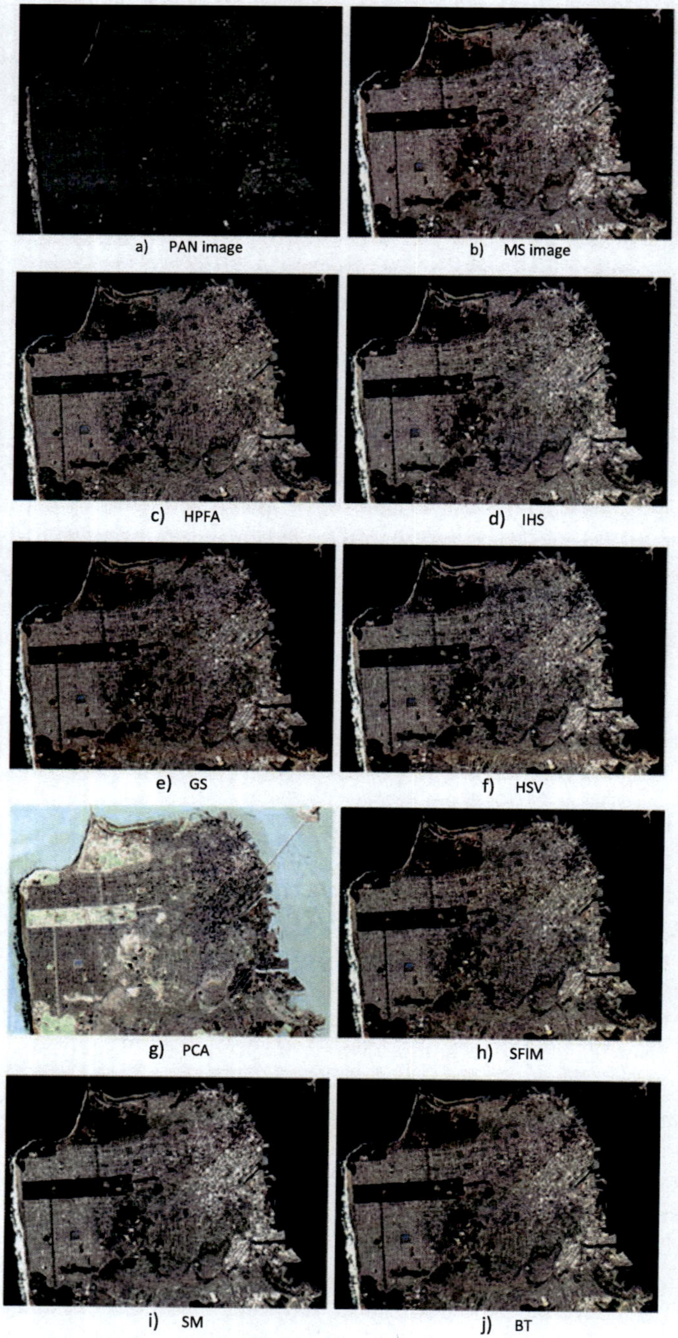

Fig. 2 MS, PAN, and pan-sharpened images using different techniques

3.1 Pan-Sharpening Algorithms

The eight PS techniques used in this research are HPFA, GS, IHS, HSV transform, SFIM, PCA, SM, and BT.

HPFA is a spatial domain image fusion technique. It fuses the HR textural and structural characteristics of an image into a lower resolution (LR) image. Thus, spectral features are mostly preserved. High-frequency information which is derived from the high-pass filter for the PAN image is fused into the LR MS image. In contrast to sophisticated PS techniques like wavelet-based techniques, HPFA uses standard square box high-pass filters [11].

IHS. Different satellite sensors make use of IHS PS technique. The major drawback of this technique is the spectral distortion [12] which results from the overusing of PAN images and also the difference in the grey values between the MS and PAN bands [13]. The IHS technique uses the RGB color model of the MS image which is up-sampled for converting it to an IHS. PAN images are used for replacing the value of I (Intensity).

GS technique was created to enhance the MS images' spatial resolution without affecting their spectral properties. This technique is explained as follows: The weighted sum of multiple bands of the MS image simulates the PAN band to the MS bands' lowest spatial resolution. The optical diffusion of the sensor and the spectral ranges of the PAN and MS bands are used to calculate the weights of the MS bands [14]. The simulated PAN band is used as the first band in the GS transformation. The GS technique is applied to the simulated PAN and MS bands. For balancing the statistics of GS transformations' first band the statistics of the high pixel-sized PAN band are altered. Then, the modified PAN band with high spatial resolution is substituted for the first band of the GS transformation [15]. Then the MS image is generated by inverse GS transforms using the PAN image's spatial resolution.

HSV is a PS technique that requires mainly four steps. Firstly, it transforms the RGB image to the corresponding HSV color space. The HR image is used to replace the value band. Nearest neighbor, cubic convolution or the bilinear technique is used for resampling the saturation and hue bands to HR pixel size. The obtained image is transformed from HSV back to RGB color space.

PCA. The MS bands are inter-correlated and are transformed into a set of new uncorrelated components using the PCA transform [16]. This is similar to a PAN image. Therefore, a PAN image with high spatial resolution is used as its replacement for PS. A reverse PCA transform is applied for combining the PAN image with the MS bands having low spatial resolution.

SFIM. The HR PAN images' spatial details can be modulated to a co-registered LR MS image by retaining its contrasts and spectral characteristics. This is achieved with the ratio between the PAN image with HR and its corresponding low-pass filtered image (using smoothing filter) [17].

SM. Using the SM PS technique, a HR multiband image is obtained by calculating the unsharpened band and the PAN-band to be the mean of each band.

BT approach multiplies each MS images' individual pixels with low spatial resolution bands the ratio of the PAN image with high spatial resolution and divides it by the sum of the MS images to create a new MS image with the spatial detail of the PAN image. The enhancement of the contrast between an image's lowest and highest histogram endpoints, this approach creates visually appealing images. It can only fuse three bands at once, and it is unable to eliminate the radiometric data from the source MS image from appearing in the resulting pan-sharpened image [14].

3.2 *Image Quality Evaluation of Pan-Sharpening Techniques*

In PS techniques, the MS image is resampled to the PAN images' spatial resolution. The spectral distortion can be evaluated using the source MS image with the dimensions corresponding to the PAN image [18].

RMSE uses statistical summaries like mean, standard deviation, etc., along with the pixel values from the entire image are used in the straightforward computation of the RMSE metrics. F and R are the fused and reference image respectively. The total number of pixels are represented by N.

$$\text{RMSE} = \sqrt{\frac{\sum_{i=1}^{N}(F_i - R_i)^2}{N}}$$

The pan-sharpened image's level of correctness is evaluated using RMSE. Since the errors are squared before being averaged, high weights are given for bigger errors by using RMSE metrics. RMSE may therefore be more useful in instances where large errors need to be avoided.

Bias between the pan-sharpened and the MS image shows the amount of distortion from the MS image's original. Calculation of relative bias is done with the division of the difference of the variances and means of similar bands of the source MS and pan-sharpened images at the same resolution by the variance or means of the corresponding main MS band [19].

The difference in variance suggests the quantity of data gained or lost during the image fusion process. Both positive bias and negative bias values respectively, represent the loss and gain of information.

PCC calculation is done between the reference image and the fused image, it illustrates how comparable are the spectral characteristics [20]. The mean values $\overline{R}$ and $\overline{F}$ are of the images R and F respectively. Spectral performance of pan-sharpened image is ideal at the value 1.

$$\mathrm{PCC}\left(\frac{R}{F}\right) = \frac{\sum_{i=1}^{N}\left(R_i - \overline{R}\right)\left(F_i - \overline{F}\right)}{\sqrt{\sum_{i=1}^{N}\left(R_i - \overline{R}\right)^2 \sum_{i=1}^{N}\left(F_i - \overline{F}\right)^2}}$$

Q. The distortion between a reference image and an altered version of that image, which is the result of variations in correlation, brightness, and contrast, is measured by the universal image quality index, or Q [20]. Covariance between R and F are denoted by σ_{RF}, the standard deviations of the images R and F are represented using σ_R^2 and σ_F^2 respectively. Values vary from −1 to 1, with 1 denoting the least distortion possible (identical images) [21].

$$Q(R, F) = \frac{4\sigma_{RF}\overline{RF}}{\left(\sigma_R^2 + \sigma_F^2\right)\left(\overline{R}^2 + \overline{F}^2\right)}$$

DIV. Zero is the optimum value for the best pan-sharpened image, and it measures the variation in the image which is pan-sharpened divided by the variance of the source MS image.

CMC. A measure of contrast between a reference image and an altered version of that image is called Change in Mean Contrast, or CMC. A value of 1 indicates no modification (identical images) [21].

CML. A measure of luminance change between a reference image and an altered version of that image is called Change in Mean Luminance, or CML. A value of 1 indicates no modification (identical images) [21].

MSE. The difference between a reference image (such as an unsharpened multiband image) and a pan-sharpened image is measured by MSE (such as a pan-sharpened multiband image). Low MSE values often indicate less distortion between the fused and reference image.

$$\mathrm{MSE} = \frac{\sum_{i=1}^{N}(F_i - R_i)^2}{N}$$

RASE is derived from RMSE, its effectiveness improves as the numerical value decreases. Here the L spectral bands' mean radiance is denoted by μ and ith band of R is represented by B_i.

$$\mathrm{RASE} = \frac{100}{\mu}\sqrt{\frac{1}{L}\sum_{i=1}^{L}\mathrm{RMSE}^2(B_i)}$$

ERGAS metric also considers the spatial resolution ratios of the MS images and PAN images. The MS image having higher spatial resolution is denoted using h and the lower spatial resolution MS image is denoted by l.

$$\text{ERGAS} = 100\frac{h}{l}\sqrt{\frac{1}{L}\sum_{i=1}^{L}\frac{\text{RMSE}^2(B_i)}{\mu_i^2}}$$

4 Results and Discussion

4.1 *Qualitative Analysis*

The images that are pan-sharpened using eight PS techniques along with the original MS image of Landsat 8 use RGB color mode for displaying the corresponding bands with the respective color. Figure 2 illustrates the pan-sharpened images generated from the eight different PS techniques, the source MS bands (i.e. R, G, and B) and the PAN band.

The spatial details were enhanced by the PCA technique, color distortion was generated. In contrast to the MS imagery used, the highways, water bodies, and different urban structures, such as stadiums, can be readily distinguished from the pan-sharpened image.

Based on the visual assessment, it is determined that the pan-sharpened imagery generated using the HPFA and GS technique is effective for identifying urban characteristics such as highways, buildings, and other structures.

The BT pan-sharpened images, best retained the spectral information and improved contrast, resulting in an image that was visually appealing. Although the edges of the features were not as crisp, the color intensity was best retained in the pan-sharpened image created using the GS technique which is shown in Fig. 2e. It appears that the pan-sharpened images produced by HPFA, and GS techniques are comparable.

4.2 *Quantitative Analysis*

The spectral and spatial quality of the pan-sharpened Landsat-8 images were assessed using the aforementioned 10 quality metrics in order to confirm the findings of the qualitative analysis. For these metrics to choose the best pan-sharpened image with the highest level of spatial features and spectral information, the following requirements must be met: In addition to bias being zero, the following values should be near to one: PCC, Q, and SSIM; DIV, RMSE, RASE, and ERGAS must be close to zero (see Tables 1, 2, 3, and 4).

The bias values for the B2-Blue, B3-Green, B4-Red bands and the average of the three bands of fused images created using various PS algorithms are displayed in Tables 3 and 4. The best pan-sharpened results are the ones acquired from GS, SM, HPFA, and SFIM PS techniques when compared to other techniques. There is a clear variation in the spectral band performance with respect to the different PS techniques used for the study.

The positive values being high for bias in the HSV, IHS, and BT (in the descending order of values) techniques is a clear indication of information loss. For the different PS techniques, the RMSE values obtained are as shown in Table 3 revealing that the least RMSE value was observed for SFIM and then GS techniques.

The PCA techniques yielded the highest RMSE value. The Q and PCC values derived from Table 1 indicate that the highest correlation between the image that is pan-sharpened and the reference MS image is acquired for the GS technique and the lowest for PCA. The DIV is observed to be highest for GS and the lowest value is obtained by IHS technique. The Table 2 shows ERGAS metrics for the different PS techniques. The best values for ERGAS are as shown by SFIM and PCA gave poor results.

From the quantitative evaluation, it is clear that the image which is pan-sharpened using the HPFA PS technique outperformed the other pan-sharpened images. The HPFA technique performed better for most of the image quality metrics. This shows that, in contrast to the other images that are pan-sharpened, the spectral characteristics of the image generated using HPFA PS are extremely similar to the MS imagery used as reference.

Table 1 Image Quality evaluation metrics for HPFA, IHS, GS and HSV pan-sharpened images

Image quality metrics	HPFA	IHS	GS	HSV
RASE	12.9991	32.35	**10.1582**	31.9312
ERGAS	1.8453	4.331	**1.4790**	4.2712

Table 2 Image Quality evaluation metrics for PCA, SFIM, SM, and BT pan-sharpened images

Image quality metrics	PCA	SFIM	SM	BT
RASE	96.3217	**9.1099**	14.0992	16.9677
ERGAS	14.0261	**1.2826**	1.8378	2.2951

Table 3 Image Quality evaluation metrics for HPFA, IHS, GS, and HSV pan-sharpened images corresponding to R, G, B bands of Landsat 8 and their average

Image quality metrics	Band	HPFA	IHS	GS	HSV
RMSE	Blue	0.0099	0.0157	0.0091	0.0153
	Green	0.0101	0.0223	0.0075	0.0220
	Red	0.0101	0.0339	0.0068	0.0335
	Avg	0.0100	0.0239	0.0078	0.0236
Bias	Blue	−0.00008	0.2344	−0.0858	0.2362
	Green	−0.00006	0.2713	−0.0559	0.2721
	Red	−0.00005	0.3064	−0.0390	0.3068
	Avg	−0.00006	0.2707	−0.0602	0.2717
PCC	Blue	0.9839	0.9517	0.9898	0.9574
	Green	0.9521	0.9755	0.9769	0.9791
	Red	0.9744	0.8523	0.9913	0.8612
	Avg	0.9701	0.9265	0.9860	0.9325
Q	Blue	0.9555	0.7291	0.9882	0.7483
	Green	0.9460	0.9135	0.9747	0.9209
	Red	0.9810	0.9179	0.9857	0.9235
	Avg	0.9608	0.3202	0.9828	0.8642
DIV	Blue	−0.1652	−0.0874	0.0747	−0.0354
	Green	−0.2544	−0.4376	0.0778	−0.3564
	Red	−0.4872	−1.3738	0.1276	−1.1874
	Avg	−0.3022	−0.6329	0.0933	−0.5264
CMC	Blue	0.9970	0.9991	0.9992	0.9998
	Green	0.9936	0.9837	0.9991	0.9884
	Red	0.9806	0.9133	0.9976	0.9280
	Avg	−0.1007	0.9653	0.9962	0.9720
CML	Blue	0.9999	0.9653	0.9966	0.9647
	Green	0.9999	0.9519	0.9985	0.9515
	Red	0.9999	0.9365	0.9992	0.9364
	Avg	0.9999	0.9512	0.9981	0.9508
MSE	Blue	0.0001	0.0011	0.00004	0.0011
	Green	0.0001	0.0005	0.00005	0.0004
	Red	0.0000	0.0002	0.00008	0.0002
	Avg	0.00006	0.0006	0.00005	0.0005

Table 4 Image Quality evaluation metrics for PCA, SFIM, SM, and BT pan-sharpened images corresponding to R, G, B bands of Landsat 8 and their average

Image quality metrics	Band	PCA	SFIM	SM	BT
RMSE	Blue	0.0870	0.0067	0.0066	0.0092
	Green	0.0710	0.0068	0.0054	0.0118
	Red	0.0646	0.0075	0.0169	0.0171
	Avg	0.0742	0.0070	0.0096	0.0127
Bias	Blue	−1.0033	−0.0005	−0.0701	0.0972
	Green	−0.6531	0.0005	0.0454	0.1161
	Red	−0.4576	0.0011	0.1499	0.1346
	Avg	−0.7046	0.0003	0.0417	0.1159
PCC	Blue	−0.0395	0.9922	0.9540	0.9820
	Green	−0.9336	0.9748	0.9916	0.9760
	Red	0.2393	0.9849	0.8862	0.9512
	Avg	−0.2446	0.9839	0.9439	0.9697
Q	Blue	0.2099	0.9790	0.8488	0.9046
	Green	−0.8064	0.9734	0.9890	0.9594
	Red	−0.0297	0.9915	0.9489	0.9752
	Avg	−0.2087	0.9813	0.9289	0.9464
DIV	Blue	0.5113	−0.0783	0.1442	−0.1244
	Green	0.3624	−0.1147	−0.1160	−0.3189
	Red	−1.0361	−0.2450	−0.6371	−0.7652
	Avg	−0.0541	−0.1460	−0.2029	−0.4028
CMC	Blue	0.9991	0.9992	0.9969	0.9982
	Green	0.9751	0.9985	0.9984	0.9904
	Red	0.9399	0.9940	0.9703	0.9609
	Avg	0.9713	0.9972	0.9885	0.9831
CML	Blue	0.7992	0.9999	0.9977	0.9947
	Green	0.8857	0.9999	0.9989	0.9924
	Red	0.9329	0.9999	0.9869	0.9896
	Avg	0.8726	0.9999	0.9945	0.9922
MSE	Blue	0.0041	0.00005	0.00028	0.00029
	Green	0.0050	0.00004	0.00002	0.00014
	Red	0.0075	0.00004	0.00004	0.00008
	Avg	0.0055	0.00004	0.0001	0.00017

5 Conclusion

In this study, the PAN band of Landsat 8 satellite with high spatial resolution and the MS bands of Landsat 8 satellite with low spatial and high spectral resolution are combined with the goal of evaluating the image quality of eight PS techniques: HPFA, GS, IHS, HSV transform, SFIM, PCA, SM, and BT. An urban neighborhood in California state called San Francisco with a high population density was chosen as the test site. The quantitative and qualitative analysis was carried out on the obtained pan-sharpened images. For the quantitative analysis, ten image quality metrics namely RMSE, RASE, bias, ERGAS, PCC, Q, DIV, CMC, CML, and MSE was used.

The PS techniques used in this study have enhanced certain spatial details in the images that are obtained using PS, according to the qualitative analysis of the fused images. Although fewer spatial features were preserved by the BT technique, it better conserved the spectral information. The spatial details were enhanced by the PCA technique, color distortion was generated. The pan-sharpened images were created using all eight techniques offering a better spatial resolution. The spatial improvement was verified qualitatively. From the whole pan-sharpened images, urban characteristics including streets, buildings, water bodies, and the stadium could be easily recognized. Qualitative analysis revealed that the GS technique generated the best pan-sharpened image while poor results were obtained by PCA technique. According to the characteristics of the input images, the results obtained by using different PS techniques vary in image quality. PS technique that performs well on images captured by a single remote sensing sensor may not do so on images captured by multiple sensors. Therefore, it is important to consider each remote-sensing sensor while evaluating PS techniques. This study also demonstrates that a PS technique would work better when input images have characteristics that are similar to the corresponding PS design principles.

References

1. Ghamisi P, Rasti B, Yokoya N, Wang Q, Hofle B, Bruzzone L et al (2018) Multisource and multitemporal data fusion in remote sensing. http://arxiv.org/abs/1812.08287
2. Govind NR, Rishikeshan CA, Ramesh H (2019) Comparison of different pan sharpening techniques using Landsat 8 imagery. In: 2019 IEEE 5th international conference for convergence in technology, I2CT 2019. Institute of Electrical and Electronics Engineers Inc.
3. Campbell JB, Wynne RH (2011) Introduction to remote sensing, 5th ed. Google Books. [cited 2022 Sep 4]
4. Gilbertson JK, Kemp J, van Niekerk A (2017) Effect of pan-sharpening multi-temporal Landsat 8 imagery for crop type differentiation using different classification techniques. Comput Electron Agric 134:151–159

5. Shelestov A, Lavreniuk M, Kussul N, Novikov A, Skakun S (2017) Exploring google earth engine platform for big data processing: classification of multi-temporal satellite imagery for crop mapping. Front Earth Sci (Lausanne) 5:1–10. [cited 2022 Aug 1]
6. Gao Z, Ai J, Gao W, Shi R, Zhang C, Liu C (2016) Integrating pan-sharpening and classifier ensemble techniques to map an invasive plant (Spartina alterniflora) in an estuarine wetland using Landsat 8 imagery. 10:026001. https://doi.org/10.1117/1JRS10026001. [cited 2022 Sep 4]
7. Raj A, Minz S (2021) A scalable unsupervised classification method using rough set for remote sensing imagery. Int J Softw Sci Comput Intell 13:65–88. [cited 2022 Dec 2]
8. Raj A, Minz S (2020) Spatial clustering using neighborhood for multispectral images. 14:038503. https://doi.org/10.1117/1JRS14038503. [cited 2022 Dec 2]
9. Korhonen L, Hadi, Packalen P, Rautiainen M (2017) Comparison of Sentinel-2 and Landsat 8 in the estimation of boreal forest canopy cover and leaf area index. Remote Sens Environ 195:259–74
10. Zhang Y (2004) Understanding image fusion. Photogramm Eng Remote Sens 70:657–661
11. Gangkofner UG, Pradhan PS, Holcomb DW (2007) Optimizing the high-pass filter addition technique for image fusion. Photogramm Eng Remote Sens 73:1107–18
12. Trijayanto DP, Tjandrasa H (2019) Improving spectral quality of IHS-pansharpening result by integrating equalization process using SVE-DWT for satellite imagery data. In: 12th international conference on information & communication technology and system (ICTS). IEEE
13. Xu Q, Zhang Y, Li B (2014) Recent advances in pansharpening and key problems in applications 5:175–95. https://doi.org/10.1080/194798322014889227. [cited 2022 Nov 10]
14. Pushparaj J, Hegde AV (2017) Comparison of various pan-sharpening methods using Quickbird-2 and Landsat-8 imagery. Arab J Geosci 10:1–17
15. Sarp G (2017) Spectral and spatial quality analysis of pan-sharpening algorithms: a case study in Istanbul 47:19–28. https://doi.org/10.5721/EuJRS20144702. [cited 2022 Nov 20]
16. Du Q, Younan NH, King R, Shah VP (2007) On the performance evaluation of pan-sharpening techniques. IEEE Geosci Remote Sens Lett 4:518–522
17. Liu JG (2010) Smoothing filter-based intensity modulation: a spectral preserve image fusion technique for improving spatial details 21:3461–72. https://doi.org/10.1080/014311600750037499. [cited 2022 Nov 10]
18. Alparone L, Aiazzi B, Baronti S, Garzelli A, Nencini F, Selva M (2008) Multispectral and panchromatic data fusion assessment without reference. Photogramm Eng Remote Sens. 74:193–200
19. Jalan S, Sokhi BS (2012) Comparison of different pan-sharpening methods for spectral characteristic preservation: multi-temporal CARTOSAT-1 and IRS-P6 LISS-IV imagery 33:5629–43. https://doi.org/10.1080/014311612012666811. [cited 2022 Nov 10]
20. Panchal S, Thakker R (2015) Signal & image processing. Int J (SIPIJ) 6
21. Wang Z, Bovik AC (2002) A universal image quality index. IEEE Signal Process Lett 9:81–84

Statistical Machine Learning in Loan Analysis from Financial Institutions

Bikram Pratim Bhuyan, Ravi Tomar, and Amar Ramdane Cherif

1 Introduction

By evaluating and using vast volumes of data, machine learning (ML) aims to create more effective methods of completing a specified list of tasks [1]. It is considered an element of artificial intelligence. It's possible for algorithms that employ machine learning to learn from examples, often known as training data, and then use the model they've learned to make conclusions or judgments even when they haven't been given any particular instructions to do so [2]. Machine learning algorithms are utilized in various fields, including those in which it is difficult or impossible to design traditional algorithms to carry out the necessary tasks [3]. Some examples of these fields include medicine, email filtering, voice recognition, agriculture, and computer vision. Machine learning is not synonymous with statistical learning; nonetheless, several subfields of machine learning have significant ties to computational statistics, a branch of statistics that emphasizes the use of computers to generate predictions [4]. The discipline of machine learning may benefit from new methodologies, theoretical frameworks, and prospective application areas made available by mathematical optimization. Data mining is an interconnected field of study that conducts exploratory data analysis via unsupervised learning. Data and neural networks are used in particular types of machine learning in ways that are theoretically comparable to how the brain of a living human being operates. Machine learning, often known as predictive analytics, is a technique that is frequently used in the corporate world to resolve a broad range of problems [5].

B. P. Bhuyan (✉)
School of Computer Science, University of Petroleum and Energy Studies, Dehradun, India
e-mail: bikram23bhuyan@gmail.com

B. P. Bhuyan · A. R. Cherif
LISV Laboratory, University of Paris Saclay, 10-12 Avenue of Europe, 78140 Velizy, France

R. Tomar
Persistent Systems, Pune, India

A. Ramdane-Cherif et al. (eds.), *Machine Intelligence and Data Science Applications*, Algorithms for Intelligent Systems, https://doi.org/10.1007/978-981-99-1620-7_32

Financial organizations often undertake loan assessments not just to guarantee that loans are provided under fair terms and circumstances but also to ascertain whether or not borrowers have the means and the will to repay the loans [6]. The information provided by the applicant is weighed against the criteria for acceptable lending to establish eligibility. Examining the loan provides the creditor with some degree of comfort by evaluating the possibility that the borrower will return the obligation (principal and interest). Through follow-up meetings with the potential borrower, especially for individual loans, the lender can get further information about the profitability of the investments or assets that would give money to aid in paying back the loans [7].

A process must be followed when applying for a loan, starting with verification of eligibility and ending with the disbursement of the loan money. Initial steps include providing the consumer with a loan application form, which must be completed before being submitted for review. Accurate details on the customer's identity, address, income, assets, etc., are required [8]. The credit officer has to be able to verify the applicant's claims without any help from the applicant. You may help the company maintain track of loan repayments through the sophisticated security procedures for loans by giving comprehensive and correct information. The information is used in the loan's preliminary appraisal and analysis, where it is considered alongside other factors, including risk, the value of collateral, and the predictability of repayments. If payments are missed, the lender may get in touch with you using the details you provided [9].

This research allows the lender to go through the statistical details of the borrower and allows for a safe loan to be issued with the help of machine learning algorithms. The next section explores the literature review, followed by machine learning techniques. The data preprocessing, statistical, and machine learning results are analyzed in the results. Finally, we conclude the paper in the conclusion section.

2 Literature Review

In the past, determining whether or not to provide a loan required human experts and intricate statistical models to make an informed choice. Credit scoring was one of these methods. Researchers and institutions have recently chosen training classifiers based on machine learning algorithms and neural networks to automatically anticipate an individual's credit score based on their previous data and weed out credit defaulters from the bunch before the loan is issued. This is done to improve accuracy [10].

Researchers have employed various statistical approaches, including linear regression, probit analysis, and discriminant analysis, to predict loan default [10]. Nave Bayesian method was combined with the Support Vector machine model to make predictions about Loan defaulting in the study's hybrid model. A dataset consisting of 21 characteristics was compiled from German sources by UCI researchers. Although

Naive Bayesian is recognized for its speed and effectiveness in classification problems, it takes a long time to install. However, support vector machines are favored for classification because of their rapid processing and high accuracy. The study's authors experimented with a wide variety of configurations, including ones based solely on Naive Bayes, solely on support vector machines, and a combination of the two.

Researchers proposed a technique to forecast "Korean Student Aid Foundation" loan defaulters (KOSAF) [11]. They identified loan defaulters using logistic regression and the Cox proportional hazard technique. The study identified the top reasons students default on their loans. KOSAF provided all data from 2012 to 2014. 125,291 good loans and 2,141 poor loans were analyzed. Like training data, the test set has excellent and bad loans (83,560 and 1,480, respectively). Student loan default was linked to age, family income, monthly repayable amount, and field of study. A logistic regression study found that 20-year-olds are less likely to default on college debts than teenagers and their early 20s. A strong and accurate logistic regression model worked well on testing data (AUC = 0.697).

In Ghana, where 30% of the population is poor, a logistic regression model detected bank loan defaulters. The study sought to identify Ghanaian loan default factors. The research employed Ghanaian microfinance data [12]. It supplied a client database. Data were collected on approximately 500 consumers in 2013 and 2014. Instead of linear regression, the researcher utilized logistic regression to predict loan failure. Logistic regression accurately identified loan defaulters at 86%.

The research proposed a two-stage risk assessment method using a decision trees-based classifier to establish creditworthiness [13]. The study employed an industry-standard method. Support vector machine improved data from IDBI, syndicate banks, and AXIS midsize Indian banks. The dataset included 1140 occurrences and contained numerous dependent characteristics, including age, sex, marital status, loan type, and credit authorization. The researchers predicted loan acceptance and default risk using Weka's C4.5 decision tree system. Their classifier uses a customer's credit score and borrowing history to decide whether to lend.

Recent study [14] uses artificial neural networks as classifiers to predict loan defaults. India's leading financial institution, "lending club bank," provided data. Due to the massive unstructured data, the researchers used principle component analysis (PCA) to reduce dimensionality and decreased the number of columns to 19. 9500+ rows and 14 columns comprised the original dataset. The data was selected to preserve 90% of the dataset for training and 10% for testing the model. The study employed a multi-layered perceptron neural network to predict loan defaulters. The multi-layered perceptron's input layer links the 18 main perceptrons to the 20 secondary perceptrons. The study's final layer is directly connected to the output, and its two previous layers each include 20 perceptrons. The program predicted 93% of loan defaulters. The model performed better than traditional models, although adding hidden layers might increase its performance.

In a similar research [15], the authors proposed a neural network-based approach to classify borrowers as defaulters or non-defaulters based on explanatory features.

The research used Bandora's dataset. The dataset comprised 16037 default and non-default loans. 15% of observations were defaulted, whereas 85% were not. After splitting the data 70:30 into training and testing sets, the neural network input layer received the 15 characteristics. 14 neurons in the input layer communicated, with 5 in the hidden layer and 1 in the output layer. The model predicted 74% of defaulters.

In the study [16], authors developed a novel credit-scoring approach utilizing genetic programming and deep neural networks. The study employed Australian and German credit databases. 307 of 690 Australian dataset observations were not loan defaults. 14 bits of information. Their concept enabled the evolutionary algorithm to uncover latent correlations between variables used to define categories. During testing, these rules would filter the neural network's first layer. They built a stacked Autoencoder neural network using the greedy layer strategy, which automatically generates a deep neural network with many hidden layers based on output. It outperformed models that employed more standard statistical approaches.

3 Machine Learning Techniques in Loan Analysis

The various machine-learning techniques are discussed in this section.

3.1 *Logistic Regression*

Logistic regression is technically a classification model, even though its name includes the word "regression" [17]. Logistic regression offers an easy and better solution to problems involving binary and linear classification. It's a model for classifying things that don't need a lot of effort to implement. Yet, it nevertheless manages to come up with quite good outcomes when applied to groups of things that can be partitioned along a linear axis. The algorithm is a useful classification tool used extensively in commercial settings. Multinomial logistic regression [18] is an approach that may be used in situations with more than two unique discrete outcomes. Logistic regression is an analytical method for classification challenges, such as selecting which category is the best fit for a freshly obtained data set. One application of logistic regression is determining which category is the best match for newly gathered data [19]. Logistic regression is a useful analytical method in cyber security since many aspects of the field involve classification challenges, such as threat detection. Consequently, logistic regression may be used to analyze data. The logistic function used is formally defined as:

$$Logistic\ Function = \frac{1}{1+e^{-x}} \tag{1}$$

3.2 K-Nearest Neighbour (KNN)

To categorize data, KNN is a non-parametric approach. Additionally, it is one of the most well-known methods that may be used to categorize data [20]. The fundamental concept is to arrange the available data in a space that is determined by the features that have been selected. The method will determine the class of the new data with the assistance of the classes of the k data located closest to it. The examination process includes noting how changes in distance and other factors impact the findings of classification [21]. The primary advantage of the KNN classification system, in addition to its effectiveness, is the simplicity with which it may be used. Although it is a useful method, it is susceptible to outliers, has lengthy computation times for large datasets, and requires trial and error to establish the ideal number of neighbors to use (k).

3.3 Naive Bayes

The Naive Bayes method is a simple approach to predictive modeling, yet it yields accurate results [22–25]. The model contains two different kinds of probabilities, both of which can be computed from the training data: the likelihood of each class and the conditional probability for each class given each value. I the likelihood of each class and (ii) the conditional probability for each class given each value. With Bayes' theorem, it is possible to generate predictions for data that have not yet been seen after the probability model has been designed and implemented. It is common practice to assume that real-valued data follow a Gaussian distribution, sometimes known as a bell curve, to assist the estimation of probabilities like these. Naive Bayes gets its name because it functions on the premise that each of the variables used in the input process may be handled independently. Even though this is not a fair assumption for the data in question, the method is very helpful for addressing a broad range of difficult problems.

Mathematically we need to maximize the following:

$$P(C_i|T) = \frac{P(T|C_i)P(C_i)}{P(T)} \tag{2}$$

where C_i is the class information, T represents the tuples. Now, $P(T)$ being a constant term, we need to maximize:

$$P(C_i|T) = P(T|C_i)P(C_i) \tag{3}$$

or in other words,

$$P(C_i|T) = P(T_0|C_i)P(T_1|C_i)P(T_2|C_i)\ldots P(T_n|C_i)P(C_i) \tag{4}$$

3.4 Decision Tree

The decision tree is a kind of supervised learning approach that may predict both discrete (using a classification tree) and continuous (using linear regression) input and output variables (regression tree). Understanding and making decisions are made easier due to the visual aspect of the information. A structure analogous to a flowchart and uses the if-then statement is referred to as a decision tree. The first question is asked at the top of the tree, which is the root node, and from there, the questions go gradually lower down the tree [26].

Entropy is a measurement that determines how much detail there is in a given piece of data. When all of the data's components are identical, the entropy value is equal to zero, often known as "pure," and increases closer to one as the data gets more scattered (that is impure). Mathematically,

$$Entropy = -\sum_{i=1}^{n} P_i * log(P_i) \tag{5}$$

4 Results Analysis

This section is dedicated to understanding the data and performing preprocessing and finally performing machine learning algorithms for predictive analysis.

4.1 Data Preprocessing

The data set consists of 500 entries with 14 features. The feature set $\Delta = \{$ID, Age, Experience, Income, ZIP Code, Family, CCAvg, Education, Mortgage, Personal Loan, Securities Account, CD Account, Online, CreditCard$\}$. The data set was first tested for NULL values in the attributes. There were no occurrences of such values.

As 'ID' is not important, the feature is deleted. Negative values were observed in the 'Experience' feature. The absolute value is taken to convert the negative to positive values. The pairwise distribution of attributes is plotted as shown in Fig. 1.

The categorical features of the data set are extracted, and a histogram plot is used to count the attribute type as shown in Fig. 2.

The numerical features of the data set are extracted, and a box plot is used to find the median, maximum value, minimum value, and quartile information for the attribute type as shown in Fig. 3.

4.2 Statistical Analysis

After the initial description of the data set, we perform a basic statistical analysis of it. Correlation analysis is performed in the data as shown in Fig. 4. It is seen that

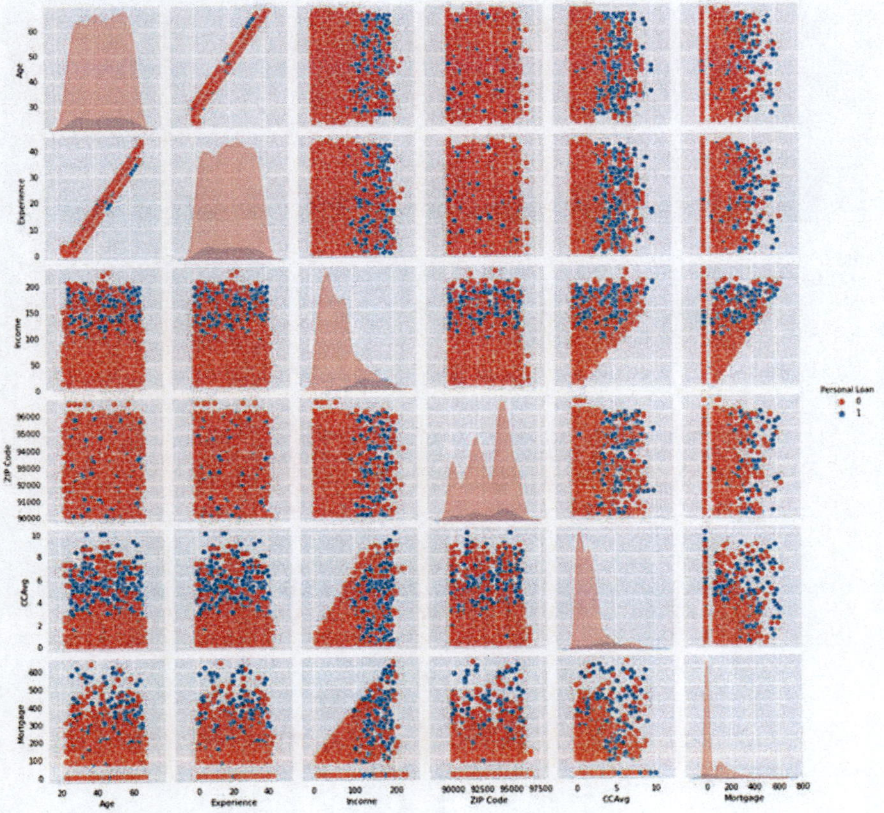

Fig. 1 Pair wise distribution of attributes

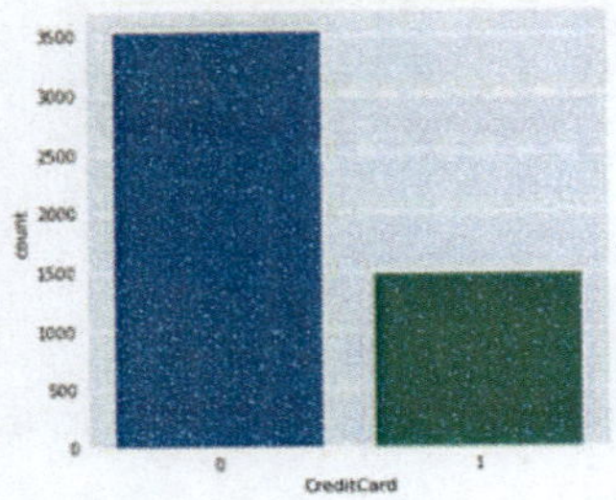

(a) Histogram plot of 'CreditCard' feature.

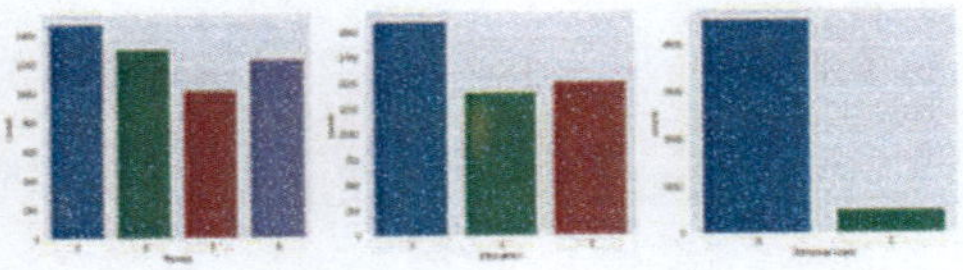

(b) Histogram plot of 'Family', 'Education' and 'Personal Loan' features.

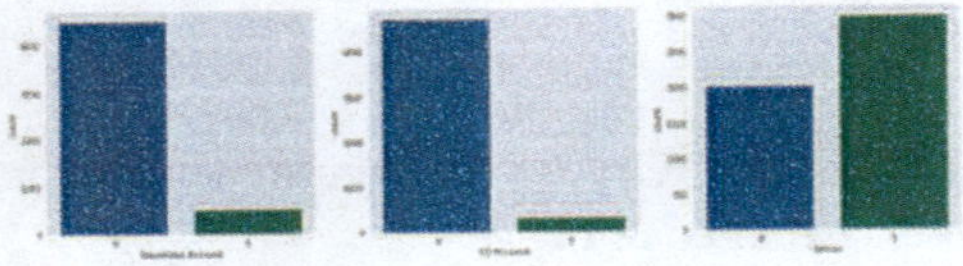

(c) Histogram plot of 'Securities Account', 'CD Account' and 'Online' features.

Fig. 2 Histogram plot of categorical features

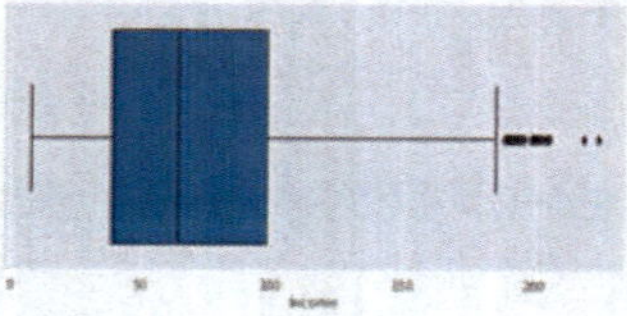

(a) Box plot of 'Income' feature.

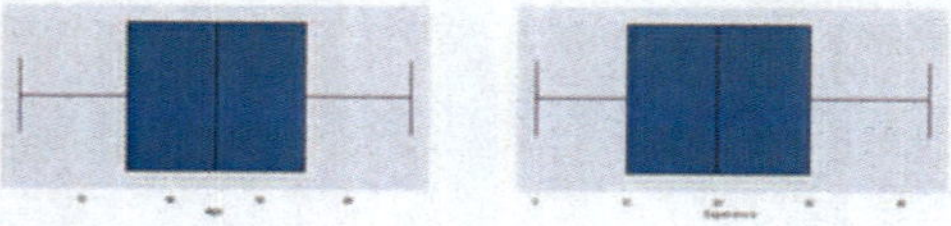

(b) Box plot of 'Age', 'Experience' features.

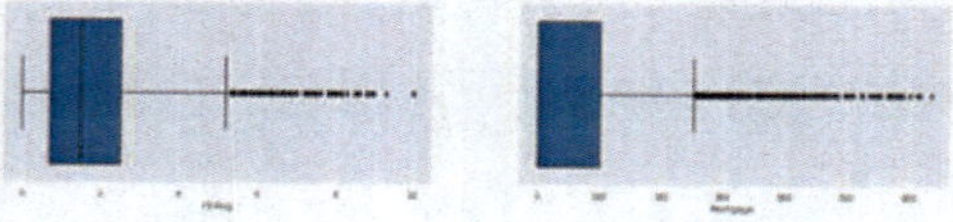

(c) Box plot of 'CCAvg', 'Mortgage' features.

Fig. 3 Box plot of categorical features

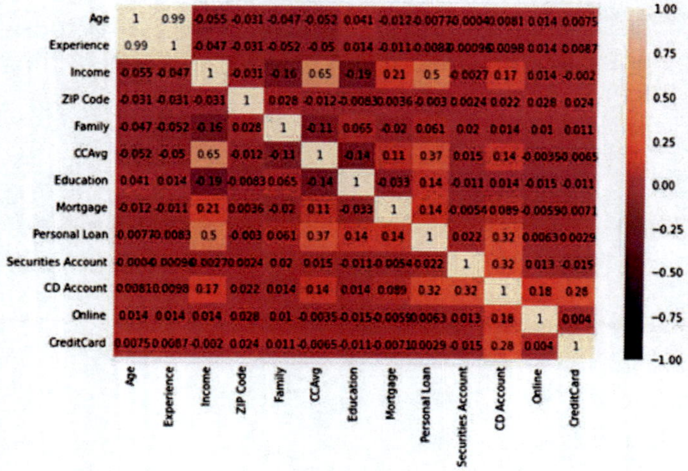

Fig. 4 Correlation matrix

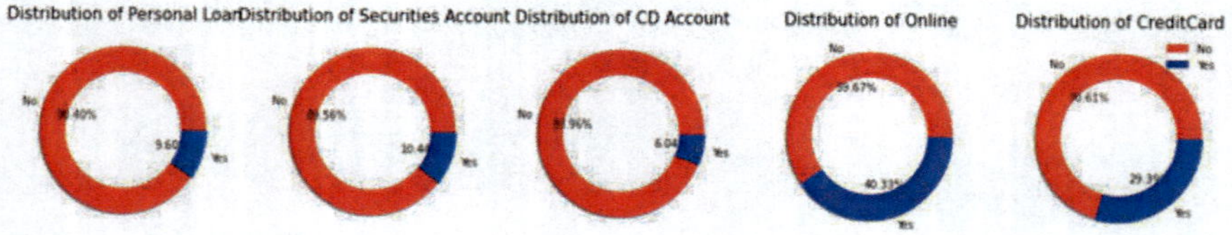

Fig. 5 Percentage distribution

'Income' and 'CCAvg' are bearing a high correlation. Percentage distribution as a plot is explored in Fig. 5.

Pie plots for mortgage and spending on credit cards per month are shown in Fig. 6.

Acceptance of loan based on the pair-wise feature distribution is explored in Fig. 7.

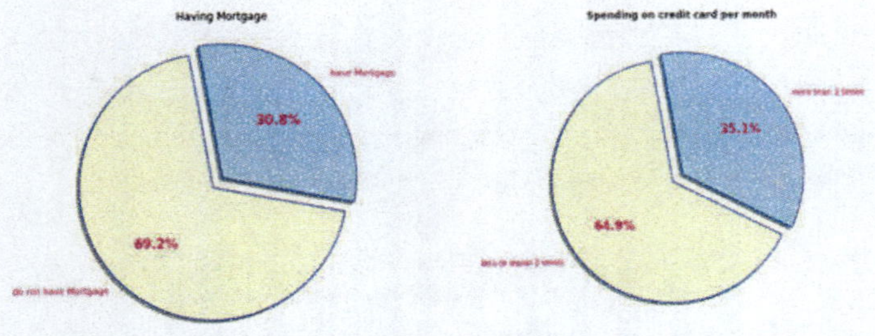

(a) Mortgage pie plot (b) Credit card spending pie plot

Fig. 6 Pie plots for mortgage and spending on credit cards per month

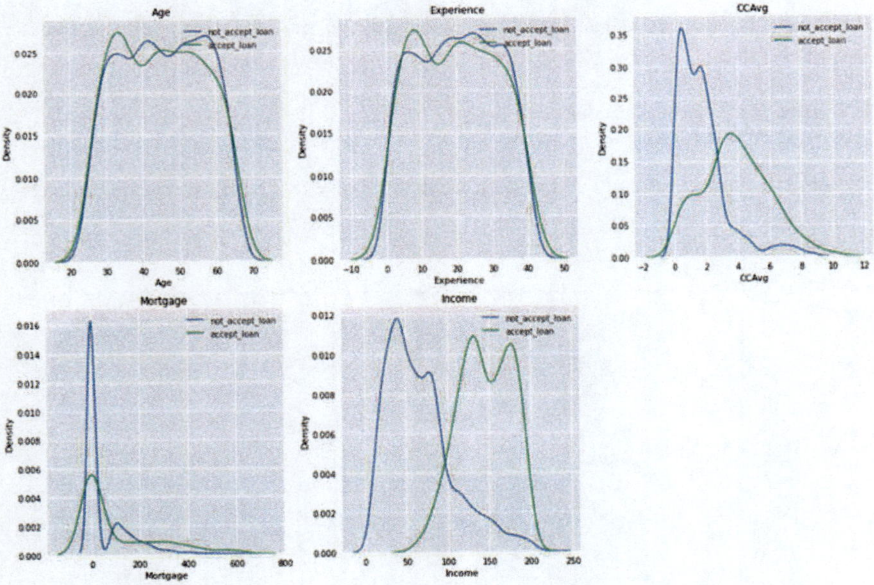

Fig. 7 Pair distribution for loan status information

The personal loans taken with respect to age are explored in Fig. 8. It is seen that people who have just entered the age of 30 and people with age 60 generally do not take a personal loan.

The personal loans taken with respect to income are explored in Fig. 9. It is seen that people who have higher incomes generally take a personal loan and not vice versa.

The personal loans taken with respect to experience are explored in Fig. 10. It is seen that people who have less experience generally take a personal loan and with experience, the loan-taking ratio lessens.

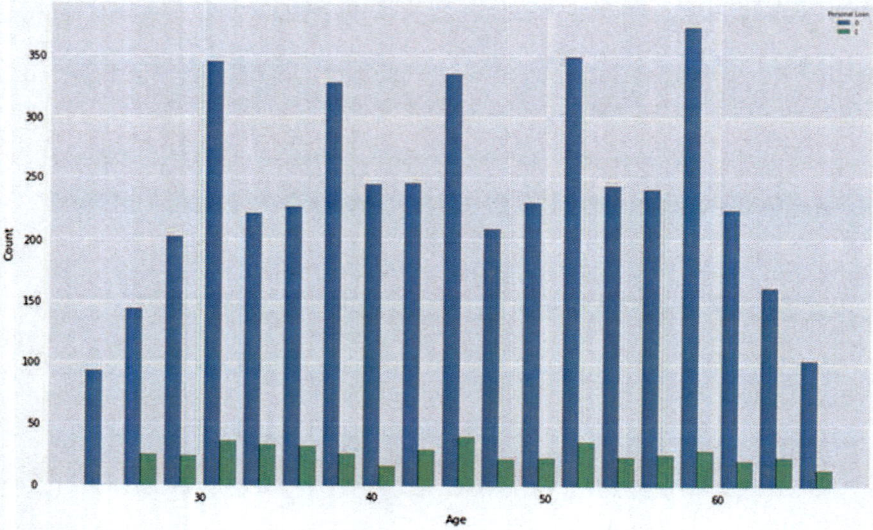

Fig. 8 Age-wise personal loan distribution

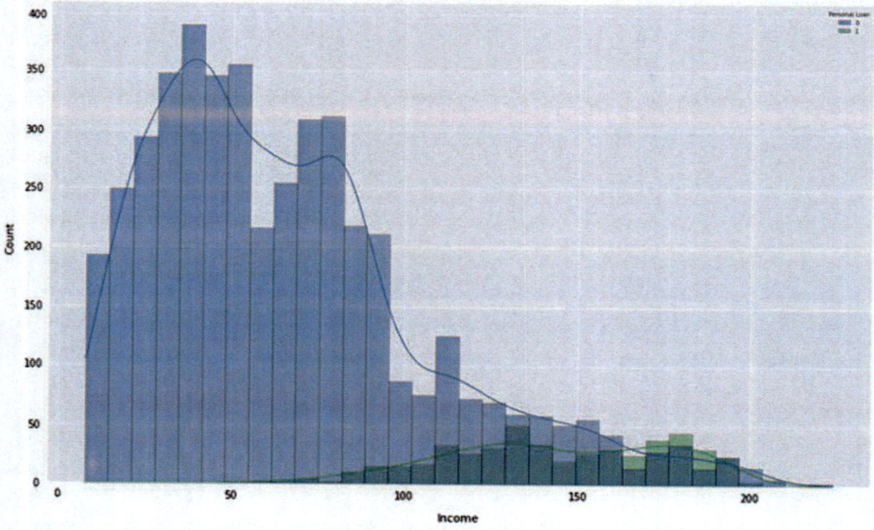

Fig. 9 Income-wise personal loan distribution

The personal loans taken with respect to family size are explored in Fig. 11. It is seen that people who are alone taking a personal loan and once the family size grows to 4, i.e. have 2 children, personal loans are reapplied for at a faster rate.

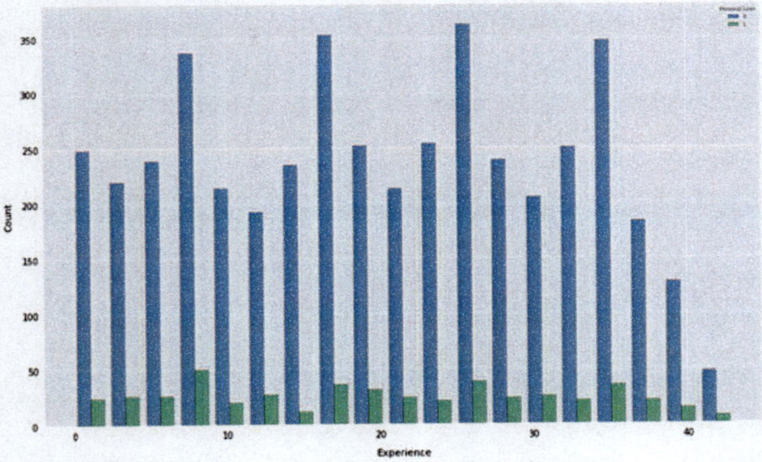

Fig. 10 Experience wise personal loan distribution

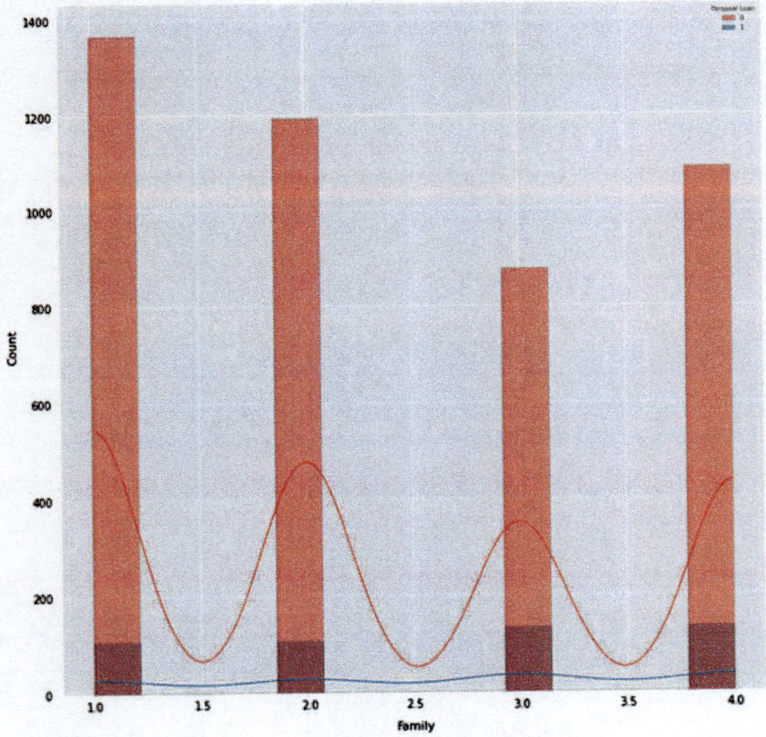

Fig. 11 Family size wise personal loan distribution

4.3 Machine Leaning Results

After the preprocessing steps and gathering statistical information about the data, the data set is subjected to the machine learning algorithms discussed in the previous section.

Tables 1, 2 and 3 show the confusion matrix [27] computed for the Logistic Regression, KNN, and Naive Bayes. After a 10 cross-validation, the average precision, recall, accuracy, and F1-score for every model is shown in Table 4. It is seen that the Decision Tree clearly outperforms the other models.

Table 1 Confusion matrix for logistic regression

Actual	Predicted	
	Yes	No
Yes	4392	127
No	331	149

Table 2 Confusion matrix for KNN

Actual	Predicted	
	Yes	No
Yes	4496	23
No	452	28

Table 3 Confusion matrix for Naive Bayes

Actual	Predicted	
	Yes	No
Yes	4222	297
No	189	291

Table 4 Performance assessment using tenfold cross-validation

Model	Avg. precision	Avg. recall	Avg. accuracy	F1-score
LR	0.55	0.33	0.92	0.41
KNN	0.64	0.07	0.90	0.13
NB	0.49	0.60	0.90	0.54
DT	0.60	0.66	0.98	0.63

5 Conclusion

Credit scoring, used to determine risk, is based on artificial intelligence models created by experts and academics from all over the globe. The study surveyed previously published materials. Various machine learning techniques have been utilized in the past, including decision trees, KNN, logistic regression, and Naive Bayes. The accuracy and usefulness of machine learning algorithms vary depending on the task at hand. It was also pointed out that although many studies focused on improving accuracy in loan default prediction, relatively few looked at the repercussions of false negatives, which might have severe effects on financial organizations. Therefore, the next generation of academics should tackle the problem of loan lending prediction by placing more weight on the possibility of false negatives.

References

1. Roh Y, Heo G, Whang SE (2019) A survey on data collection for machine learning: a big data-ai integration perspective. IEEE Trans Knowl Data Eng 33(4):1328–1347
2. Ngiam KY, Khor W (2019) Big data and machine learning algorithms for health-care delivery. Lancet Oncol 20(5):e262–e273
3. Sarker IH (2021) Machine learning: algorithms, real-world applications and research directions. SN Comput Sci 2(3):1–21
4. Ray S (2019). A quick review of machine learning algorithms. In: 2019 International conference on machine learning, big data, cloud and parallel computing (COMITCon). IEEE, pp 35–39
5. Gambella C, Ghaddar B, Naoum-Sawaya J (2021) Optimization problems for machine learning: a survey. Eur J Oper Res 290(3):807–828
6. Gilchrist D, Yu J, Zhong R (2021) The limits of green finance: a survey of literature in the context of green bonds and green loans. Sustainability 13(2):478
7. Sanchís Pedregosa C, Berenguer E, Albort-Morant G, Anton Sanz J (2020) Guaranteed crowdlending loans: a tool for entrepreneurial finance ecosystem sustainability. Amfiteatru Economic 22(55):775–791
8. Kordzadeh N, Ghasemaghaei M (2022) Algorithmic bias: review, synthesis, and future research directions. Eur J Inf Syst 31(3):388–409
9. Naim A (2022) Role of artificial intelligence in business risk management. Amer J Bus Manag Econ Bank 1:55–66
10. Mhlanga D (2021) Financial inclusion in emerging economies: the application of machine learning and artificial intelligence in credit risk assessment. Int J Financ Stud 9(3):39
11. Chitty R, Gunawikrama K, Fernando H (2022) Development of loan default prediction model for finance companies in Sri Lanka-a case study. In: 2022 international conference on data science and its applications (ICoDSA). IEEE, pp 103–108
12. Dosalwar S, Kinkar K, Sannat R, Pise N (2021) Analysis of loan availability using machine learning techniques
13. Faisal SM (2021) Overview of the ARIMA model average crude oil price forecast and its implications on the Indian economy post-liberalization. Int J Multidiscip: Appl Bus Educ Res 2(2):118–127
14. Mor S, Aneja R, Madan S, Gupta S, Artificial intelligence and loan default: the case of commercial banks in India. Strategic Change
15. Aslam U, Tariq Aziz HI, Sohail A, Batcha NK (2019) An empirical study on loan default prediction models. J Comput Theor Nanosci 16(8):3483–3488

16. Pérez-Martín A, Pérez-Torregrosa A, Rabasa A, Vaca M (2020) Feature selection to optimize credit banking risk evaluation decisions for the example of home equity loans. Mathematics 8(11):1971
17. Shipe ME, Deppen SA, Farjah F, Grogan EL (2019) Developing prediction models for clinical use using logistic regression: an overview. J Thorac Dis 11(Suppl 4):S574
18. Bayar Y, Sezgin HF, Öztürk ÖF, Şaşmaz MÜ (2020) Financial literacy and financial risk tolerance of individual investors: multinomial logistic regression approach. SAGE Open 10(3):2158244020945717
19. Bhuyan BP, Singh TP (2022). Artificial intelligence in financial portfolio management. In: Revolutionizing business practices through artificial intelligence and data-rich environments. IGI Global, pp 108–139
20. Bhuyan BP, Prasad A (2022). Statistical distribution and socio-economics in accordance with the Indian stock market in the COVID19 scenario. In: International conference on intelligent vision and computing. Springer, Cham, pp 193–206
21. Zhang S (2020) Cost-sensitive KNN classification. Neurocomputing 391:234–242
22. Chen S, Webb GI, Liu L, Ma X (2020) A novel selective naïve Bayes algorithm. Knowl-Based Syst 192:105361
23. Kumar P, Choudhury T, Rawat S, Jayaraman S (2016) Analysis of various machine learning algorithms for enhanced opinion mining using twitter data streams. In: 2016 international conference on micro-electronics and telecommunication engineering (ICMETE), Ghaziabad, India, pp 265–270. https://doi.org/10.1109/ICMETE.2016.19
24. Jayaraman S, Choudhury T, Kumar P (2017) Analysis of classification models based on cuisine prediction using machine learning. In: International conference on smart technologies for smart nation (SmartTechCon), Bengaluru, India, pp 1485–1490. https://doi.org/10.1109/SmartTechCon.2017.8358611
25. Choudhury T, Dangi G, Singh TP, Chauhan A, Aggarwal A (2018) An efficient way to detect credit card fraud using machine learning methodologies. In: 2018 second international conference on green computing and internet of things (ICGCIoT), Bangalore, India, pp 591–597. https://doi.org/10.1109/ICGCIoT.2018.8753077
26. Charbuty B, Abdulazeez A (2021) Classification based on decision tree algorithm for machine learning. J Appl Sci Technol Trends 2(01):20–28
27. Luque A, Carrasco A, Martín A, de Las Heras A (2019) The impact of class imbalance in classification performance metrics based on the binary confusion matrix. Pattern Recogn 91:216–231

Security-Aware Efficient Multi-tenant Cloud Environment

Anuj Kumar Yadav, M. L. Garg, and Ritika

1 Multi-tenancy in Cloud Computing

Multi-tenancy is a significant aspect of the service-oriented computing paradigm, including infrastructure, platform, and application as a service [1]. The idea of sharing virtual cloud infrastructure with system servers supports access control and resource availability. Multi-tenancy-based architecture includes single-tenant and multiple tenants that use unique resources or share the storage, computing, and network resources [2]. An individual user is considered a tenant of cloud infrastructure. In a multi-tenant environment, sharing of resources relates to software sharing and shared hosting on servers inside the data center. In a multi-tenant supported data center node, each tenant is physically connected but cognitively isolated [3]. It enables a system to work in a shared environment in which multiple users can access the cloud services at the same time with minimum SLA violations and minimum power consumption with high resource availability. The concept of virtualization supports the cloud computing environment [4] and acts as a backbone in a multi-tenant environment with high resource reliability and availability.

1.1 Types of Tenancy

Cloud services can be accessed in a single-tenant as well as a multi-tenant environment. In a single-tenant environment, resources are assigned to the end users uniquely. In contrast, in a multi-tenant environment, multiple users share the resources which

A. K. Yadav (✉) · M. L. Garg
DIT University, Dehradun, India
e-mail: anujbit@gmail.com

Ritika
DBUU University, Dehradun, India

A. Ramdane-Cherif et al. (eds.), *Machine Intelligence and Data Science Applications*, Algorithms for Intelligent Systems, https://doi.org/10.1007/978-981-99-1620-7_33

include storage, computing, and network resources [5]. Cloud computing environment with a single tenancy, each tenant has their own database and instance of the software, but in multi-tenancy, tenants are free to share their software applications and databases. The application segregates the configuration and data in a multi-tenant architecture to ensure each tenant works using a tailored virtual application instance [6].

The functionality of the banking system is an example of multi-tenancy. Money can be stored by multiple people in the same bank, but the customers cannot access another's account information, and various customers are unaware of each other's account information.

1.2 Benefits of Multi-tenancy

Multi-tenancy provides various benefits to cloud users, including developers and end users; key benefits of multi-tenancy are as follows:

Cost-Effectiveness: While developing a SaaS module in a service-oriented architecture, development is a major concern. The multi-tenancy supports the cost-effectiveness and operational efficiency improvement by allowing multiple clients or customers to share databases, resources (storage, computing, and network), and applications [7].

Ease in maintaining applications: The multi-tenant application architecture is intended to make software application maintenance easier. The modules can be customized with little effort, and no changes to the code are required. This also dramatically lowers the maintenance costs borne by tenants during the software update.

Quick deployment: Multi-tenancy makes it simple to add customers by reducing the amount of manual workforce required to get the service up and running and accelerating the time-to-value for the end-user [8].

Effective resource utilization: One of the most prominent features of a multi-tenant application is its ability to make optimal use of available resources. Multi-tenant has the advantage of being able to share infrastructure and resources, which allows the maintenance and utilization of resources automated. In addition, the downtime and idle time of the cloud resources are significantly minimal.

Efficiency and flexibility: A SaaS delivery model can run one instance of its program on one database instance and provides several customers with online access. The data of each tenant is isolated from other tenants.

Ease in customization: Multi-tenancy provides an additional layer that allows adapting the changes while preserving an underlying codebase. It is consistent for all users, including new clients, with multi-tenancy-based applications.

1.3 Limitations of Multi-tenancy

The Cloud has several multi-tenancy issues. The following section focuses on some significant limitations of the Multi-tenant environment.

Complex architecture: A multi-tenant environment follows a more complex architecture than a single-tenant application. In a single-tenant architecture, one does not have to write any code to detect which tenant a web request is intended for or prevent data leakage among the tenants. Some challenging issues are more critical in multi-tenant architectures.

Performance-Related Issue: The system's performance slows down, and resources may be over-utilized. It happens because the multi-tenant users share the environment excessively using unlimited computing power, which may slow down the performance of the virtual cloud environment.

Security-Related Issues: In a multi-tenancy-supported cloud environment, data is stored in third-party systems at a remote location. The security concern is there because multiple tenants put the requests on the same instance inside the data center node. The data may have requested by the users may be in flight mode or rest mode. Furthermore, security issues or corrupted data from one tenant could propagate to other tenants on the same server [9]. The cloud service providers need to ensure the isolation of instances in a multi-tenant environment. The isolated instance needs to deal with security principles at the service layers. Our presented work focuses on the efficient use of cloud environments with multi-tenant. Section 2 describes the complete details of the secure multi-tenant environment using legal consignment parameters between tenants and cloud service providers.

2 SLA Aware Multi-tenant Secure Cloud Environment

Cloud computing provides a shared pool of storage, computing, and network resources. The users put their requests from different geographical regions. Resource availability is the primary concern of the service providers at the end of Cloud computing consumers. The resource availability at the user end can be measured using legal agreement among the entities [10]. The legal consignment supports the secure multi-tenant cloud environment. The secure multi-tenant environment supports the quality of service at the level of the data center.

2.1 SLA and Role of SLA in Multi-tenant Cloud Environment

SLA is a contract between the cloud provider and the customer. It specifies the services that the cloud provider will deliver the services with high-quality metrics, time deadlines, security, and remuneration plan for the data centre utilization. It includes the parameters, such as uptime, downtime, and number of virtual machines, service assurance, and SLA violation compensation mechanism [11]. In case of SLA violations, the compensation needs to be paid. Some of the main qualities for SLA specification are service monitoring, billing method, network connection bandwidth, data security, data privacy, storage capacity, processing speed, etc. In order to provide consumers with a high level of service, SLA violations should be minimum [12, 13].

2.2 Constituents of Cloud-Based SLA

The Cloud-based SLA is created using several metrics and other essential elements. General Cloud-based SLA has the below-given components:

Purpose: It describes the reason behind the creation of SLA.

Service Guarantee: It describes all the facilities and their contracts that are provided by the Cloud service provider to the client.

Validity Period: It describes the validity period of the SLA specified at the time of the SLA negotiation.

Scope: It describes the scope of the service level agreement.

Limitation: This section of the SLA processes the limitations revealed in the SLA.

Service Level Objectives: service level objectives measure the objectives of the service level agreement.

Service Level Indicators: service level indicators are the components of the SLA which measure the current standing of the accomplishment of service level objectives mentioned in the SLA.

Exclusions: Exclusions are the constraints that are not covered under the Service Level Agreement.

Security Measures: This component deals with all the security primitives that are used to provide security services with high resource availability without any interruptions.

Figure 1 shows the layered architecture for implementing SLA measurement for optimal utilization in a multi-tenant environment. The SLA-aware policy is implemented at the level of the virtual machine, which ensures a better quality of service. The layered architecture is simulated using an integrated environment with Eclipse IDE and Cloudsim 4.0.

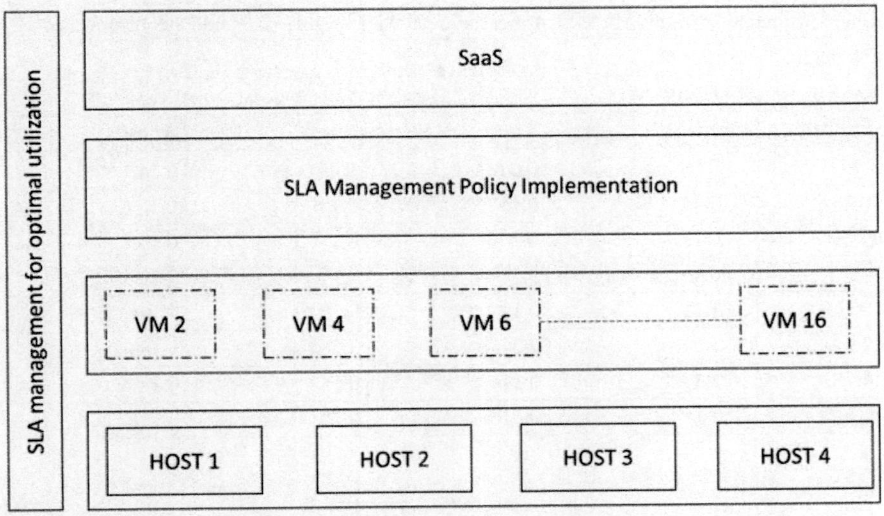

Fig. 1 Layered architecture of SLA-aware virtual cloud environment

2.3 *Experimental Setup for SLA-Aware Multi-tenant Environment*

The experimental setup includes an open-source simulator Cloudsim 4.0 integrated with Eclipse IDE. Cloudsim 4.0 provides a scalable scenario for implementing the process for a multi-tenant cloud environment. It supports the creation of different data centers, hosts, and virtual machines on available hosts. Table 1 illustrates the configuration parameters. It includes the four different configuration scenarios at the level of IaaS. Each scenario usages the PaaS layer with virtual machines 2–16 with the step 2. The virtual machine contains the homogeneous configuration parameters for all eight configurations with respect to the 4 IaaS configuration scenarios. Table 1 shows a clear picture of the configuration with the optimal value of service level agreement. There are a total of 8 scenarios that incorporate the variation of nodes at the level of IaaS and PaaS. The objective of the work includes finding the optimal configuration of IaaS PaaS, which have minimum violations of service level agreement.

Figure 2 illustrates the variation of the SLA violations with respect to the eight scenarios. The optimal value of SLA violation is 0.04% using scenario 1, in which the IaaS model includes 4 hosts and the PaaS model includes 2 virtual machines with homogeneous storage, computing, and network resources.

Optimality can be measured using the availability measurement parameter, i.e., SLA. This approach used not only provides QoS but also helps to secure the multi-tenant environment inside the cloud datacentre node. Further enhancement of the security in multi-tenant cloud environment virtual machines consolidation approach has been explored and implemented using an integrated environment with Cloudsim

Table 1 Simulation results of the SLA measurement with varying IaaS and PaaS parameters

Overall SLA violation %				
VM	Hosts			
	2 host	4 host	6 host	8 host
2	0	0.04	0.04	0.04
4	5.44	0.15	0.15	0.15
6	0.17	0.2	0.1	0.1
8	0.6	0.2	0.11	0.29
10	1.13	0.19	0.19	1.1
12	1.5	0.24	3.58	0.18
14	2.35	0.31	0.13	0.18
16	3.1	0.31	0.22	0.5

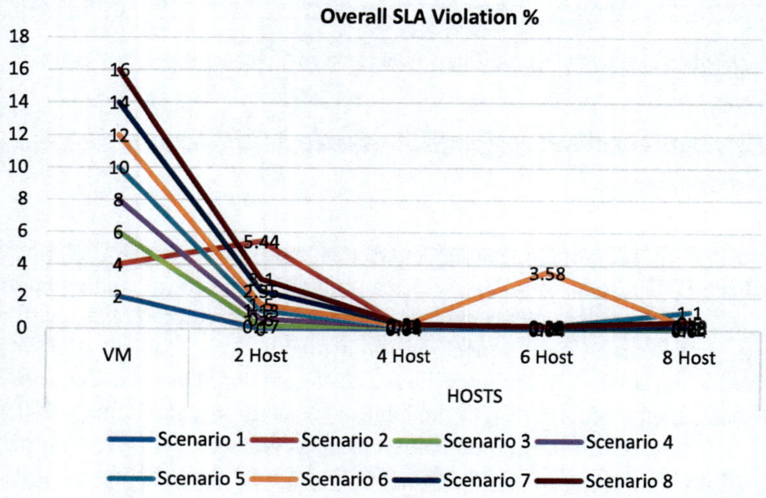

Fig. 2 Simulation results of the SLA measurement with varying IaaS and PaaS parameters

4.0 and Eclipse IDE. A detailed explanation of the virtual machine consolidation is shown in Sect. 3.

3 Virtual Machine Consolidation to Provide a Secure Cloud Environment

Virtual machine consolidation indicates minimum power consumption, maximum resource utilization, and improvement in the security level of the virtual machine instance with respect to power consumption. The functional relationship between power consumption security level enhancements with virtual machine consolidation is shown in Eq. 1.

$$\text{Power Consumption} = F(\text{Security Level}) \tag{1}$$

3.1 *Security-Aware Virtual Machine Migration Using Performance Matrices*

In large-scale Cloud Data Centers, virtualization effectively handles the increased demand for computing, storage, and communication resources. Through Virtual Machine (VM) migration, it assists in achieving various resource management [14] objectives such as power management and resource sharing. The virtual machine constantly demands suitable CPU cycles, cache memory, memory capacity, and communication bandwidth. The VM migration is a resource-intensive process that depends on the availability requirement of the services.

VM migration is famous because it keeps application services running while the migration takes place. The state-of-the-art methods of VM migration examines various methodologies. The review process results give some observations, i.e., security is one of the primary concerns when talking about the multi-tenant cloud computing environment. The rapid adoption of cloud computing has resulted in many challenges, including high energy consumption in Cloud Data Centers (CDCs) and new security risks to Virtual Machines (VMs) as a result of co-residency of user requests on VMs running on the same Physical Machine (PM). The virtual machine migration creates more vulnerability which tends toward security risks. The significant findings of the simulation environment are that we can improve cloud computing security in a multi-tenant environment with the help of some performance matrices, including power consumption and virtual machine migration. The security risk at the level of the virtual machine can be mitigated by using the minimum number of migrations. Cloud computing simulation related to multi-tenant security has been simulated based on cold and hot VM migration techniques. The cold migration technique migrates the powered-off virtual machines or adjourned virtual machines between different infrastructural elements. Other infrastructural elements include hosts, clusters, and data centers.

During cold workload migration, the virtual machine remains powered off for the entire migration duration. Hot migration is a different migration technique in which virtual machines remain powered during the migration process. Hot migration is also called live migration [15]. Hot migration is the migration technique that works during an active workload. The cold migration technique is less complex than the hot migration technique, but it can lead to more downtime as the machine needs to start up again to initiate the process. Hot migration also has limitations that include the risk of data corruption during migration due to hardware or software failure. Also, we can't perform hot migration between two computers having different CPUs.

Experimental simulation has been done using the following parameters for both scenarios. We can easily create different data centers and hosts and deploy virtual machines over available hosts with the minimum number of migrations.

To support a multi-tenant environment, optimal configuration parameters at the various level are as follows:

- We have used a minimum of 2 hosts and maximum of 8 hosts at the infrastructural level.
- The number of virtual machines varies from 2 to 16, which provides a PaaS module for successfully executing the tasks in a scalable manner.

The configuration parameters used at the host and virtual machines are as follows:

3.2 *Host Machine Configuration Parameters*

HP ProLiant ML110 G4 Xeon 3040 1860 MHz.

HP ProLiant ML110 G5 Xeon 3075 2660 MHz.

Host machines RAM Size = 4 GB.

Host MIPS range = 1660–2460.

3.3 *Virtual Machine Configuration Parameters*

Virtual machine MIPS range 500–2500.

Virtual machine RAM range 800–1700 MB.

Simulation has been done for the following two configuration scenarios:

- Comparison of Power consumption with Virtual Machine Migration
- Comparison of Vulnerability level with the number of Virtual Machine Migration.

3.4 *Comparison of Power Consumption with Virtual Machine Migration*

Table 2 given below shows the total energy consumption in kWh (Watt Hour) with variations in the number of hosts and number of virtual machines.

Hence Table 2 and Fig. 3 shown above exhibit that the minimum number of hosts power consumption matrices has been improved, but there is a maximum number of migrations that leads to security risks and increase the vulnerability level, compromising security. The same process can be repeated for an increasing number of hosts and virtual machines, but it improves the simulation time depending on the number of robust infrastructural configurations.

The simulation process is repeated 10–15 times with the mean values of the performance metrics.

Table 2 Power consumption in kWh with respect to the eight scenarios

Total energy consumption in kWh				
VM	Hosts			
	2 host	4 host	6 host	8 host
2	30	30	30	30
4	50	60	60	60
6	50	60	60	60
8	50	60	70	80
10	50	60	70	90
12	50	70	80	90
14	50	60	90	110
16	50	60	120	130

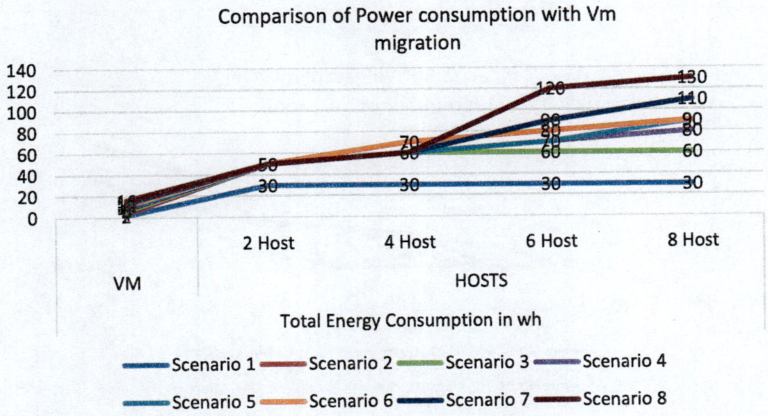

Fig. 3 Power Consumption in kWh with respect to the Eight Scenarios

3.5 *Comparison of Vulnerability Level with Number of Migration*

Table 3 and Fig. 4 show the total number of migrations that occurred with the variations in the number of hosts and the number of virtual machines in a scalable scenario.

Table 4 shows the parameters which are used to calculate the average number of migrations with respect to each of the host and virtual machine scenarios. The average number of migrations with respect to vulnerability levels (0, 1, 2, and 3).

Figure 5 illustrates that the level 0 vulnerability level is high due to the average maximum number of migrations. During the migration process, Infrastructure as a service configuration includes 2 host machines. At level 2, the vulnerability level is minimum due to the minimum number of migrations of virtual machines from one node to another node inside the data center. The simulation results show that there is a high probability of vulnerability when the number of migrations increases, i.e.,

Table 3 Number of migrations with eight scenarios

No of VM migration				
VM	Hosts			
	2 host	4 host	6 host	8 host
2	0	1	1	1
4	4	6	6	6
6	10	5	4	4
8	11	5	8	13
10	15	5	5	15
12	17	17	9	16
14	30	31	18	21
16	34	36	17	27

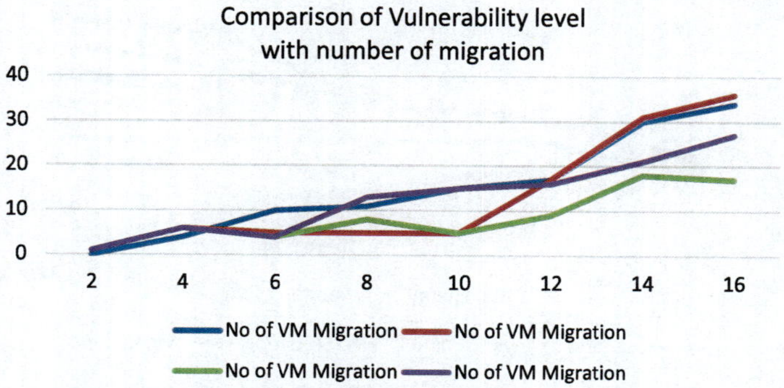

Fig. 4 Vulnerability versus number of migrations

Table 4 Infrastructure configuration variations with average migrations

Vulnerability level	Host configuration	Average migrations
0	2	8
1	4	7
2	6	4
3	8	6

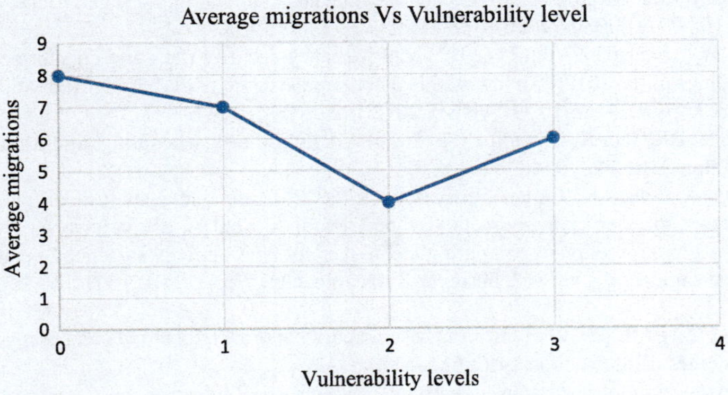

Fig. 5 Average migrations with vulnerability level

the movement of the virtual machines increases. So, virtual machine consolidation considers the power consumption and status of the source node and destination node before migration. Hence, this configuration setup performs better in all the tested scenarios after repeating the process 10–15 times.

4 Conclusion and Future Scope

Simulation results show that using the proposed approach results in security improvement, and the overall risk was decreased positive impact on energy consumption acting as the quality of service metrics. For the simulation, instead of using fix number of hosts and virtual machines, variations at the host and VM level have been performed as the number of hosts varies from 2 to 8, and the number of virtual machines varies from 2 to 16 in different scenarios. Hence, the outcomes indicate that the SLA Violation is minimum while using 4 hosts and 2 virtual machines, the Vulnerability level is minimum while using 6 hosts, and Energy Consumption is minimum in the case of 2 host machines. In future work, the above scenarios can be evaluated and tested using EC2 services.

References

1. AlJahdali H, Albatli A, Garraghan P, Townend P, Lau L, Xu J (2014) Multi-tenancy in cloud computing. In: IEEE 8th international symposium on service oriented system engineering, pp 344–351. https://doi.org/10.1109/SOSE.2014.50
2. Singh BK, Danish M, Choudhury T, Sharma DP (2021) Autonomic resource management in a cloud-based infrastructure environment. In: Choudhury T, Dewangan BK, Tomar R, Singh BK, Toe TT, Nhu NG (eds) Autonomic computing in cloud resource management in Industry 4.0. EAI/Springer Innovations in Communication and Computing. Springer, Cham. https://doi.org/10.1007/978-3-030-71756-8_18
3. Brown W, Anderson V, Tan Q (2012) Multi-tenancy-security risks and countermeasures. In: Proceedings of the 2012 15th international conference on network-based information systems, pp 7–13. https://doi.org/10.1109/NBiS.2012.142
4. Dewangan BK, Jain A, Choudhury T (2020) AP: hybrid task scheduling algorithm for cloud. Rev d'Intell Artif 34:479–485
5. Bhardwaj P, Lohani K, Tomar R, Srivastava R (2022) Comparative analysis of traditional and cloud-based disaster recovery methods. In: Tripathi A, Soni A, Shrivastava A, Swarnkar A, Sahariya J (eds) Intelligent computing techniques for smart energy systems. Lecture notes in electrical engineering, vol 862. Springer, Singapore. https://doi.org/10.1007/978-981-19-0252-9_11
6. Yadav A, Garg ML (2019) Monitoring based security approach for cloud computing. Ingénierie des Systèmes d'Information 24(6):611–617
7. https://www.cloudflare.com/en-in/learning/cloud/what-is-Multi-tenancy/. Accessed 12 May 2022
8. https://rickscloud.com/multi-tenancy-in-the-cloud-understanding-its-benefits/. Accessed 15 June 2022
9. https://www.ibm.com/docs/en/cognos-analytics/11.1.0?topic=guide-multitenantenvironments. Accessed 18 June 2022
10. Li L, Dong J, Zuo D, Wu J (2019) SLA-aware and energy-efficient VM consolidation in cloud data centers using robust linear regression prediction model. IEEE Access 7. https://doi.org/10.1109/ACCESS.2019.2891567,9490-9500
11. Singh S, Chana I, Buyya R (2020) STAR: SLA-aware autonomic management of cloud resources. IEEE Trans Cloud Comput 8(4):1040–1053. https://doi.org/10.1109/TCC.2017.2648788
12. Singhal A, Sarishma, Tomar R (2016) Intelligent accident management system using IoT and cloud computing. In: 2016 2nd international conference on next generation computing technologies (NGCT), Dehradun, India, pp 89–92. https://doi.org/10.1109/NGCT.2016.7877395. (2015) Technology Convergence (ICTC), pp 500–502. https://doi.org/10.1109/ICTC.2015.7354595
13. Choi Y, Lim Y (2015) Resource management mechanism for SLA provisioning on cloud computing for IoT. In: International conference on information and communication
14. Choudhury T, Dewangan BK, Tomar R, Singh BK, Toe TT, Nhu NG (2021) Autonomic computing in cloud resource management in Industry 4.0. Spring
15. Yadav AK, Ritika MG (2021) Cryptographic solution for security problem in cloud computing storage during global pandemics. Int J SafSecur Eng 193–199

Hybrid Approach for Classification of Medical Imaging Data Using Machine Learning Techniques

Arun Kumar Gautam and Shikha Verma

1 Introduction

In recent scenario the impact of COVID-19 has put the whole medical system on verge of developing algorithm which may combine the medical terminology as well as the Machine Learning technique and Artificial Intelligence. In the context of these the processing of medical imaging has been an immense task. Chest X-Ray images has played a vital role in detecting the presence of COVID-19 virus. These medical imaging require meticulous approach to correctly classify in the corresponding classes. Recently, classification models that can categorize and automatically interpret x-ray pictures have been created using deep learning techniques. These techniques identify the complex patterns of the x-rayed objects before classifying them as normal or abnormal [1].

2 Related Work

A two-step classification approach has been described in Ref. [2]. The approach were implemented with two set of Dense Net Classifier. These two sets of dense networks are formed in a stepwise fashion to address the issues of body part identification and forecasting the normality or abnormality of the x-ray image.

In Ref. [3] to establish a baseline for the final proposed custom model, CovCNN 4, three deep learning architectures—CovCNN 1, CovCNN 2, and CovCNN 3—were used in the development of the models. A variety of convolution models were combined with effective activation layers in the suggested model to facilitate enhanced learning for identifying and classifying elite patterns. To attain the best

A. K. Gautam · S. Verma (✉)
Department of Computer Science, Ram Lal Anand College, New Delhi, India
e-mail: shikhaverma@rla.du.ac.in

A. Ramdane-Cherif et al. (eds.), *Machine Intelligence and Data Science Applications*, Algorithms for Intelligent Systems, https://doi.org/10.1007/978-981-99-1620-7_34

accuracy, the data has also been examined using various transfer learning models that have already been trained. The ROC curve, training loss, and validation loss for both bespoke and pre-trained models are used to display the results.

In Ref. [3], the utilization of several layers, shared weights, local connections, and pooling are the four criteria used to categorize the data processing carried out by CNN. Pre-trained models have been utilized using transfer learning. Two convolutional layers, two pooling layers, a flatten layer, and two dense output layers make up the straightforward structure the authors utilized in this paper.

In Ref. [4], three key experiments make up the technique that has been utilized to gauge the effectiveness of the models and gauge the impact of the various stages of the process. Every experiment uses the same photos for COVID-19 positive cases in all instances when they were used. Three separate datasets for negative cases were employed in this methodology. Experiments 1 and 2 evaluate positive vs. negative case datasets, and Experiment 3 uses photos from the Pre-COVID era.

In Ref. [5], using deep learning techniques, automatic chest X-ray analysis has advanced significantly. Deep models' data driven characteristic necessitates training data with a wide distribution in order to function. Therefore, it is important to combine information from several sources, particularly for medical imaging. Learning a disease classification model with additional Chest X-ray (CXR) data is still difficult, though. Recent studies have shown that joint training on several CXR datasets suffers from a performance bottleneck, but few have attempted to overcome the problem. We contend in this study that using an external CXR dataset produces inaccurate training data, which increases the difficulties. In particular, there are two types of incomplete data: domain discrepancy, which arises from the fact that picture appearances vary among datasets, and label discrepancy, which results from the partial labelling of various datasets. To achieve this, we design the multi-label classification issue for thoracic diseases as a set of weighted independent binary tasks based on the categories. To reduce feature disparities for common categories that are shared across domains, we use task-specific adversarial training. To further mine the data from the missing labels for categories that exist in a single dataset, we provide uncertainty-aware temporal ensembling of model predictions. By modelling and addressing the domain and label conflicts simultaneously, our approach enhances knowledge mining capabilities. On three da-tasets containing more than 360,000 chest X-ray pictures, we undertake extensive tests. Our approach performs better than other rival models.

3 Proposed Work

a. Dataset description

The Chest X-ray dataset has been deployed from Official Website i.e. Kaggle website (Table 1). Source:-https://www.kaggle.com/datasets

https://www.kaggle.com/paultimothymooney/chest-xray-pneumonia

Table 1 Description of the dataset

	Testing data	Training data	Validation data
Normal	234 Images	1341 Images	8 Images
Pneumonia	390 Images	3905 Images	8 Images
Dataset	*Features*		*Class*
Chest X-ray Phenomena	270		Binary Class Values Normal Pneumonia 2 Categories
	Histogram features—256 Contrast-1 Other features (MFCC)-13		

b. **Phases of proposed work**

The proposed model architecture aims to classify the medical imaging data. With a hybrid approach. There are two phases in the model architecture. Phase one: where the X-Ray images are classified using the supervised machine learning algorithms. The next phase comprises of improving the prediction results by re classifying it using 5 different supervised machine learning algorithms. This approach seeks to make sure that the classes predicted incorrectly are further tackled in the subsequent phase and are classified again using a different model.

Phase 1

The classification of the entire medical image dataset is the first step of phase one, and it is done using the KNN, Random Forest, Ensemble Learning, and Decision Tree classification algorithms. The dataset is first given to four distinct classification algorithms, which partition it into training and testing module. The algorithm then uses the test dataset to apply the model, which is then used to make predictions about the test set. The accuracy may be determined once the test set is predicted, and using this accuracy, a prediction can be generated for the entire dataset. The entire dataset receives the new label as a result of this process.

Phase 2

In Phase-2 the output of the predictions or the learnings from these phase-1 predictions are carried in this phase where the output of the earlier one serves as the input here. All the X-Ray images for both normal and pneumonia are divided into 70:30 ratio for training and testing. In this phase the Random forests, ensemble learning, Nave Bayes, and KNN, The classes have been divided into training and test groups, respectively for both the phases. By using the four machine learning algorithms provided, a model was created, predictions were made using test data, and accuracy was calculated as a result. The model's accuracy and prediction have increased compared to phase one's accuracy and prediction. In comparison to the base classifier, the hybrid classification model consistently produces results with higher accuracy.

c. **Algorithm for hybrid approach**

Phase 1:

1. Let i={i1, i2, i3……i270}

   where i1, i2, i3 and so on are the defines the features or the attributes of the dataset
2. Let J be the output class

   J= {0, 1}

   Where pneumonia=1 and normal=0
3. S={{i1,j1},{i2,j2} {i3,j3}…….{im,jm}} where m=1….5216 x-ray images.
4. Train(T1)= {{i1,j1}…….{ik,jk}} where k are randomly selected records from S using the method of Train test split of Decision Tree Algorithm/Random Forest/Ensemble Learning and K-Nearest neighbor(KNN)
   a. ITrain={i1…..ik}
   b. JTrain={j1….jk}
5. Model(T1)

   where model is the implementation of all the four Classification Algorithm.
6. Test(T2)= {{i1,j1}…….{in,jn}} where n are randomly selected records from S using the method of Train test split of implemented Algorithms
   a. ITest={i1…..in}
   b. JTest={j1….jn}
7. out=Predict (ITest) where out is the output of the prediction applied on ITest Data.
8. Calculation of Performance Measure Accuracy

   Accuracy= {out, Test}
9. out=Predict(I)

   New set of prediction for whole dataset with above accuracy.

Phase 2:

1. Let Class 1:- Pneumonia =(S1 for all J1}and Class 2:-Normal=(S2 for all J0}
2. Repeat step 5 to 8 for each classified model applied on class 1 and class 2 separately with Decision Tree Random forest, Ensemble Learning, Naïve Bayes and K-Nearest Neighbor (KNN) Algorithms.

4 Results and Observations of Chest X-ray Pneumonia Dataset

Model-1					
Step 1	Decision Tree	Decision Tree	Decision Tree	Decision Tree	Decision Tree
	97.194%	97.194%	97.194%	97.194%	97.194%
Step 2	Decision Tree	Random Forest	Ensemble	Naïve Bayes	KNN
	99.322%	97.509%	95.782%	85.344%	96.421%

Model-2					
Step 1	Ensemble	Ensemble	Ensemble	Ensemble	Ensemble
	95.463%	95.463%	95.463%	95.463%	95.463%
Step 2	Decision Tree	Random Forest	Ensemble	Naïve Bayes	KNN
	99.003%	96.264%	99.041%	88.122%	95.757%

Model-3					
Step 1	KNN	KNN	KNN	KNN	KNN
	95.421%	95.421%	95.421%	95.421%	95.421%
Step 2	Decision Tree	Random Forest	Ensemble	Naïve Bayes	KNN
	99.520%	98.371%	98.913%	52.490%	98.921%

Model-4					
Step 1	Random Forest	Random Forest	Random Forest	Random Forest	Random Forest
	97.605%	97.605%	97.605%	97.605%	97.605%
Step 2	Decision Tree	Random Forest	Ensemble	Naïve Bayes	KNN
	99.175%	98.754%	98.913%	86.590%	95.901%

5 Conclusion and Future Work

When the proposed model was applied to the Medical Imaging dataset, it was found that phase I produced a good classification results, with a maximum accuracy of 97.6% when the Random forest algorithm was used. The classification outcomes were improved further when other classification algorithms were applied to them. This hybrid approach enabled the succeeding phase to fix the incorrectly classified data from the original phase. The decision tree algorithm's Phase-2 iteration improved the outcome to 99.5%.

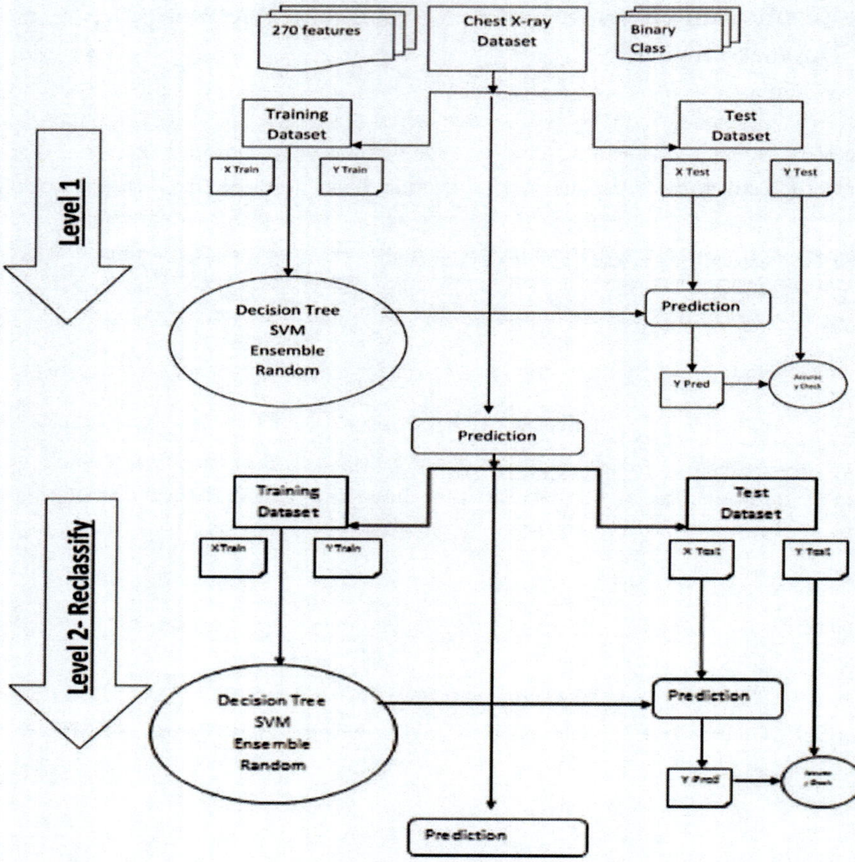

Fig. 1 Proposed architecture of the hybrid model

In the future, experiments might be set up using many phases in place of the hybrid approach, and alternative supervised learning techniques could be tested. The dataset sizes can be extended as well, allowing this technique to be applied to larger datasets in big data applications (Fig. 1).

References

1. Lu Y, Vincent N, Yuen PC, Zheng W-S, Cheriet F, Suen CY (Eds.), Pattern Recognition and Artificial Intelligence: International Conference, ICPRAI 2020, Zhongshan, China, October 19–23, 2020, Proceedings, vol. 12068. Cham: Springer International Publishing. doi: https://doi.org/10.1007/978-3-030-59830-3

2. "X-ray Image Classification Using Two-step DenseNet Classifiers|The 14th PErvasive Technologies Related to Assistive Environments Conference." https://dl.acm.org/doi/abs/https://doi.org/10.1145/3453892.3461632 (accessed 10 November 2022)
3. Sanket S, Vergin Raja Sarobin M, Jani Anbarasi L, Thakor J, Singh U, Narayanan S (2022) Detection of novel coronavirus from chest X-rays using deep convolutional neural networks. Multimed Tools Appl 81(16):22263–22288. doi: https://doi.org/10.1007/s11042-021-11257-5
4. Arias-Garzón D et al (Dec.2021) COVID-19 detection in X-ray images using convolutional neural networks. Mach Learn Appl 6:100138. https://doi.org/10.1016/j.mlwa.2021.100138
5. Luo et al. Deep mining external imperfect data for chest X-ray disease screening" ArXiv, Jun. 06 2020. doi: https://doi.org/10.48550/arXiv.2006.03796
6. "Machine Learning Techniques for Classification of Spambase Dataset. In: Proceedings of the 2019 3rd International Symposium on Computer Science and Intelligent Control. https://dl.acm.org/doi/abs/https://doi.org/10.1145/3386164.3389089 (accessed 01 December 2022)

Deep Learning Based World Coin Currency Detection

Zarak Jahan, Nazia Parween, Arun Prakash Agrawal, Ankur Choudhary, Gaurav Raj, and Aziz Deraman

1 Introduction

Identification of coin currency and a variety of applications, like the detection of fake coins, identifying ancient coins for collectors using smartphone cameras, etc. Coins can be recognized using 3 different techniques: electromagnetic recognition, mechanical recognition, and image processing. Electromagnetic recognition identifies the coin based on the material properties of the coin, on the other hand mechanical recognition utilizes the physical characteristics of the coin like its diameter, thickness, and width for recognition.

In image processing, coin recognition is done based on the engraving on the coins that are taken as the input image and compared with the already established coin currency database. The database includes images of coin currency from various countries captured at different angles and in different lighting environments using smartphone cameras or any other scanning instruments. Image processing is done using various digital image processing techniques like edge detection, fast Fourier transformation, discrete cosine transformation, feature extraction, image segmentation, and scale-invariant feature transformation.

The objective of this paper is to utilize Deep convolutional neural networks as they are highly successful in the domains of image classification [1], object tracking [2], object detection [3], and semantic image segmentation [4, 5]. In addition to CNN, we used an improved network model established on depth-wise dissociable filters called MobileNet-v2. The MobileNet-v2 network is used to extend the neural network and

Z. Jahan (✉) · N. Parween · A. P. Agrawal · A. Choudhary · G. Raj
Sharda University, Greater Noida, Uttar Pradesh, India
e-mail: zkhalid25@gmail.com

N. Parween
e-mail: naziaparween633@gamil.com

A. Deraman
University Malaysia Terengganu, 21030 Kuala Terengganu, Terengganu, Malaysia

A. Ramdane-Cherif et al. (eds.), *Machine Intelligence and Data Science Applications*, Algorithms for Intelligent Systems, https://doi.org/10.1007/978-981-99-1620-7_35

decrease the parameters and computation time and cost to be implemented in smart devices with low memory and lower computational complexity.

2 Literature Survey

In recent years, numerous researchers have donated to the field of recognition. Coin recognition was first given by Abdolah Chalechake in 2007, in which he used Spiral decomposition and Image abstraction.

Tajane et al. [6] proposed a system to identify Indian coin currency using deep learning. The methodology was employed using MATLAB and Raspberry Pi, where the system captured the image of the implanted coin and processed it using Raspberry Pi. A total of over 1600 sample images were used from 4 categories: 1, 2, 5, and 10 rupee coin currency. The model was trained on a variety of datasets using Alex-network transfer training, and the achieved accuracy was used to optimize the output system. The results obtained outshined the existing traditional methods.

In Zhang et al. incorporated MLP, SSD, and CNN for the recognition of currency. The 5, 10, and 20 NZD denominations were used for the research. 300 images were captured at a size of 1280 $\times$ 720 pixels, which was insufficient for deep learning. In order to overcome this, they used dataset augmentation and generated more novel datasets for training and testing. When comparing CNN, Ada-Boost, FNN, and PRFNN, it was discovered that FNN had the highest detection accuracy, achieving 96.6 per cent.

Laavanya et al. [8] employed the deep neural network method along with Alex-Net to identify and recognize fake currency apart from the original. They collected 100 images for individual currencies and increased the dataset count using augmentation. Transfer training of Alex-net with the Adam optimizer was proposed using MATLAB. Out of 100 images, 50 were captured, 50 were made through augmentation, and 50, 200, 500, and 2000 rupees were taken. The average accuracy obtained was 81.5% for original notes and 75% for fake notes.

Jadhav et al. [9] used deep learning to discriminate between fake currency notes from Saudi Arabia and India. A smartphone camera was used to obtain the images, and features were extracted using the proposed methodology to identify fake currency like discontinuity and dissimilarity features. The methods proposed were employed using MATLAB tools and were more efficient and inexpensive than the existing ones.

Ali et al. [10] utilized state-of-the-art Generative Adversarial Networks to distinguish counterfeit currency notes from real currency notes. In this process, Pakistani currency was used, which was augmented to create fake copies in order for to demonstrate high accuracy in identifying real paper currency. The GAN applied achieved an accuracy of 80%.

Van and Hoang [11] developed a process to distinguish currency notes by utilizing deep learning for the extraction of features and recognition. The dataset is comprised of images of Vietnamese dong, euro, and dollar paper currencies taken from fake notes. A total of 9,736 original notes and 1,083 fake notes were taken with a resolution of 200 × 450 pixels. The deep learning technique of convolutional neural networks was used for feature extraction and recognition and Support vector machines for classification. This model achieved a real-life application accuracy of 99.97%.

Navya et al. [12] created an automatic fake currency recognition system using CNN and VGG 16 to identify fake currency notes and detect their originality. The model was made in Python using Keras' deep learning library, and the data was taken through screenshots for the experiment, where it was discovered that there was a noticeable difference between counterfeit and real paper currency. To make it available to the people they advised to use it as a smartphone application.

Pham et al. [13] aimed at detecting the difference between fake bank paper currencies which were created by re-generating using general-purpose imaging instruments for people with perception impairments. The approach was developed using CNN, and the dataset samples were recorded using smartphone cameras in good-lighting environments. The dataset included Jordanian dinar, USD, Korean won, and Euro currencies, and the model outshined other deep learning methods in the detection of counterfeit (Table 1).

3 Deep Learning Fundamentals

Deep learning is a subgroup of artificial intelligence and is fundamentally used for the identification recognition of objects, image processing and cooperation rated with computer vision is used to develop algorithms for self-driving cars. Deep learning's architecture and functionality are encouraged by the biological brain, called artificial neural networks. Deep learning performs better than other learning algorithms as it excels at processing data that is analogue but is in large quantities [14], like pixel data of images, files of text data, and analogue data of audio.

3.1 Convolutional Neural Network

CNN is a deep neural network algorithm that contains a single or multiple convolutional layers that are proceeded by multiple fully connected layers. The 2D images are taken as input in CNN, and the features are extracted using weights and connections, followed by pooling layers with the help of kernels. A convolutional neural network has a lot of applications, and the most commonly used ones are object detection and image processing, as it learns directly from the input dataset without any human input for the extraction of features [15]. The architecture of CNN is shown in Fig. 1 [16], which is made up of multiple convolutional layers [17] that take images as input, followed by pooling layers, and then a fully connected neural network. The size of input images ranges from 100 KB to 1 MB, but in terms of pixel size, it depends upon the height and width of input images. In CNN, each convolutional layer has distinct functions associated with it.

Table 1 Summarized analysis of existing work

Authors	Year	Dataset	Analysis
Tajane et al. [6]	2018	Toy dataset	i. Utilized CNN i.e. AlexNet for training more than 1600 images and categorize them into 4 types
Zhang et al. [7]	2019	Standard as well as Toy dataset	i. Presented a summary of currency recognition ii. ii. Worked to increase the recognition accuracy already in place
Laavanya and Vijayaraghavan [8]	2019	Standard dataset	i. Developed a deep learning model to detect fake money ii. Used photos of Indian cash worth 50, 200, 500, and 2000 to spot false currency
Megha Jadhav et al. [9]	2019	Toy dataset	i. Developed a technique to identify counterfeit banknotes from various nations ii. The method operated on two pictures, one of which was the front side of the original image of the cash that was being checked and the other was the back
Toqeer Ali et al.	2019	Toy dataset	i. Deepmoney architecture was proposed as a way to discern between false and real currency ii. Used GAN to train the model on Pakistani currency iii. iii. Reached an 80% accuracy rate
Hoang and Vo	2018	Toy dataset	i. Created a model based on face values to distinguish between real and fraudulent currency ii. Used data augmentation and deep neural networks iii. Obtained a 99.97% accuracy rate
Navya et al. [12]	2019	Toy dataset	i. Developed a system to automatically distinguish between bogus and genuine money ii. The dataset consisted of images from the children's bank churan label

(continued)

Table 1 (continued)

Authors	Year	Dataset	Analysis
Pham et al. [13]	2020	Toy dataset	i. Proposed a technique to use the cameras on smartphones to identify counterfeit money, particularly for visually handicapped persons ii. Self-developed a dataset of a few nations' currencies to test the model's performance

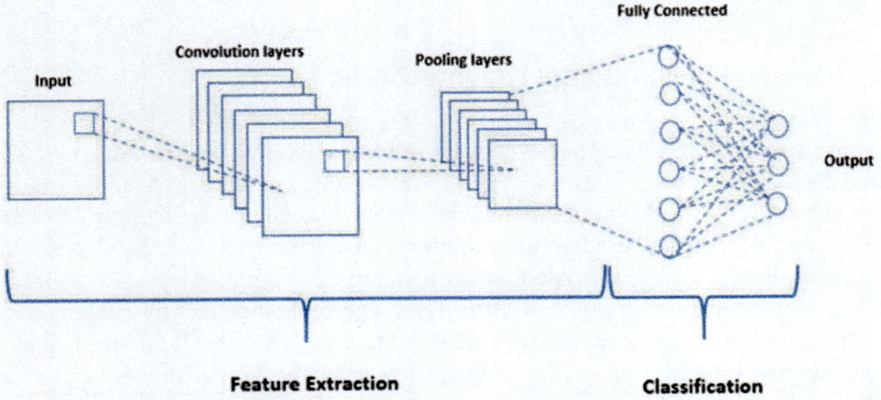

Fig. 1 CNN architecture [16]

3.2 *MobileNet V2*

MobileNet V2 is an up-to-date framework that utilizes depthwise divisible convolutions to develop lightweight deep convolutional neural networks which stipulate powerful models for smartphones and embedded computer vision utilization [18]. The framework of MobileNet V2 is given in Fig. 2 [19]. MobileNet V2 framework is built on two types of filters: depthwise convolution filters and point convolution filters, where the former executes a single convolution on dual input channel and the latter puts together the output of the previous filter in a continuous manner with 1*1 convolutions.

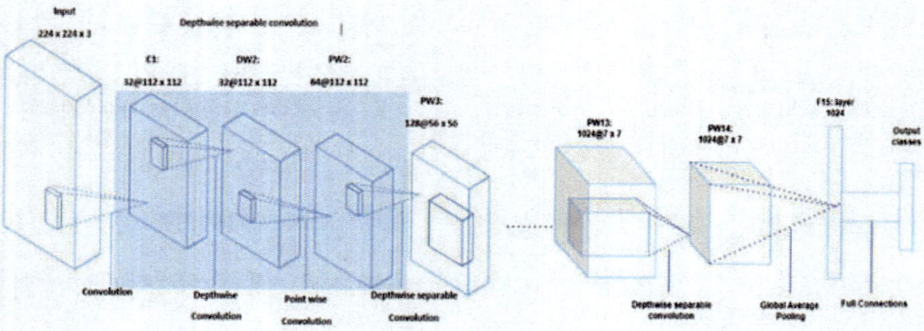

Fig. 2 MobileNet V2 architecture [19]

4 Methodology and Experimental Setup

In this paper, a convolutional neural network and MobileNet V2 classifiers are trained on the dataset in order to process images of the dataset and perform feature extraction accordingly. The deep learning implementation was done using Google's Colab GPU, and the techniques used were convolutional neural networks and MobileNetV2, which are discussed in Sect. 3. Along with this, many other libraries were also utilized in this project, like Keras and Tensorflow for image generation and for creating the layers of the neural network. The importance of using CNN is that it doesn't require manual input for image processing, which helps in saving computational costs and time.

5 Dataset

The dataset used in the model consists of images of coin currency of 211 types from multiple countries, like the British Pound, Canadian Dollar, Australian Dollar, Brazilian Real, etc., as shown in Fig. 3. The memory size of each coin image is 65 KB, and in terms of resolution, each pixel is 499 * 499.

The dataset was divided into a training dataset, a validation dataset, and a testing dataset, each containing 6494, 6494, and 7302 images, respectively, belonging to 211 classes. The input shape for the datasets was taken as 244*244*3. In addition to the dataset images, a JSON file was used to give names to each category of coin currency, as shown in Fig. 4.

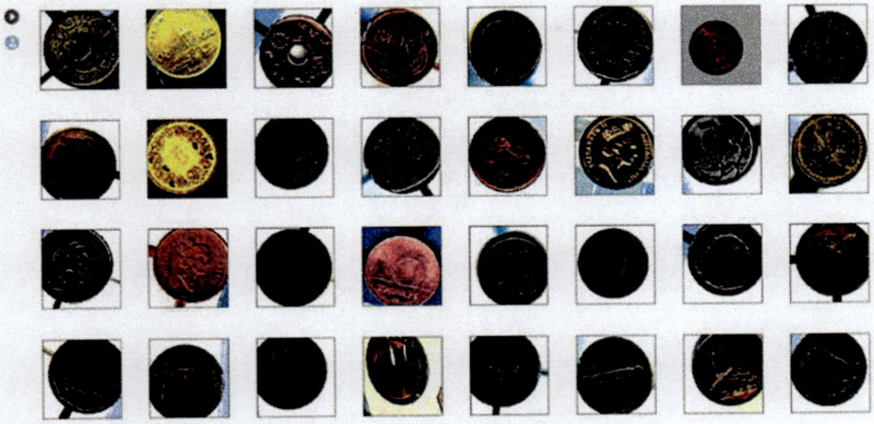

Fig. 3 Coin currency of different countries

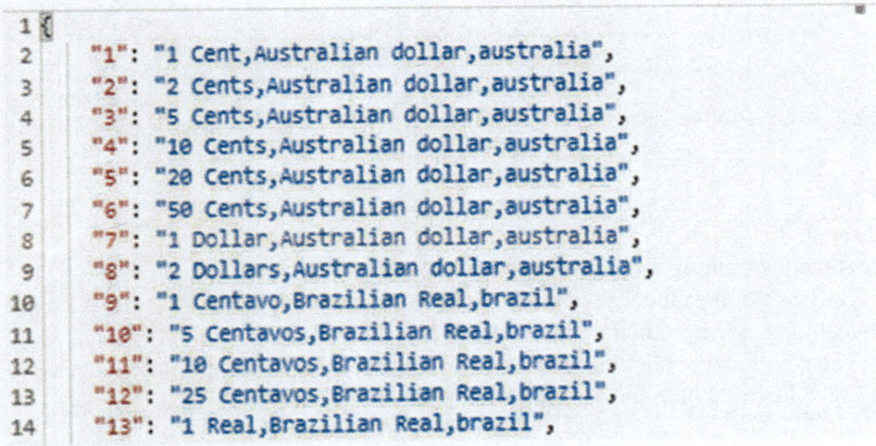


```
{
    "1": "1 Cent,Australian dollar,australia",
    "2": "2 Cents,Australian dollar,australia",
    "3": "5 Cents,Australian dollar,australia",
    "4": "10 Cents,Australian dollar,australia",
    "5": "20 Cents,Australian dollar,australia",
    "6": "50 Cents,Australian dollar,australia",
    "7": "1 Dollar,Australian dollar,australia",
    "8": "2 Dollars,Australian dollar,australia",
    "9": "1 Centavo,Brazilian Real,brazil",
    "10": "5 Centavos,Brazilian Real,brazil",
    "11": "10 Centavos,Brazilian Real,brazil",
    "12": "25 Centavos,Brazilian Real,brazil",
    "13": "1 Real,Brazilian Real,brazil",
```


Fig. 4 Image of JSON file

6 Result Analysis

The database was trained and tested with a batch size of 64 for a target size of 244 * 244. For the input layer in the neural network, a dense layer of 512 was utilized, along with Relu as an activation function and a dropout function of 0.8 for the input layer. Similarly, for the output layer, a dense layer was used along with Softmax as an activation function. The model was compiled using the Adam optimizer for 100 epochs, which resulted in a training accuracy of 86.87%, validation accuracy of 82.65%, and testing accuracy of 84.24%, as shown in Fig. 5.

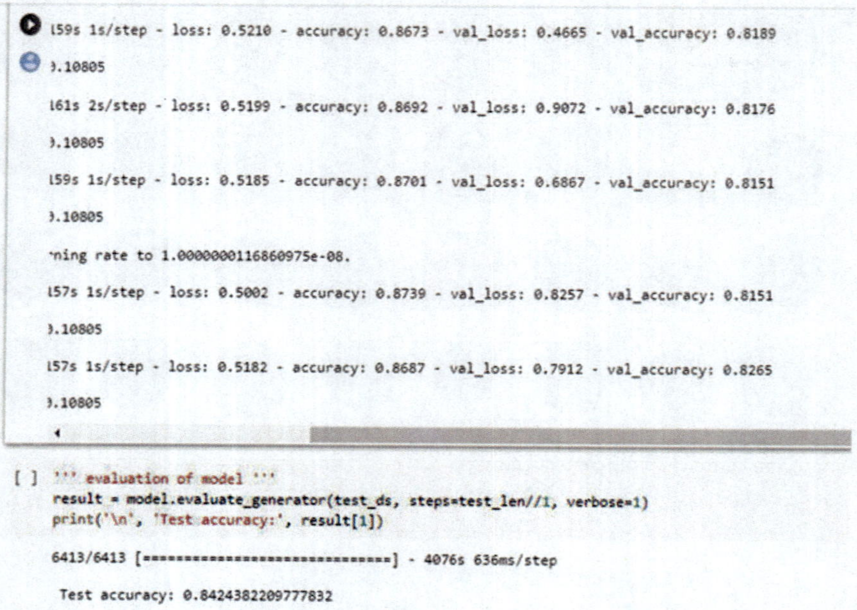

Fig. 5 Accuracy of the model

The graphs of training accuracy and validation accuracy are shown in Fig. 6. Through this graph, it can be observed that the validation accuracy came out to be less than the training accuracy.

Only one of the model's ten predictions, as shown in Figs. 7 and 8, was incorrect, indicating that the prediction data was largely accurate.

The additional predictions and the picture of the coin from the dataset are shown in Fig. 8 below, which shows how accurate the model is.

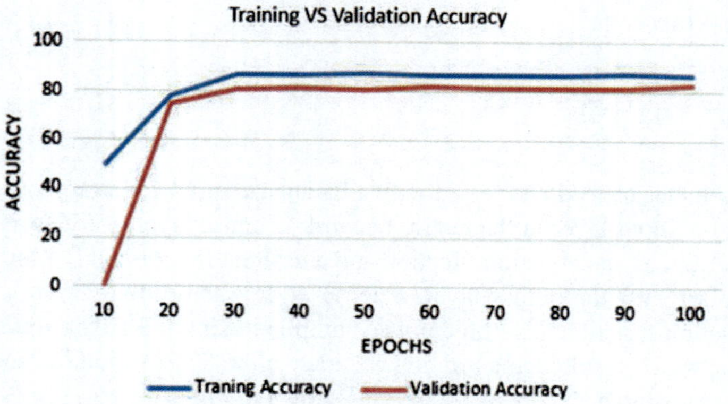

Fig. 6 Training accuracy versus validation accuracy

```
Predicted: 5 Cents,South African Rand,south_africa
Real:      5 Cents,South African Rand,south_africa

Predicted: 100 Forint,Hungarian Forint,hungary
Real:      20 Forint,Hungarian Forint,hungary

Predicted: 50 Satang,Thai Baht,thailand
Real:      50 Satang,Thai Baht,thailand

Predicted: 20 Kroner,Norwegian Krone,norway
Real:      20 Kroner,Norwegian Krone,norway

Predicted: 50 Cents,Canadian Dollar,canada
Real:      50 Cents,Canadian Dollar,canada
```

Fig. 7 Prediction versus real

```
Predicted: 1 Piso,Philipine peso,philippines
Real:      1 Piso,Philipine peso,philippines

Predicted: 1 Baht,Thai Baht,thailand
Real:      1 Sen,Malaysian Ringgit,malaysia

Predicted: 5 Kronor,Swedish Krona,sweden
Real:      5 Kronor,Swedish Krona,sweden

Predicted: 5 Rupees,Pakistan Rupee,pakistan
Real:      5 Rupees,Pakistan Rupee,pakistan

Predicted: 200 Forint,Hungarian Forint,hungary
Real:      200 Forint,Hungarian Forint,hungary
```

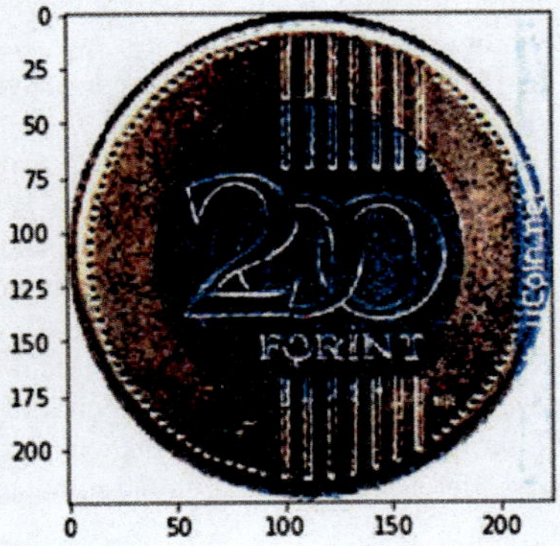

Fig. 8 Prediction versus real

7 Conclusion and Future Scope

Traditional image processing models need a lot of computational power and training data to achieve better prediction accuracy on the other hand deep learning models have higher operational speed and decreased computation time which makes them the most versatile and better suited for image processing tasks. In our model, we utilized more than 6000 images where deep learning proved to have an advantage over other models in response time and the ability to utilize GPU for quick training, testing and validation of the neural network. In the coming future, we can incorporate this model along with computer vision to develop an application for smartphones where real-time detection of coin currency from around the world could be achievable.

References

1. Krizhevsky A, Sutskever I, Hinton GE (2017) ImageNet classification with deep convolutional neural networks. Commun ACM 60(6):84–90
2. Wang N, Yeung DY (2013) Learning a deep compact image representation for visual tracking. In: Advances in neural information processing systems, Cambridge, MA, USA. MIT Press, pp. 809–817
3. Wang W, Tang C, Wang X, Luo Y, Hu Y, Li J (2019) Image object recognition via deep feature-based adaptive joint sparse representation. In: Computational intelligence and neuroscience, vol. 2019
4. Wang W, YangY, Wang X, Wang W, Li J (2019) 1e development of convolution neural network and its application in image classification: a survey. Opt Eng 58(4)
5. Li F, Wang C, Liu X, Peng Y, Jin S (2018) Corrigendum to 'A composite model of wound segmentation based on traditional methods and deep neural networks.' Comput Intell Neurosci 2018:4967290
6. Tajane AU, Patil JM, Shahane AS, Dhulekar PA, Gandhe ST, Phade GM (2018) Deep learning based Indian currency coin recognition. In: 2018 International conference on advances in communication and computing technology (ICACCT)
7. Zhang Q, Yan WQ, Kankanhalli M (2019) Overview of currency recognition using deep learning. J Bank Financial Technol 3(1):59–69
8. Real time fake currency note detection using deep learning. Int J Eng Adv Technol 9(1S5): 95–98
9. Jadhav M, Sharma YK, Bhandari GM (2019) Currency identification and forged banknote detection using deep learning. In: 2019 International conference on innovative trends and advances in engineering and technology (ICITAET)
10. Ali T, Jan S, Alkhodre A, Nauman M, Amin M, Siddiqui MS (2019) DeepMoney: counterfeit money detection using generative adversarial networks. PeerJ Comput Sci 5(e216):e216
11. Hoang VD, Vo HT (2018) Hybrid discriminative models for banknote recognition and anti-counterfeit. In: 2018 5th NAFOSTED conference on information and computer science (NICS), 2018.8606900
12. Navya Krishna G (2019) Recognition of fake currency note using convolutional neural networks. Int J Innov Technol Exploring Eng 8(5)
13. Pham TD, Park C, Nguyen DT, Batchuluun G, Park KR (2020) Deep learning-based fake-banknote detection for the visually impaired people using visible-light images captured by smartphone cameras. IEEE Access 8:63144–63161

14. Dhulekar PA, Gandhe ST, Shewale A, Sonawane S, Yelmame V (2017) Motion estimation for human activity surveillance. In: IEEE international conference on emerging trends & innovation in ICT (ICEI), pp. 82–85
15. Dhulekar PA, Aher P, Choudhari S, Khairnar Y (2016) Design of IR based image processing technique for digital camera deactivation. In: IEEE International conference on global trends in signal processing, information computing and communication (ICGTSPICC), pp. 83–86
16. Balaji S (2020) Binary Image classifier CNN using TensorFlow – Techiepedia – Medium," Techiepedia, 29-Aug-2020. [Online]. Available: https://medium.com/techiepedia/binary-image-classifier-cnn-using-tensorflow-a3f5d6746697
17. Hussein R (2010) E cient coin recognition using a statistical approach, pp. 1–04
18. Howard AG, et al. (2017) MobileNets: efficient convolutional neural networks for mobile vision applications. arXiv [cs.CV]
19. Medium.com. [Online]. Available: https://miro.medium.com/max/1384/1*7R068tzqqK-1edu4hbAVZQ.png

Hybrid Framework for Genomic Data Classification Using Deep Learning: QDeep_SVM

Girija Rani Karetla, Daniel R. Catchpoole, and Quang Vinh Nguyen

1 Introduction

Machine learning applications are widely adopted for healthcare data analysis, including disease detection, gene analysis, image processing, and precision medicine [1]. Gene expression variation can greatly influence phenotypic variety within species and species differentiation [2]. Advances in functional and structural knowledge from genome sequencing analysis have increased both the opportunities and challenges for machine learning in recent years [3]. More specifically, advances in RNA sequencing gene expression have made it possible to test the expression levels of 100 s, if not hundreds or thousands, of genes in a single analysis [4]. Recently, precision medicine, which uses genetic information systems to enhance therapy and transform care, has been using the prediction capability of machine learning methods. Many researchers have developed a series of analyses that can be used to identify specific biomarkers associated with different kinds of tissue, tumor or non-tumor [5].

"RNA sequencing (RNA-Seq) and DNA microarray" are two of the most used modern tools for implementing high-throughput gene expression evaluation. The DNA microarray technique uses an array of complementary probes to calculate the

G. R. Karetla (✉) · Q. V. Nguyen
School of Computer, Data and Mathematical Sciences, Western Sydney University, Penrith, Australia
e-mail: 19791744@student.westernsydney.edu.au

Q. V. Nguyen
e-mail: q.nguyen@westernsydney.edu.au

Q. V. Nguyen
MARCS Institute, Western Sydney University, Penrith, Australia

D. R. Catchpoole
Biospecimen Research Services, Children's Cancer Research Unit, The Kids Research Institute, The Children's Hospital at Westmead, Westmead, Australia
e-mail: daniel.catchpoole@health.nsw.gov.au

A. Ramdane-Cherif et al. (eds.), *Machine Intelligence and Data Science Applications*, Algorithms for Intelligent Systems, https://doi.org/10.1007/978-981-99-1620-7_36

proportion of a specific set of transcripts [6]. Various classification techniques based on DNA microarray gene expression data have been developed in recent years, and they have made significant contributions to cancer research [7]. Nevertheless, this approach has significant limitations since background hybridization reduces the efficiency of expression metrics, especially for low-abundance transcripts. Moreover, hybridization characteristics of probes vary significantly, thus arrays are confined to probing just those genes for which probes are developed [8]. RNA-Seq, on either side, offers information on a cell's transcriptome. Because of the rapid evolution of the transcriptome, it gives significantly more protection and precision than DNA microarray. As a result, it has evolved into a robust tool that might eventually replace DNA microarrays for transcriptome analysis [2].

RNA-Seq technique, frequently utilized to analyze gene expression data and reveal different features of the transcriptome. "RNA-seq is a high-throughput sequencing (HTS) technique for quantifying cDNA transcripts." Transcript abundance must also be correlated with gene/isoform expression [9]. Considering this, HTS for diagnostic testing, monitoring, and treatment is not frequently used in molecular diagnostics. HTS implementation in the clinic has formerly been limited by the availability of empirical standardization and validation. Furthermore, RNA-Seq provides insight into formerly unknown variations emerging in disease processes, treatments, and other investigations under various environmental settings. Although RNA-Seq is more expensive than DNA microarray, it may produce a vast amount of sequencing data.

As a result, RNA-Seq generates gene expression data with optimal feature selection, such as the Quantum Inspired Immune Clone Optimization (QIICO) algorithm, which would also be effective for deep convolutional neural networks (DCNN) and support vector machines (SVM) that categorize RNA-Seq gene expression data. The classifications of RNA-Seq gene expression data have offered valuable information for cancer diagnosis and medication research [9]. Gene expression is a result of one or more environmental, behavioral, and genetic variables. When compared to DNA microarray technology, the RNA-Seq technique has become a popular method for quantifying gene expression and is predicted to provide superior answers to a variety of biological and biomedical concerns. Several steps are involved in RNA-Seq gene expression analysis to build a data matrix.

Due to the massive dimension of gene expression data with a restricted quantity of information, the problem creates statistical and analytical difficulties, as also traditional statistical procedures that produce incorrect results. "Support vector machines (SVM), neural networks, *k* nearest neighbors, decision trees, random forests, random forests of oblique decision trees, bagging and boosting, and random ensemble oblique decision stumps" are just a few of the classifiers that were used to recognize gene expression data over the last decade. Even though much research has been conducted to categorize gene expression data, there is still a vital requirement for improved accuracy. DCNN and SVM are two effective pattern recognition techniques.

The SVM has several benefits when it comes to categorizing large amounts of data. This algorithm's primary goal is to increase the margin while lowering the upper bound on the generalization error. To identify globally optimum solutions, SVM uses a multi-objective optimization problem. An SVM is a basic architectural model with a single adjustable layer that includes other techniques. Therefore, a deep convolutional neural network (DCNN) is a deep model that trains hidden representations. For classification, traditional DCNN design utilizes multinomial logistic regression at the outer layer. SVM is a convenient approach to soft max for improving classification accuracy.

This paper proposes a "new learning algorithm for precise classification of RNA-Seq gene expression data of optimal feature selection based SVM using features extracted by DCNN". Feature extraction, Optimal Feature Selection, and Classification" are the three key steps of the proposed classification algorithm. Benchmark datasets are first collected. The feature extraction is done using the gene expression data that has been collected. To reduce the length of the feature vectors, a new meta-heuristic approach called the quantum-inspired immune clone optimization algorithm (QIICO) is applied. After the necessary features have been chosen, a classification method entitled hybrid SVM-DCNN is used to classify the data. Finally, the experimental results show that the proposed "QIICO-based feature selection model outperforms other heuristic-based feature selection models, as well as hybrid SVM-DCNN, outperforms other classification techniques."

The following is the structure of this paper: Sect. 2 addresses the related work. The optimal feature selection is described in Sect. 3. SVM, DCNN, and the proposed technique are briefly discussed in Sect. 4. The results are provided in Sect. 5 and the conclusions are provided in the final session.

2 Related Works

With the huge trend of RNA-Seq technology, many tools and software have been developed for the analysing gene expression to study the correlations between genes and diseases [10]. RNA-Seq is an exceptionally useful expression analysis technology that can be of assistance in a variety of situations relating to specific conditions [11]. Under many studies, the feature selection method is used to select the most essential features, and then the SVM or RF classification method is used on these features. Guyon et al. [6] conducted a study in which they classified samples based on a selection of genes that were relevant to the problem.

Marioni [12] the purpose of this paper is to evaluate the official variance linked with illumina sequencing in this setting and evaluate its capacity to find out specific genes with previous array technologies. We used several sequencing replicates to quantify gene interpretation differences between RNA models of liver and kidney then compared the arraying data to Affymetrix array results through similar RNA models. We discovered that Illumina arraying information is extremely reproducible, with tiny technical deviation, and that sequencing each mRNA sample only once may be sufficient for many uses.

Deep convolutional neural network (DCNN) has produced impressive results recently in text classification and computer vision [13]. Additionally, deep neural networks are also applied in medical appliances and genomic signal processing [14]. Using the information derived from a support vector machine, Huynh et al. [5] proposed a new classification model of gene appearance DCNN-Deep Convolutional Neural Network. "RNA sequencing is a strong approach for gene expression profiling in model organisms that can provide bimolecular information for cancer diagnosis." In this paper, the DCNN derives hidden patterns from gene expression data, which combined with SVM to categorize RNA-Seq gene expression information effectively. Huynh et al. [5] present a hybrid model that combines DCNNSVM and non-linear SVM to classify extremely high-dimensional RNA-Seq gene expression data. This model performs quite well on tasks tested on The Cancer Genome Atlas (TCGA).

(Chen 2020) presented the use of two (rPCA) resilient principal component analysis techniques, PCAHubert and PCAGrid, to identify outlier prototypes in simulated and actual genetic RNASeq data sets with standard control samples. rPCA and cPCA methods were applied on an RNASeq dataset, both rPCA approaches identified the two similar outlier samples but cPCA does not detect any. The authors of this paper used quantitative reverse transcription PCR to confirm gene expression variations and the results are added as a benchmark to evaluate the performance of data analysis methodologies. Stupnikov et al. [9] provided a complete and in-depth assessment of five DGE models for RNASeq by a controlled quantitative method. The tested DGE pipelines were based on count and observed expression at the gene stage. Using unbiased metrics, the standardization of RNASeq exactness medicine is improved. The performance of these three classifiers is evaluated on certain features. Källberg [15] Clustering should have wisely chosen features for high-dimensional RNA sequencing data. Many feature selection methods are already proposed, but still, there is a lack of knowledge on how and when to use these approaches. This paper provided the evaluation of thirteen feature selection methods with known subtypes. All methods were examined by taking the mean square expression and standard deviation of the sample genes. The clustering performance is obtained for all methods.

Gakii [16] used a graph-based feature selection approach for the classification of RNASeq datasets from datasets of lung cancer. Recursive feature removal and principal component analysis approach are also used in this paper to choose features for classification. The outcome of this research showed that this feature selection method developed the sequential minimal optimization and also multilayer perception classifier's performance in all datasets. The graph-based feature selection method is joined with rule mining for accurate selection and association finding between features in great dimensional RNASeq data.

3 The Quantum-Inspired Immune Clone Optimization (QIICO) Algorithm

The functioning of the quantum-inspired immune clonal algorithm [17] is as follows. Thus, every quantum antibody is represented by a qubit. During iteration, each qubit phase of the antibody in the population is related to the current greatest one, and then its occurring probability is modified, to evolve towards some kind of fitter antibody with a higher probability.

The basic specifications for the generic optimization algorithm are first given. The systematic procedure may be used to express a basic optimization method:

$$\min g(a) a = (a_1, a_2, \ldots, a_n) \in 0 \tag{1}$$

where $g(a)$ represents the objective function, O specifies the space search which is an m-dimensional space. This signifies
$O = [\underline{a}, \overline{a}], \underline{a} = \left(a_1, a_2, \ldots, a_m \right), \overline{a} = \overline{a_1}, \overline{a_2}, \ldots, \overline{a_m})$.

The following is how the affinity for the generic optimization method is determined.

Note: A proposed method for the optimization model represented by the antibody, k. The negative value of the objective function is equivalent to its affinity level.

$$k \in O \text{and} W(k) = -g(k) \tag{2}$$

Let's look at the quantum antibody's current status in the generation:

$$p^n = \begin{pmatrix} \alpha_1^n \alpha_2^n \ldots\ldots \alpha_z^n \\ \beta_1^n \beta_2^n \ldots\ldots \beta_z^n \end{pmatrix} \tag{3}$$

where z denotes the length of the qubit antibody p^n, α_u^n and β_u^n ($u = 1,2, \ldots, z$) are randomly generated between −1 and +1, satisfying $|\alpha_u^n|^2 + |\beta_u^n|^2 = 1$.

The quantum antibodies then generate common antibodies for the monitoring operator to evaluate their affinity.

Including the affinity function, the clone operator is used to copy a qubit antibody in the population into the identical qubit antibodies in the optimal solution, $() = J$. and J denotes the identity matrices of dimensionality C expressed as:

$$C = \left[X_c * \frac{W(p^n)}{\sum W(p^n)} \right] \tag{4}$$

Which may be self-adjusted by the affinity W (*). $\sum W(p^n)$ Denotes the total affinity of all the qubit antibodies in a population, and X_e is a specified value related to the clone scale.

Proliferative qubit antibodies emerge after the clone operator:
$p^{n'} = \left\{ p^n, p_1^n, p_2^n, \ldots, p_j^n, p_{c-1}^n \right\}$,

$$\left(p_j^n = p^n\right) = \left\{ \begin{pmatrix} \alpha_1^{n'} \alpha_2^{n'} \ldots \alpha_z^{n'} \\ \beta_1^{n'} \beta_2^{n'} \ldots \beta_z^{n'} \end{pmatrix}, \begin{pmatrix} \alpha_1^{n'} \alpha_2^{n'} \ldots \alpha_z^{n'} \\ \beta_1^{n'} \beta_2^{n'} \ldots \beta_z^{n'} \end{pmatrix} \ldots \begin{pmatrix} \alpha_1^{n'} \alpha_2^{n'} \ldots \alpha_z^{n'} \\ \beta_1^{n'} \beta_2^{n'} \ldots \beta_z^{n'} \end{pmatrix} \right\} \tag{5}$$

It is also modified by

$$\begin{bmatrix} \alpha_1^{n''} \\ \beta_1^{n''} \end{bmatrix} = (\theta_u^n)^* \begin{bmatrix} \alpha_1^{n'} \\ \beta_1^{n'} \end{bmatrix} u = 1, 2, \ldots, z \tag{6}$$

The modified operator is

$$\mathrm{M}(\theta_u^n)^* = \begin{bmatrix} \cos(\theta_u^n) & -\sin(\theta_u^n) \\ \sin(\theta_u^n) & \cos(\theta_u^n) \end{bmatrix} \tag{7}$$

where denotes the rotate step and is described as $= x^* g()$, and the function $g()$ focuses on guidance by the current best antibody in the population. If the termination criterion is true, output the best solution and stop the process; if not, execute the clonal selection and quantum recombination operations.

4 Classification Methods

The "support vector machine (SVM) and the deep convolutional neural network (DCNN)" are both briefly described in this section. We concentrate on examining their basic frameworks to get knowledge into their different strengths and weaknesses regarding the current objective. The analysis that follows explains why our approach should be proposed.

4.1 *Support Vector Machine (SVM)*

The author [18] proposed the Support Vector Machine (SVM), which is based on a statistical learning approach and is comprehensive. SVM is the most well-known method in the kernel substitution class of learning algorithms. For classification and regression, SVM and kernel-based approaches are useful. The SVM method determines the optimal separation plane between the various classes. The SVM method needs to enhance the margin while minimizing the error in pursuit of this objective.

With z data points ($j = 1, \ldots, z$) in the m-dimensional input space Q^m and related labels $= S_j = \pm 1$, for binary classification issue (see Fig. 1). The SVM algorithms look for the optimal hyper-plane that is the furthest away from both classes + 1 and class −1. It may calculate for each class by maximizing the distance or margin between the enabling planes (a.d−f = for class +1 and a.d−f = for class −1).

The distance between these supporting planes is 2/||D|| (where w is the vector's 2-norm). Any point xi that is on the incorrect edge of its hyperplane is deemed an error. As a result, SVM has improved the margin and reduced error. The following quadratic Eq. (8) yields the conventional SVM with a linear kernel:

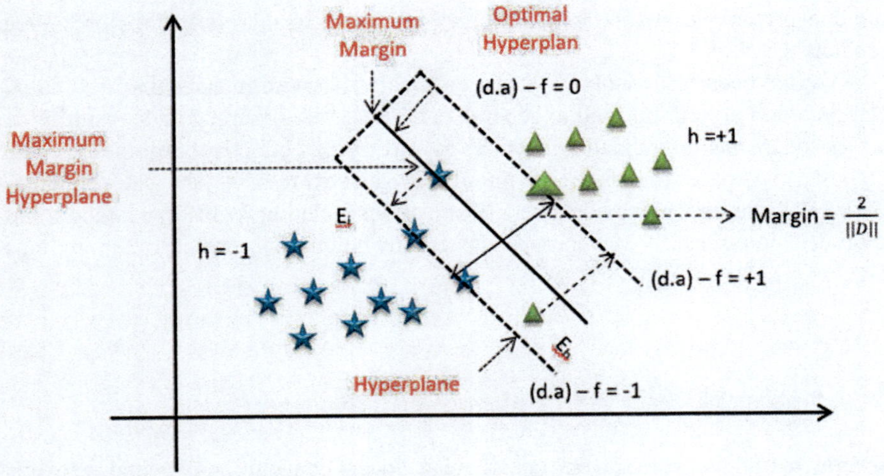

Fig. 1 SVM classifier

$$min_{\mu}(1/2)\sum_{j=1}^{z}\sum_{h=1}^{z}s_j s_h \mu_j \mu_h X\left(A, A_j\right) - \sum\nolimits_{j=1}^{z}\mu_j \tag{8}$$

$$s.t.\begin{cases}\sum_{j=1}^{z} A_j\mu_j = 0\\ 0 \le \mu_j \le T\forall_j = 1, 2, \ldots, z\end{cases} \tag{9}$$

Here, "T is a positive constant that is used to modify the margin and error, and X(A_j, A_h) = X(A_j, A_h) is a linear kernel function."

The solution of the quadratic program (8) determines the support vectors (for which an μ_j>0), and the differentiating surface and scalar f are then defined by the support vectors. The following is how a novel data object A is categorized using the SVM classifier:

$$\text{Predict(a, SVM Classifier)} = \text{sign}\left(\sum_{j=1}^{\#sv} s_j\mu_j X\left(A_j A_h\right) - f\right) \tag{10}$$

Other classification functions that SVM can utilize include a "polynomial function of degree w, a radial basis function (RBF), and a sigmoid function. Cristianini and Shawe-Taylor [18] provide further information regarding SVM and other kernel-based learning algorithms."

Due to their simplicity, one-versus-all and one-versus-one have to be the most common multiclass approaches. Find the probability of x classes ($x > 2$). The one-versus-all technique creates k distinct classifiers, each of which distinguishes the ith class from the others. Using some of the binary pairwise variations of the k classes, the one-versus-one method creates x * (×1)/2 classifiers. A simple majority then predicts the class.

SVM has been used to solve high-dimensional issues with great results. Traditional classification algorithms tend to generalize the system by training data, resulting in lesser accuracies on validation data. SVM was trained effectively on the absolute highest input spaces in the area of "microarray gene expression data" categorization [19]. SVM was used alongside other classifications including k closest neighbors and C4.5 decision trees. SVM improves classical techniques, according to the findings [4].

4.2 Deep Convolutional Neural Network

Several neural network layers constitute a "deep convolutional neural network (DCNN)". The consecutive layers of a network system are intended to know sufficiently high-level information until the final layer provides classifications [4]. All layers of the network are trained at the same time throughout the training phase, and feature extraction is a key component of the classification method. The final layer, which is a linear classification working on the features extracted by the preceding layers, is applied when the training procedure is over. Since such attributes are the product of the integrated training method and have been tuned to meet the needs and limits of the final layer, they are the outcome of the integrated training technique [20].

DCNN was developed to handle a variety of data formats, particularly two-dimensional pictures that have been directly influenced by the human brain's visual cortex. Simple cells and complex cells are the two fundamental cell types found in the human brain. On the one hand, basic cells respond to basic structures in visual stimulator sub-regions. Complex cells, on the other hand, combine information from basic cells to detect higher complex structures. As a result, the visual cortex is a strong and natural mechanism for visual processing. DCNN is used to simulate three fundamental concepts: local connection, location invariance, and local transition invariance [21].

Feature extraction may be thought of as a "stack of convolution and sub-sampling layers" being built up. Over the preceding layers, "the convolution layers compute convolutions" with a minimal set of trainable convolution kernels x:

$$p_o = g\left(\sum p_i * x_n + y\right) \tag{11}$$

"A hyperbolic tangent (HT), rectified linear unit, or sigmoid are examples of non-linear functions. Y is biased. The convolution process may construct a local edge detector or a low-pass filter for every value of the kernel coefficients x. Multiple convolution kernels can be used on each layer to generate a set of feature maps."

The sub-sampling layers multiply the average or maximum of a $m \times m$ pixel blocks by a trainable scalar, add a bias, and then feed the output via a sigmoid:

$$p_o = g\left(\sum_{p_i}^{m*m} + y\right) \tag{12}$$

These layers have the effective impact of extracting a feature vector r from an input x, which is expressed as $r = t$. (a). The last layer is a linear classification that uses the characteristics retrieved by the preceding layers to make decisions. This layer includes a bias vector to the product of the feature vector v and the weight matrix D, and then runs the output via activation functions. At the topmost network, the conventional DCNN employs the soft-max activation function. During the training of a network, gradient descent methods are applied for optimization.

4.3 Hybrid SVM Based DCNN for Gene Expression

The proposed method is a hybrid of the DCNN and SVM algorithms. The training work is divided into two stages by the algorithm. Figure 2 shows the process of our approach.

DCNN was used in our system to learn latent features from very high-dimensional input fields. Data is projected from a higher-dimensional realm to a lower-dimensional space in this procedure. These additional characteristics may enhance the dissimilarity power of gene expression models, resulting in a greater accuracy rate than before. While the non-linear SVM can not learn complicated invariances, it

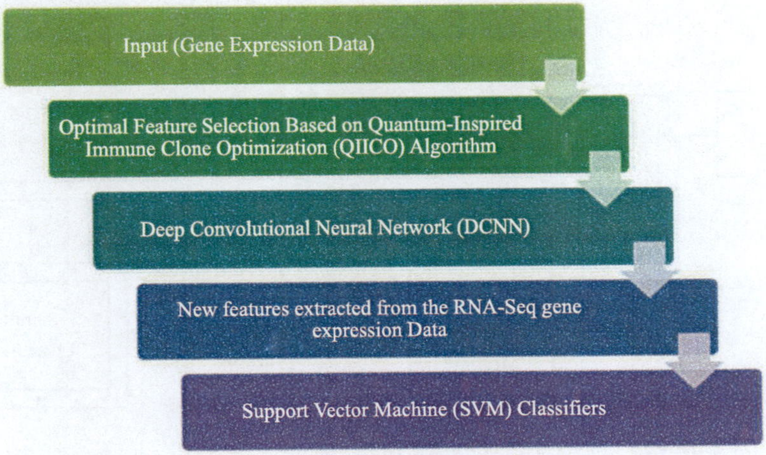

Fig. 2 Proposed workflow model for classification

may provide effective decision surfaces by utilizing soft-margin techniques to maximize margins [20]. Suggested the use of DCNN and SVM together in the previous as components of a deep learning process [22].

First, it develops a novel DCNN that extracts new features from RNA-Seq gene expression data at the source. To extract latent characteristics from RNA-Seq gene expression data creates a novel DCNN model. They discovered that pooling global averages improves model stability. In Fig. 3, two convolutional layers, two pooling layers, and a fully connected layer that make up the framework are shown. "Conv1, Pooling1, Conv2, Pooling2, and output are the terms of the layers." The gene expression is sent to the input layer as a 2-D matrix. By applying some zeros to the image's last edge we're able to incorporate the high-dimension expression data (20,531 $\times$ 1) into a 2-D image (142 $\times$ 142). The initial Conv1 layer includes four feature maps as well as the kernel size (3 $\times$ 3). The average pooling output of the first layer is sent into the second layer, Pooling1, which filters it using a (2 $\times$ 2) sub-sampling layer. Conv2 outputs two feature maps with a convolution kernel size of (3 $\times$ 3). Pooling2 is a subsampling layer with a size of (2 $\times$ 2). As neurons, it suggests using Tanh activation. The last layer has a configurable number of maps that mix inputs from all of Pooling2's maps. A 2312-dimensional vector is generated via networks. The characteristics retrieved by the Conv1 to Pooling2 layers are represented by these vectors.

Finally, the SVM is trained to categorize gene expression data effectively at this stage using the new feature generated by DCNN. This paper suggests using the RBF kernel type in the SVM classifier in our method. It is generic and inexpensive.

DCNN is utilized to extract features in our method. The training and testing samples are sent into the trained network, and the features are taken from the output of the final layer. The training samples' feature sets are then utilized as input to train an SVM in the normal way. It discovered that this hybrid model not only considerably improves classification results but also computes quite quickly.

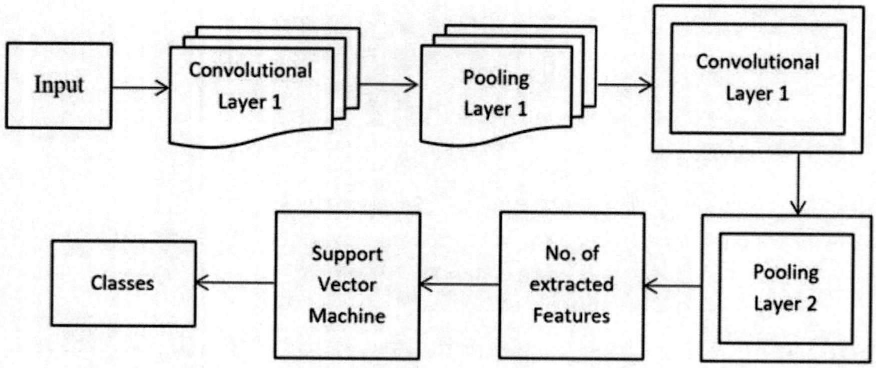

Fig. 3 The framework of hybrid DCNN and SVM

5 Experimental Results and Discussion

The classification results of our approach for classifying RNA-Seq gene expression data are of significance to us. As a result, it compares the classification results of **QDeep_SVM** with those of the finest state-of-the-art methods, such as support vector machines (SVM), random forests (RF),DCNN and DCNN–SVM. And evaluate several DCNN–SVM models with SVM and RF. These findings are performed to analyze the efficiency of classifiers.

This analysis applies RNA-Seq gene expression statistics from the public website to test the effectiveness of our technique on various dataset sizes. In terms of datasets for binary class classification, we employ 5 datasets with sample sizes ranging from 1131 to 20,000 and sample sizes ranging from small to medium. The entire dataset, which comprises over 10,000 data samples covering 5 cancer types, is also utilized. It includes all RNA-Seq level 3 expression data from 9027 patients with 5 different abnormal tissue types, as well as normal samples. The huge dataset includes newly updated abnormal samples from RNA-Seq gene expression data.

This paper used Tensor-Flow and Scikit packages to develop DCNN–SVM and its Python version to test its efficacy in classification models. The Scikit package is used to create random forests from C4.5 decision trees. For larger classes, it may utilize the Lib-SVM standard SVM method, which is quite effective.

This paper employs the tenfold cross-validation method for evaluating the performance of 5 datasets, which is still the most commonly used. The classification results are compared using the overall classification accuracy measure. Using the huge dataset, we dynamically split the data into an 70% training set and a 30% testing set, with samples selected proportionally from each cancer type.

The gene expression matrix is normalized to the region [−0.9 to +0.9] in the DCNN–SVM arrangement. We utilize Adam for optimization to train the algorithm, with batch sizes ranging from 250 to 1500. It started training at a learning rate of 0.00002 for those layers and automatically enhanced it when the validation error rate stopped increasing. The loss function is cross-entropy, and the epochs are 250. The network structure hyper-parameters on either side can never be maintained between workloads. As a result, DCNN, can evaluate several topologies on the validation set and select the best one, then retrain the whole network on the training set and provide test accuracy.

Thus, QDeep_SVM outperformed LSVM, RF, and DCNN in the coupled Student ratio test. When evaluated against SVM, in column 3 of Table 1, **QDeep_SVM** has 17 triumphs, 6 splits, and 3 losses (p-value = 0.0051). In addition, when compared to RF in column 2, **QDeep_SVM** has 17 triumphs, 7 splits, and 2 defeats (p-value = 0.0009). Furthermore, our proposed system outscores in 4 out of 5 datasets.

The time it takes to train the deep convolutional networks to extract features and the time it takes to train the classifier on the newly extracted feature set to make up the hybrid model's running time. The first half takes an average of 79.54 s across 5 datasets. SVM and RF in the hybrid model take 0.08 s and 0.19 s on average for the second layer. Linear SVM and RF take 11.73 and 5.28 s to execute, respectively.

Table 1 The findings of classifying algorithms on five binary-class gene expression datasets

S.No	RF	SVM	DCNN	DCNN-SVM	QDeep_SVM
	87.58	77.58	88.46	87.25	89.28
	78.42	88.42	78.72	86.27	88.46
Lung cancer	67.24	77.24	89.27	89.87	89.01
Covid	88.76	68.76	76.2	89.03	91.05
RMS	77.02	87.02	84.41	88.24	93.46

Table 2 Overall results on RNA-Seq Gene Expression Datasets

Dataset	RF	SVM	DCNN	DCNN- SVM	QDeep_SVM
RNA-Seq_Gene Expression Datasets	79.804%	74.96%	75.91%	80.45%	90.42%

The results of Q-DCNN–SVM, different variants of DCNN–SVM, SVM, RF, and DCNN on the huge RNA-Seq gene expression dataset are shown in Table 2. The DCNN–SVM classifier has the highest accuracy, followed by DCNN–RF. Because it uses DCNN and SVM, DCNN–SVM has greater accuracy than other models on this dataset. Furthermore, due to the efficiency of the RBF kernel, QDeep_SVM has the best classifier performance (90.42%) among the different DCNN–SVM versions. Our suggested DCNN and SVM combination improves the DCNN, SVM, and RF separately. On large RNA-Seq gene expression datasets, our results suggest that DCNN–SVM is the optimum classification method.

6 Conclusion

To classify highly dimensional RNA-Seq gene expression data, we propose a hybrid model integrating Quantum Immune clone optimization with DCNN and SVM. With the use of optimum feature selection algorithms like the quantum-inspired immune clonal algorithm (QIICO), the DCNN extracts latent features from gene expression data, which then utilized in conjunction with SVM to successfully categorize RNA-Seq gene expression data. The latent features are learned by a convolutional procedure and then fed into the RBF kernel-based SVM classifier. This algorithm performs quite well on the task performed on RNA-Seq gene expression datasets after changes by provided hyperparameters. When compared to standard DCNN, SVM, DCNN–SVM and RF, numerical test results on RNA-Seq gene expression datasets reveal that our suggested technique, **QDeep_SVM**, is the most accurate.

References

1. Bist AS (2018) A Survey of deep learning algorithms for malware detection
2. Skelly DA et al (2011) A powerful and flexible statistical framework for testing hypotheses of allele-specific gene expression from RNA-seq data. Genome Res 21(10):1728–1737
3. Xing E, Jordan M, Karp R (2001) Feature selection for high-dimensional genomic microarray data. Proc 18th International Conf on Machine Learning
4. Costa-Silva J, Domingues D, Lopes FM (2017) RNA-Seq differential expression analysis: an extended review and a software tool. PLoS ONE 12(12):e0190152
5. Huynh P-H, Nguyen V-H, Do T-N (2019) Novel hybrid DCNN–SVM model for classifying RNA-sequencing gene expression data. J Inform Telecommun 3(4):533–547
6. Guyon I et al (2002) Gene selection for cancer classification using support vector machines. Mach Learn 46(1):389–422
7. Guyon I (2002) Sene selection for cancer classification using support vector machines. Mach Learn 46(1):389–422
8. Zhao S et al (2014) Comparison of RNA-Seq and microarray in transcriptome profiling of activated T cells. PLoS ONE 9(1):e78644
9. Stupnikov A et al (2021) Robustness of differential gene expression analysis of RNA-seq. Comput Struct Biotechnol J 19:3470–3481
10. Zhang ZH et al (2014) A comparative study of techniques for differential expression analysis on RNA-Seq data. PLoS ONE 9(8):e103207
11. Oshlack A, Robinson MD, Young MD (2010) From RNA-seq reads to differential expression results. Genome Biol 11(12):1–10
12. Marioni JC, Mason CE, Mane SM, Mane SM, Gilad Y (2008) RNA-seq: an assessment of technical reproducibility and comparison with gene expression arrays. Genome Res. https://doi.org/10.1101/gr.079558.108
13. Krizhevsky A, Sutskever I, Hinton GE, Imagenet classification with deep convolutional neural networks. Advances in neural information processing systems, 2012, **25**
14. Min S, Lee B, Yoon S (2017) Deep learning in bioinformatics. Brief Bioinform 18(5):851–869
15. Källberg D, Vidman L, Rydén P (2021) Comparison of methods for feature selection in clustering of high-dimensional RNA-sequencing data to identify cancer subtypes. Front Genet. https://doi.org/10.3389/fgene.2021.632620
16. Gakii C, Mireji PO, Rimiru R (2022) Graph based feature selection for reduction of dimensionality in next-generation RNA sequencing datasets. Algorithms 15(21). https://doi.org/10.3390/a15010021
17. Wu Q et al (2009) A novel quantum-inspired immune clonal algorithm with the evolutionary game approach. Prog Nat Sci 19(10):1341–1347
18. Vapnik V (1999) The nature of statistical learning theory. Springer science & business media
19. Pirooznia M et al (2008) A comparative study of different machine learning methods on microarray gene expression data. BMC Genomics 9(1):1–13
20. Huang, F.J. and Y. LeCun. Large-scale learning with svm and convolutional for generic object categorization. In 2006 IEEE Computer Society Conference on Computer Vision and Pattern Recognition (CVPR'06). 2006. IEEE
21. LeCun Y, Bengio Y, Hinton G (2015) Deep learning. Nature 521 (7553):436–444. Google Scholar Google Scholar Cross Ref Cross Ref, 2015
22. Huynh P-H, Nguyen V-H, Do T-N (2018) A coupling support vector machines with the feature learning of deep convolutional neural networks for classifying microarray gene expression data. Modern approaches for intelligent information and database systems. Springer, pp 233–243

Agricultural Drought Assessment of a Canal Command Area Using Space and Ground Observation. A Case Study of Tons Pump Canal Command

Mohd Sufiyan, Derrick Denis, Himanshu Mishra, and Mukesh Kumar

1 Introduction

Water serves as one of nature's most basic needs for plants, animals, and people to thrive because life cannot exist without it. Vegetation is vital in providing a range of ecosystem services and benefits that aid in coping with and reducing the impacts of global climate change.

Agriculture plays a significant role in India's economy, contributing to the country's Gross Domestic Product (GDP). Agriculture has contributed more than 20% of the Indian GDP in the fiscal year 2020–2021. Over 57 percent of Indians rely on agriculture as their primary source of income, and it is the foundation of the national economy. India is making progress in agriculture; however, there is a pressing need for monitoring in this sector, particularly due to the country's large population. Therefore, vegetation indices play a prominent role in vegetation monitoring.

Drought indices are created to aid in decision-making in the event of a drought. Drought can be quantified and measured at various scales, such as at the watershed, regional, provincial, national, or global level, to determine the severity of the event [1]. Drought indices are classified based on their input data requirements, intended use, or both. Examples include indices that use hydrological, meteorological, agricultural and remote sensing data. Remote sensing techniques allow for the quick acquisition and dissemination of information across broad regions by utilizing

M. Sufiyan (✉)
Irrigation and Drainage Engineering, VIAET, SHUATS, Prayagraj, India

D. Denis
Irrigation and Drainage Engineering Department, VIAET, SHUATS, Prayagraj, India

H. Mishra
VIAET, SHUATS, Prayagraj, India

M. Kumar
Centre for Geospatial Technologies, VIAET, SHUATS, Prayagraj, India

A. Ramdane-Cherif et al. (eds.), *Machine Intelligence and Data Science Applications*, Algorithms for Intelligent Systems, https://doi.org/10.1007/978-981-99-1620-7_37

sensors operating in multiple spectral bands and mounted on aircraft or satellites [2]. While in orbit, the satellite can examine the earth's surface, identifying, monitoring, and assessing drought conditions in a short period of time. Using this technology, one can analyze the effects of droughts not just on vegetation but as well as on groundwater, surface temperature, and precipitation [3]. Since 1980, several distinct vegetation indices have been developed. These indices are classified into the following categories: vegetation greenness index, vegetation water index, vegetation temperature index, and combination index. Vegetation greenness indices are calculated by combining ratios of predominantly red, infrared, green, and short-wave spectral bands. Several indicators are used to evaluate vegetation, but one commonly used method is the Normalized Difference Vegetation Index (NDVI). This method has been utilized for over 20 years to assess vegetation stress and the most recognized vegetation index [3]. A composite of spectral bands is employed to determine the vegetation water indices. In moisture detection, the green, short-wave, and infrared spectrums are most normally employed [4]. The Land Surface Water Index (LSWI) is a widely used method for determining vegetation moisture levels and assessing drought conditions.

The primary goals of this research are to measure and quantify the Normalized Difference Vegetation Index (NDVI) and Land Surface Water Index (LSWI) using satellite data during the Rabi (November- March) and Zaid (March -June) seasons, Linear Regression model will be developed using the correlations between Normalized Difference Vegetation Index (NDVI) and Land Surface Water Index (LSWI) and Validation of the result.

2 Related Works

In 1998, the research aim was to demonstrate that relatively straightforward methods can be used to map evaporation at a broad scale using limited information from Advanced Very High-Resolution Radiometers (AVHRR) and meteorological data. These methods can be linked to moisture stress and are directly related to concepts and components of water balance approaches for determining regional soil moisture and moisture deficits [5]. These water balance methods are often used to infer drought; however, they can be challenging to calibrate and are typically spatially limited.

In 2004, the author demonstrated the existence of relatively simple methods for mapping evaporation at a regional scale, which can be linked to moisture stress, and are directly related to concepts and components of water balance approaches for determining regional soil moisture and moisture deficits. These methods utilize the limited information that is typically available from Advanced Very-High-Resolution Radiometers (AVHRR) and meteorological data. Although these water balance approaches are commonly used to predict drought, they can be geographically limited and challenging to calibrate [6].

In 2010, the research primarily centered on using the Normalized Difference Moisture Index (NDMI) and Normalized Difference Vegetation Index (NDVI) generated from the Moderate Resolution Imaging Spectro-Radiometer (MODIS) to gather geographical information on drought status. A Pearson correlation analysis was conducted to investigate the correlation between NDVI and NDMI, which revealed a strong positive relationship [7]. Drought indices have been used to identify drought conditions and determine the geographic scope of droughts. A comparison of normal and drought years shows that the amplitude analysis, which takes into account both moisture and vegetation conditions, is an efficient tool for identifying drought conditions.

In 2019, the research was mainly formulated to analyze NDVI for vegetation cover change. Four separate monitoring locations were used to examine changes in vegetation cover [8]. The minimum and highest NDVI values were computed from June 2015 to May 2016. The research discovered the minimum and maximum Normalized Difference Vegetation Index (NDVI) values for monitoring points 1, 2, 3, and 4 from June 2015 to May 2016. The findings indicated that the average cover of vegetation decreased in December (Rabi season) while strong vegetation was observed in September (Kharif Season).

In 2021, the research was executed to examine land surface temperature (LST and study its link with the normalized difference water index (NDWI), land use and land cover, the normalized difference built-up index (NDBI), and the normalized difference vegetation index (NDVI) in the larger Arba Minch, Rift Valley, Ethiopia [9]. The study was conducted over two time periods, specifically December 8th, 2013, and December 27th, 2020. The research found that the Land Surface Temperature (LST) values of 38 °C, 34 °C, 8 °C, and 7 °C were low and high, respectively, for the years 2013 and 2020. To map the various LULC units in the current research region, a supervised image classification approach was used in ArcGIS software. Furthermore, it was indicated that the water bodies and forests have decreased slightly, but shrubland has decreased significantly. The current study's normalized difference vegetation index (NDVI) data indicate a significant decrease in forest area, and normalized difference water index (NDWI) values suggest a decrease in water bodies throughout 2020. The Normalized Difference Built-up Index (NDBI) value increased significantly from 2013 to 2020.

3 Materials

3.1 Study Area

Tons Pump Canal cultural command area is situated in the Prayagraj district of Uttar Pradesh, India. Tons Pump Canal originates at 81°2′ 34.44″ E longitude and 25°2′ 34.44″ N latitude. It falls within UTM Zone 44N and is located between the parallels of 25°16′46″ and 25°23′59″ north latitudes and 81°44′03″ and 81,056′58″ east

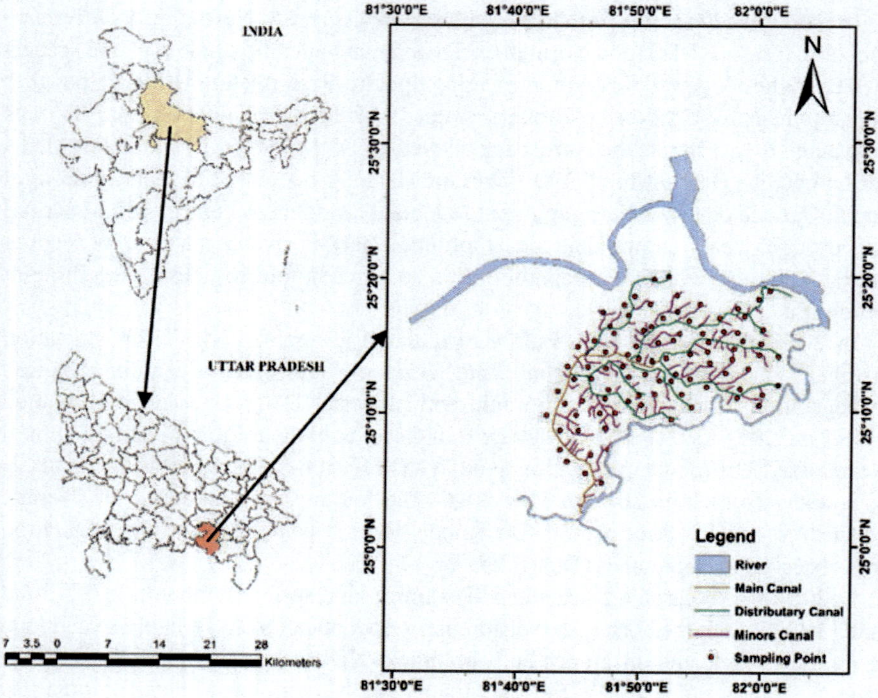

Fig. 1 The study area location map

longitudes. The Tons River is a significant source of water for the regions where irrigation is practiced. Prayagraj climate is tropical, with typical maximum temperatures varying from 43 °C to 45 °C, reaching 46 °C during peak summers, and a minimum average temperature of 8–9 °C; however, it may drop as low as 4 °C during peak winter months (Fig. 1).

3.2 Data Available

In this agricultural drought analysis research, data from the Operational Land Imager (OLI), LANDSAT-8, with a geographical image resolution of 30 m, as well as ground survey data, will be used. The path or row was identified as 143/43. The Operational Land Imager (OLI) measures the spectrum's near-infrared, visible, short-wave infrared, and thermal infrared bands [10]. Satellite data for the following dates will be used to meet the study's goal. Table 1 shows the details of the LANDSAT-8 satellite data in addition to the date of acquisition.

Table 1 Operational Land Imager (OLI) Landsat 8 data from several pass days

Type of satellite data	Season	Date of pass
Landsat 8 (OLI)	Rabi	22-Nov-2021 08-Dec-2021 25-Jan-2022 26-Feb-2022 14-Mar-2022
	Zaid	30-Mar-2022 15-April-2022 01-May-2022 06-June-2022

4 Methodology

This section includes the LSWI and NDVI indices that are going to be used to extract distinct characteristics that are easy to get to in the 11-band satellite image of the Landsat 8 data of the study area.

Land Surface Water Index (LSWI) is calculated by using Eq. 1.

$$\text{LSWI} = (\text{NIR} - \text{SWIR})/(\text{NIR} + \text{SWIR}) \tag{1}$$

where SWIR indicates near-infrared reflectance and SWIR indicates short-wave infrared reflectance. The wavelength spectrum wavelength of the SWIR region is (1.57–1.65 m), while the spectrum wavelength of the NIR region is (0.85–0.88 m) [11]. We calculated the LSWI for the Tons Pump Canal Command region using Landsat-8 data. Several sample points were used to determine the vegetation moisture conditions of the area.

Normalized Difference Vegetation Index (NDVI) is computed using Eq. 2.

$$\text{NDVI} = (\text{NIR} - \text{RED})/(\text{NIR} + \text{RED}) \tag{2}$$

where RED indicates visible red reflectance, whereas NIR indicates near-infrared reflectance. The spectrum wavelength of the NIR band is 0.85–0.88 m, whereas the spectrum wavelength of the red band is 0.64–0.67 m [12]. We calculated the NDVI for the Tons Pump Canal command region using Landsat-8 data. Different sampling points were used to determine if the vegetation condition was low, high, or not.

4.1 Methodology Adopted for the Study and Software Used

See Fig. 2.

In data analysis, the following software programs were used:

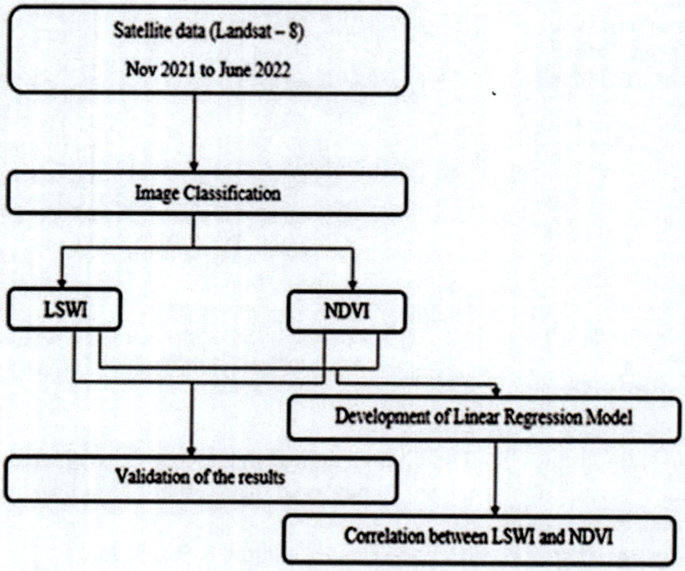

Fig. 2 Flow chart of methodology

- ArcMap 10.8 is used for database construction and analysis, especially image processing.
- Microsoft Excel was used for creating a spreadsheet, creating scatter plots, and finding correlation.

4.2 Remote Sensing Data

LANDSAT—8 satellite images can be accessed and downloaded from a variety of sources. The USGS Earth Explorer website has the most comprehensive collection of satellite data. The present study used the LANDSAT 8 sensors from November 2021 to June 2022. The satellite imagery has a resolution of 30 m, excluding the thermal and PAN bands [13]. Satellite pictures were turned to the Universal Transverse Mercator (UTM) projection in Zone 44N (Fig. 3).

The Land Use and Land Cover (LULC) map shown in Fig. 4 depicts the land surface's vegetation, water bodies, and natural surface as well as cultural elements. Tons Main Canal is 37.4 km long, and the Tons Pump Canal Command area comprises water bodies, vegetation, barren land, built-up areas, and settlements [13].

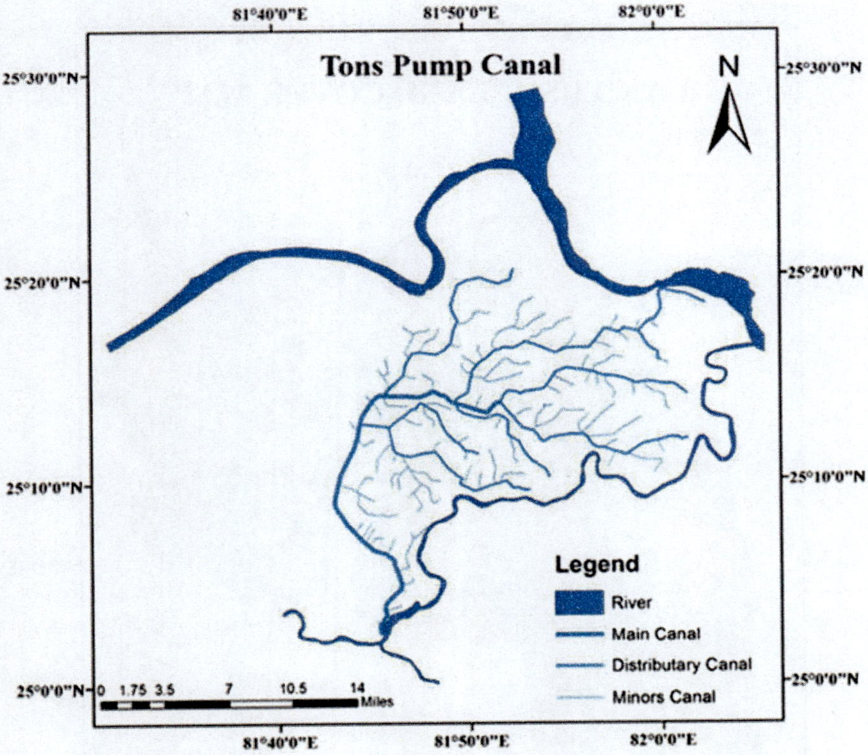

Fig. 3 Tons pump canal map

4.3 Multiple Linear Regression Models (MLR) and Square of Correlation Coefficient

MLR is a method for modeling the linear connection between one or more varying dependent and independent variables. Every independent variable value is associated with a dependent variable value. There are two variables in the multiple regression analysis, one of which is known as an independent variable or predictor. The values of the second variable are forecasted based on the values of this variable. The dependent variable is the other variable whose values are forecasted.

R^2 is the correlation coefficient. It defines the relation of a two-variable linear association. The coefficient of determination (R^2) is an important outcome of a regression analysis. It is defined as the amount of modification in the dependent variable that the explanatory variables can predict.

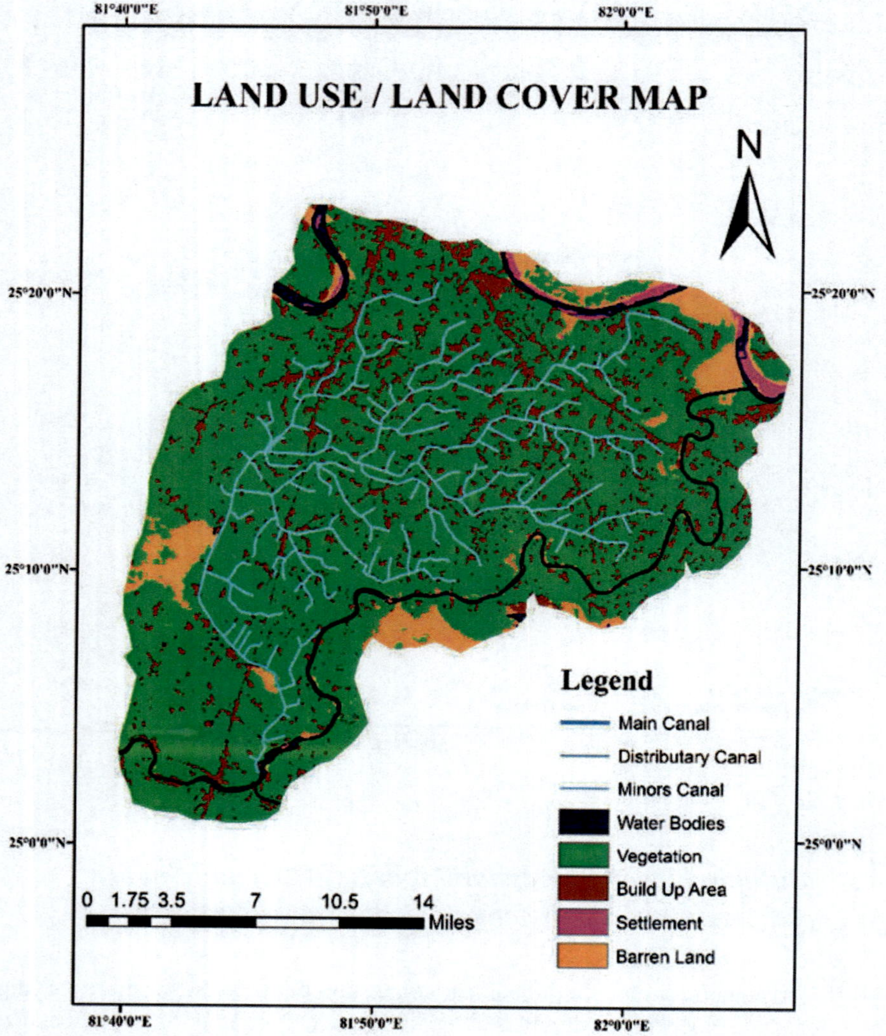

Fig. 4 The study area land use land cover map

4.4 *Drought Range of Land Surface Water Index (LSWI) and Normalized Difference Vegetation Index (NDVI)*

The value ranges of the Land Surface Water Index (LSWI) and the Normalized Difference Vegetation Index (NDVI) are shown in Tables 2 and 3. We can quickly understand the vegetation and vegetation moisture status of the area based on that Tables [14, 15].

Table 2 Classification of normalized difference vegetation index (NDVI)

NDVI ranges	Drought
(<0)	Extreme dry/water body
(0.2–0)	Dry condition
(0.4–0.2)	Moderate condition
(0.6–0.4)	Wet condition
(0.6–1.0)	Extreme wet condition

Table 3 Classification of Land Surface Water Index (LSWI)

LSWI ranges	Drought
(1.0 to 0.1)	No drought
(0.1 to 0)	Abnormally dry
(0 to −0.1)	Moderate
(−0.1 to −0.5)	Severe
(−0.5 to −1.0)	Extreme

5 Results and Discussion

5.1 Vegetation Moisture Detection Employing Satellite Indices

As vegetation moisture indices seem to be more prevalent and susceptible to agricultural drought conditions. It detects moisture using near- and short-wave infrared bands. The shortwave band exhibits variations in the water content of vegetation as well as the mesophyll microstructure in canopy vegetation, whereas near-infrared reflectance is modulated by leaf. Water content is not affected by intrinsic structure or dry matter content. This band combination lowers changes caused by internal leaf structure as well as dry matter leaf proportion, thus enhancing the precision of monitoring the internal water content of the vegetation and its soil background. The moisture present in the leaf's organic structure substantially controls reflectance in the electromagnetic spectrum in the infrared short-wave region. As a result, SWIR reflectance is significantly correlated to the water content of the leaf.

5.2 Detection of Vegetation from Satellite-Based Indices

The amount of vegetative cover and vegetation condition were quantified using vegetation indices. Leaf pigment and structure tightly regulate the interactions of incident sunlight with green plants. The leaf's primary pigment, chlorophyll, absorbs light strongly in the visible blue and red spectral regions while bouncing green wavelengths, bringing in the color of the green leaf we witness.

Table 4 Showing Multiple Linear Regression and Square of Correlation Coefficient (R^2) on monthly basis from Nov 2021 to June 2022

MLR equation	Square of correlation coefficient (R^2)	Month
$y = 0.6451x + 0.1764$	($R^2 = 0.5489$)	22-Nov-21
$y = 0.8022x + 0.1641$	($R^2 = 0.6269$)	8-Dec-21
$y = 0.8786x + 0.1714$	($R^2 = 0.9024$)	25-Jan-22
$y = 0.7708x + 0.195$	($R^2 = 0.9592$)	26-Feb-22
$y = 0.7582x + 0.1884$	($R^2 = 0.9526$)	14-Mar-22
$y = 0.5461x + 0.1657$	($R^2 = 0.5487$)	30-Mar-22
$y = 0.666x + 0.1568$	($R^2 = 0.6995$)	15-Apr-22
$y = 0.6586x + 0.1604$	($R^2 = 0.7095$)	1-May-22
$y = 0.943x + 0.1587$	($R^2 = 0.8377$)	2-Jun-22

5.3 *Development of Multiple Linear Regression Models (MLR)*

MLR models are developed using the parameters acquired from satellite data, notably the Land Surface Water Index (LSWI) and Normalized Difference Vegetation Index (NDVI) values. MLRs were utilized for investigating the links between the two satellite data sets. The satellite data values of the Land Surface Water Index (LSWI) are deployed as independent variables, while the Normalized Difference Vegetation Index (NDVI) values are employed as dependent variables. The R^2 values between two indices are computed using Remote Sensing Parameters (Table 4).

5.4 *Square of Correlation Coefficient*

For each month of the satellite pass, Landsat-8 OLI data were retrieved. I.e. 22-Nov-2021, 8-Dec-2021, 25-Jan-2022, 26-Feb-2022, 14-Mar-2022, 30-Mar-2022, 15-Apr-2022, 1-May-2022, and 02-Jun-2022, correlations have been made by taking ground parameters and space observations, i.e., satellite data. Figures 5, 6, 7, 8, 9, 10, 11, 12 and 13 shows the correlation graphs that reveal the correlation between the Land Surface Water Index (LSWI) and Normalized Difference Vegetation Index (NDVI) for each date of pass for several months.

Each point on the plot represents a separate observed sample point. Between the two indices, a correlation and regression equation is developed. It indicates that in the early stages of vegetation, Land Surface Water Index (LSWI) and Normalized Difference Vegetation Index (NDVI) do not have a good relationship, but when the vegetation reaches the harvesting stage, the relationship improves.

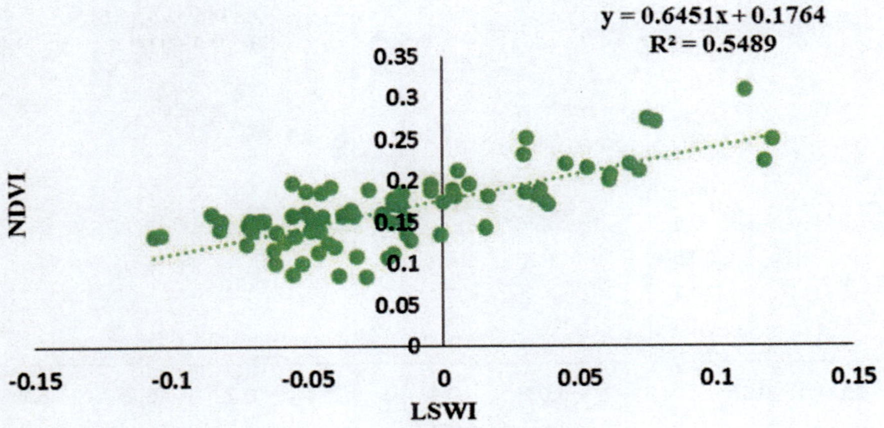

Fig. 5 Co-relation between LSWI and NDVI for 22-Nov-2021

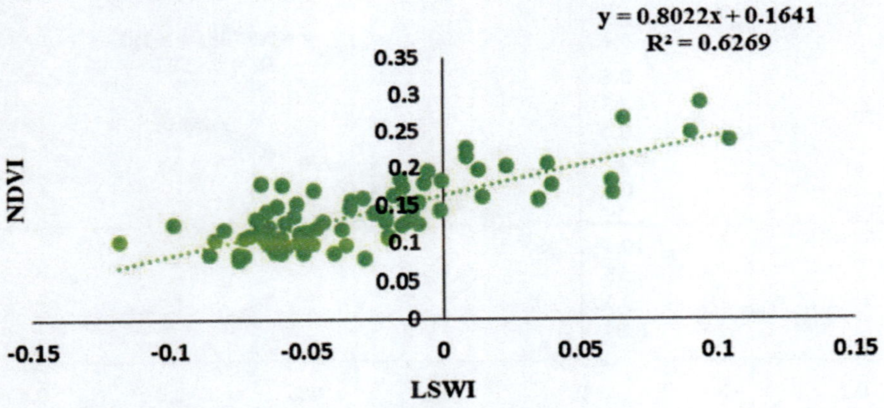

Fig. 6 Co-relation between LSWI and NDVI for 8-Dec-2021

5.5 *A graphical portrayal of drought threshold of Land Surface Water Index (LSWI) and Normalized Difference Vegetation Index (NDVI) for the year 2021–2022 in Tons Pump Canal*

Figures 14, 15, 16, 17, 18, 19, 20, 21, and 22 depict the land surface water index (LSWI) and normalized difference vegetation index (NDVI) drought thresholds for the period November 2021–June 2022. The normalized difference vegetation index is commonly used to measure vegetation health. Normalized Difference Vegetation Index values above 0.2 indicate healthy vegetation, while values below 0.2 indicate agricultural drought conditions in an area. The Land Surface Water Index is used to measure moisture content and drought conditions. A value of 0 or above indicates

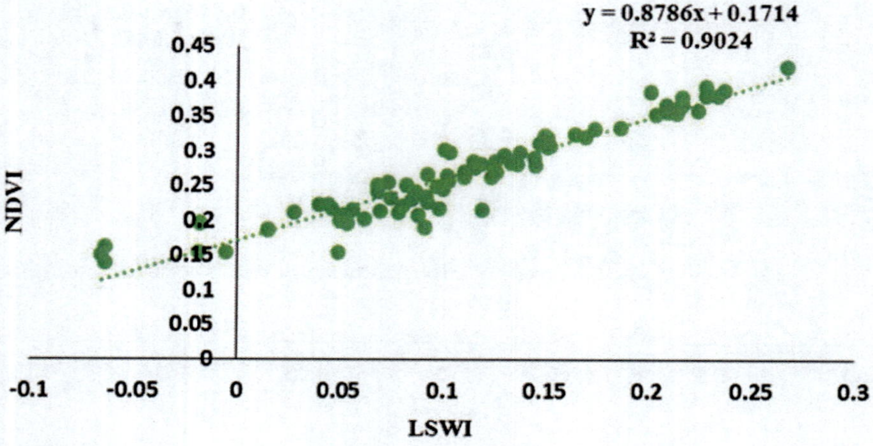

Fig. 7 Co-relation between LSWI and NDVI for 25-Jan-2022

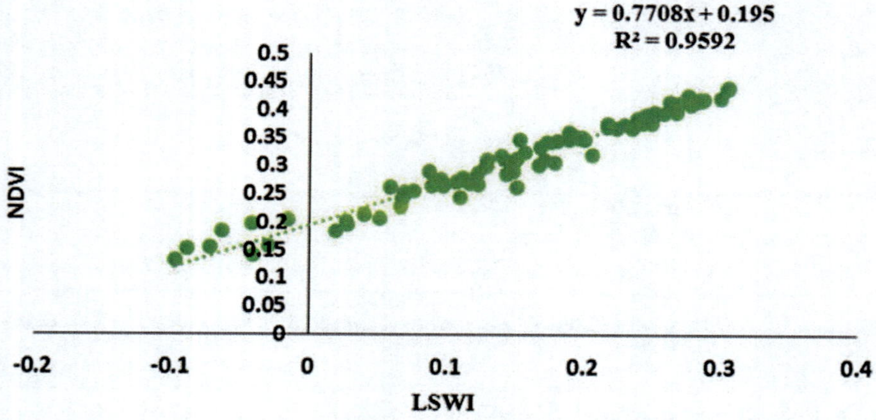

Fig. 8 Co-relation between LSWI and NDVI for 26-Feb-2022

that the vegetation is not in a drought state or is in the early stages of growth, while a value below 0 suggests that the area is experiencing agricultural drought in Tons Pump Canal Command during the months of November 2021 to June 2022.

5.6 *Drought Classification of NDVI & LSWI for Each Month*

The maps of the Drought Classification of Land Surface Water Index (LSWI) and Normalized Difference Vegetation Index (NDVI) are shown in the figures (Figs. 23,

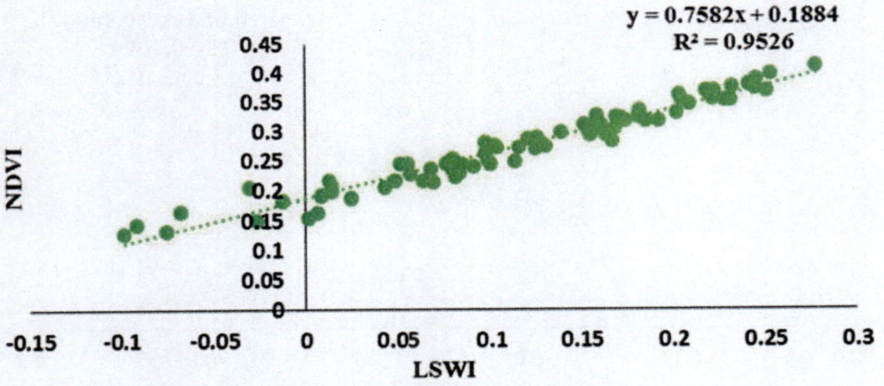

Fig. 9 Co-relation between LSWI and NDVI for 14-Mar-2022

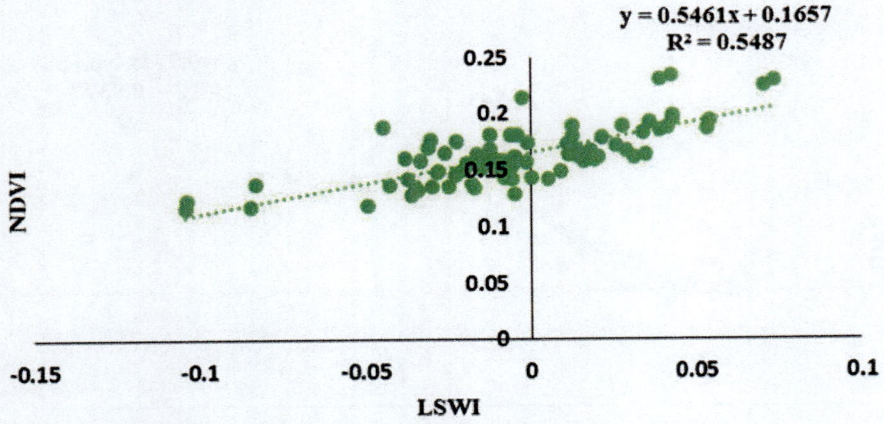

Fig. 10 Co-relation between LSWI and NDVI for 30-Mar-2022

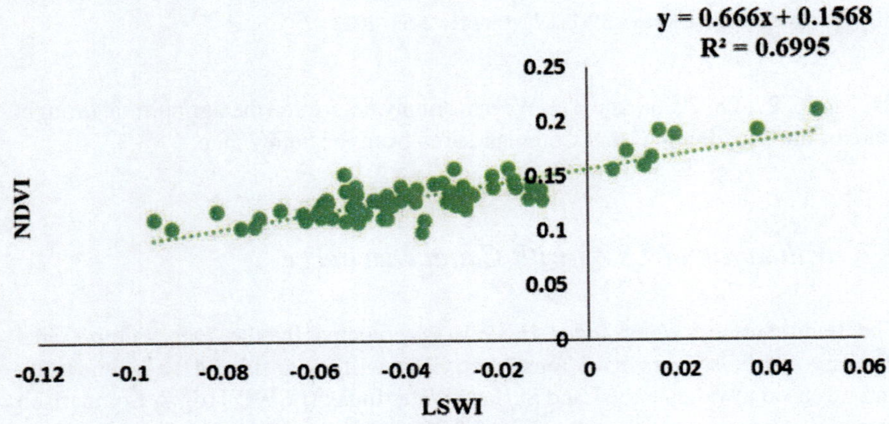

Fig. 11 Co-relation between LSWI and NDVI for 15-Apr-2022

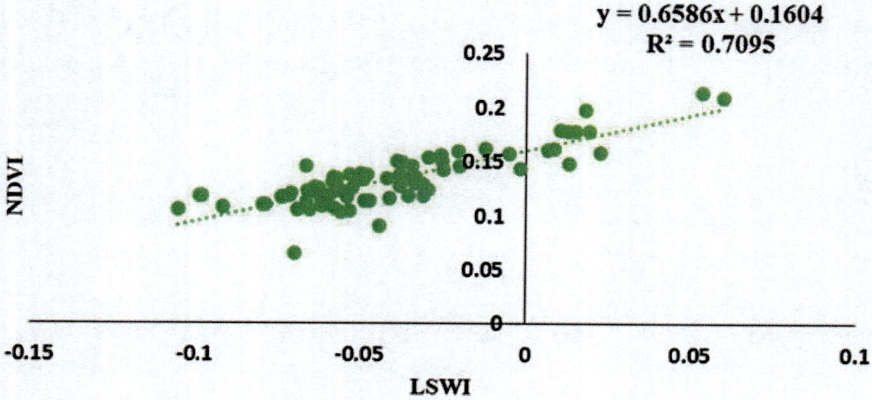

Fig. 12 Co-relation between LSWI and NDVI for 1-May-2022

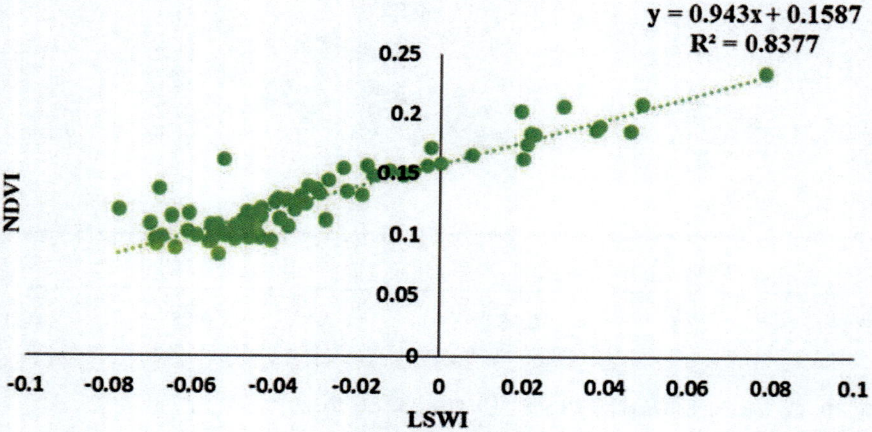

Fig. 13 Co-relation between LSWI and NDVI for 2-Jun-2022

24, 25, 26, 27, 28, 29, 30 and 31). We can simply determine the agricultural drought state of the Tons Pump Canal Command area from the below map.

5.7 *Validation of LSWI with Canal Discharge*

The Land Surface Water Index (LSWI) is validated by the Tons Pump Canal discharge. The discharge in the Tons Pump Canal will be considered the ground truth, and it is used to validate the Land Surface Water Index (LSWI) [16]. A Comparison between percentages of the area under LSWI with that of Tons Canal Discharge is

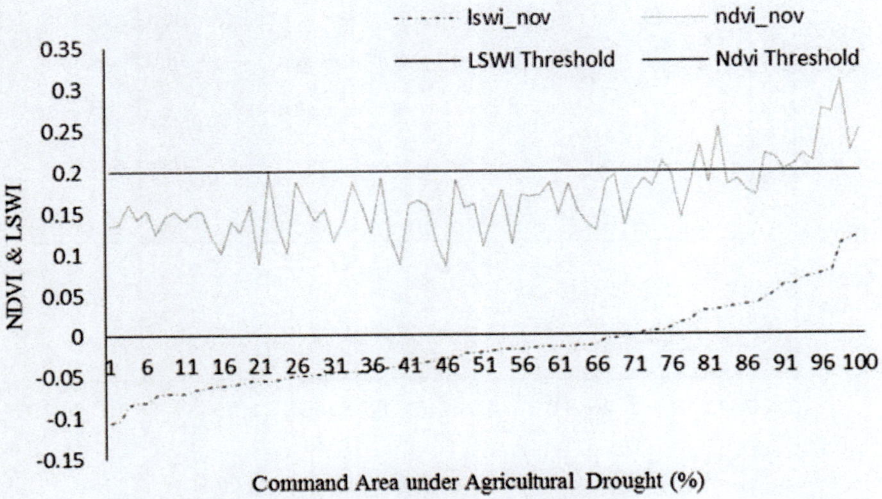

Fig. 14 A graphical portrayal of drought threshold of LSWI and NDVI for 22-Nov-2021

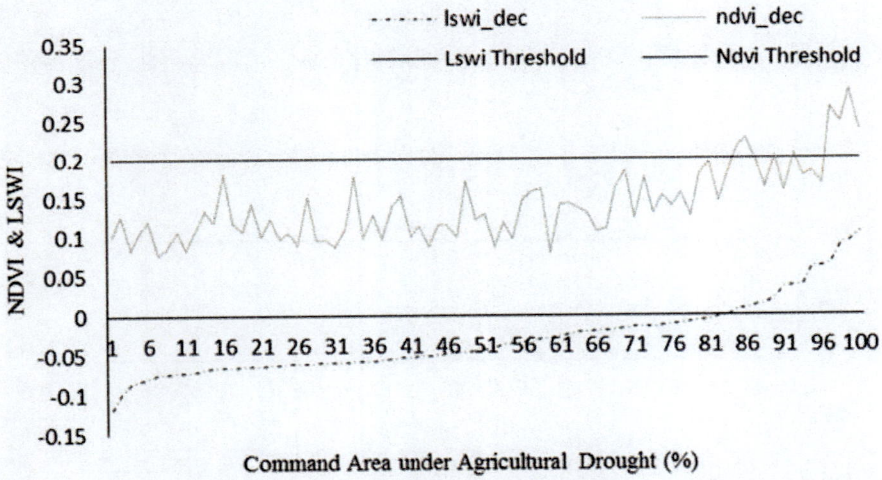

Fig. 15 A graphical portrayal of drought threshold of LSWI and NDVI for 8-Dec-2021

done to understand the validity of LSWI with discharge. It was found that whenever there is discharge in Tons Pump Canal Command, the percentage of area under moisture stress is lower. The validation is shown in Fig. 32.

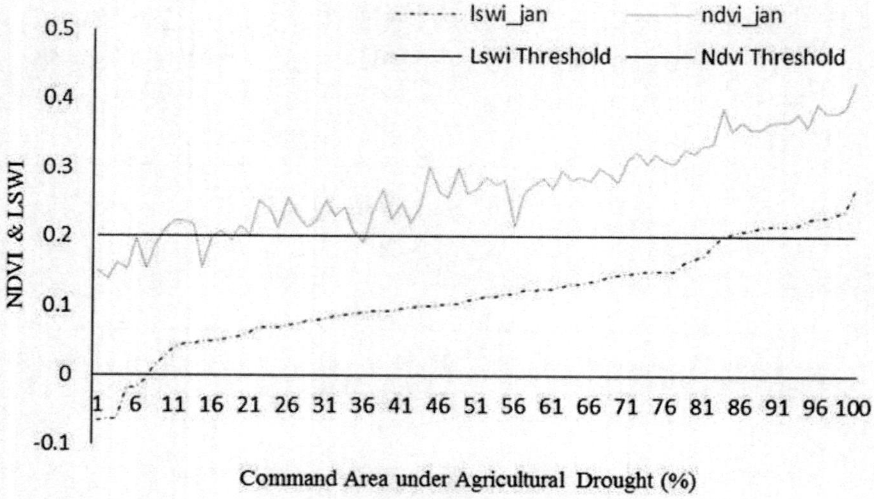

Fig. 16 A graphical portrayal of drought threshold of LSWI and NDVI for 25-Jan-2022

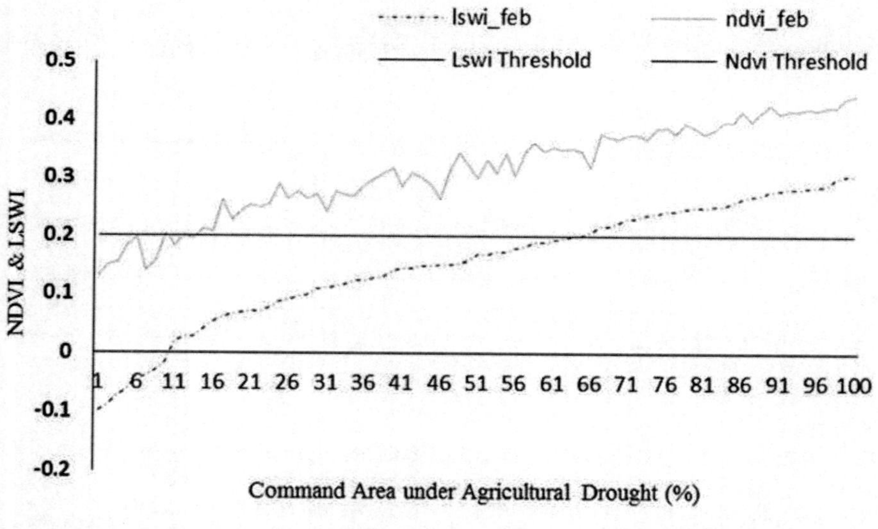

Fig. 17 A graphical portrayal of drought threshold of LSWI and NDVI for 26-Feb-2022

6 Conclusion

The study was focused on identifying vegetation moisture and condition using satellite-based indices. The research was carried out at the Tons Pump Canal Command area in Uttar Pradesh, in the Prayagraj district. This study used Landsat-8 data from November 2021 until June 2022, between the Rabi and Zaid seasons.

From this study, the following conclusion was made-

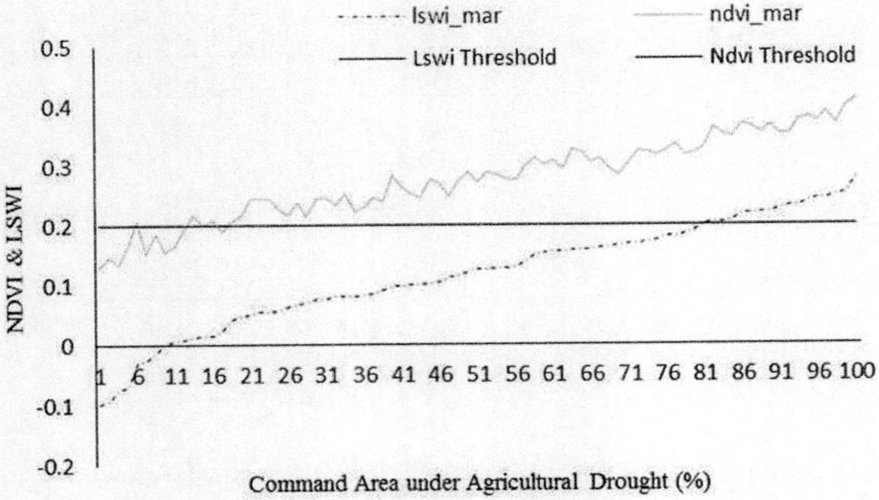

Fig. 18 A graphical portrayal of drought threshold of LSWI and NDVI for 14-Mar-2022

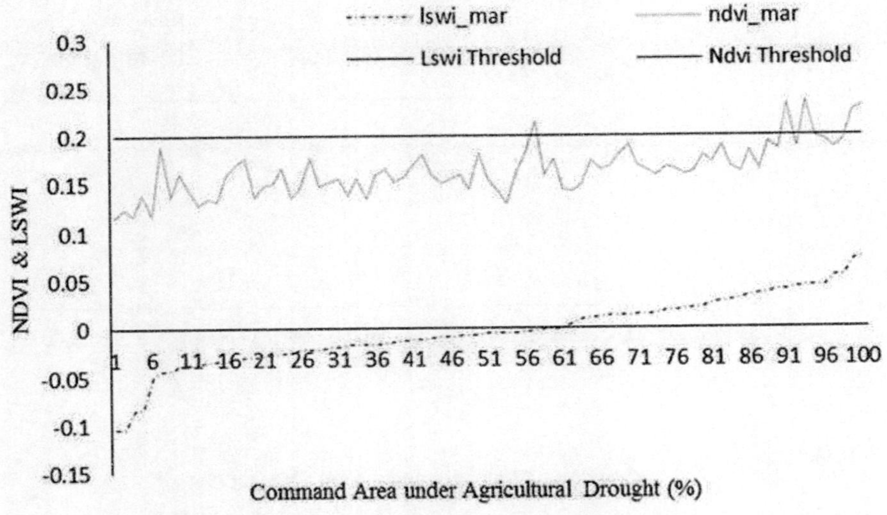

Fig. 19 A graphical portrayal of drought threshold of LSWI and NDVI for 30-Mar-2022

The co-relationship between the Land Surface Water Index (LSWI) and the Normalized Difference Vegetation Index (NDVI) is not always clear or linear. However, during the vegetation's growth stages, it is observed that both indices' values generally increase together, until the vegetation reaches its harvesting stage. It has been revealed that during the Rabi and Zaid seasons, the value of the square of the correlation coefficient (R^2) shows an increasing trend, which reveals that both indices have a strong relationship. The normalized difference vegetation index is commonly used to measure vegetation health. Normalized Difference

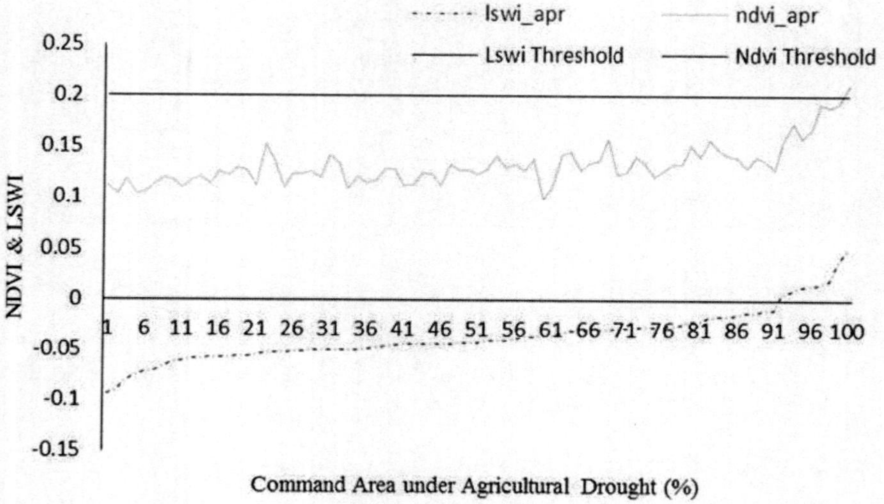

Fig. 20 A graphical portrayal of drought threshold of LSWI and NDVI for 15-Apr-2022

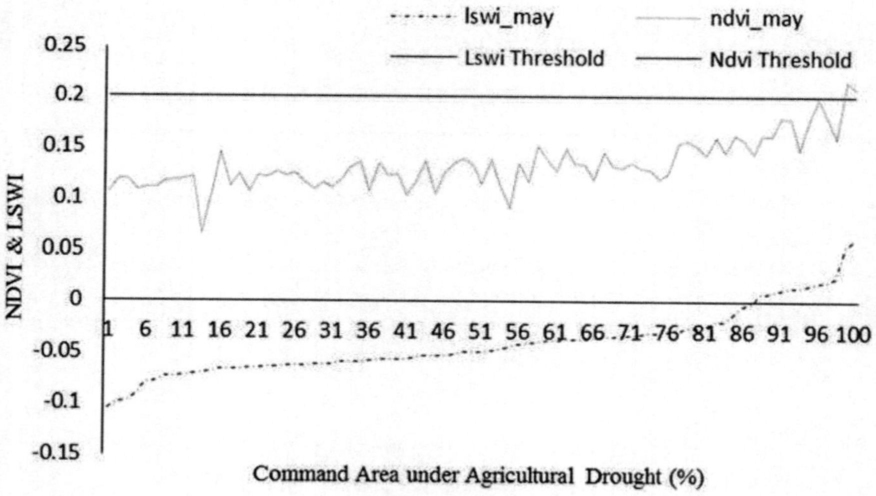

Fig. 21 Graphical representation of drought threshold of LSWI and NDVI for 1-May-2022

Vegetation Index values above 0.2 indicate healthy vegetation, while values below 0.2 indicate agricultural drought conditions in an area. The Land Surface Water Index is used to measure moisture content and drought conditions. A value of 0 or above indicates that the vegetation is not in a drought state or is in the early stages of growth, while a value below 0 suggests that the area is experiencing agricultural drought. It was indicated that both drought indices (vegetation and moisture) are excellent approaches for monitoring drought conditions.

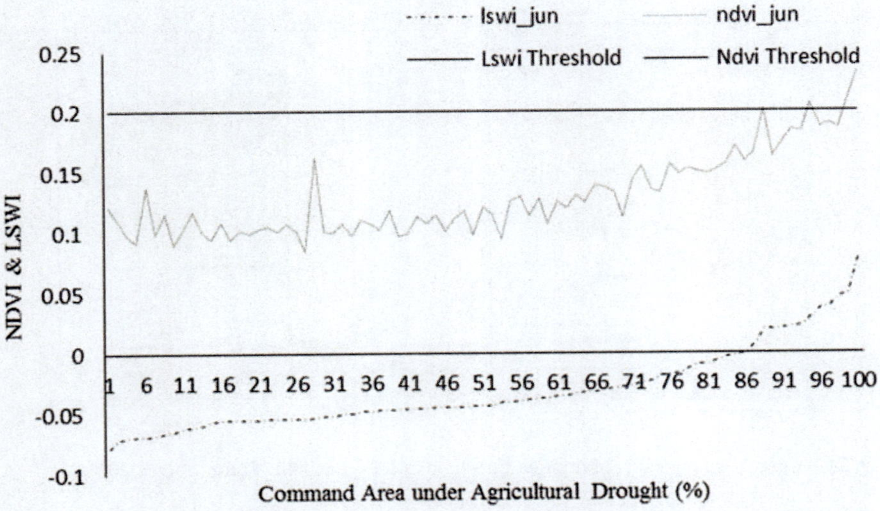

Fig. 22 A graphical portrayal of drought threshold of LSWI and NDVI for 2-Jun-2022

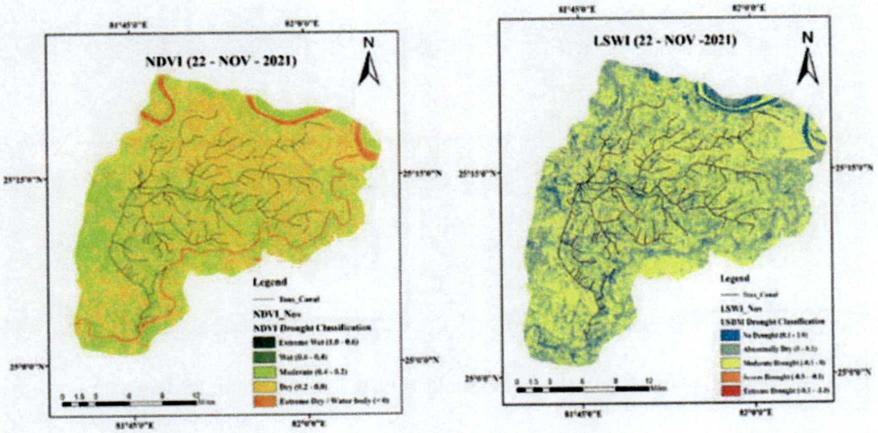

Fig. 23 A graphical portrayal of drought threshold of LSWI and NDVI for 22-Nov-2021

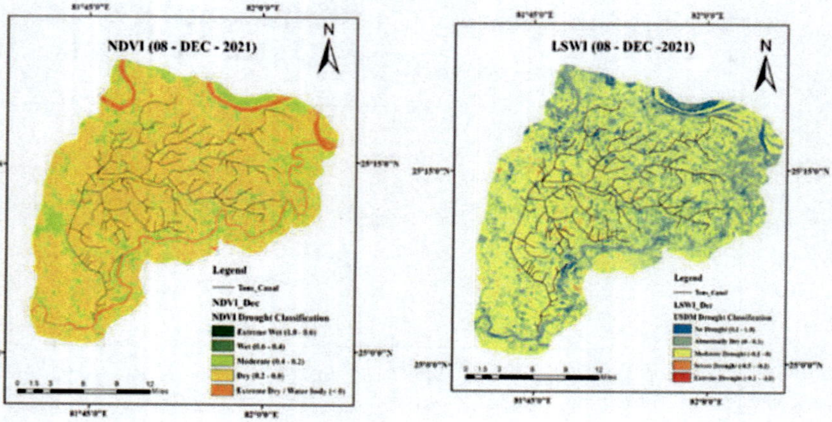

Fig. 24 A graphical portrayal of drought threshold of LSWI and NDVI for 08-Dec-2021

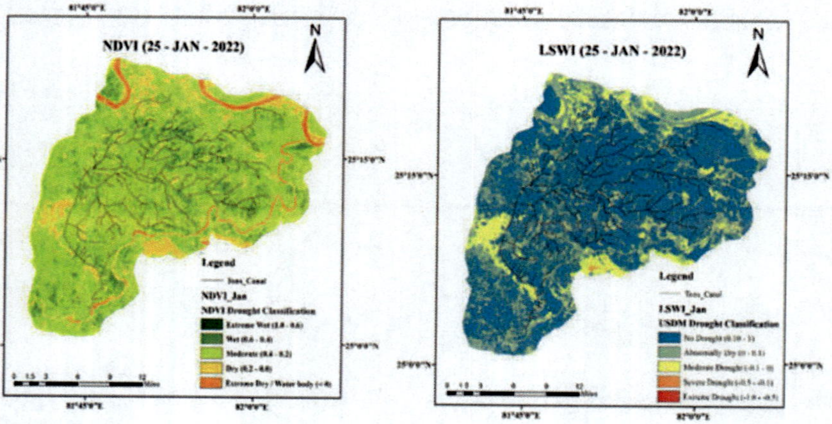

Fig. 25 A graphical portrayal of drought threshold of LSWI and NDVI for 25-Jan-2022

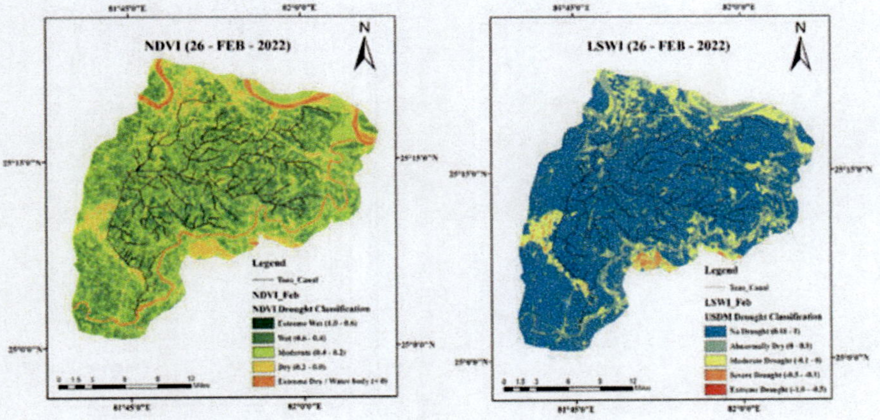

Fig. 26 A graphical portrayal of drought threshold of LSWI and NDVI for 26-Feb-2022

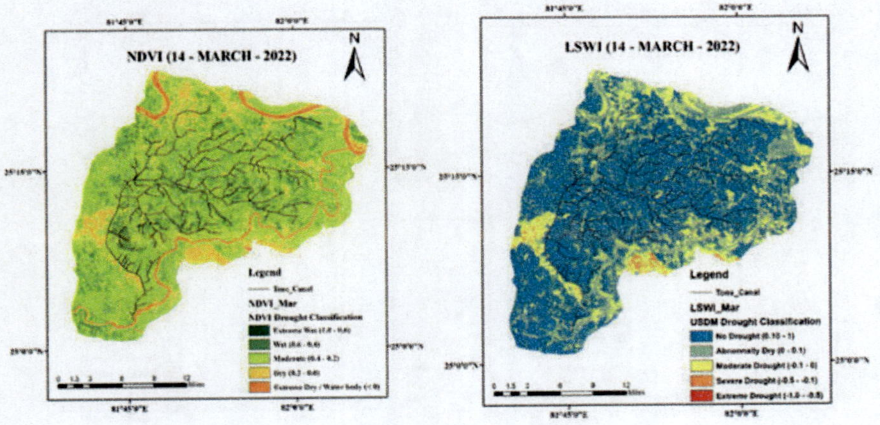

Fig. 27 A graphical portrayal of drought threshold of LSWI and NDVI for 14-Mar-2022

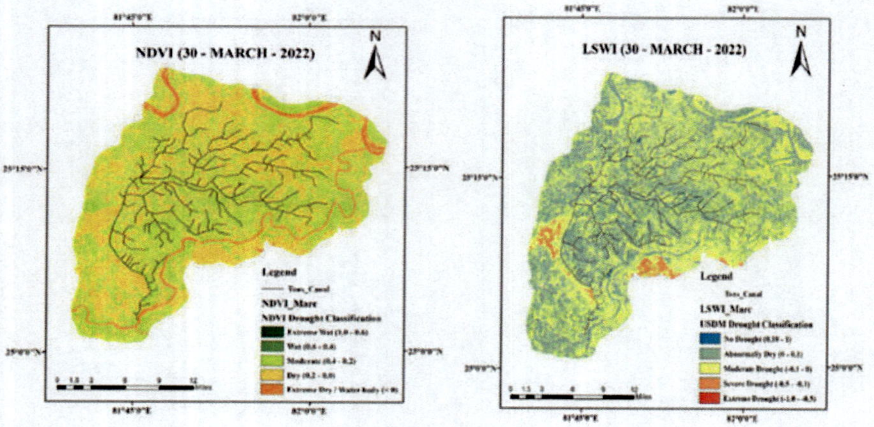

Fig. 28 A graphical portrayal of drought threshold of LSWI and NDVI for 30-Mar 2022

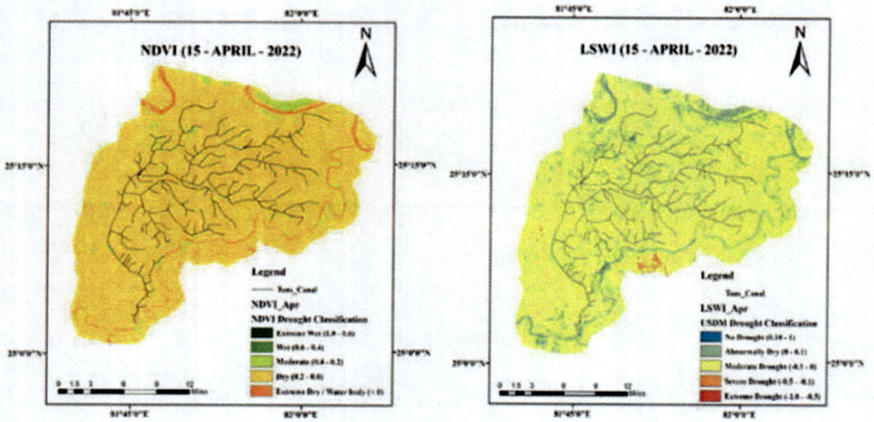

Fig. 29 A graphical portrayal of drought threshold of LSWI and NDVI for 15-Apr-2022

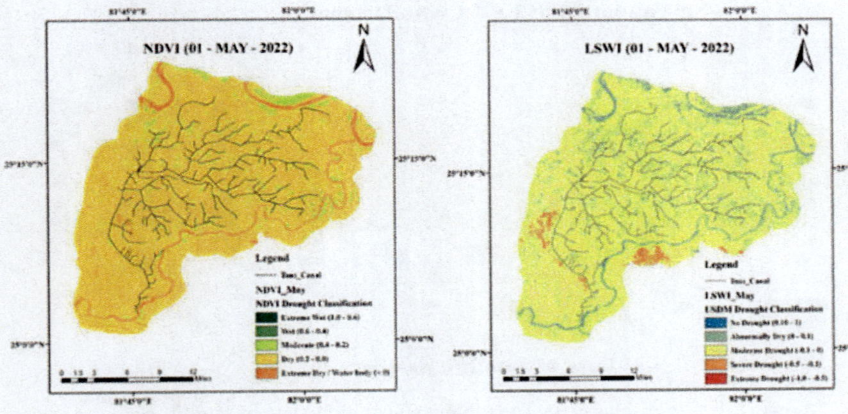

Fig. 30 A graphical portrayal of drought threshold of LSWI and NDVI for 01-May-2022

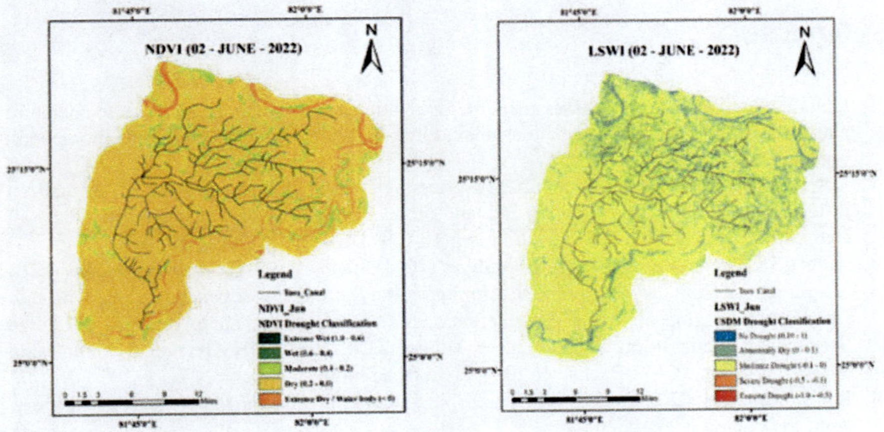

Fig. 31 A graphical portrayal of drought threshold of LSWI and NDVI for 2-Jun-2022

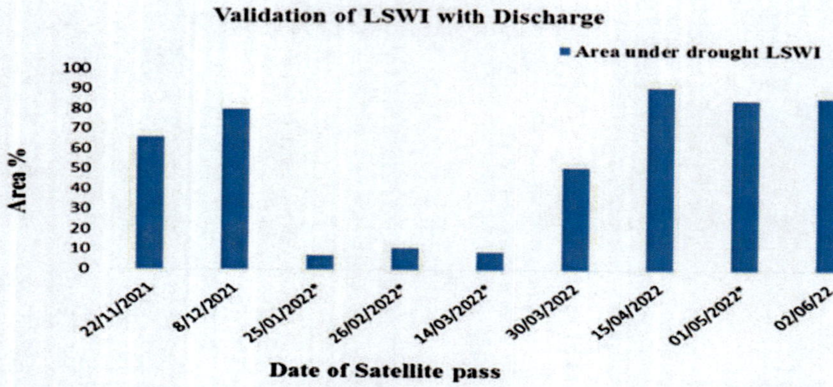

*showing date of discharge in Tons Pump Canal with Date of satellite pass

Fig. 32 Validation of LSWI with canal discharge for the year (2021–2022)

References

1. Guha S, Govil H (2021) An assessment on the relationship between land surface temperature and normalized difference vegetation index. Environ Dev Sustain 23:1944–1963. https://doi.org/10.1007/s10668-020-00657-6
2. Garai S, Khatun M, Singh R et al (2022) Assessing correlation between Rainfall, normalized difference Vegetation Index (NDVI) and land surface temperature (LST) in Eastern India. Saf. Extreme Environ. 4:119–127. https://doi.org/10.1007/s42797-022-00056-2
3. Padhee SK, Nikam BR, Aggarwal SP, Vaibhav (2014) Integrating effective drought index (EDI) and remote sensing derived parameters for agricultural drought assessment and prediction in Bundelkhand region of India. The International Archives of the Photogrammetry, Remote Sensing and Spatial Information Sciences, Volume XL-8, 2014 ISPRS Technical Commission VIII Symposium, 09–12 December 2014, Hyderabad India
4. Raj A, Minz S (2020) Game theory based pixel approximation for remote sensing imagery. Appl Soft Comput 93:106365
5. Gao B (1996) NDWI is a normalized difference water index used for remote sensing of liquid water in plants from space. Remote Sens Environ 58:257–266
6. Thenkabail PS, Gamage MSDN, Smakhtin VU (2004) The use of remote sensing data for drought assessment and monitoring in Southwest Asia. Research Report 85. Colombo, Sri Lanka International Water Management Institute. ISBN 92-9090-575-1 ISSN 1026-0862
7. Chandrasekar K, Sesha Sai MR, Roy PS, Dwevedi RS (2010) Response of the land surface water index (LSWI) to rainfall and NDVI using the MODIS vegetation index product. Int J Remote Sens 31(15):3987–4005. https://doi.org/10.1080/01431160802575653
8. https://mail.irjet.net/archives/V6/i1/IRJET-V6I136.pdf
9. Muralitharan J, Gunalan J, Duraisamy R, Abebe A (2021) Study the relationship between LULC, LST, NDVI, NDWI and NDBI in Greater Arba Minch Area, Rift Valley, Ethiopia. https://doi.org/10.2991/ahis.k.210913.023
10. Raj A, Minz S (2021) A Scalable Unsupervised Classification Method Using Rough Set for Remote Sensing Imagery. International Journal of Software Science and Computational Intelligence (IJSSCI) 13(2):65–88
11. Zhao S, Qin Q, You L, Yao Y, Yang N, Li J (2009) Application of two shortwave infrared water stress indices to drought monitoring over Northwestern China. Geoscience and Remote Sensing Symposium (IGARSS 2009), III-530–III-533

12. Gu Y, Brown JF, Verdin JP, Wardlow B (2007) A five-year analysis of MODIS NDVI and NDWI for grassland drought assessment over the central Great Plains of the United States. Vol 34, Issue 6, March 2007
13. Kurnaz B, Bayik C, Abdikan S (2020) Forest fire area detection by using Landsat-8 and Sentinel-2 satellite images: a case study in Mugla, Turkey. https://doi.org/10.21203/rs.3.rs-26787/v1
14. Bajgain R, Xiao X, Wagle P, Basara J, Zhou Y (2015) Drought sensitivity investigation of vegetation indicators at two tallgrass prairie locations. ISPRS J Photogramm Remote Sens 108(2015):151–160
15. Du TLT, Bui DD, Nguyen MD, Lee H (2015) Satellite-based, multi-indices for evaluation of agricultural droughts in a highly dynamic tropical catchment, central vietnam. licensee MDPI, Basel, Switzerland
16. Chandrasekara K, Sesha Saia MVR, Beheraa G (2011) Assessment of early season agricultural drought through land surface water index (LSWI) and Soil Water Balance Model. International Archives of the Photogrammetry, Remote Sensing and Spatial Information Sciences, Volume XXXVIII-8/W20, 2011, ISPRS Bhopal 2011 Workshop, 8 November 2011, Bhopal, India

Assessing Irrigation Performance of a Canal Command Area Using Space and Ground Observation. A Case Study of Belan Canal, Prayagraj

Vikas Mishra, Derrick Denis, Himanshu Mishra, and Mukesh Kumar

1 Introduction

In spite of climate variability, irrigation will continue to be a crucial component of food security programmes. It is the primary method of intensification. It is unlikely that the world can abandon a food production system that supplies 40% of the world's food from just 20% of the world's arable land and supports food supplies in populous nations like Pakistan, China, and India despite underperformance and past overinvestment with its environmental consequences. These three nations account for 80, 35, and 34% of the agricultural areas in Pakistan, China, and India, respectively, of the world's total irrigated land [2]. A little more than 300 million ha of the world's arable land can be irrigated [2].

Agriculture's contribution to India's Gross Domestic Product (GDP) is progressively declining. Programs for food security are successful thanks to recent significant investments in irrigation infrastructure. As a result, the local population in the Gross Command Area has a higher standard of living, which raises a country's GDP overall [5, 6, 13]. India's irrigation water management and supply are determined by government projections rather than the level of farmer requests. At the moment, these evaluations simply take into account how resources, including productivity and

V. Mishra (✉)
Irrigation and Drainage Engineering, VIAET, SHUATS, Prayagraj, India
e-mail: vikashm496@gmail.com

D. Denis
Dept. of Irrigation and Drainage Engineering, VIAET, SHUATS, Prayagraj, India
e-mail: derrick.denis@shiats.edu.in

H. Mishra
VIAET, SHUATS, Prayagraj, India

M. Kumar
Centre for Geospatial Technologies, VIAET, SHUATS, Prayagraj, India
e-mail: mukesh@shiats.edu.in

A. Ramdane-Cherif et al. (eds.), *Machine Intelligence and Data Science Applications*, Algorithms for Intelligent Systems, https://doi.org/10.1007/978-981-99-1620-7_38

profitability, are used. However, using geographical and temporal operational indicators that take into account not just productivity and profit but also service quality and agricultural water demand would be a better choice and irrigation of crops Performance evaluation can be used to aid in the diagnosing process [2]. The overarching goal of performance evaluation is to deliver pertinent input to project management at all levels of the irrigation system in order to achieve efficient and effective project performance.

The main objectives of this research are:

(I) Estimation of Actual evapotranspiration at a study area.
(II) Quantification of irrigation performance indicators at a study area.
(III) To develop an approach to select the most relevant performance indicators to assess irrigation performance.

2 Related Works

In 2001, this paper shows a method for evaluating the effectiveness of the water distribution system from the standpoint of farmers, which is what the research wants to explore. The research examines farmer replies to questions on how they view the irrigation service offered using the idea of fuzzy set theory. The study adopts the stance that it is possible to utilise farmers' assessments to evaluate the efficacy of the water delivery system in schemes where statistics on water delivery are unavailable or their integrity cannot be assured. The Dawhenya irrigation system in Ghana served as the case study for the methodology.

In 2018, According to the researcher, a case study in Southern California used remote sensing and GIS techniques for evaluating irrigation performance. This study examines the effectiveness, sufficiency, and sustainability of geographically dispersed irrigation using remotely sensed data. A southern California irrigation district's real evapotranspiration (ET) was mapped using the surface energy balance algorithm for land (SEBAL). The Priestley–Taylor method was improved to map potential ET and take horizontally transmitted energy into account when increasing or inhibiting ET. A Geographical Information System (GIS) environment was used to combine remotely sensed data with ground-based data to quantify a number of irrigation and drainage performance metrics. Actual Evapotranspiration between and between field coefficients of variation were comparable to those from earlier experiments, indicating that water consumption was consistent across the irrigation district.

In 2019, using remote sensing (RS) and the geographic information system, the Atabey Irrigation Scheme's irrigation performance was studied (GIS). To achieve this, the scheme's crop pattern maps were created using the supervised classification approach with the study area's Rapid Eye satellite data from 2012 using RS techniques, whilst the scheme's canal maps were created using techniques from the Geographic Information System. These data were used to calculate the monthly water supply ratios, one of the performance measures.

In 2020, The purpose of the study is to demonstrate how effectively Burkina Faso's irrigation plans work, which leads to labor-intensive, expensive processes. Although widely used for this type of performance assessment, Burkina Faso continues to underutilize remote sensing (RS)-based methods. We looked at the spatiotemporal performance patterns of the Kou Valley irrigation scheme (KVIS) across two successive crop seasons using multi-temporal Land sat imagery within the Python module for the Surface Energy Balance Algorithm for Land model. For rice, maize, and sweet potatoes, four performance parameters (depleted fraction, relative evapotranspiration, uniformity of water consumption, and crop water productivity) were calculated and contrasted with reference values.

2.1 Study Area

The Prayagraj district is located between latitudes 24°47′ and 25°43′ N and longitudes 81°31′ and 82°21′ E. The district is 90 to 140 m above mean sea level. The study region is between 24 48′32″ and 25 19′0″ N latitude and 81 44′38″ to 82 19′19″ E longitude. The start of the Belan canal is located at Meja Dam at 24 50′12.42″ N latitude and 82 4′41.34″ E longitude. The canal command area's length, measured from north to south is about 56 km and the width from east to west is about 55 km. The total lengths of Main canals, branch canals, distributaries and minors in the study area are 64,850 m, 33,268 m, 194,687 m, and 143,430 m respectively. The canal network is having one main and one branch canal, 14 distributaries, and 127 minors. The Belan canal system is a significant irrigation project. In the Prayagraj district. The four blocks of the district are Meja, Manda, Urwa, and Koraon having a 1674 km^2 area are covered under this Network of the Belan Canal (Fig. 1).

2.2 Climate

The climate of Prayagraj is tropical and the average maximum temperature range is between 42 °C and 46 °C which may go as high as 48 °C during peak summers. The minimum average temperature of the study area is 7–10 °C which may fall as low as 3 °C during peak winter months. The district experiences 952 mm of rainfall on average per year, and the monsoon season lasts from July to September [5, 13]. The climate is marked by high relative humidity nearly 71–82% during monsoon and it decreases during the summers by nearly 14–21%.

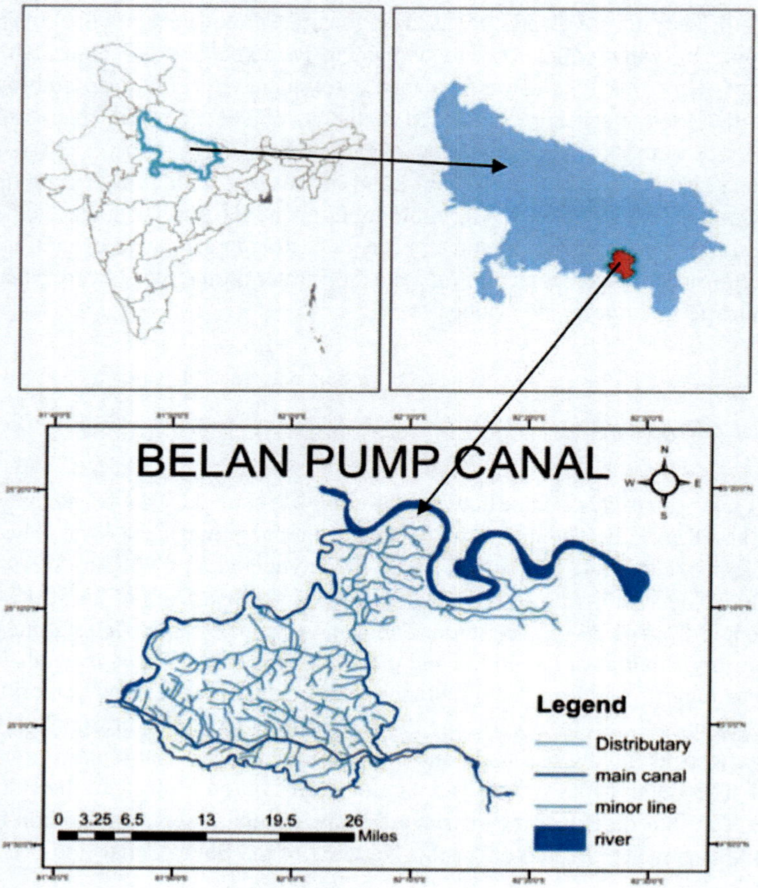

Fig. 1 The study area location map

3 Methodology

3.1 Methodology Adopted for the Study

See Fig. 2.

3.2 Fao Penman–Monteith Equation

ETo estimation using the FAO Penman–Monteith technique:-

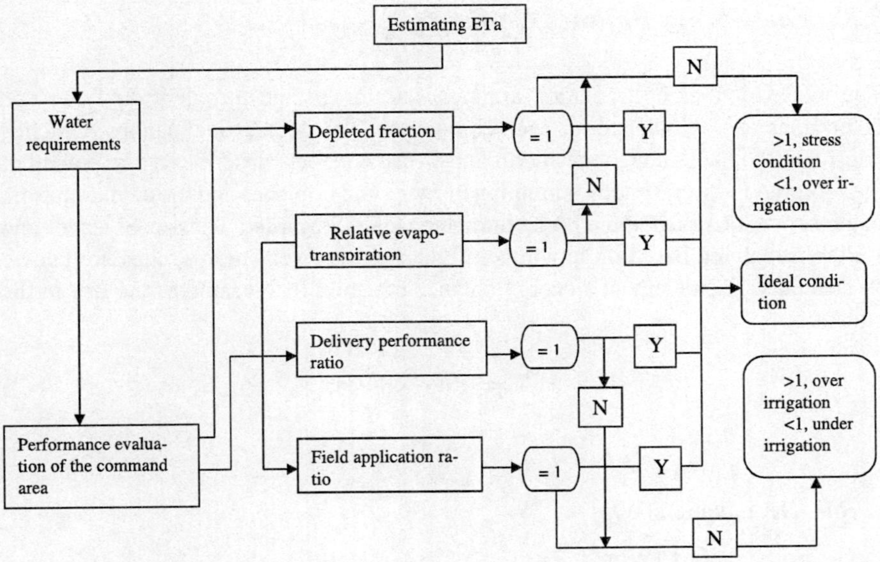

Fig. 2 Flow chart of methodology

$$ETo = \frac{0.408\Delta(Rn - G) + \gamma \frac{900}{T+273} u_2(e_s - e_a)}{\Delta + \gamma(1 + 0.34u_2)} \quad (1)$$

where,

ETo—reference evapotranspiration [mm day^{-1}],

Rn—net radiation [MJ m^{-2} day^{-1}],

G—Soil heat flux density [MJ m^{-2} day^{-1}],

T—Mean daily air temperature at 2 m height [°C],

u_2—wind speed at 2 m height [m s^{-1}],

e_s—Saturation vapors pressure [kPa],

e_a—Actual vapor pressure [kPa],

$e_s - e_a$—Saturation vapor pressure deficit [kPa],

Δ—slope vapor pressure curve [kPa °C^{-1}],

γ—Psychometric constant [kPa °C^{-1}].

3.3 The Energy Balance Equation

The model divides net radiation among heat storage and turbulent air flows and determines energy balance for each component. The geometrical canopy, which is considered to be homogenous and in one dimension, best describes a component of the prototype. As a result, the clumping of twigs and branches and the deviation from macroscopic qualities in the horizontal plane are disregarded. Classes of leaves and soil are separated based on how much light they receive. These groups are known as elements. Each element's energy balance equation is the same as the one in the equation.

$$\lambda ET = Rn - G - H. \tag{2}$$

Where,

Rn—Net radiation [W/m^2],

G—Soil heat flux [W/m^2].

λET—latent heat flux [W/m^2].

H—sensible heat flux. [W/m^2].

3.4 Evaluation of the Irrigated Areas' Performance Based on Actual Evapotranspiration

The primary goal of irrigated area performance assessment is to comprehend it's performance based on Actual evapotranspiration. To assure water supply equality and to maximize water distribution efficiency, it contains water service indicators, and criteria for evaluating an irrigated zone that are based on actual evapotranspiration. However, there are numerous signs that can be used to evaluate the accessibility of data was limited. These are the performance indicators:

(1) **Delivery performance ratio:** The water delivered as anticipated for crop growing season at any location with the irrigation system is known as the DPR [7]. The delivery performance ratio is the most straightforward but arguably most significant operational performance metric [13]. In it's basic form, it is defined as:

$$DPR = \frac{\text{Actual flow of water}}{\text{intended flow of water}} \tag{3}$$

$$DPR = \frac{V_C}{V_{C,int}} \tag{4}$$

where,

Vc—actual flow of water.

Vc, int—intended flow of water.

Amount of water to be given for irrigation from the main source (i.e., the reservoir or the river diversion) Vc, int can be expressed as [2, 8].

$$V_{c,int} = \frac{V_m - P_{e,ant}}{\varepsilon_{app}\varepsilon_{dis}\varepsilon_{con}} \tag{5}$$

where,

Vm—the amount of irrigation water needed.

Pe, ant—the effective use of anticipated rainfall.

Ɛapp—the application efficiency.

Ɛdis—the distribution efficiency.
Ɛcon—the conveyance efficiency.

(2) **Depleted fraction:** The Depleted Fraction compared the irrigated area's actual evapotranspiration to totally amount of precipitation received along with the surface water used for irrigation. The ratio between three elements of the water balance in an irrigated area is known as the depleted fraction. The diagnosis of issues in areas with a lack of water is made extremely simple by this indicator. For all of the area's rainfall as well as surface water inflows into the irrigated zone, depleted fraction compares total evapotranspiration from the specified area to the actual evapotranspiration from that site (typically irrigation water). It is defined as [7, 11, 13]:

$$\mathrm{DF} = \frac{\mathrm{ETa}}{\mathrm{Pe} + \mathrm{Vc}} \tag{6}$$

.

(3) **Relative evapotranspiration:** The dimensionless ratio can be used to assess the efficiency of irrigation water supply to particular command area as The water management can learn a lot from the function of actual evapotranspiration over prospective evapotranspiration over time. The ratio is defined as:

$$\mathrm{RET} = \frac{\mathrm{ETa}}{\mathrm{ETp}} \tag{7}$$

(4) **Crop water deficit:** The discrepancy between the potential and actual evapotranspiration caused by the cropping pattern over time in a given area is known as the CWD.

$$\text{CWD} = \text{ETp} - \text{ETa} \tag{8}$$

(5) **Field application ratio:** The Field application ratio quantifies the degree to which the requirements of crop irrigation are satisfied in the irrigated area by irrigation water [9, 10]. Assuming negligible non-irrigation water deliveries to the area, the ratio is defined as [9]:

$$\text{FAR} = \frac{\text{ETp} - \text{Pe}}{\text{Volume of water supplied to the field}} \tag{9}$$

where,

Pe—gross precipitation [mm].
Vc—surface water flowing [mm].
ETa—Actual Evapotranspiration [mm/day].
ETp—Potential Evapotranspiration [mm/day].

3.5 *Process of Selecting Performance Indicators*

See Fig. 3.

3.6 *Process of Selecting Performance Indicators for the Study Area*

A minimal number of performance indicators (PI) is selected from the lengthy list that has been published in the literature [2, 13], taking into account the cost of data collection and management. It is also necessary to confirm whether these indications can be measured quantitatively from routine fieldwork. Other measures or methods can occasionally be necessary. As a result, it is necessary to assess if doing so is feasible given the facilities and resources at hand. The application of GIS and satellite remote sensing is one such option. Additionally, it is vital to determine whether the parameters can be accurately measured or collected.

In order to quantify performance indicators, canal flow of the belan canal was available. Drainage data were collected from Irrigation Department, Govindpur, and Prayagraj (Uttar Pradesh). Rainfall data and other weather data (temperature,

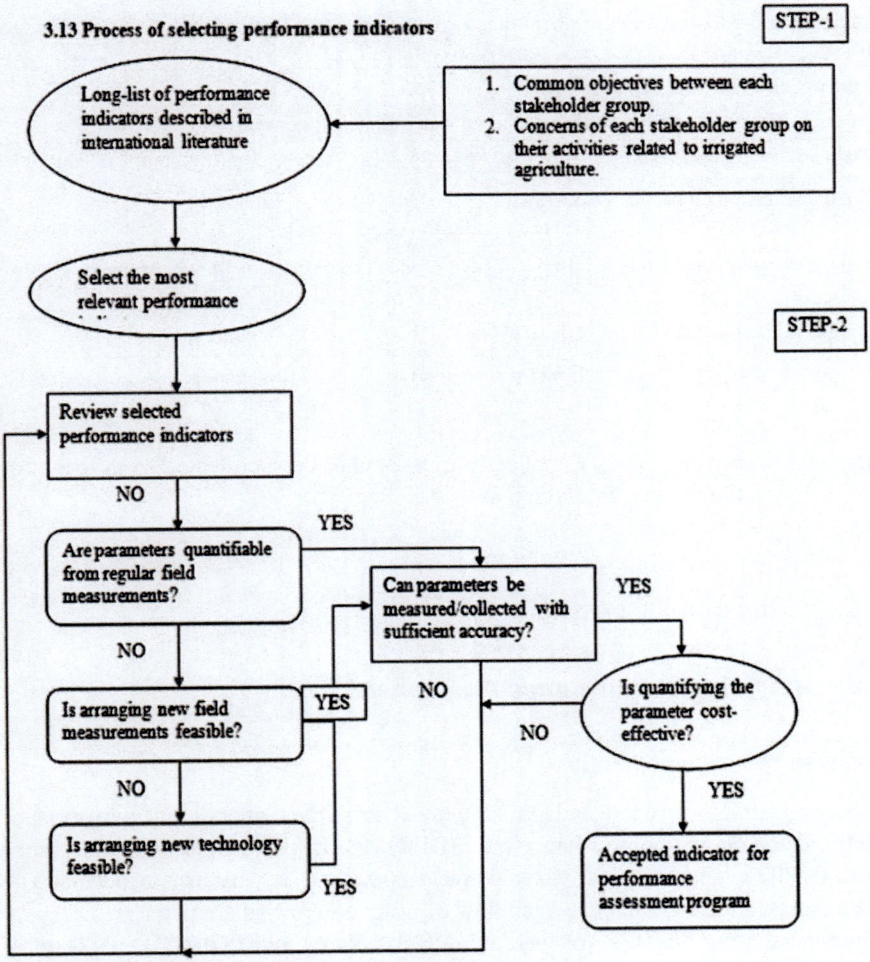

Fig. 3 The two steps in the process of choosing performance indicators are: (1) choosing the research's most pertinent indicators, and (2) determining whether the indicators are applicable to the local circumstances

humidity, wind speed, daily durations of sunshine etc.) were obtained from the site of NASA POWER (https://power.larc.nasa.gov/data-access-viewer/) (Table 1).

3.7 Required Data

The required meteorological data was collected from the site of NASA POWER (https://power.larc.nasa.gov/data-access-viewer/). This includes the Temperature (°c), Wind speed (m/s), Sunshine hours, Atmospheric pressure (hpa), and rainfall data

Table 1 Lists the most important performance metrics based on operational activity and resource use

Performance indicator definition criteria		
nQuality of service, crop water demand, crop water use, and drainage		
Delivery performance ratio	$\text{DPR} = \frac{\text{Actual flow of water}}{\text{intended flow of water}}$	Adequacy, equity, reliability
Field application ratio	$\text{FAR} = \frac{\text{ETp} - \text{Pe}}{\text{volume of water delivered to the field}}$	Efficiency
Relative evapotranspiration	$\text{RET} = \frac{ETa}{ETp}$	Reduction in ET, Adequacy, uniformity
Depleted fraction	$\text{DF} = \frac{ETa}{Pe+Vc}$	Adequacy, Equity
Crop water deficit	$\text{CWD} = \text{ETp} - \text{ETa}$	Efficiency

(mm) on daily basis. Belan Canal discharge data has been collected from Irrigation Department, Govindpur, Prayagraj, and Uttar Pradesh.

4 Results and Discussions

4.1 Irrigation Performance Assessment

Delivery Performance Ratio

The most straightforward and possibly most important operational performance metric is the delivery performance ratio (DPR) [13–15]. The Delivery performance ratio (DPR) evaluates actual water delivery compared to what was expected. The water supplied to AOI and expected flow of water entering. it are taken into account when calculating the DPR for the AOI. DPR is configured to have a goal level of 0.8 and a critical level of 0.6. An ideal scenario is one with a DPR of 1.0. Whenever the DPR exceeds 1.0. It indicates over-irrigation also indicating moisture stress if it falls below this level [18]. For the estimation of the DPR, we consider only the canal water supply, and we can see in Fig. 4 the DPR is not ideal through the growing season. This shows that the canal water supply has not been maintained according to the demand. For August, September, October, and January the DPR is too high which results from over-irrigation of the fields. This result shows that Water is not applied consistently and requires proper maintenance in this case to strengthen the DPR.

Depleted fraction

The depleted fractions are the ratio between three elements of an irrigated area's water balance which are the actual evapotranspiration, surface water supply, and precipitation. This indicator is particularly useful for the purpose of diagnosis in water-scarce areas. This indicates the water scarcity over the irrigated command area.

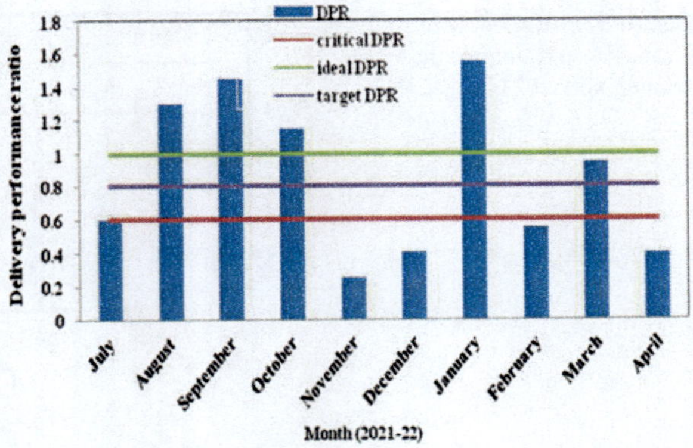

Fig. 4 The AOI delivery performance ratio for the coming year 2021–22

The depleted fraction compares the actual amount of water that evaporates from the chosen area to the total amount of precipitation in the area plus surface water flows into the irrigated area. This has an impact on the local water supply decision. It gauges how much water is available to be stored in the groundwater system. Figure 5 gives the DF of the irrigated area for the cropping period of 2021–22.

We can see that DF remains below the critical level for July, August, September, October, January, and February and is quite high during November, December, and April. A portion of the irrigated water is stored if rainfall covers the irrigated region and the ETa is less than 0.6; otherwise, groundwater depletion happens. Water may not be directed as much to the crops. If it is kept in the command area, and more water may be used for irrigation if the groundwater supply is running low. This is a result of high rainfall during these months. This water can be diverted during the dry months if it is still in the storage reservoir, which is frequently the case in arid and semi-arid countries.

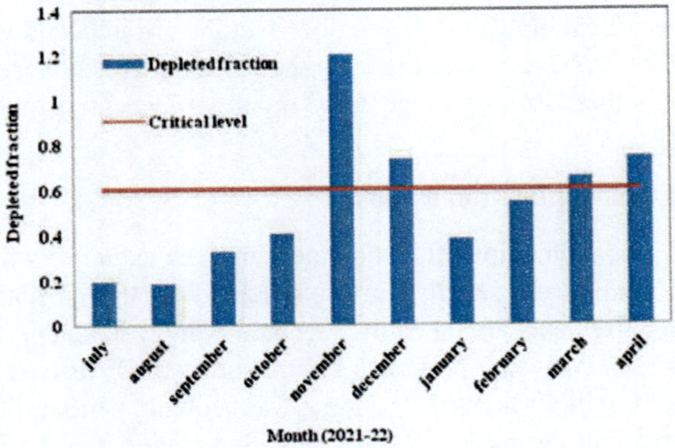

Fig. 5 The AOI depleted fraction for the coming year 2021–2022

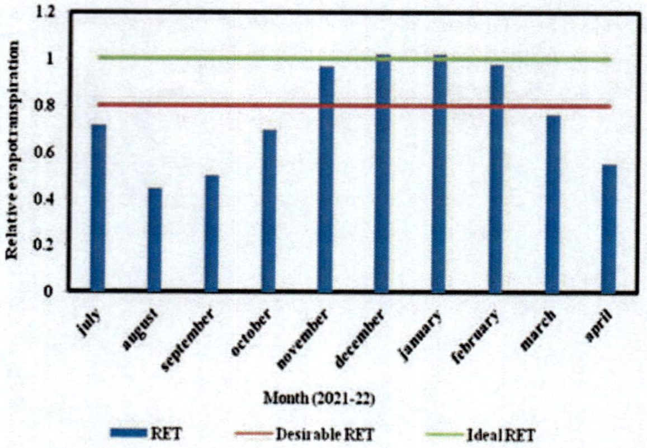

Fig. 6 The AOI Relative Evapotranspiration for the coming year 2021–22

Relative Evapotranspiration

The dimensionless ratio of actual to potential evapotranspiration provides the water manager with vital information to assess the delivery of enough irrigation water in relation to time to a particular command region. RET between 0.8 and 1.0 indicates an adequate level of water supply and below 0.8 indicates an inadequate level of water supply. In Fig. 6 we can see that for the month of November, December, January, and February RET is high, and for the rest of the months RET is below the desired level. These results show that for almost all the cropping period there was not enough water available.

Crop water deficit

The difference between the potential and actual evapotranspiration of the cropping pattern within an area as established by the water management is the crop water deficit over a time. It signifies a water shortage, which developed over time in the irrigated area. It is acceptable for CWD to average 30 mm per month [15]. The CWD is critical in August, September, October, March, and April which is due to low ETa. The CWD is acceptable from the month of November to February. Also, we can see that there is no shortage in the month of December and January which is acceptable (Fig. 7).

Field Application Ratio

Field application efficiency measures the extent to which the requirements of crop irrigation are satisfied in the irrigated area by irrigation water. The entire water requirements are taken into account while calculating the FAR for the AOI and the amount of water provided. Figure shows the FAR over the irrigated area.

Figure 8 shows that while it was growing period of 2021–22 the FAR goes down to minus during July, August, and September. It crosses the ideal level in November, December, February, March, and April. This is because of excess irrigation due to high rainfall. We can see that Water is provided even when not required. This requires maintenance in the provision according to the requirements. When the ratios are low,

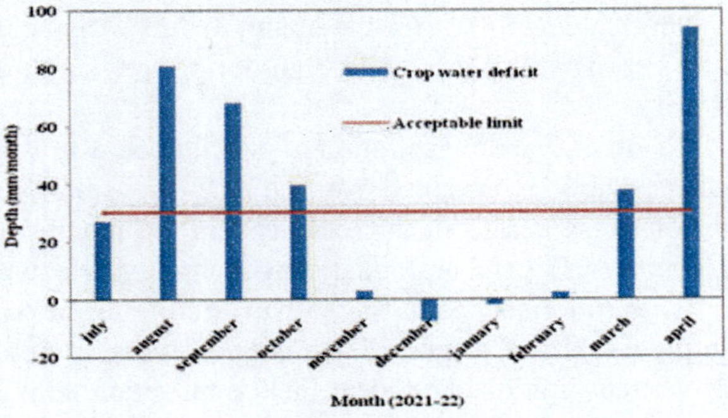

Fig. 7 The AOI crop water deficit for the coming year 2021–22

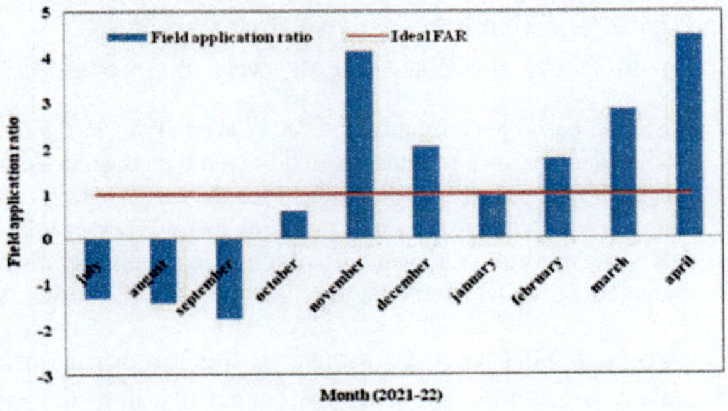

Fig. 8 The AOI field application ratio for the coming year 2021–22

the ground table will rise as a result of the water that is not used. To prevent a water scarcity, groundwater must be extracted and stored if the ratio rises above 0.6 [15].

Examining specific performance indicators

The relationship between farmers and the water authority has to do with the ecology and water balance.

- The justification for water distribution taking into account the water supply the system is expected to receive.
- The timing of agriculture, which includes regular upkeep, preparation of the land, cultivation, and harvesting.
- The type and diversity of crops grown under each irrigation plan.

5 Conclusion

The delivery performance ratio is not ideal throughout the crop growing period (Kharif and Rabi). The ratio increases and decreases above and below the ideal limit (1.0) respectively. This shows that the supply is not uniform and requires proper supply interval and quantity depending on the water demand (i.e. 3-10 mm/day for kharif crop and 2-4 mm/day for Rabi crops). The Depleted fraction remains below the critical level (0.6) for the month of July, August, September, October, January, and February and it is quite high during the month of November, December, and April.

Relative evapotranspiration of 0.8 and below this indicates that the crop is under stress. The RET obtained over the AOI shows prolonged water stress periods. The water stress period in this case covers the months almost all the months except November, December, January, and February (adequate supply of water).

The crop water deficit was seen to be crossing the acceptable limit (30 mm/month) in the months of August, September, October, March, and April due to low actual evapotranspiration. The Ideal Field application ratio (1.0) was low throughout the period and goes down to minus during July, August, and September. This means that there was more water available due to high precipitation during this month exceeding the demand which indicates over irrigated lands.

From all the above statements, we can conclude that:

> The ETa based performance indicators are helpful in the assessment of the system performance and can give reliable results if taken the consideration to mention above. The supply does not meet the irrigation demand for the crop growing season due to which water stress occurs. Assessment of an irrigated area using ETa can help in understanding the issues of adequacy and equity of water distribution. In case if water is limited, deficit irrigation should be practiced Water should be provided to match the critical levels if water is inadequate.

For each performance measure, the temporal variation of the irrigated region experiencing water stress was successfully demonstrated. Water stress is indicated by the Delivery Performance Ratio (DPR) for the first four months of the growing season (August, September, October, and January). With the exception of November to February, Water stress was present for the majority of the irrigation era, according to the Relative Evapotranspiration. However, it also displays the permissible levels for the identical two months of December and January as that of Relative Evapotranspiration, in addition to for two more months of March and April. Water stress was also evident for the bulk of the time period according to the Crop Water Deficit. The evaluation of an irrigation system is aided by the ETa-based water balance. However, there are considerable uncertainties in these ETa calculations as well.

References

1. Ram A, Tiwari SK, Pandey HK, Chaurasia AK, Chaudhary SK (2020) Assessment of groundwater behavior in Kulpahar Watershed, District Mahoba, Uttar Pradesh. India IOP Conf Ser Earth Environ Sci 597 012014
2. Allen RG, Pereira LS, Raes D, Smith (1998) Crop evapotranspiration, guidelines for computing crop water requirements. FAO Irrigation and Drainage paper 56 (Food and Agricultural Organization of United Nations (FAO) Rome, Italy), pp 300
3. Sawadogo A, Kouadio L, Traoré F, Zwart S, Hessels T, kemal sulhi gündo gdu (2020) spatiotemporal assessment of irrigation performance of the kou valley irrigation scheme in burkina faso using satellite remote sensing-derived indicators indian journal of geo information

4. Anderson MC, Allen RG, Morse A, Kustas WP (2012) Use of Land sat thermal imagery in monitoring evapotranspiration and managing water resources. Remote Sens Environ. https://doi.org/10.1016/j.rse.2011.08.025
5. Avil Kumar K, Reddy MD, Uma Devi M, Narender N, Neelima TL, Ramulu V, Praveen Rao V, Raghavaiah R (2014) Irrigation Performance Assessment of Left Bank Canal. Nagarjuna Sagar Project, India During Rabi Using Remote Sensing and GIS Agrotechnol 2014(3):1. https://doi.org/10.4172/2168-9881.1000122
6. Awan UK, Tischbein B, Conrad C, Martius & Hafeez M, (2011) Remote sensing and hydrological measurements for irrigation performance assessments in a water user association in the lower Amu Darya River Basi. Water Resour Manage 25:2467–2485
7. Elnmer A, Khadr M, Allam A, Kanae S, Tawfik A (2018) Assessment of irrigation water performance in the Nile delta using remotely sensed data, Indian J Geo Information
8. Ayşegül parladir karci and yusuf uçar (2019) use of remote sensing and geographic information systems in irrigation performance: a case study of atabey irrigation scheme türk tarım ve doğa bilimleri dergisi 6(4): 624–635.
9. Bastiaanssen WGM, Menenti M, Feddes RA, Holtslag AAM (1998) A remote sensing surface energy balance algorithm for land (SEBAL). 1: Formulation. J Hydrol 212:198–212
10. Karatas B, Akkuzu E, Unal H, Asik S, Avci M (2009) Using satellite remote sensing to assess irrigation performance in water user associations in the lower gediz basin, Turkey article in agricultural water management, June 2009
11. Çakmak B, Polat E, Kendirli B, Gökalp Z (2009) evaluation of irrigation performance of Asartepe irrigation association: a case study from Turkey. Akdeniz Üniversitesi Ziraat Fakültesi Dergisi 2009, 22(1):1–8
12. Bos MG (1997) Performance indicators for irrigation and drainage systems 11:119–137
13. Bos MG, Burton MA, Molden DJ (2005) Irrigation and drainage performance assessment: practical guidelines. IWMI Books Reports, International Water Management Institute, No, p H037064
14. Bos MG, Kselik RAL, Allen RG, Molden D (2008) Water requirements for irrigation and the environment
15. Bos MG, Wolters W, Drovandi A, Morabito JA (1991) The secondary canal of Viejo Retamo, Performance evaluation case study. Mendoza, Argentina. Irrigation and Drainage Systems 5, 77–88
16. Conrad C, Dech SW, Hafeez M, Lamers JPA, Tischbein B (2013) "Remote sensing and hydrological measurement based irrigation performance assessments in the upper Amu Darya Delta, Central Asia". Phys. Chem. Earth, Parts A/B/C 61:52–62
17. Droogers P, and Bastiaanssen W G M (2002) Irrigation performance using hydrological and remote sensing modeling. J Irrig Drain Engg
18. Bandara KMPS (2006) Assessing irrigation performance by using remote sensing. ITC Dissertation 134:86–87
19. Bastiaanssen WGM, Molden D (2017) Performance evaluation of the Bhakra Irrigation System, India. In: Using Remote Sensing and GIS Techniques R. Sakthivadivel S. Thiruvengadachari Upali Amerasinghe
20. Zwart S, Leclert, (2010) A remote sensing-based irrigation performance assessment: a case study of the Office du Niger in Mali. Irrig Sci 28(5):371–385. https://doi.org/10.1007/s00271-009-0199-3

Twitter-Based Disaster Response Framework Using ELECTRA Transformer

Pardeep Singh, Monika, Bhawna Shishodia, and Satish Chand

1 Introduction

During crises or natural disasters, people share situation-sensitive information on social media platforms about what they have seen, witnessed or heard from different sources [5]. These platforms deliver information faster than mainstream media and help government and disaster response agencies regarding aid and assistance [25]. In the 2013 Westgate mall terror attack in Kenya, Information flooded on Twitter and Facebook within a minute after the attack lacking mainstream media by half an hour [7]. The disaster response agencies examine actionable or situational information in real-time for crisis assessment and to provide relief measures accordingly.

We considered the Twitter platform for analyzing tweets in the field of disasters since its acts as a short and instant basis for communication and ease of access to data. We focus on getting situational awareness from tweets and categorize it into six classes based on the type of information. Situational awareness is when you are familiar with what is going on around you [4]. Therefore, this paper aims to develop a classifier based on the recent state-of-the-art transformer model: ELECTRA, that can accurately classify disaster-related tweets into various categories related to caution and advice, infrastructure damages etc. Our contributions in the paper are as follows:

- We propose a novel disaster response framework using ELECTRA, a transformer based pre-trained masked language model that needs significantly fewer compute resources at the pretraining stage and results in better performance on downstream disaster response classification task.
- The proposed classifier classifies disaster-related tweets into various categories related to affected individuals, Caution and advice, Infrastructure and utilities damage etc.

P. Singh (✉) · Monika · B. Shishodia · S. Chand
School of Computer & Systems Sciences, Jawaharlal Nehru University New Delhi, New Delhi, India
e-mail: pardeepsinghinfo@gmail.com

A. Ramdane-Cherif et al. (eds.), *Machine Intelligence and Data Science Applications*, Algorithms for Intelligent Systems, https://doi.org/10.1007/978-981-99-1620-7_39

- We evaluate the proposed model's performance in Macro-F1 and compare it with other competitive models.

The organization of the paper is as follows: Sects. 2 and 3 discuss the related work and methodology. The dataset and experimental results are presented in Sect. 4. Section 5 discusses the experimental results and analysis. Finally, in Sect. 6, we discuss the conclusion and future suggestions.

2 Related Work

During disasters, social media sites like Twitter and Facebook are useful sources of situational information [12, 14, 15, 21]. However, this information in the form of tweets or Facebook posts must be retrieved and analyzed in real time to gain insights into it. This section discusses existing methods that investigated and classified situational information during disaster events. The researchers have tried numerous methods to extract situational information during disaster events. We discuss below the existing state-of-the-art literature in the disaster classification field.

2.1 *Shallow Methods*

Using shallow detection techniques, we refer to disaster tweet classification that employs traditional word representation techniques to encode text into numerical representation and apply machine learning classifiers to perform the detection. Most of these methods used feature-based bag-of-words or n-grams models classifiers to differentiate between tweets that provide situational information and tweets that do not [8, 16, 19, 23]. However, the problem with these approaches is that the classifiers are trained on a specific event's vocabulary and may perform badly for a new disaster event.

2.2 *Deep Learning Methods*

The primary challenge with shallow methods is manual feature engineering. Several deep learning-based methods come into the picture that automatically extracts features from the input data [11, 18, 20]. Authors in [13] discussed CNN-SkipCNN deep learning model to categorize tweets into crisis-related information and humanitarian aid. The study of [1] examined CNN with the hashtag and user mentions features for classifying disaster-related tweets from the CrisisNLP dataset. For a better comprehensive literature review on this topic, we direct the reader to study a paper by [6].

2.3 Transformers-Based Methods

Transformers are a collection of attention-based neural network models introduced in 2017 in the paper "Attention is all you need" [22]. It has revolutionized the NLP community by achieving state-of-the-art performance for various language tasks such as question answering, text classification, and text summarization. Fine-tuning the transformer-based models such as BERT [3], DistilBERT [17], ELECTRA [2] etc., producing higher-capacity models and capturing long-range sequence features. In the field of NLP for disaster response, [10] proposed CrisisBERT which utilizes a pre-trained BERT model for crisis event identification. The common limitation of the above discussed works is that either the classifiers used more time for training or needed manual feature engineering. To overcome each of these limitations, we propose to use ELECTRA, a pre-trained masked language model that needs significantly fewer compute resources at the pretraining stage and results in better performance on downstream disaster response classification tasks.

3 Methodology

This section discusses our proposed framework for multi-label classification of disaster related tweets.

3.1 Problem Statement

We define the disaster response framework $DRF = (T, \phi)$, which is defined by $T = \{t_1, t_2, t_3 \ldots, t_n\}$ a finite set of tweets having size n. Each sample t_i is a series of tokens at timestamp T, i.e. $t_i = \{t_i^1, t_i^2, t_i^3, ..., t_i^T\}$. ϕ represents the set of labels having the same arrangement as the sample set, $\phi = \{\phi_1, \phi_1, \ldots, \phi_n\}$ and $\phi_i \epsilon \{1, 2, 3, 4, 5, 6\}$ represents six different class labels. A multi-class classifier $MC : T \rightarrow \phi$ defines the mapping from representative tweets to their class label. Our prime objective is to train a disaster response classifier that predicts the tweet is related to situational awareness especially in disaster domain.

3.2 Efficiently Learning an Encoder That Classifies Token Replacements Accurately (ELECTRA)

Transformer, a deep learning model architecture first explained in the paper "Attention is all you need" overwhelms the field of Natural Language Processing. The first transformer model is BERT (Bidirectional Encoder Representations from Transform-

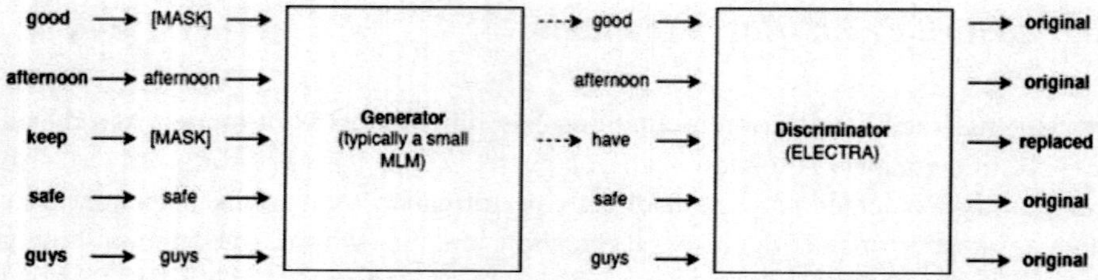

Fig. 1 Fine-tuning ELECTRA model

ers), a masked language model achieving state-of-the-art results in many challenges and tasks. Various models have been developed in the last couple of years that have improved on the success of BERT - like RoBERTa, DistilBERT and XLNet. Still, the major concern with these architectures is relying on bigger networks and larger datasets.

To handle the above challenges, the ELECTRA model was released by Google in 2020 in the paper [2] and has been demonstrated as potent transformers architecture for its smaller architecture and higher performance. This model follows a new pretraining procedure that trains two transformer models: the generator and the discriminator as shown is Fig. 1.

The generator component generates fake input samples during the pre-training process. The generator is a neural network trained to produce examples similar to the actual input examples but with some tokens replaced by other tokens. The generator is trained to create samples that are difficult for the discriminator to distinguish from real examples, with the goal of learning a good representation of the input data. The discriminator component is typically used in fine-tuning or pre-training tasks to improve the model's ability to distinguish between real and generated text. It is trained to predict if the input text is real or fake by comparing it with real text and generated text by the generator component. It is trained using the binary cross-entropy loss function. Together, the generator and discriminator components work in a way that the generator learns to generate realistic text and the discriminator learns to distinguish between real and generated text. This process is known as adversarial training. This improves the overall performance of the model in generating realistic text for the downstream task.

4 Dataset and Experimental Setup

In this section, we discuss the dataset and experimental setup that we use for the evaluation.

4.1 Dataset

To evaluate our proposed work, we need a labeled dataset that contains tweets related to the crisis. We choose CrisisNLP dataset [9] that includes tweets from 19 natural and human-induced disasters. Our study only chooses tweets related to four events: India Floods, Nepal Earthquake, California Earthquake and Hurricane Odile. The Tweets are categorized based on their information types:

1. Caution and advice: Tweets related to warnings and advice by concerned authorities and people.
2. Affected individuals: Tweets concerning injuries, trapped, missing, death as well as other updates in the vicinity of the disaster.
3. Sympathy and emotional support: Tweets related to prayers and well wishes.
4. Other useful information: Tweets related to the disaster.
5. Infrastructure and utilities damage: Tweets about either buildings, bridges, or infrastructures damaged by the disaster.
6. Donations and volunteering: Tweets about donating clothes, medicines, and other relief material.

For preprocessing the tweets, we have used the tweet-preprocessor library[1] that removes Hashtags, mentions and URLs from the text. Figure 2 shows the tweets before and after preprocessing step.

4.2 Experimental Setup and Hyperparameter Tuning

To fine-tune the ELECTRA transformer model, we use the Hugging-face Transformers library [24] on NVIDIA TESLA P100 GPU. We used categorical cross entropy as a loss function and AdamW as an optimizer. To overcome the model from overfitting, we add small proportion of dropout. We used class weights for imbalanced classes also. The value of hyperparameters is shown in Table 1.

During the classifier's training, we used categorical cross entropy as a loss function to compute gradients and train the model via backpropagation. The training loss is plotted as a moving average over 35 epochs are shown in Figure 4. Table 2 presents the twenty most frequent Bigrams and Trigrams identified in crisis NLP dataset. Figure 3 depicts a word cloud representation showcasing most frequent words within the crisis NLP dataset. It is observed that the most frequent words are "Nepal flooding", "India flood", and "death toll", and there are other words such as "monsoon hits", "people" and "killed" (Fig. 3).

[1] https://pypi.org/project/tweet-preprocessor/.

Tweet	CleanTweet
RT @ABSCBNNews: Philippines braces for Typhoon...	: Philippines braces for Typhoon Pablo's return
Philippine Typhoon Bopha death toll passes 300...	Philippine Typhoon Bopha death toll passes :
RT @MiaFarrow: Haunting image - read what it m...	: Haunting image - read what it means to the p...
City Room: New York Today: After Derailment, B...	City Room: New York Today: After Derailment, B...
NSW on fire, VIC saturated. What is the #price...	NSW on fire, VIC saturated. What is the to you...
@Leonuhhh @cupsaucerblog @VickiHigham sounds l...	sounds like Australia! But with simultaneous b...
RT @WOOPIGSOOOIE: BREAKING: Experts have dete...	: BREAKING: Experts have determined Colorado F...
The Bow from our deck! She's not happy! #yycfl...	The Bow from our deck! She's not happy!
RT @gmanews: PGH not in emergency situation. B...	: PGH not in emergency situation. But relief g...
RT @_SGkid: Haze, I just met you, And this is ...	: Haze, I just met you, And this is crazy, But...
RT @Krenvrro: @MatamaanKa ASAP RESCUE in Villa...	: ASAP RESCUE in Villa Monteverde, Brgy.Nagbal...
RT @ItsDileyJonas: RT si eres de Guatemala, El...	: RT si eres de Guatemala, El Salvador o Mexic...
@edwyplenel bravo pour votre reflexion sur les...	bravo pour votre reflexion sur les petites tra...
RT @MiaFarrow: We can contact complicit compan...	: We can contact complicit companies. We can r...
Premier Newman: weather situation has deterior...	Premier Newman: weather situation has deterior...
RT @gmanews: Residents of Kasiglahan Village i...	: Residents of Kasiglahan Village in Montalban...
RT @Fraintesa: Se interessa a qualcuno di #San...	: Se interessa a qualcuno di e dintorni: ho po...
#LAX Shooting #MANNEQUIN strolld out on CBS cl...	Shooting strolld out on CBS claimed was wounded ?
RT @CuteLovesMsgs: Before anything else, #Pray...	: Before anything else,
@rhm1947 @JoseM_Palacios Esta tarde quise dona...	Esta tarde quise donar sangre haban ido tantos...

Fig. 2 Data preprocessing

Table 1 Hyperparameters of fine-tuning ELECTRA model

Hyperparameters	Values
Dropout rate	0.2
Max sequence length	80
Batch size	16
Epochs	25
Learning rate (AdamW)	1e-6

5 Experimental Results and Analysis

Our proposed framework employ the ELECTRA transformer model for classifying situational tweets during disasters. However, other mainstream transformer models like BERT, XLNet and deep learning RNN-based models like LSTM and GRU achieve state-of-the-art performance on various NLP tasks. Hence, we compare the performance of our proposed work with these methods to validate our work.

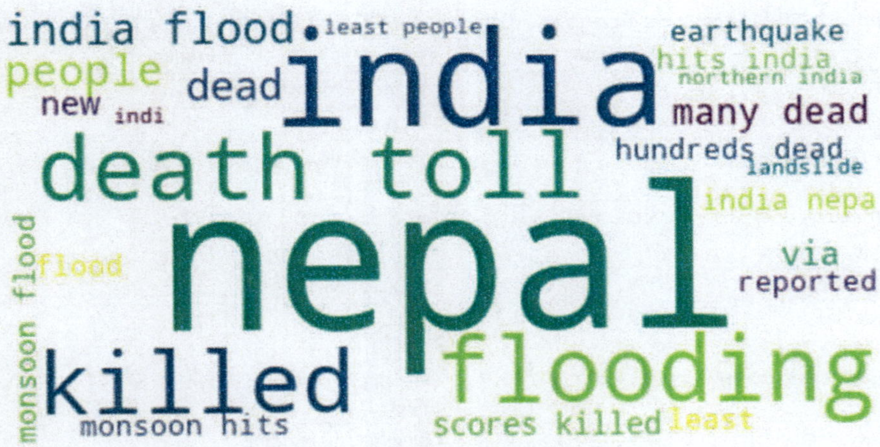

Fig. 3 Word cloud of most frequent disaster keywords from the CrisisNLP dataset

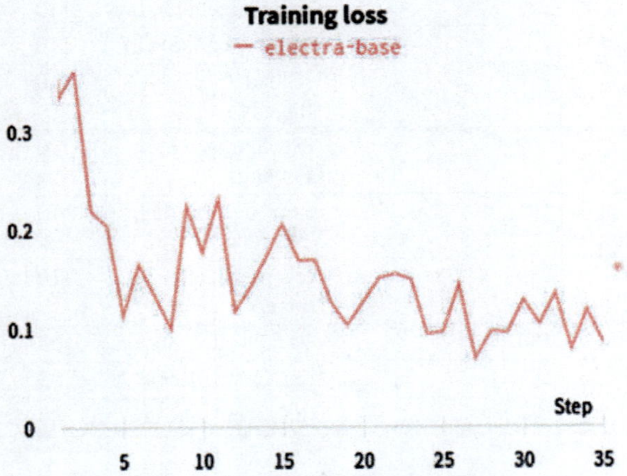

Fig. 4 Training loss over 35 epochs using ELECTRA transformer

Table 3 presents the evaluation results in terms of Precision, Recall, Accuracy and macro F1-score. It is clear from the results that The ELECTRA model outperforms the other methods in all four evaluation metrics. If the data is imbalanced or skewed, metrics like accuracy is not enough. Hence we used macro F1 score that calculate the harmonic mean of precision and recall and works well for imbalanced distribution of data. Hence in our study, we used these four evaluation metrics to compare the results. In Table 4, we also report class-wise results of to see how well the model performed in each class.

Table 2 20-Most frequent bigrams and trigrams in the CrisisNLP Dataset

Bigram	Frequency	Trigram	Frequency
(Costa, Rica)	325	(accidente, de, tren)	221
(New, York)	320	(de, tren, en)	191
(train, derailment)	316	(Santiago, de, Compostela)	178
(Santiago, de)	302	(en, Santiago, de)	159
(de, tren)	290	(tren, en, Santiago)	130
(tren, en)	279	(So, sad, hear)	113
(de, la)	277	(nightclub, fire, Brazil.)	112
(helicopter, crash)	271	(Praying, everyone, affected)	111
(accidente, de)	249	(Bangladesh, building, collapse)	110
(death, toll)	239	(sad, hear, nightclub)	110
(At, least)	231	(hear, nightclub, fire)	110
(building, collapse)	223	(fire, Brazil., Such)	110
(train, crash)	214	(Brazil., Such, tragedy.)	110
(nightclub, fire)	209	(Such, tragedy., Praying)	110
(en, el)	195	(tragedy., Praying, everyone)	110
(de, Compostela)	189	(everyone, affected, there.)	110
(en, la)	186	(@ddlovato:, So, sad)	108
(NYC, train)	184	(en, Costa, Rica)	93
(terremoto, en)	174	(del, accidente, de)	88

Table 3 Comparision of all Model

Model → / Metrics ↓	ELECTRA	XLNet	BERT	Bi-LSTM	Bi-GRU
Precision	0.68	0.64	0.59	0.78	0.58
Recall	*0.68*	0.67	0.63	0.55	0.57
Accuracy	*0.70*	0.67	0.64	0.55	0.57
F1 score	*0.68*	0.65	0.60	0.59	0.57

Table 4 Class-wise result of ELECTRA Transformer

Classes↓	Precision	Recall	Macro-F1
Caution and advice	0.56	0.62	0.59
Affected individuals	0.77	0.80	0.79
Sympathy and emotional support	0.82	0.70	0.75
Other useful information	0.66	0.69	0.67
Infrastructure and utilities damage	0.53	0.52	0.52
Donations and volunteering	0.75	0.75	0.75

6 Conclusion

This paper presents a transformer-based memory efficient framework for classifying situational tweets during disasters. We empirically demonstrate our approach and achieve leading performance with an F1 score of 68% in comparison to other baseline and state-of-the-art methods. In class-wise performance, some classes are well classified while others may not be due to subjectivity and ambiguity. This finds applications in various disaster scenarios and helps crisis agencies and the government for aid and assistance. The present study is confined to in-domain (i.e. model trained in one dataset) but can be extended to cross-domain with different disaster datasets without any feature engineering approach.

References

1. Aipe A, Mukuntha N, Ekbal A, Kurohashi S (2018) Deep learning approach towards multi-label classification of crisis related tweets. In: Proceedings of the 15th ISCRAM conference
2. Clark K, Luong MT, Le QV, Manning CD (2020) Electra: pre-training text encoders as discriminators rather than generators. arXiv:2003.10555
3. Devlin J, Chang MW, Lee K, Toutanova K (2018) Bert: pre-training of deep bidirectional transformers for language understanding. arXiv:1810.04805
4. Gautam AK, Misra L, Kumar A, Misra K, Aggarwal S, Shah RR (2019) Multimodal analysis of disaster tweets. In: 2019 IEEE fifth international conference on multimedia big data (BigMM), pp 94–103
5. Hughes AL, Palen L (2009) Twitter adoption and use in mass convergence and emergency events. Int J Emerg Manag 6(3–4):248–260
6. Imran M, Castillo C, Diaz F, Vieweg S (2015) Processing social media messages in mass emergency: a survey. ACM Comput Surv (CSUR) 47(4):1–38
7. Imran M, Castillo C, Lucas J, Meier P, Rogstadius J (2014) Coordinating human and machine intelligence to classify microblog communications in crises. In: ISCRAM
8. Imran M, Castillo C, Lucas J, Meier P, Vieweg S (2014) AIDR: artificial intelligence for disaster response. In: Proceedings of the 23rd international conference on world wide web, pp 159–162

9. Imran M, Mitra P, Castillo C (2016) Twitter as a lifeline: human-annotated twitter corpora for NLP of crisis-related messages. In: Proceedings of the tenth international conference on language resources and evaluation (LREC 2016). European language resources association (ELRA), Paris, France
10. Liu J, Singhal T, Blessing LT, Wood KL, Lim KH (2021) Crisisbert: a robust transformer for crisis classification and contextual crisis embedding. In: Proceedings of the 32nd ACM conference on hypertext and social media, pp 133–141
11. Manderna A, Kumar S (2022) Effective long short-term memory based-driver identification in its. In: 2022 international conference on inventive computation technologies (ICICT), pp 464–469. IEEE
12. Neubig G, Matsubayashi Y, Hagiwara M, Murakami K (2011) Safety information mining-what can NLP do in a disaster. In: Proceedings of 5th international joint conference on natural language processing, pp 965–973
13. Paul NR, Sahoo D, Balabantaray RC (2022) Classification of crisis-related data on twitter using a deep learning-based framework. Multimed Tools Appl 1–21
14. Purohit H, Castillo C, Diaz F, Sheth A, Meier P (2014) Emergency-relief coordination on social media: automatically matching resource requests and offers. First Monday
15. Qu Y, Huang C, Zhang P, Zhang J (2011) Microblogging after a major disaster in China: a case study of the 2010 Yushu earthquake. In: Proceedings of the ACM 2011 conference on computer supported cooperative work, pp 25–34 (2011)
16. Rudra K, Ghosh S, Ganguly N, Goyal P, Ghosh S (2015) Extracting situational information from microblogs during disaster events: a classification-summarization approach. In: Proceedings of the 24th ACM international on conference on information and knowledge management, pp 583–592
17. Sanh V, Debut L, Chaumond J, Wolf T (2019) Distilbert, a distilled version of bert: smaller, faster, cheaper and lighter. arXiv:1910.01108
18. Singh P, Chand S (2019) Pardeep at semeval-2019 task 6: identifying and categorizing offensive language in social media using deep learning. In: Proceedings of the 13th international workshop on semantic evaluation, pp 727–734
19. Singh P, Chand S (2020) Predicting the popularity of rumors in social media using machine learning. In: Social networking and computational intelligence, pp 775–789. Springer
20. Singh YP, Lobiyal D (2022) A comparative study of deep learning algorithms for epileptic seizure classification. In: 2022 international conference on computing, communication, security and intelligent systems (IC3SIS), pp 1–6. IEEE
21. Varga I, Sano M, Torisawa K, Hashimoto C, Ohtake K, Kawai T, Oh JH, De Saeger S (2013) Aid is out there: looking for help from tweets during a large scale disaster. In: Proceedings of the 51st annual meeting of the association for computational linguistics (Volume 1: Long Papers), pp 1619–1629
22. Vaswani A, Shazeer N, Parmar N, Uszkoreit J, Jones L, Gomez AN, Kaiser Ł, Polosukhin I (2017) Attention is all you need. Adv Neural Inf Process Syst 30
23. Verma S, Vieweg S, Corvey W, Palen L, Martin J, Palmer M, Schram A, Anderson K (2011) Natural language processing to the rescue? "Extracting situational awareness tweets" during mass emergency. In: Proceedings of the international AAAI conference on web and social media, vol 5
24. Wolf T, Debut L, Sanh V, Chaumond J, Delangue C, Moi A, Cistac P, Rault T, Louf R, Funtowiczet M et al (2019) Huggingface's transformers: State-of-the-art natural language processing. arXiv preprint arXiv:1910.03771
25. Xu J (2020) Does the medium matter? A meta-analysis on using social media vs. traditional media in crisis communication. Public Relations Rev 46(4):101947

Tumor-Type Detection Using Machine Learning Algorithm on Gene Expression Cancer RNA-Seq Dataset

Rajkumar Patra, Debajyoti Dutta, Mitadru Datta, and Anupam Ghosh

1 Introduction

With the growth in technologies, improvements, and research happening in the field of Machine Learning, in the form of new algorithms, tools, and software, research scientists and medical teams have been able to achieve significantly high accuracy for predicting tumors in the early stages for proper diagnosis and treatment. Nowadays getting hold of medical data like reports and findings is quite easy as a large number of datasets containing thousands of data entries are easily available publicly for free on government and medical facilities websites. The real challenge lies in how early tumors can be detected accurately for proper and immediate treatment. Because of this, ML techniques are gaining popularity among medical researchers. These methods can be used to discover and identify new patterns and relationships between them, extracted from new and complex datasets, while they are able to effectively predict future outcomes of a specific cancer type or category.

Nowadays, we have witnessed significant growth in personalized medicine and rising demand and implementation of ML concepts and techniques. We are hereby using those methods and techniques that have been discovered in a review of studies that are currently using them in the domains of tumor prediction and prognosis. In these studies, predictive and prognostic elements have been taken into account to be either completely specific for a certain treatment or integrated in order to guide proper therapeutic treatment for patients suffering with tumors, respectively. Apart from these, we have also elaborately explained the types of ML techniques being used, the types of data we have integrated, and the overall performance of each proposed scheme and stating the pros and cons associated with it.

In recent years, we have witnessed a mixture of both clinical and genomic data in the proposed works in this field. But, we have observed a deficiency in terms of

R. Patra (✉) · D. Dutta · M. Datta · A. Ghosh
Netaji Subhash Engineering College, Kolkata, India
e-mail: rajkr.patra@gmail.com

A. Ramdane-Cherif et al. (eds.), *Machine Intelligence and Data Science Applications*, Algorithms for Intelligent Systems, https://doi.org/10.1007/978-981-99-1620-7_40

validating the accuracy of the various predictive models by an external source. There is no doubt at all regarding the fact that the ML models can achieve accuracy values for the detection of cancer, recurrence, and predicting chances of survival. We have seen that after applying ML models, the accuracy for the detection of cancer in its early stages has achieved a significant improvement by 15–20% in the last decade.

Recently, different methods and strategies are getting implemented and reports are generated for early cancer diagnosis and prognosis. The studies that are conducted are based on approaches related to improving the circulation of RNA gene expressions that have been witnessed to be a promising class for tumor type detection and identification. However, these new methods and strategies have some drawbacks such as suffering from low sensitivity when they are being used for screening purposes at the premature stages and lacking the proper ability to differentiate benign from malignant tumors. Various aspects regarding the use of gene expression signatures for the prediction of the tumor are constantly discussed. These studies list the potential as well as the limitations of microarrays for the prediction of cancer outcomes. Though gene signatures have the full potential for improvement in the ability for prognosis in cancer patients, there has been a lack of progress or growth in the implementation on clinical grounds. However, first, the utmost necessity lies in studying and building models by training with large data samples for more accurate prediction and validation before gene expression profiling can be tested in clinical practice.

In our current work, we have only incorporated studies from research work that have deployed ML techniques for modeling tumor diagnosis and prognosis. As we are witnessing an increasing demand in the application of machine learning techniques and methods in the domain of tumor research, we therefore, present the most up to date publications that have incorporated these methods with the aim to build a better model to get superior accuracy over a large dataset to predict better results in the future.

2 Related Works

Machine learning is gaining popularity in cancer research. To identify potential cancer biomarkers, machine learning algorithms are used to correctly predict the cancer types/subtypes. Machine learning methods have the capability to identify key features [1, 2] from complex datasets, which makes them ideal in cancer studies. Some previous works [3] were done on cancer detection using machine learning. One of the previous works was done on the [4] Gene Expression Cancer RNA-Seq dataset, which is obtained from UCI Repository. The dataset is composed of five different types of tumors (BRCA, KIRC, COAD, LUAD, and PRAD), which are expressed by the gene expressions of the patients.

The dataset [4] is an 801 patient samples collection where 136 samples are patients with PRAD tumor, 141 samples are patients with LUAD tumor, 300 samples are patients with BRCA tumor, 146 samples are patients with KIRC tumor and the

Table 1 RNA-seq dataset details

Tumor classes	Samples per class
PRAD	136
LUAD	141
BRCA	300
KIRC	146
COAD	78

remaining 78 samples are patients with COAD tumor. So, all samples are divided into five classes as five tumor types (Table 1).

The dataset [4] contains 20,531 features of each sample. For each sample, the features are RNA-seq gene expression levels, which are measured with the Illumina Hiseq platform.

In this study, the Naive Bayes method [5] on the WEKA tool was used to classify the tumor types with 98.7516% accuracy in 2.69 s by tenfold cross-validation, 98.5% accuracy in 2.47 s by 50–50% train-test data split, 98.7526% accuracy in 2.63 s by 40–60% train-test data split, and 98.5294% accuracy on 2.65 s by 66–34% train-test data partition.

For diagnosing brain tumors interest in designing tools has been increasing in recent years. In the work of Gopal and Karnan [6], image processing clustering algorithm is used to classify images into a group that has brain tumor and another group that does not. In this work, we have used 42 MRI images for analysis. The authors remove the film artifacts (labels and X-ray marks) in the preprocessing phase. A Median filter is used by them to remove high-frequency components in the MRI image. The algorithm used here is Fuzzy C Means (FCM) as an image clustering algorithm, in addition to using Genetic Algorithm (GA) as an intelligent optimization tool. From this experiment, the result said that the classification accuracy of the classification algorithm FCM is 74.6% with less than 0.4% error rate. The authors used an optimization technique called Particle Swarm Optimization (PSO) in order to enhance the accuracy. They achieved an accuracy level of 92%.

In [7], a new system for diagnosis of brain tumor automatically is proposed by Othman and Ariffanan. For pattern classification problems, the Probabilistic Neural Network (PNN) provides a good solution. A dataset from University Teknologi Malaysia (UTK) is used here in this paper. The dataset goes through a preprocessing phase as follows. The MRI images are converted into matrices by using MATLAB. To classify the MRI images, the classification algorithm PNN is used then. The result said that more than 73% accuracy is achieved by the proposed system. Depending upon a smoothing factor [7], the accuracy level can even be higher.

Finally, a classifier to detect abnormalities in CT brain images caused by the diseases/cases—atrophic, hemorrhage, hematoma, infarct, and craniotomy—is designed by Najadatat al. [8].

3 Proposed Work

Based on the previous work [4] conducted using Naive Bayesian, we have tried to improvise first to get a better result by using MultinomialNB [5]. First, we have encoded the categorical data into classes. Then we have applied GaussianNB [5] and MultinomialNB with all the features. We have observed a significant increase in accuracy level while using all the features available. But to make the model more simpler and to reduce execution time, we have tried to reduce features and consider only the crucial ones contributing the most by using chi2 and RFECV (Recursive Feature Elimination with Cross-Validation) [9]. But as chi^2 has some limitations, we have proceeded with RFECV [9]. We have used StratifiedKFold with cv values 5, 10, and 15. Within those cv values, 10 is found to give the best result. So, we have proceeded with cv value 10. We have used StratifiedKFold so that in each fold we get almost an equal percentage of samples belonging to every class level. After the feature selection process, we got 2661 features, which are the most important for our prediction. Here we have noticed that the number of features has significantly reduced. The accuracy level is expected to be >98% so that we can conclude that the selection of features is successful according to the dataset.

Naïve Bayes Classifier (2) that is also called Simple Bayes, Idiot's Bayes, and Independence Bayes. This is an efficient supervised learning algorithm that is used for both binary and multiclass classification. Also, it is the statistical method based on Bayes' Theorem (1). Posterior probability is calculated in Bayes theorem using the prior probability.

$$P(c/x) = P(x/c)P(c)/P(x) \tag{1}$$

The assumption of the Naive Bayes classifier is that a particular feature present in a class is independent of any other features present in that class.

$$P(y|x_1, \ldots, x_n) = P(y)P(x_1, \ldots, x_n|y) / P(x_1, \ldots, x_n) \tag{2}$$

K-Nearest Neighbor is a simple Machine Learning algorithm that is based on supervised learning technique. It finds out the similarity between the new data point and the available class's data points. The value of K is to be determined by the trial and error method. K-N-N method finds the K nearest data points from the new unknown data point. Generally, the distance can be any metric measure but most commonly used metric is standard Euclidean distance. Then it labels the new data point as a class in which the maximum nearest points are found. As it simply remembers all of its training data, so it is a non-generalized machine learning method.

Support vector machine algorithm is a supervised learning algorithm. It is used to find a hyperplane in an N-dimensional space (N is the number of features), which is used to distinctly classify the classes using the data points. There can be many hyperplanes that can classify two classes. The objective of this algorithm is to find the hyperplane, which has the maximum distance between data points of both classes. Maximizing distance gives us more confidence to classify future data points.

Decision tree algorithm is a supervised learning algorithm, which is used for both classification and regression problems. It is a tree-like structure. Internal nodes represent the attributes, the branches represent decision rules and the leaf nodes represent the outcome. It first splits the data according to the attribute selection measure. Then it finds the best possible root node. There are two methods for doing this, the Gini index and Information gain. The impurity of a node according to the incorrectly classified results is measured by the Gini index. And it tries to reduce the Gini impurity in each split of the data. The attribute that has the lowest impurity becomes the root node. On the other hand, information gain finds out the purity of an attribute and according to that, it decides the root node. In each and every step, it finds the root node for each split until the leaf node. We can reach to the leaf node or to the outcome starting from the root node by asking a series of questions. Each time when we receive the answer another follow-up question is asked until we reach the leaf node which can classify the data at a class level. The series of questions and the possible answers form a tree-like structure, which is a hierarchical structure made of nodes and direct edges.

After using RFECV [9] to get the proposed working features, the selected dataset is split in a 70:30 ratio using a train test split. But as this splitting function splits the dataset randomly, the accuracy value may differ according to the split dataset. So, it is better to use cross-validation techniques to get the average accuracy. We have moved on and used MultinomialNB and KNeighboursClassifier to fit the data into the model and used Cross-Validation with fold value 10 to get the average accuracy value. We also experimented by using the Support Vector Machines and Decision Tree method to check which one is more suitable and fits the most. We have also tried the Random Forest model. All of these results were cross-validated by keeping the same fold value as a benchmark for a standard and unbiased comparison and noted the accuracy value. We have also tested various types of error values for each of the models used. Following is the flow chart of our work (Fig. 1).

We have extended the previous research work on tumor type detection and to broader the research area, we have used various types of classification algorithms so that we can compare the result among those and choose the best one. We are expecting a higher accuracy score than the previous one. Apart from that to reduce the computation time for early detection, we have reduced the complexity of the algorithm by using different feature selection methods.

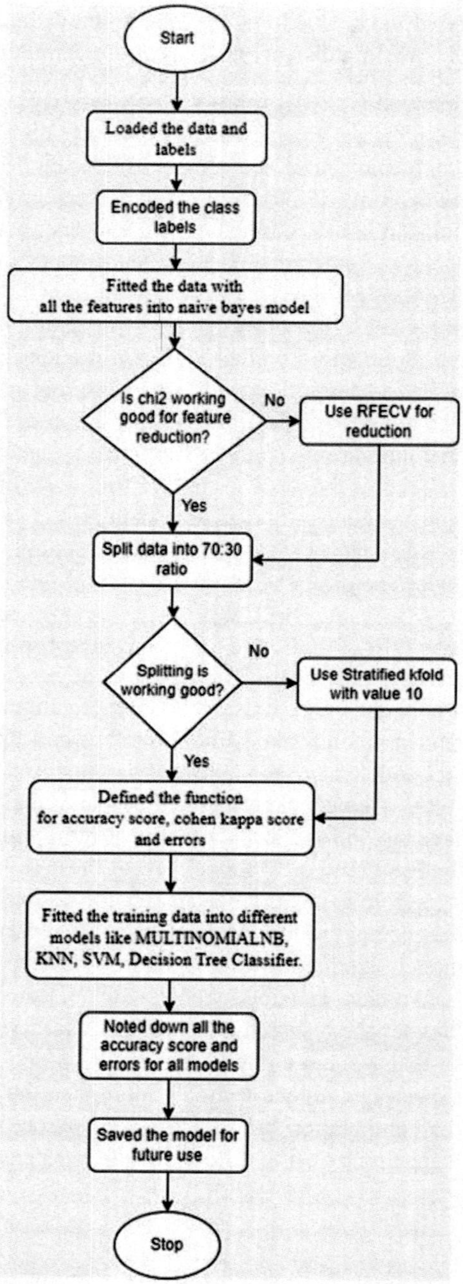

Fig. 1 Flowchart of the proposed method

4 Result

4.1 Naïve Bayes Algorithm

From a report presented by [4] Gemci et al. on Tumor Type Detection using Naïve Bayes Algorithm on Gene Expression Cancer RNA-Seq Dataset, we can compare our result with this previous work. We have used the following metrics for the comparison:

Kappa Value: Cohen kappa gives us a scale to measure the accuracy of algorithms where 1.0 is the highest value.

Mean Absolute Error (MAE): This is the mean of all absolute errors. The less the value of MAE, the more the accuracy is.

Root Mean Square Error (RMSE): This gives us a more accurate error without deviation (Table 2).

4.2 Other Classifiers

When using K-Nearest Classifier, SVM, and Decision Tree, we expect a higher success rate of the model such as 95–99%. So, if the model gives us an accuracy level of 99% then we can conclude that the selection of features is successful according to the dataset (Table 3).

Table 2 Result of previous work versus result of our work using Naïve Bayes algorithm

The performance measurement of Naive Bayes	Result of previous work	Result of our work
The classification accuracy	98.7526	99.6265
Kappa value	0.9834	0.9945
MAE	0.005	0.0082
RMSE	0.0706	0.1288

Table 3 Accuracy percentage and different errors value using different algorithms

The performance measurement	Using KNN	Using SVM	Using decision tree	Using random forest
The classification accuracy	99.3796	99.3780	95.3796	99.6265
Kappa value	0.9837	0.9837	0.9296	0.9945
MAE	0.0373	0.0248	0.1535	0.0082
RMSE	0.3347	0.2410	0.6967	0.1288

Apart from that, we have also tried with Random Forest classifier, which has given an accuracy score of 99.62% with tenfold cross-validation.

5 Discussions

From the experiment and result, we can understand that there are many classification algorithms to work with but the *Naïve Bayes Algorithm* gives us the best result with less amount of error (Naïve Bayes accuracy 99.6265% and error 0.008).

Other algorithms also give us satisfactory results but the *Decision Tree* is not so suitable for this kind of dataset as it gives us significantly less accuracy (Decision tree accuracy 95.3796). *Support Vector Machine and KNN* algorithm give us 99.37% of accuracy, which is less than Naïve Bayes. Though the accuracy score of the *Random Forest* algorithm is very good and comparable to the *Naive Bayes Algorithm* with tenfold cross-validation still we have not considered it to be a good model because the Random Forest algorithm is based on the Decision Tree algorithm and we have seen that the accuracy score of Decision Tree algorithm is very low as compared to the other methods. So, there is a high chance that the Random Forest Algorithm will predict the tumor type wrongly which can deviate the decision of the physician. As we have got our expected accuracy score from some of the algorithms that means our feature selection is successful.

Here, with the kappa value, we can also understand What is the best classification algorithm for this kind of dataset. Cohen kappa gives us a scale to measure the accuracy of algorithms where 1.0 is the highest and Naïve Bayes has got the best kappa value among the other classifiers.

6 Conclusion

It is very important to detect cancer early in a cancer control plan. So that the case can be detected in an earlier stage for more effective treatment which has a greater chance of cure.

In this work, we learned about the various techniques and algorithms used in tumor-type detection (like Naïve Bayes), feature selection techniques (like Recursive Feature Elimination and Cross-Validation Selection algorithm), and classification techniques (like K nearest neighbors, Support Vector Machine, Decision Tree). We implemented our project using classification algorithms and using various feature selection methods for data trimming. Along with them, we used a confusion matrix (a table that describes the performance of a classification model) as a tool to ensure that our classification algorithms are working and to provide features, like the Cohen kappa score, to justify the accuracy of our algorithms over any provided dataset.

It is certain that these accurate predictions will help us to predict future tumor-based problems and how to get desired results with better accuracy using classification algorithms of machine learning.

References

1. Renato Cordeiro de Amorim (2007) Computational Methods of feature selection, huan liu, hiroshi motoda. CRC Press, Boca Raton, p 440
2. Pal M, Giles M (2010) Foody: feature selection for classification of hyperspectral data by SVM. IEEE Trans Geosci Rem Sens 48(5):2297–2307
3. Saba T (2020) Recent advancement in cancer detection using machine learning: systematic survey of decades, comparisons and challenges. J Infect Public Health 13(9):1274–1289
4. Gemci F, Ibrikci T (2017) Tumor type detection using Naïve Bayes algorithm on gene expression cancer RNA-seq data set. In: International conference on engineering technologies (ICENTE'17)
5. Lewis DD (1998) Naive (Bayes) at forty: the independence assumption in information retrieval. In: Nédellec C, Rouveirol C (eds) Machine learning: ECML-98. ECML 1998. Lecture notes in computer science, vol 1398. Springer, Berlin, Heidelberg
6. Gopal NN, Karnan M (2010) Diagnose brain tumor through MRI using image processing clustering algorithms such as fuzzy C means along with intelligent optimization techniques. In: IEEE International conference on computational intelligence and computing research, pp 1–4 (2010)
7. Othman MFB, Basri MAM (2011) Probabilistic neural network for brain tumor classification. In: Second international conference on intelligent systems, modelling and simulation, pp 136–138
8. Najadat H, Jaffal Y, Darwish O, Yasser N (2011) A classifier to detect abnormality in CT brain images. In: International multiconference of engineers and computer scientists: IMECS 2011, vol 1, pp 374–377, Hong Kong, China
9. Wang C, Xiao Z, Wu J (2019) Functional connectivity-based classification of autism and control using SVM-RFECV on rs-fMRI data, pp 99–105, Phys Med

Proactive Fault Prediction and Tolerance in Cloud Computing

P. Pabitha, N. Sandheep, K. Nivitha, and R. Praveen

1 Introduction

Cloud computing delivers services like tools and applications through the Internet. Cloud computing is a popular option for people and businesses for a number of reasons. Some of the cloud services are Email, Storage, backup, and data retrieval, Creating and testing apps, and Delivering software. Various types of cloud computing are Software-as-a-service (SaaS), Infrastructure-as-a-service (IaaS) and Platform-as-a-service (PaaS) [25].

1.1 Fault Tolerance in Cloud Computing

The types of faults in cloud setting system are The Transient faults, Intermittent faults, Permanent faults, and Byzantine faults. Some other Critical Problems faced in cloud computing are Need to Automate, Halt in the clouds, and Workloads in the cloud. Parameters that are mainly used in Fault Tolerance are Performance, Response time, Scalability, Throughput, Reliability, Availability, Usability, Overhead associated, and the Cost-effectiveness [36].

Fault Tolerance in Cloud Computing is a system's ability to perform its task accurately in spite of inner defects that are present or the downtime. The scheme classification of the faults is Static fault tolerance scheme and Dynamic fault tolerance

P. Pabitha · K. Nivitha · R. Praveen (✉)
Department of Computer Technology, Anna University, Madras Institute of Technology Campus, Chennai 600044, Tamil Nadu, India
e-mail: Praveenram2510@gmail.com

N. Sandheep
Department of Information Technology, Anna University, Madras Institute of Technology Campus, Chennai 600044, Tamil Nadu, India

A. Ramdane-Cherif et al. (eds.), *Machine Intelligence and Data Science Applications*, Algorithms for Intelligent Systems, https://doi.org/10.1007/978-981-99-1620-7_41

scheme. They are further classified as Forward recovery, Backward recovery, and Checkpoint [27].

The techniques of fault tolerance are Reactive Methods, Proactive Methods, and Resilient Methods. In Reactive Fault Tolerance Mechanisms, the measures are only reaction to the faults that have already occurred. After a fault has taken place, it will be tried to rectify those faults and try to recover the system from those particular faults [28]. This is the Reactive Mechanism in the Fault Tolerance. Proactive Fault Tolerance in Cloud Computing means the faults are predicted in prior and measures are taken to prevent those to provide uninterrupted cloud services to the customers. The Resilient Fault Tolerance Mechanisms are similar to the Proactive Fault Tolerance Mechanisms. It continuously monitors the system and gains tolerance to faults from various outcomes. It is a process of Intelligent Learning via Interaction. The Resilient Fault Tolerance Mechanisms are Machine learning and Fault induction.

1.2 Objectives

- To schedule the input tasks in cloud computing environment
- To predict the fault in the system
- To evaluate the existing algorithm with the proposed algorithm
- To find optimal Virtual Machine

2 Literature Survey

A comprehensive literature review is performed, which is summarized here.

In [24], one of the most common meta-heuristic strategies for tackling optimal task scheduling problems is particle swarm optimization. To start the PSO search process, the suggested technique uses heuristic scheduling methods. Heuristic methods called MCT-PSO and LJFP-PSO to used to create a hybridized task scheduling algorithm [37]. Degree of Imbalance, Makespan, Amount of Execution Time, and Consumed Energy are the four variables that influence the fitness value. The proposed algorithms are compared to numerous recent meta-heuristic algorithms [30]. All virtual machines' idle are set at random between this interval. For simulations, two different scenarios are considered, the number of tasks varies between 200 and 300 or the number of tasks remains constant.

In [23], used hybrid particle swarm optimization HPSO to incorporate the classification strategy of artificial fish swarm optimization into particle swarm optimization and introduces the strategy of pre-generating search particles. There are optimal solution areas Dbest and bad solution areas Dbad in the two-dimensional solution space. Because of this, the following judgment can be made for the (x, y) position after each iteration of particle optimization, The position (x, y) is better when it is closer to Dbest [20]. The position (x, y) becomes worse as it gets closer to Dbad. Before updating the position of the particles, the field of view of each population and the crowding degree of each particle must be calculated. As experimental data, two task sets are built of DAG by hand synthesis based on distinct task quantities and partial order relationships. HPSO has been shown to be superior to existing algorithms in tests. When scheduling different DAG task sets, the results of HPSO are always better than those of PSO and IA. The benefit of HPSO becomes more apparent as the complexity of problems to be solved increases [21]. HPSO ensures that each virtual machine has a consistent execution time.

In [3], the ability of a system to run continuously in case of failure of system components is fault tolerance. ACO is not updated for each iteration. As a result, a novel concept, reinforced ant colony optimization, is introduced in this study work to tackle job scheduling. This examines fault tolerance as the most central feature in the cloud to make task scheduling more effective [39]. A fault tolerant system will locate the issue and make an effort to resolve it. This method prevents the system from getting caught in a failure trap while perhaps extending the lifespan of the process. In order to validate the suggested phenomena, cloud data has been taken into account. All of the tasks in the priority queue, which range from 10 to 100, will be carried out by 2–10 virtual computers. In [32], used Chemical reaction optimization (CRO) and Particle Swarm Optimization [31] (PSO) as a combined approach to schedule tasks in the cloud computing domain is described in this section. CRO is a new evolutionary computing approach that has a quick convergence speed [8]. The original CRO is first adjusted by eliminating the number of such phases that increase the additional computation and time complexity for solution development in order to generate the updated CR-PSO algorithm. In order to improve the quality of the optimal solution, the standard PSO is adjusted in a second step. This step determines the best task scheduling strategy to lower computation costs and shorten the make-span time for scheduling the work on a cloud-based virtual machine.

In [41], framework contains data centers with physical machines or servers. Each server consists of a virtual machine hypervisor layer and resource pool. Power analyzer which is used for measuring CPU utilization and resource estimator for measuring the status of the server. To predict failures ensemble models are used. Preparation, prediction, and output generation are three different operations to be performed by the Failure Prediction Unit (FPU). Preparation involves selecting the attributes. Prediction uses the ensemble learning approach which involves different predictors, namely, support vector machine, feedforward neural network, and linear regression. The final output is predicted by a voting process from all the base learners.

In [40], error-prone virtual machines are evaluated against the Resource Provisioning Matrix in the Failure Tolerance Unit and the maximum value of each resource capacity is selected in the select box to determine the effective capacity of x resources. We evaluate the performance of the proposed work using two real benchmark workloads with CPU, memory, disk I/O requirements, and Google Cluster Data (GCD) providing usage information [11]. OFP-TM is compared to three splitting works that predict failure in order to provide high accessibility, Prediction-based Energy Aware

Fault Tolerant Scheduling Scheme (PEFS), Hadoop Distributed Computing Clusters for Fault Prediction (HDCC), Cloud Disk Error Forecasting (CDEF).

In [5], the scheduler is built on the MAPE-k model. The control system is made up of four components: monitor, analyzer, planner, and executor, all of which share a knowledge base. The monitor can be used to collect the supervised data. The analyzer forecasts the rate of resource load [16, 38]. An IDE algorithm for workflow scheduling has been proposed in the planner component. The basic goal of proactive fault tolerance is to prevent anomalous Virtual Machines from entering the problem zone and from receiving new load. A live Virtual Machine migration is when a Virtual Machine's present load is transferred to another Virtual Machine without halting it [14]. The proposed method is compared with three existing algorithms: Enhanced Differential Evolution (IDE) and multi-objective scheduling methods based on IDE heuristic features of Fault Tolerant Workflow Scheduling (ICFWS) and Ant Colony Optimization (ACO) are evaluated increase.

In [7], the PCFT method's goal is to keep track of and predict declining Physical Machines. If the physical machine is degraded, our approach looks for the best target physical machine for virtual machines hosted on the degraded physical machine. To predict which physical machines will degrade, the failure characteristics of physical machines are used to select which physical machines will degrade based on CPU temperature [2]. Shutting down physical machines due to overheating and improving data center energy efficiency. Finding the best physical target machine is considered an optimization problem. They propose an adaptive heuristic algorithm based on the improved PSO algorithm to solve the optimization problem. The 3D coding scheme aims to solve the virtual machine reassignment problem [35]. Through simulation experiments, the efficiency is assessed and effectiveness of our approach. In particular, the approach is compared to five others in terms of overall transmission overhead while running a set of parallel applications.

In [10], the Replica Set Controller in the described system is used to control a service that can be mapped to many pod instances in Kubernetes, a popular container orchestration system. We need the ability to predict a node's vulnerability to cloud service system failures before a failure occurs, and a migration mechanism that can migrate a Kubernetes container application to another healthy worker node [26]. Prometheus, a popular monitoring framework, is used as a search block in the proposed work to collect real-time data for prediction and store it in an external database for retraining strategies. The Bi-LSTM model is the main contribution of this paper. The task executor determines whether each node should fail by setting a threshold on the model's output. A worker node that consumes more than 80% CPU or memory in the next time step will be marked unhealthy in the architecture and Task Executor will migrate the container service from that node to another healthy node using migration API increase [12]. A live migration pod strategy with checkpoints by CRIU (checkpoint restore in user space) is used to migrate running containers.

In [19], they make the cloud environment more energy efficient, and a software solution such as Virtual Machine consolidation is required. Virtual Machine consolidation is when a virtual machine is mapped to a real machine and no more migrations are required. The process of choosing the best Virtual Machine that can complete the work without failure is known as Virtual Machine selection. Using the cloudsim simulator, they analyze the efficacy of the VUOD [Virtual Machine utilization-based Virtual Machine overload detection algorithm] and MVU [Maximum Virtual Machine Utilization based selection policy] algorithms. They suggested a modified maximum utilization-based selection criteria as well as a modified overload detection approach based on virtual machine CPU usage [42]. This algorithm determines if a virtual machine host or server will be overloaded or oversubscribed. It shows the total number of Virtual Machine migrations that occur as a result of overloaded or underloaded hosts. In comparison to the former ways, this strategy resulted in no Virtual Machine migration. The number of hosts shut down in the new approaches decreased by 22%, demonstrating that our suggested algorithm outperforms the previous methods. Cloud computing is becoming increasingly popular. High power is required to keep the data center running, which increases operating costs and generates user unhappiness. Energy usage also results in the production of carbon dioxide, which contributes to global warming. When compared to the traditional methodology, the proposed method minimizes power usage, ASLA violations, the number of Virtual Machine migrations, and host shutdown.

In [4], Qaiser et al., discussed system, the live migration of Virtual Machines from over-utilized hosts is required for Dynamic Consolidation of Virtual Machines. The efficient selection of Virtual Machines is a crucial but often overlooked aspect of the consolidation process. This paper describes two policies for virtual machine selection: Capacity Based Selection (CBS) and Threshold Based Selection (TBS). These policies are based on the idea of reducing several elements that contribute to the degradation of service quality as a result of consolidation at the same time [13]. Criterias used in TBS are (I) The migration of the Virtual Machine reduces the total resource request to the host's resource consumption threshold. (II) Of all the Virtual Machines that meet the criteria, the selected one requires the least amount of time to migrate. CBS uses the following criteria to bring over-utilized hosts below their resource consumption capacity, preferably in a single Virtual Machine migration. (I) Moving the Virtual Machine brings the overall resource request to the host's resource consumption capacity. (II) Of all the Virtual Machines that meet the criteria, the selected one requires the least amount of time to migrate. It may be necessary to migrate multiple Virtual Machines from an overburdened host. TBS and CBS employ the same technique for selecting Virtual Machines for migration but with different overload levels as parameters.

In [22], the Cloud infrastructure meets the demand for big computing. As a result, a prominent ongoing study focus is how to reduce energy use in data centers. Our study field is Dynamic Virtual Machine consolidation, which involves intelligent decision-making to monitor server load and optimize datacenter use. The proposed (MC MC) Maximum Correlation with Minimum Virtual Machine Selection by Modified Knapsack technique handles extremely loaded servers by relocating the most relevant Virtual Machine to the suitable destination server. The authors of [1, 33] are proposed novels' methods to address the fault in the cloud computing environment [17]. CloudSim is used to model the implementation of our strategy. Our findings show that energy consumption and the number of Virtual Machine migrations have decreased, while cloud data center performance has improved. The minimum selection knapsack is a modified version of the basic knapsack, with the goal of 1. To achieve migration control, the bare minimum of Virtual Machines should be used. 2. With a given weight, obtain the most profit (maximum correlation coefficient) (overutilization) Profit is less significant than the amount of things chosen in the recommended approach. As a result, it is concentrated on selecting the smallest number of Virtual Machines with the highest potential correlation coefficient in order to achieve migration control. The Maximum correlation is suggested with migration control (MCMC) strategy in our work, and the smallest number of Virtual Machines is chosen with the highest possible correlation with Host. First, the parameter Over-Utilized Host is linked to modified knapsack parameters, with the value of overutilization assumed to be "Knapsack Weight," the correlation coefficient of Virtual Machines on the host assumed to be "Value of Items," and the current requested MIPS of Virtual Machines assumed to be "Weight of Items". The suggested approach provides a solution for Virtual Machine consolidation by selecting the smallest number of Virtual Machines, resulting in fewer migrations and lower datacenter host energy consumption.

2.1 The Major Contribution of the Work

- Predicting the fault even before it occurs in the virtual machines using efficient machine learning models.
- Round Robin algorithm is initially used to schedule tasks efficiently without the need of waiting for so long unattended.
- Choosing the most contributing parameters has improved the accuracy of predicting the occurrence of fault.
- Optimal virtual machine is recommended to migrate the processes from Faulty Virtual Machines using the Best Fit algorithm to which the migration can be done with less compensation.

3 Proposed System

In our proposed system, the tasks are scheduled using the Round Robin algorithm to respective VM's. Fault's are identified in its earlier stage using Multi-layer perceptron based on the parameters like CPU temperature, utilization, disk storage, and memory usage. After fault detection in a VM, the job is allocated to another optimal VM using the Best fit algorithm.

3.1 Architecture Diagram

We use CloudSim to initialize cloud environments like hosts, virtual machines, and cloud lists. It explains how the tasks are being allocated to the VMs based on its inputs like host details, cloud details, and VMs details. Machine learning models are trained simultaneously to predict the fault and the model with highest accuracy is used in the proposed system.

3.2 Methodology

The system takes the Host Details, Cloudlet Details, and the Virtual Machine Details as input to the Proactive Fault Prediction and Tolerance System as shown in Fig. 1. It then creates the Hosts, Virtual Machines, Cloudlets according to the required specifications, and quantities, creates lists for each of them, and adds them to the lists. All the lists are submitted to the broker and are scheduled according to their resource requirements using the Round Robin algorithm and allocated to the respective Virtual Machines in the hosts in a datacenter.

Simultaneously, to predict fault the model is trained in the system using the Multi-Layer Perceptron Neural Networks as shown in Fig. 2. Other models are also to be trained similarly and the accuracy of the trained models is compared and the model with maximum accuracy is to be selected. For the predicted fault, an optimal Virtual Machine is to be found using the Best fit algorithm to which the faulty ones had to be migrated. Initially, the search for the optimal Virtual Machine is to be done within the same host as the Faulty Virtual Machine for minimizing the effort, time for migration, etc. If there is no such Virtual Machine within the same host, then an optimal Virtual Machine is to be found in a neighboring host. If there is no such Virtual Machine within any of the hosts in the datacenter, the search begins in the nearby datacenters. If not found, the cloud service providers are alerted regarding this.

3.3 Workflow Diagram

After scheduling the task, if the VM didn't enter into the fault stage, then it resumes its execution. If the trained model predicts the fault in any VM it will search for another optimal VM in its vicinity to migrate that specific job running in that VM.

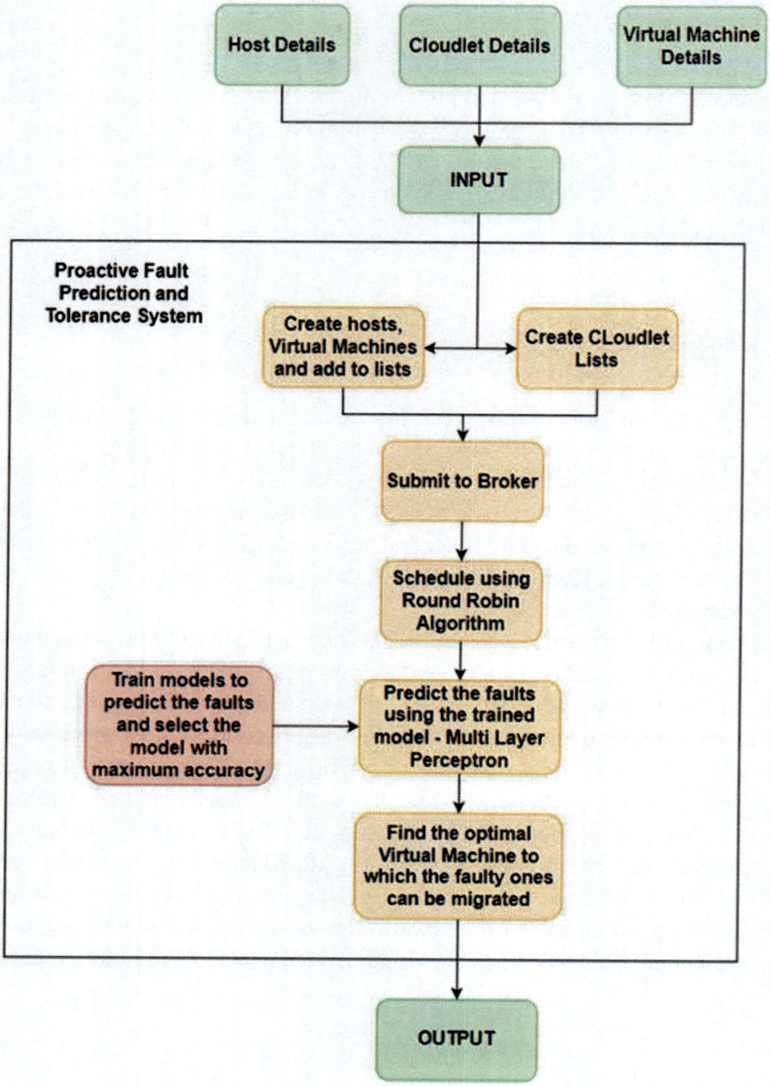

Fig. 1 Proactive fault tolerance

3.4 Modules Involved

The set of modules involved in proactive fault predictions are

- Module 1-Task Scheduling Round Robin algorithm.
- Module 2-Fault prediction in Virtual Machine using Multi-Layer Perceptron considering different parameters like CPU temperature, utilization, disk storage, and memory usage.

Fig. 2 Workflow diagram

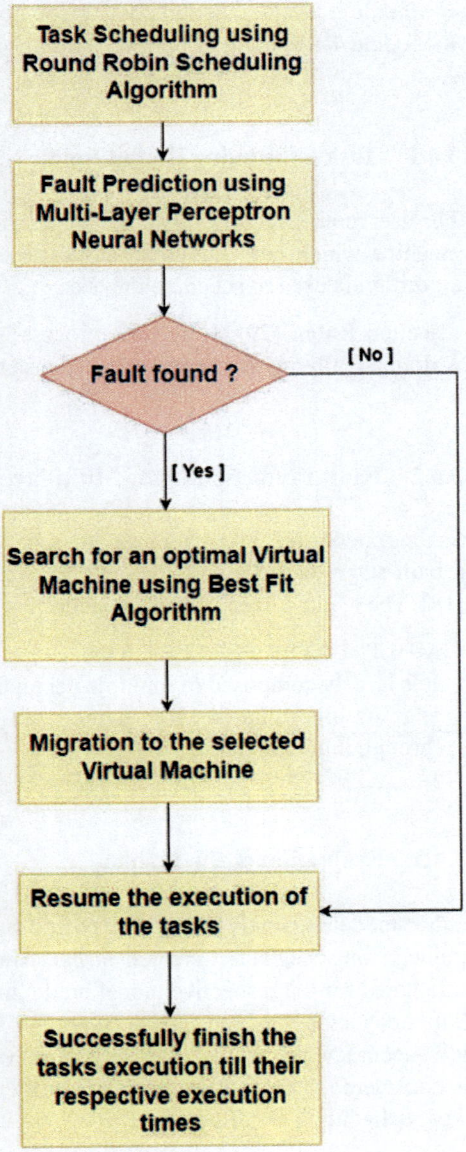

- Module 3-Evaluation and comparison between existing and proposed system.
- Module 4-Selection of optimal Virtual Machine using the Best Fit algorithm

3.4.1 Task Scheduling Round Robin Algorithm

The user requests for the resources are queued, processed, and allocated to the virtual machine which has enough resources to execute the task [6]. The Round Robin algorithm is used to schedule the task.

- Round Robin [29]: Here each process is assigned in a fixed time slot hence it dramatically increases average response time and not to waste time.

3.4.2 Fault Prediction Using Multi-layer Perceptron

Saxena and Singh [34]: After tasks are scheduled to Virtual Machines, they may enter a fault stage due to high CPU temperature, memory exhaustion, and other factors [18]. The Multi-Layer Perceptron neuron network model is used to predict the fault.

- MULTI-LAYER PERCEPTRON
  It is largely composed of multiple perceptron layers. Backpropagation is common learning algorithm in MLP. It produces output from the given inputs that pass through input and hidden layers.

3.4.3 Evaluation of the System

Other models, namely, Naive Bayes and Support Vector Machine have also been trained using the dataset created earlier. After testing the model, accuracy has been calculated for the respective model prediction as 77% for Naive Bayes and 70% for Support Vector Machine. The accuracy for Multi-Layer Perceptron is calculated as 99% accuracy. So, Multi-Layer Perceptron with highest accuracy for fault prediction is considered. The predicted value is written into the fault column of the provided Task Scheduling Details page.

1. **Naive Bayes:** It determines whether or not there is a fault in the virtual machine. This is a two-class issue. The two classes identified are fault presence (Yes) and fault absence (No). The columns represent parameters such as CPU temperature, CPU utilization, disk storage, and memory usage, while the rows represent individual entries.
2. **Support Vector Machine (SVM):** It seeks a hyperplane in an N-dimensional space that distinguishes between data points. It finds the boundary that separates the space. A hyperplane is the best decision boundary. The loss function aids in increasing the margin.

Four input parameters are taken into account. The CPU temperature, CPU utilization, disk storage, and memory usage are all taken into account. The hyperplane is neither a line nor a plane in two dimensions. This is a problem of two classes. The two classes identified are fault presence (Yes) and fault absence (No).

3.4.4 Selection of Optimal Virtual Machine Using Best Fit Algorithm

When a Virtual Machine is predicted with fault, it will search for the Virtual Machine with the required resources and CPU utilization to run the task within the same host [9]. If it is found, it will migrate the task from the faulty to the found Optimal system. If there is no such Virtual Machine within the same host, it'll check the neighboring and then the remaining hosts in the datacenter [15]. Then it will search in the nearby and remaining data centers. Even then if no optimal Virtual Machine is found, it will be reported to the Cloud Service Provider about the shortage of resources.

4 Implementation

The cloud environment is initialized in CloudSim and the cloudlet details are stored for future references. After the VMs are in a runnable state, the tasks are allocated and their parameters are stored in .csv format which was later used to find fault using the trained Multi-layer Perceptron. For the VMs where the fault is predicted its job is migrated to another optimal VM.

4.1 Dataset Description

A dataset has been generated using the reference from the published papers in this domain consisting of 500 records as shown in Fig. 3. The features that have been considered are CPU temperature, CPU utilization, Disk Storage, and Memory. The threshold for CPU temperature is set as 60 °C. i.e., if the CPU temperature of the Virtual Machine crosses 60 °C, then there is a chance that the machine may enter into the fault state by shutting down. Similarly, for other parameters like CPU utilization, Disk storage, and Memory a threshold has been set as 85%. If any of the parameters has the value greater than the threshold value then the system may enter into the fault stage.

1	CPU Temperatur	CPU Utilization	Disk Storage	Memory Usage	Fault
2	89.27	87.96	89.47	87.69	Yes
3	80.96	70.84	75.2	53.64	No
4	88.89	95.24	95.91	87.99	Yes
5	54.58	64.52	57.64	71.95	No
6	93.15	91.51	89.46	88.34	Yes
7	53.7	78.49	79.19	77.4	No
8	93.36	88.06	93.91	91.15	Yes
9	59.19	51.72	80.4	52.7	No
10	94.46	94.87	88.61	92.31	Yes
11	57.65	68.67	71.41	54.75	No
12	95.99	93.88	96	89.7	Yes
13	65.45	77.25	51.26	52.37	No
14	94.07	94.82	93.37	89.67	Yes
15	55.05	62.79	71.43	80.64	No
16	93.89	88.1	92.48	95.24	Yes
17	76.59	57.45	64.59	54.66	No

Fig. 3 Dataset for fault prediction

4.2 Task Scheduling Using Round Robin Algorithm

The set of input tasks is taken and stored in a separate file as public static final variables. The input details consist of Number of Hosts, Number of Host PEs, Host MIPS, Host RAM, Host Storage, Number of VMs, Number of VM PEs, VM MIPS, VM Size, VM RAM Number of Cloudlets referring to tasks, Number of Cloudlet Pes, Cloudlets Length, Cloudlet File size, Cloudlet Output Size, Scheduling Interval, and some other requirements as shown in Fig. 4.

A host list, Virtual Machine list, and the cloudlet lists are created and submitted to the broker along with specifying the Round Robin algorithm. After task allocation, the Finished Cloudlets list is generated by the Cloudlet Table Builder. It is displayed in the output and also stored in a CSV file as shown in Fig. 5.

All the details of the task scheduling like VM ID, CPU Temperature, CPU Utilization, Disk Storage, Memory Usage, VM Power Consumption, Host ID, Host CPU Utilization, and Host Power Consumption are all written to a CSV as shown in Fig. 6. As no fault prediction has taken place yet, the Fault column has been left blank.

4.3 Fault Prediction Using Multi-layer Perceptron

A dataset is generated using a reference of the threshold values of CPU Temperature, CPU Utilization, Disk Storage, and Memory Usage along with the target value of whether the fault is present or not (Yes or No). MLP is trained using 70% of data and the remaining 30% is used to test. Then the accuracy is calculated for the model

```
public static final int HOSTS = 8;
public static final int HOST_PES2[] = {8, 9, 10, 11, 12, 13, 14, 15};
public static final int VMS = 30;
public static final int VM_PES = 2;
public static final int CLOUDLETS = 16;
public static final int CLOUDLET_PES = 2;
public static final int CLOUDLET_LENGTH = 10000;
public static final int SCHEDULING_INTERVAL = 1;
public static final double HOST_START_UP_DELAY = 5;
public static final double HOST_SHUT_DOWN_DELAY = 3;
public static final double HOST_START_UP_POWER = 5;
public static final double HOST_SHUT_DOWN_POWER = 3;
public static final double STATIC_POWER = 35;
public static final int MAX_POWER = 50;
public static final int VM_PES2[]   = {2, 2, 2, 1};
public static final long   CLOUDLET_LENGTH2 = 20_000;
public static final int SCHEDULING_INTERVAL2 = 10;
public static final double HOST_UNDER_UTILIZATION_THRESHOLD_FOR_VM_MIGRATION = 0.1;
public static final double HOST_OVER_UTILIZATION_THRESHOLD_FOR_VM_MIGRATION = 0.7;
public static final int HOST_SEARCH_RETRY_DELAY = 60;
public static final long HOST_BW = 16_000L;
public static final int HOST_MIPS = 1000;
public static final long HOST_RAM[] = {15_000, 500_000, 25_000, 15_000, 500_000, 25_000, 15_000, 500_000};
public static final long HOST_STORAGE = 1_000_000;
public static final int     VM_MIPS = 1000;
public static final long    VM_SIZE = 1000;
public static final int     VM_RAM = 10_000;
public static final double VM_BW = HOST_BW/(double)VM_PES2.length;
public static final long    CLOUDLET_FILESIZE = 300;
public static final long    CLOUDLET_OUTPUTSIZE = 300;
```

Fig. 4 Sample input of tasks and system details

	B	C	D	E	F	G	H	I	J	K	L	M
1	Status	DC	Host	Host PEs	VM	VM PEs	CloudletLen	CloudletPEs	StartTime	FinishTime	ExecTime	
2		ID	ID	CPU cores	ID	CPU cores	MI	CPU cores	Seconds	Seconds	Seconds	
3	SUCCESS	1	0	8	2	2	10000	2	5	15	10	
4	SUCCESS	1	1	9	3	2	10000	2	5	15	10	
5	SUCCESS	1	2	10	4	2	10000	2	5	15	10	
6	SUCCESS	1	3	11	5	2	10000	2	5	15	10	
7	SUCCESS	1	4	12	6	2	10000	2	5	15	10	
8	SUCCESS	1	5	13	7	2	10000	2	5	15	10	
9	SUCCESS	1	6	14	0	2	10000	2	5	15	10	
10	SUCCESS	1	7	15	1	2	10000	2	5	15	10	
11	SUCCESS	1	0	8	10	2	20000	2	5	25	20	
12	SUCCESS	1	1	9	11	2	20000	2	5	25	20	
13	SUCCESS	1	2	10	12	2	20000	2	5	25	20	
14	SUCCESS	1	3	11	13	2	20000	2	5	25	20	
15	SUCCESS	1	4	12	14	2	20000	2	5	25	20	
16	SUCCESS	1	5	13	15	2	20000	2	5	25	20	
17	SUCCESS	1	6	14	8	2	20000	2	5	25	20	
18	SUCCESS	1	7	15	9	2	20000	2	5	25	20	
19												
20												

Fig. 5 Finished cloudlet list

trained. After satisfactory accuracy is attained, the model can be used to predict the faults.

As output from the previous module, the Task Scheduling Details are stored in a file. This is given as the input to the trained model to predict the faults if any and it is written in the Fault Column of the Task Scheduling Details file as shown in Fig. 7.

VM ID	CPU Temperature	CPU Utilization	Disk Storage	Memory Usage	VM Power	Host ID	Host CPU	Host Power	Fault
2	67.12	41.67	55.4	68.53	10.31	0	25	38.75	
10	94.33	95.24	92.77	88.02	12.32	0	25	38.75	
18	35.62	0	35.66	38	8.75	0	25	38.75	
26	40.58	0	38.49	37.48	8.75	0	25	38.75	
3	52.7	41.67	69.63	78.2	10.31	1	22.22	38.33	
11	94.77	95.24	92.96	94.31	12.32	1	22.22	38.33	
19	35.23	0	43.98	39.14	8.75	1	22.22	38.33	
27	41.09	0	37.06	38.8	8.75	1	22.22	38.33	
4	57.1	41.67	67.56	63.15	10.31	2	20	38	
12	93.74	95.24	95.21	94.15	12.32	2	20	38	
20	39.57	0	43.16	45.42	8.75	2	20	38	
28	40.78	0	43.24	37.99	8.75	2	20	38	
5	56.36	41.67	76.62	54.72	10.31	3	18.18	37.73	
13	93.38	95.24	93.77	89.57	12.32	3	18.18	37.73	
21	36.21	0	38.52	40.67	8.75	3	18.18	37.73	
29	44.91	0	38.01	42.86	8.75	3	18.18	37.73	
6	78.64	41.67	75.34	67.44	13.75	4	16.67	37.5	
14	91.89	95.24	90.17	93.25	16.43	4	16.67	37.5	
22	45.33	0	35.57	44.42	11.67	4	16.67	37.5	
7	76.95	41.67	62.66	56.61	13.75	5	15.38	37.31	
15	91.15	95.24	91.7	95.35	16.43	5	15.38	37.31	
23	36	0	45.76	39.58	11.67	5	15.38	37.31	
0	66.98	41.67	73.61	55.17	10.31	6	14.29	37.14	
8	89.5	95.24	89.57	92.28	12.32	6	14.29	37.14	
16	43.95	0	41.04	45.6	8.75	6	14.29	37.14	
24	44.87	0	37.46	38.46	8.75	6	14.29	37.14	
1	64.5	41.67	55.04	63.92	10.31	7	13.33	37	
9	91.28	95.24	92.85	91.88	12.32	7	13.33	37	
17	40.77	0	44.04	39.86	8.75	7	13.33	37	
25	38.44	0	45	40.89	8.75	7	13.33	37	

Fig. 6 Task scheduling details

VM ID	CPU Temperature	CPU Utilization	Disk Storage	Memory Usage	VM Power	Host ID	Host CPU	Host Power	Fault
2	67.12	41.67	55.4	68.53	10.31	0	25	38.75	0
10	94.33	95.24	92.77	88.02	12.32	0	25	38.75	1
18	35.62	0	35.66	38	8.75	0	25	38.75	0
26	40.58	0	38.49	37.48	8.75	0	25	38.75	0
3	52.7	41.67	69.63	78.2	10.31	1	22.22	38.33	0
11	94.77	95.24	92.96	94.31	12.32	1	22.22	38.33	1
19	35.23	0	43.98	39.14	8.75	1	22.22	38.33	0
27	41.09	0	37.06	38.8	8.75	1	22.22	38.33	0
4	57.1	41.67	67.56	63.15	10.31	2	20	38	0
12	93.74	95.24	95.21	94.15	12.32	2	20	38	1
20	39.57	0	43.16	45.42	8.75	2	20	38	0
28	40.78	0	43.24	37.99	8.75	2	20	38	0
5	56.36	41.67	76.62	54.72	10.31	3	18.18	37.73	0
13	93.38	95.24	93.77	89.57	12.32	3	18.18	37.73	1
21	36.21	0	38.52	40.67	8.75	3	18.18	37.73	0
29	44.91	0	38.01	42.86	8.75	3	18.18	37.73	0
6	78.64	41.67	75.34	67.44	13.75	4	16.67	37.5	0
14	91.89	95.24	90.17	93.25	16.43	4	16.67	37.5	1
22	45.33	0	35.57	44.42	11.67	4	16.67	37.5	0
7	76.95	41.67	62.66	56.61	13.75	5	15.38	37.31	0
15	91.15	95.24	91.7	95.35	16.43	5	15.38	37.31	1
23	36	0	45.76	39.58	11.67	5	15.38	37.31	0
0	66.98	41.67	73.61	55.17	10.31	6	14.29	37.14	0
8	89.5	95.24	89.57	92.28	12.32	6	14.29	37.14	1
16	43.95	0	41.04	45.6	8.75	6	14.29	37.14	0
24	44.87	0	37.46	38.46	8.75	6	14.29	37.14	0
1	64.5	41.67	55.04	63.92	10.31	7	13.33	37	0
9	91.28	95.24	92.85	91.88	12.32	7	13.33	37	1
17	40.77	0	44.04	39.86	8.75	7	13.33	37	0
25	38.44	0	45	40.89	8.75	7	13.33	37	0

Fig. 7 Fault prediction

4.4 Evaluation Metrics

The parameters considered for the prediction of the presence of the fault are CPU temperature, CPU Utilization, Disk storage, and Memory usage. The CPU temperature is to be considered as excessive heat can cause CPU to fail. Some startup issues and random crashes may be faced. It may also result in the CPU to crash and all the ongoing tasks will be terminated. This results in the halt of the processes.

If disk usage is at high capacity and if it continues for several days, performance may degrade. If the computer has a high memory usage problem, it will freeze. It leads to temporarily pausing the execution of tasks.

All these are considered, fed to a model, trained, and then tested. Various metrics recall, f1-score, precision, support, and accuracy are calculated. The model's ability is mainly judged based on the accuracy, and how effectively the system is able to predict the occurrence of the fault.

4.5 Selection of Optimal Virtual Machine Using Best Fit Algorithm

As the output from the previous module, the fault is predicted in the Virtual Machines. This is taken as input for the current module as shown in Fig. 8.

After the faulty Virtual Machines are found, search for an optimal Virtual Machine begins. At first, CPU Utilization of the Virtual Machines within the same host is checked. If there is any Virtual Machine that is capable of accepting all the tasks from the faulty Virtual Machine, it is declared as the optimal Virtual Machine as shown in Fig. 9.

```
Number of rows in the table :  30

Fault detected at VM with ID : 10
Fault detected at VM with ID : 11
Fault detected at VM with ID : 12
Fault detected at VM with ID : 13
Fault detected at VM with ID : 14
Fault detected at VM with ID : 15
Fault detected at VM with ID : 8
Fault detected at VM with ID : 9

The tasks should be migrated from a VM to another VM
with respective VM IDs as follows to avaoid fault
```

Fig. 8 Faulty virtual machines

Fig. 9 Optimal virtual machine selection

	From VM	From Host	To VM	To Host
0	10	0	18	0
1	11	1	20	2
2	12	2	22	4
3	13	3	1	7
4	14	4	26	0
5	15	5	28	2
6	8	6	7	5
7	9	7	9	7

Else, the searching process is extended to the neighboring hosts and then to all of the hosts in the current datacenter and then to the others. If no Virtual Machine is found to be optimal. Physical Machine is alerted about the shortage of virtual machines, hosts, and the datacenters.

5 Results and Analysis

It is preferable to use cloud resources as there will be no need of maintaining any personal resources, storage, etc. for the customer. In order to uninterruptedly use resources, there should be no shortage of resources, no faults or halt of the processes. So, during task scheduling the required quantity of resources, time for completion and estimated utilization are all estimated. Round Robin is used to look for a Virtual Machine which has these necessary requirements and the task are scheduled for that Virtual Machine. Classifiers, namely, MLP, NB, and SVM are trained and used to predict the Faults in the system. The loss corresponding to the number of iterations in MLP has been visualized in Fig. 10. Then, the accuracy for the models is also calculated and was found to be 99% for Multi-layer Perceptron as shown in Fig. 11, 70% for Support Vector Machine as shown in Fig. 12 and 77% for Naive Bayes as shown in Fig. 13.

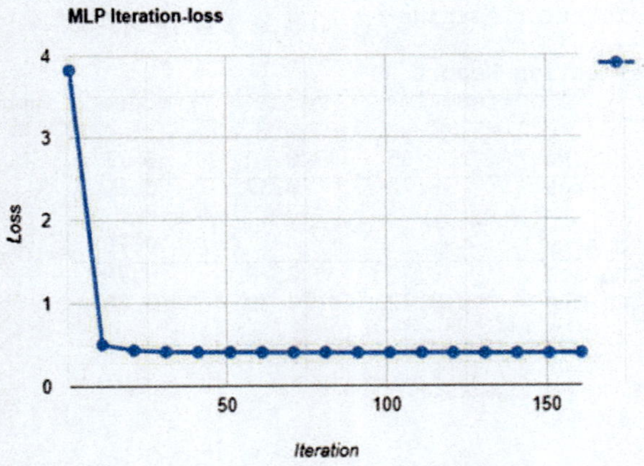

Fig. 10 Graphical representation of loss with each iteration in Multi-Layer Perceptron

```
Multi Layer Perceptron
---------------------------
Classification Report
              precision    recall  f1-score   support

           0       1.00      0.99      0.99        77
           1       0.99      1.00      0.99        73

    accuracy                           0.99       150
   macro avg       0.99      0.99      0.99       150
weighted avg       0.99      0.99      0.99       150

---------------------------------------------------------
Confusion Matrix is:
[[76  1]
 [ 0 73]]
---------------------------------------------------------
Accuracy is 0.9933333333333333
```

Fig. 11 Performance summary of Multi-layer Perceptron

Table 1 Accuracy comparison of various models

Model	Accuracy percentage
Support vector machine	70
Naive Bayes	77
Multi-layer perceptron	99

The summary of trained and tested models so far is compared as shown in Table 1, and the best model with maximum accuracy is selected for the prediction, i.e., the Multi-Layer Perceptron.

```
Support Vector Machine
-----------
Classification Report
              precision    recall  f1-score   support

          No       0.65      0.84      0.73        49
         Yes       0.78      0.57      0.66        51

    accuracy                           0.70       100
   macro avg       0.72      0.70      0.70       100
weighted avg       0.72      0.70      0.69       100

-----------------------------------------------------
Confusion Matrix is:
[[41  8]
 [22 29]]
-----------------------------------------------------
Accuracy is 0.7
```

Fig. 12 Performance summary of Support Vector Machine

```
Naive Bayes
-----------
Classification Report
              precision    recall  f1-score   support

          No       0.77      0.81      0.79        54
         Yes       0.77      0.72      0.74        46

    accuracy                           0.77       100
   macro avg       0.77      0.77      0.77       100
weighted avg       0.77      0.77      0.77       100

-----------------------------------------------------
Confusion Matrix is:
[[44 10]
 [13 33]]
-----------------------------------------------------
Accuracy is 0.77
```

Fig. 13 Performance summary of Naive Bayes

The graphs in Fig. 14 show the representation of VM ID with their corresponding CPU temperature. The graphs in Fig. 15 show the representation of VM ID with their corresponding CPU Utilization. The graphs in Fig. 16 show the representation of VM ID with their corresponding Disk Storage. The graphs in Fig. 17 show the representation of VM ID with their corresponding Memory Usage. The graphs in Fig. 18 show the representation of VM ID with their corresponding Power Consumption. By comparing these graphs, the VM with ID ranging from 6 to 15 has relatively higher value of CPU Temperature, CPU Utilization, Disk storage, and Memory usage compared to other VMs. Figure 19 represents the Host ID with their corresponding CPU utilization. Figure 20 represents Host ID with their corresponding Power Consumption.

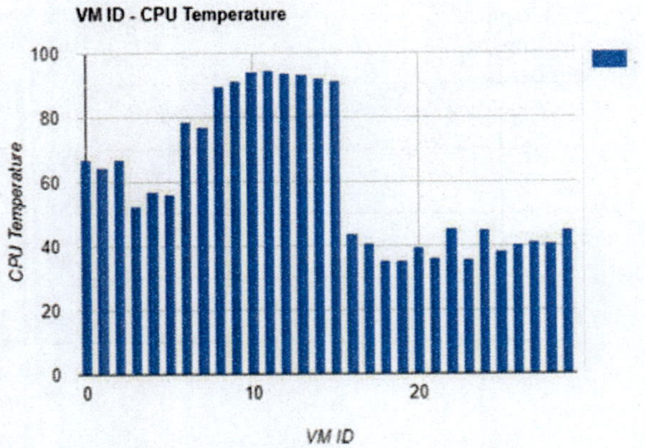

Fig. 14 Graphical representation of VM ID and CPU temperature

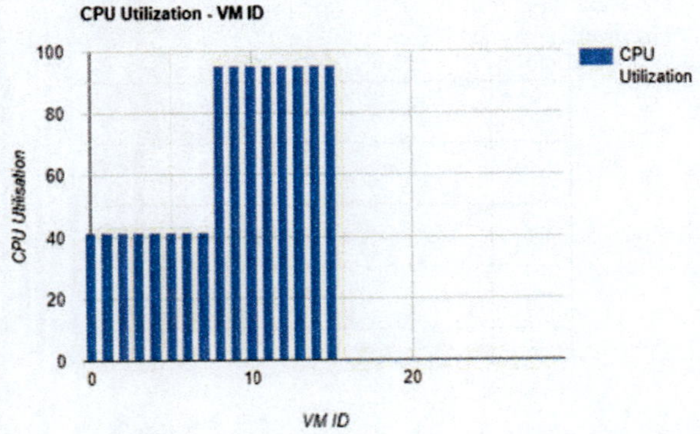

Fig. 15 Graphical representation of VM ID and CPU Utilization

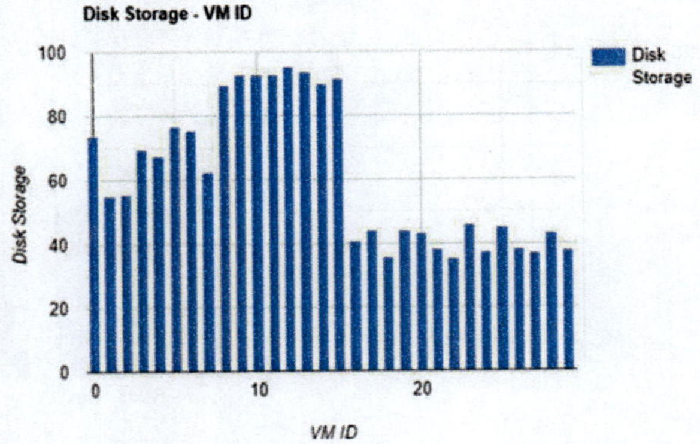

Fig. 16 Graphical representation of VM ID and Disk Storage

This way most shortage problems can be avoided. The Virtual Machine may enter the fault stage due to many factors like hardware faults, aging problems, software faults, response faults, timing faults, and omission faults. Some of these faults occur due to hardware problems. So, a Multi-Layer Perceptron model is used to predict these faults in prior, so that the system can avoid failure as shown in Fig. 21.

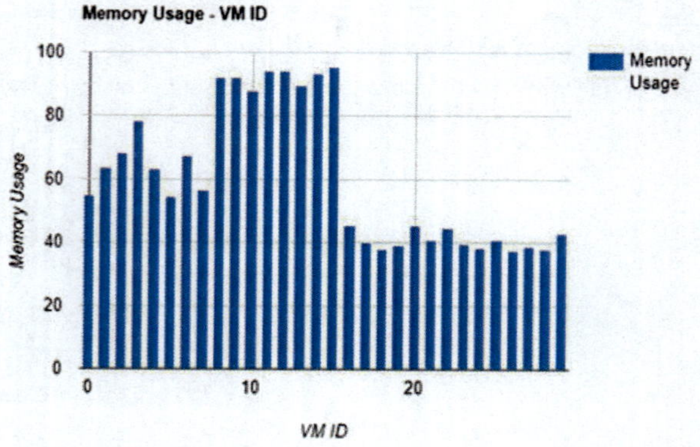

Fig. 17 Graphical representation of VM ID and Memory Usage

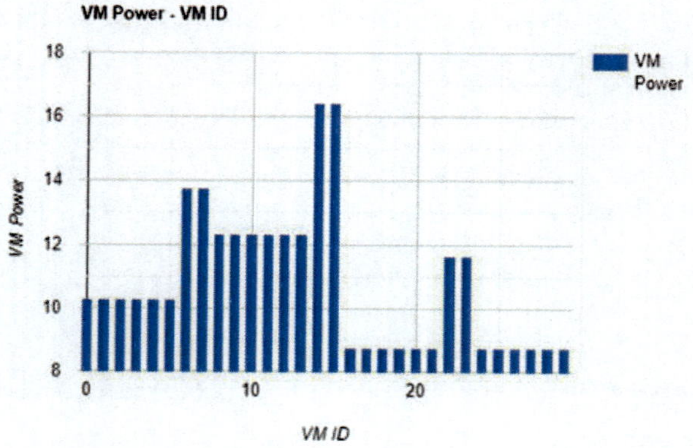

Fig. 18 Graphical representation of VM ID and VM Power

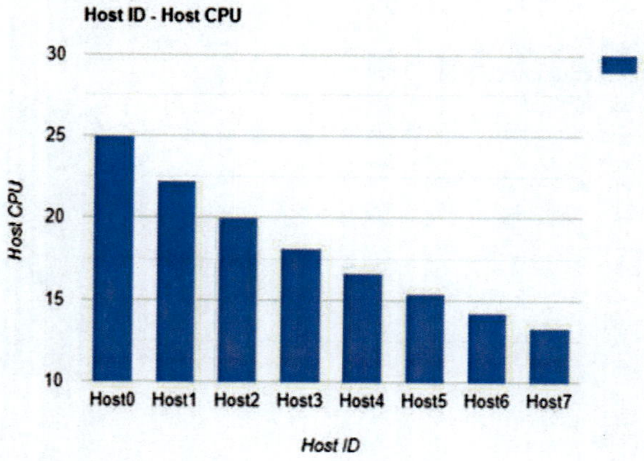

Fig. 19 Graphical representation of Host ID and Host CPU

Fig. 20 Graphical representation of Host ID and Host power

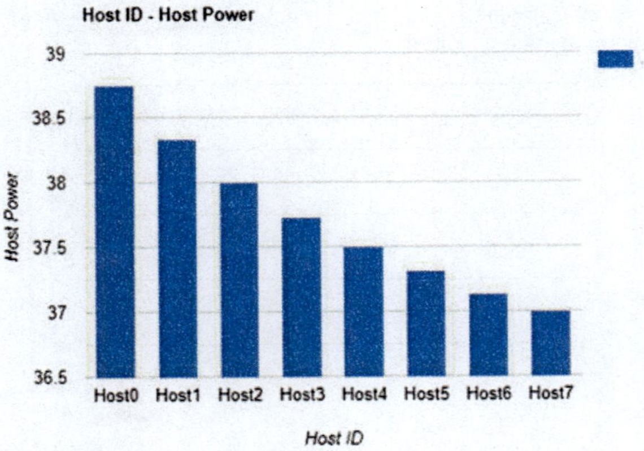

Fig. 21 Graphical representation of VM ID and their fault

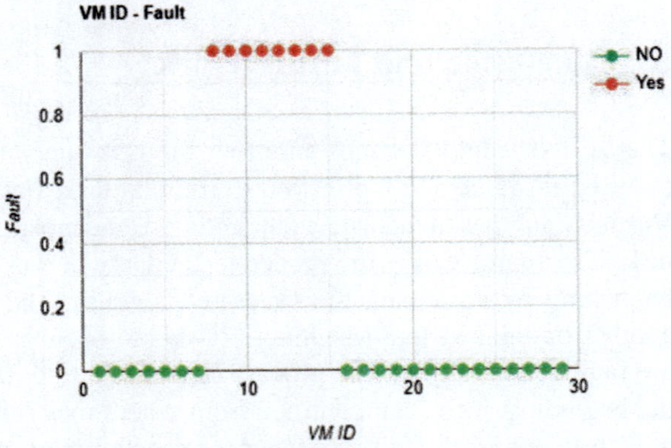

The parameters like CPU temperature, Utilization, Memory, and Disk Storage are considered in the dataset which is used to train Multi-Layer Perceptron. Two other classifier models have been implemented and their accuracy is compared to the Multi-Layer Perceptron model as shown in Fig. 22. Since Multi-Layer Perceptron has the highest accuracy of all models, it has been used for the fault prediction. Also, two hidden layers have been used in the Multi-Layer Perceptron model to improve the prediction stability.

After finding fault in the Virtual Machine the Cloud Service Provider can abort the task scheduled. But it makes the user start the task from the first which will decrease the dependability and stability of the cloud service. So, the optimal Virtual Machine is found with required resources and Utilization to resume the task execution. First, it is tried to migrate within the same host, if not possible then the other hosts, other datacenters, etc. This type of approach will increase the stability, availability of resources, and dependability of the cloud services.

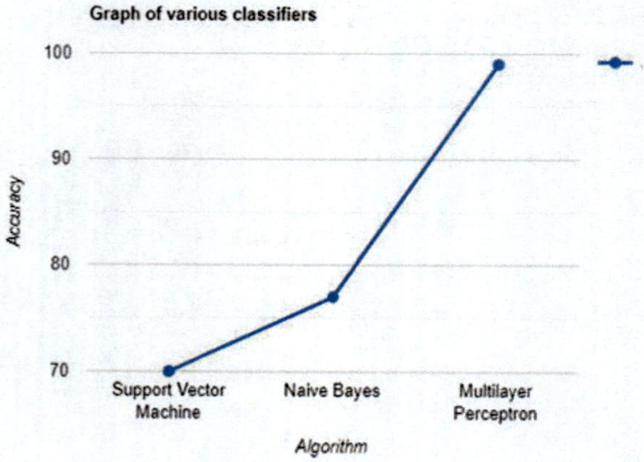

Fig. 22 Graphical representation of various classifiers

6 Conclusion and Future Work

The task scheduling makes sure that the task scheduled Virtual Machine has the required resource and Utilization to run that task. For scheduling, it checks all the Virtual Machines in the datacenter and finds the most suitable one for that particular task. The main reason for proactive fault prediction is to prevent the fault from happening by correcting the cause by migrating the task from the faulty one to another optimal Virtual Machine. For this to occur it is needed to predict whether the fault occurs or not. The process or task need not be aborted and restarted from the beginning. They can continue from where they left. Evaluating many classifier models to predict the failure widens the possibility of choosing the best model which predicts the fault more accurately. By finding an optimal Virtual Machine with the necessary resources, the task can be migrated from the faulty Virtual Machine to the healthy Virtual Machine which ensures the completion of the task.

In the future, a model which can predict faults based on many other parameters can be built. The type of faults the model can predict like network faults, omission faults, response faults, software faults, etc. can also be widened. Different models can be trained to predict faults with different types of datasets with more features and can compare the accuracy. A combined model can also be used with many different models at the classifier level and predict the fault by the voting system. Migration of the task from the faulty Virtual Machine to the optimal Virtual Machine in real time can be done without interrupting the process of the task at minimum cost and time.

References

1. Abro JH, Li C, Shafiq M, Vishnukumar A, Mewada S, Malpani K, Osei-Owusu J (2022) Artificial intelligence enabled effective fault prediction techniques in cloud computing environment for improving resource optimization. Sci Progr (2022)

2. Ajmal MS, Iqbal Z, Khan FZ, Ahmad M, Ahmad I, Gupta BB (2021) Hybrid ant genetic algorithm for efficient task scheduling in cloud data centers. Comput Electr Eng 95:107,419
3. Alaei M, Khorsand R, Ramezanpour M (2021) An adaptive fault detector strategy for scientific workflow scheduling based on improved differential evolution algorithm in cloud. Appl Soft Comput 99:106,895
4. Ali H, Qureshi MS, Qureshi MB, Khan AA, Zakarya M, Fayaz M (2020) An energy and performance aware scheduler for real-time tasks in cloud datacentres. IEEE Access 8:161,288–161,303
5. Alsadie D (2021) A metaheuristic framework for dynamic virtual machine allocation with optimized task scheduling in cloud data centers. IEEE Access 9:74218–74233
6. Alsaidy SA, Abbood AD, Sahib MA (2020) Heuristic initialization of pso task scheduling algorithm in cloud computing. J King Saud Univ-Comput Inf Sci
7. Dai Vu D, Vu XT, Kim Y (2021) Deep learning-based fault prediction in cloud system. In: 2021 International conference on information and communication technology convergence (ICTC). IEEE, pp 1826–1829
8. Dubey K, Sharma SC (2021) A novel multi-objective cr-pso task scheduling algorithm with deadline constraint in cloud computing. Sustain Comput: Inf Syst 32:100,605
9. Duong-Ba T, Tran T, Nguyen T, Bose B (2018) A dynamic virtual machine placement and migration scheme for data centers. IEEE Trans Serv Comput 14(2):329–341
10. Farahnakian F, Pahikkala T, Liljeberg P, Plosila J, Hieu NT, Tenhunen H (2016) Energy-aware vm consolidation in cloud data centers using utilization prediction model. IEEE Trans Cloud Comput 7(2):524–536
11. Gao M, Zhu Y, Sun J (2020) The multi-objective cloud tasks scheduling based on hybrid particle swarm optimization. In: 2020 eighth international conference on advanced cloud and big data (CBD). IEEE, pp 1–5
12. Guo P, Liu M, Wu J, Xue Z, He X (2018) Energy-efficient fault-tolerant scheduling algorithm for real-time tasks in cloud-based 5g networks. IEEE Access 6:53671–53683
13. Gupta P, Sahoo PK, Veeravalli B (2021) Dynamic fault tolerant scheduling with response time minimization for multiple failures in cloud. J Parallel Distrib Comput 158:80–93
14. Hasan M, Goraya MS (2019) Flexible fault tolerance in cloud through replicated cooperative resource group. Comput Commun 145:176–192
15. He T, Toosi AN, Buyya R (2021) Camig: concurrency-aware live migration management of multiple virtual machines in sdn-enabled clouds. IEEE Trans Parallel Distrib Syst 33(10):2318–2331
16. Jagatheswari S, Ramalingam P, Chandra Priya J (2022) Improved grey relational analysis-based topsis method for cooperation enforcing scheme to guarantee quality of service in manets. Int J Inf Technol 14(2):887–897
17. Jena S, Sahu LK, Mishra SK, Sahoo B (2021) Vm consolidation based on overload detection and vm selection policy. In: 2021 11th international conference on cloud computing, data science & engineering (Confluence). IEEE, pp 252–256
18. Kanwal S, Iqbal Z, Al-Turjman F, Irtaza A, Khan MA (2021) Multiphase fault tolerance genetic algorithm for vm and task scheduling in datacenter. Inf Proc & Manag 58(5):102,676
19. Khichar GS, Gupta G, Singh R, Rathi R (2018) Maximum correlation with migration control based on modified knapsack (mc_mc) approach for vm selection for green cloud computing. In: 2018 8th international conference on cloud computing, data science & engineering (Confluence). IEEE, pp 1–6
20. Liu H (2022) Research on cloud computing adaptive task scheduling based on ant colony algorithm. Optik 258:168,677
21. Liu J, Wang S, Zhou A, Kumar SA, Yang F, Buyya R (2016) Using proactive fault-tolerance approach to enhance cloud service reliability. IEEE Trans Cloud Comput 6(4):1191–1202
22. Mahmoud H, Thabet M, Khafagy MH, Omara FA (2022) Multiobjective task scheduling in cloud environment using decision tree algorithm. IEEE Access 10:36140–36151
23. Marahatta A, Pirbhulal S, Zhang F, Parizi RM, Choo KKR, Liu Z (2019) Classification-based and energy-efficient dynamic task scheduling scheme for virtualized cloud data center. IEEE Trans Cloud Comput 9(4):1376–1390

24. Marahatta A, Xin Q, Chi C, Zhang F, Liu Z (2020) Pefs: Ai-driven prediction based energy-aware fault-tolerant scheduling scheme for cloud data center. IEEE Trans Sustain Comput 6(4):655–666
25. Moorthy RS, Pabitha P (2020) A novel resource discovery mechanism using sine cosine optimization algorithm in cloud. In: 2020 4th international conference on intelligent computing and control systems (ICICCS). IEEE, pp 742–746
26. Nalini J, Khilar P (2021) Reinforced ant colony optimization for fault tolerant task allocation in cloud environments. Wireless Pers Commun 121(4):2441–2459
27. Nivitha, K., Pabitha, P.: Fault diagnosis for uncertain cloud environment through fault injection mechanism. In: 2020 4th international conference on intelligent computing and control systems (ICICCS). IEEE, pp 129–134
28. Nivitha K, Parameshwaran P (2022) C-drm: coalesced p-topsis entropy technique addressing uncertainty in cloud service selection. Inf Technol Control 51(3):592–605
29. Pang S, Li W, He H, Shan Z (2019) Wang X (2019) An eda-ga hybrid algorithm for multi-objective task scheduling in cloud computing. IEEE Access 7:146,379–146,389
30. Qaiser HUR, Shu G (2018) Efficient vm selection heuristics for dynamic vm consolidation in cloud datacenters. In: 2018 IEEE international conference on parallel & distributed processing with applications, ubiquitous computing & communications, big data & cloud computing, social computing & networking, sustainable computing & communications (ISPA/IUCC/BDCloud/SocialCom/SustainCom). IEEE, pp 832–839
31. Ramasamy V, Srirangan J, Ramalingam P (2021) Fuzzy and position particle swarm optimized routing in vanet. Int J Electr Comput Eng Syst 12(4):199–206
32. Ray B, Saha A, Khatua S, Roy S (2020) Proactive fault-tolerance technique to enhance reliability of cloud service in cloud federation environment. IEEE Trans Cloud Comput
33. Saxena D, Gupta I, Singh AK, Lee CN (2022) A fault tolerant elastic resource management framework towards high availability of cloud services. IEEE Trans Netw Serv Manag
34. Saxena D, Singh AK (2022) Ofp-tm: an online vm failure prediction and tolerance model towards high availability of cloud computing environments. J Supercomput 78(6):8003–8024
35. Sharma NK, Reddy GRM (2016) Multi-objective energy efficient virtual machines allocation at the cloud data center. IEEE Trans Serv Comput 12(1):158–171
36. Shenbaga Moorthy R, Pabitha P (2019) Optimal provisioning and scheduling of analytics as a service in cloud computing. Trans Emer Telecommun Technol 30(9):e3609
37. Singh H, Tyagi S, Kumar P, Gill SS, Buyya R (2021) Metaheuristics for scheduling of heterogeneous tasks in cloud computing environments: analysis, performance evaluation, and future directions. Simul Model Pract Theory 111:102,353
38. Sudhakar T, Ramalingam P, Jagatheswari S (2022) An improved proxy-vehicle based authentication scheme for vehicular ad-hoc networks. Int J Inf Technol 14(5):2441–2449
39. Wang JV, Ganganath N, Cheng CT, Chi KT (2019) Bio-inspired heuristics for vm consolidation in cloud data centers. IEEE Syst J 14(1):152–163
40. Yang L, Yang D, Cao J, Sahni Y, Xu X (2020) Qos guaranteed resource allocation for live virtual machine migration in edge clouds. IEEE Access 8:78441–78451
41. Zade BMH, Mansouri N, Javidi MM (2022) A two-stage scheduler based on new caledonian crow learning algorithm and reinforcement learning strategy for cloud environment. J Netw Comput Appl 202:103,385
42. Zhang P, Zhou M (2017) Dynamic cloud task scheduling based on a two-stage strategy. IEEE Trans Autom Sci Eng 15(2):772–783

Object-Based Vegetation Mapping in the Sundarbans Using Machine Learning Techniques and Earth Observation Data

Archit Jain, Anindita Nath, Bappaditya Koley, Tanupriya Choudhury, and Jung-Sup Um

1 First Section

Land use change analysis gives important insight about the vegetation of area. Rapidly changing climate affects UNESCO's Sundarbans area too, not only in climatic change but also adverse natural calamity [1, 2]. Melting of glacier has

Archit Jain contributed equally in this chapter.

A. Jain
Informatics Cluster, SoCS, University of Petroleum and Energy Studies (UPES), Dehradun 248007, Uttarakhand, India

A. Nath (✉) · B. Koley (✉)
Department of Geography, Bankim Sardar College, South 24 Parganas, Tangrakhali, West Bengal, India
e-mail: aninditan286@gmail.com

B. Koley
e-mail: bappadityakoley2012@gmail.com

J.-S. Um
Department of Geography, College of Social Sciences, Kyungpook National University, Daegu, South Korea

T. Choudhury (✉)
CSE Department, Symbiosis Institute of Technology, Symbiosis International University, Pune, Maharashtra 412115, India
e-mail: tanupriya.choudhury@sitpune.edu.in; tanupriya@ddn.upes.ac.in; tanupriya1986@gmail.com

Informatics Cluster, School of Computer Science, University of Petroleum and Energy Studies (UPES), Dehradun, Uttarakhand 248007, India

CSE Department, Daffodil International University, Daffodil Smart City, Birulia 1216, Bangladesh

Director Research (Honorary), The AI University, Cutbank Montana 59427, US

A. Ramdane-Cherif et al. (eds.), *Machine Intelligence and Data Science Applications*, Algorithms for Intelligent Systems, https://doi.org/10.1007/978-981-99-1620-7_42

increased water levels all around the world. Such a rise in water level causes loss of land, coastal erosion [3] and destroys all manly activities [4]. 36% of Mangrove is already lost in the past 20 years. Mangroves are found in about 125 tropical countries, and Sundarbans is the largest Mangrove vegetation that stretches over both India and Bangladesh. Mangrove needs special attention for its conversation and restoration because it provides a significant role in ecologically as well as economically. Generally, these areas are submerged due to tides, hence making it difficult for traditional field monitoring. Thus Landsat-8 satellite data are extremely useful because of their high accuracy and up to date data [5]. LULC mapping and analysis can use many different indices like NDVI (normalized difference vegetation index) [6], SAVI (soil-adjusted vegetation index) [7], NDBI (normalized difference built-up index) [8] and IBI (indexed-based built-up index) [9]. Image classification methods such as ML (maximum likelihood) technique, RF (Random Forest), KNN (k-Nearest Neighbour) and SVM (Support Vector Machine) can also be used [27–29].

Mangrove vegetation has been needed to urgent concentration for restoration and preservation for the balance in global climatology. Many places in Sundarbans mangrove have been submerged with tides and are located in remote areas of Sundarbans. Due to this reason, conventional method for mangrove vegetation survey is very time consuming and cost effective. To reduce the time and cost, remote sensing method has been required in the current content of research. Remote sensing is very significant for vegetation mapping and identification with their spectral visibility and spatial condition that are detectable in satellite images [10]. GIS (Geographical Information System) has been widely used to monitor the vegetation condition [11–13]. NDVI is one of the applied significant and effective techniques for assessing the vegetation condition [14]. It is (Vegetation indices) a highly applied technique for assessing the various land cover features among several groups of vegetation. Vegetation attributes are directly delineated from the high-resolution satellite images. The current research aims to identify the land cover features and mangrove conditions with their area under Sundarbans mangrove area in part of India. Machine learning is playing an important role over Geo-spatial and object observation techniques [17–19].

2 Study Area

Sundarbans is famous for its largest mangrove area, which is 40,000 sq. km. 25% of land is coved with mangrove vegetation. The area under study area lies between 20°50′N and 88°44′E, 22°42′N and 89°27′E, 22°50′N and 88°44′E, 22°01′N and 89°18′E, which is above the Bay of Bengal. Sundarbans delta especially part of India is mainly located in West Bengal (north and south 24 pgs) (Fig. 1). The area of Sundarbans is highly prone to cyclonic calamities. Malta River is the boundary between two parts of Sundarbans, India and Bangladesh, respectively. It has high fluvial activities and sedimentation of rivers like Padma, Hooghly and Brahmaputra, which make it suitable home to many rare and worldwide threatened wildlife species such as estuarine crocodile (Crocodilus porosus), royal Bengal tiger (Panthera tigris), Water monitor lizard (Varanus salvator), Gangetic dolphin (Platinista gangetica) and olive ridley turtle (Lepidochelys olivacea). The Indian Sundarban is also home to 4.5 million people, even though its area is fragile and climatically unstable. Climate change around the world is already affecting Sundarbans. Annual rainfall is about 1600–1800 mm as it is subtropical climate. The elevation is between 0.9 and 2.1 m [4]. Due to the enhancement of tourism, the land cover identification is very essential subject to assessing the mangrove area and heath of vegetation in Sundarbans. Sagar, Jharkhali and Bakkhali are prominent islands, which are observed as rapid growth of population due to the development of transportation system.

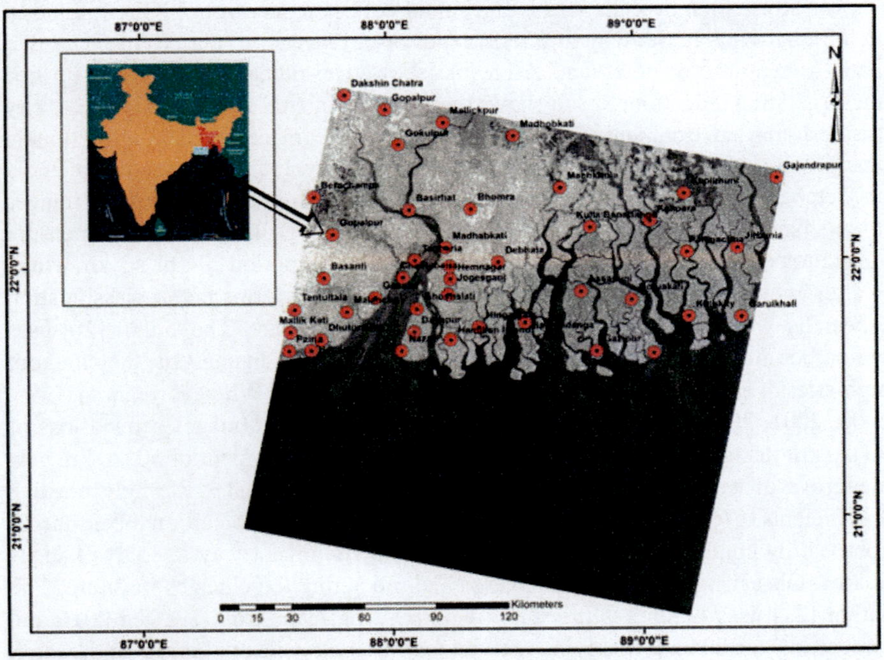

Fig. 1 Location of the study area

3 Related Work

Remote sensing imaginary provides a lot of important facts and changes about real-world area without doing any traditional physical survey. Improved feature ex-traction showed insights about Jabalpur City [16]. The NDVI image has been segmented into land cover classification, vegetation index, road structure, rocky areas, concrete structure, urban areas and remaining areas. Visible area surface features are well separated by NDVI. Density of vegetation can be estimated more precisely by varying threshold index. Change in land usage pattern can be observed using satellite images of different years. Once such paper [30] observed that changes in Sundarbans area by taking data of alternative years from 2011 to 2019. They observed changes in four categories: vegetation, high vegetation, no vegetation and water. Results show an increase of 0.02% of high vegetation and 19.22% of vegetation. Whereas, 84.14% to 54.28% water has decreased initially but finally 54.28% to 66.84% water increased. One another paper [31] observed that changes in Bangalore city transformation over the years. The author collected data from MODIS, Landsat and Global Human Settlement (GHS) layer over 17 years to classifying data into water, vegetation and other land. All results showed an enhancement of built-up area and vegetation reduction. Result shows that such unplanned development makes city unsafe living environment and urges for sustainable development. Rapid climatic changes are producing calamity occasions. Hudhud cyclone hit east coast of India on October 12, 2014. The author [32] used pre- and post-cyclone Landsat-8 images to understand damage caused. It uses NDVI to classify LULC classes and measure vegetation coverage. Vegetation covered in pre-cyclonic period 234.68 sq. km which is after cyclone was only 161.66 sq. km. The NDVI performs better classification especially when there are pre- and post-images to compare. The author [20] uses bio-carbon flux and Sundarbans forest from 1990 to 2020. In analysis, they find out high-stress years were 1990, 2006, 2007, 2009 and 2011. Whereas years of 1999, 2000, 2001, 2015, 2016 and 2019 show healthy condition. Images show a loss of 90 sq. km mangrove because of coastal erosion and development of 50 sq. km new mangrove in western parts. The study shows correlation that is strongly positive (coefficients 0.76 at $p = 0.01$) between mangrove areas and bio-carbon fluxes during non-healthy condition of mangroves. LULC is a significant way to study changes in area. Observing the pattern in change and predicting the changes in future. The author [21] uses Lahore as the study region. Using maps from 1988 to 2016 and forecasting for 2040. Classifying area into six classes built-up, barren land, water, sparse vegetation, and dense vegetation shrub and grassland. LULC map of 1988 and 2001 was used to form LULC transformation. Accuracy was compared with 2016 image. The changes were visible when compared 2040 to 1988, water was reduced from 25.37 km^2 to 5.32 km^2; 625.16 km^2 from 168.29 km^2 built-up is expected to increase; decrease to 137.53 km^2 of barren land from 514.13 km^2; change observe in grass field to 297.68 km^2 from 539.46 km^2; change observed in vegetation from

539.46 km^2 to 297 km^2; an increase trend noticed in dense vegetation to 409.65 km^2 from 191.51 km^2. Most of coastal region is covered by mangrove and it's important to tackle challenges. Studies show [22] satellite provides cost-efficient, instant, real-time assessment of coastal region. With the changes in the Sundarbans vegetation, a change in temperature can also be noticed [23].

4 Methodology

NDVI is a majorly applied method of identifying the health condition of vegetation and crops in agricultural field. NDVI value and vegetation is obtained by using geospatial techniques in remote sensing and GIS application. Satellite data have been used by the spatial and spectral resolution. The spectral and spatial resolution has been shown in the pixel with an image that shows the geometric interaction of one another and represents the variation of wavelength within a pixel, respectively. In satellite images, vegetation has been identified for the absorption in red band and blue band among the other objects or land features of the visible spectrum [24–26].

Landsat-8 is the eighth American satellite in Landsat series, which was introduced on February 11, 2013. It aims to measure in, near infrared, visible and shortwave infrared spectrum portions as it images from 30 to 100 m. It takes 99 min to revolve one orbit of Earth in which it takes 400 scenes a day. It has Thermal Infrared Sensor and Operational Land Imager sensor. Satellite is combined effort of NASA and US Geological Survey. Data for this paper have been taken from USGS site (https://earthexplorer.usgs.gov/). Using high resolution satellite image of Sundarbans of 8 spectrum band which has spatial resolution of 30 m. Used Arc Map 10.8 to visualize data and create LULC map by using Raster Calculator. Map was then covered with Layer of locations with the help of Google Earth. NDVI value and NDVI range for each location have been filled in table. The difference between near infrared (NI) and red light gave NDVI. Vegetation reflects near infrared strongly and red light absorbs by vegetation. NDVI always ranges from -1 to $+1$. For negative pixel values, it's most likely water and value near $+1$ are strong vegetation. A value near zero can be classified as urban area or waste land. Thus, there is no strict boundary of NDVI values. NDVI analysis is the most applied method in remote sensing for assessing the vegetation indices and land cover mapping (Thakur et al. 2021). For the calculation of the NDVI Red (RED) and near-infrared (NIR) band has been used and the light reflected on the vegetation cover. Calculating NDVI uses a simple formula NDVI = NIR-RED/NIR + RED, where NIR represents Near Infrared and RED presents red light. For data processing, Arc Map software tool has been used. Raster Calculator with six classes has been used from Arc Map Toolbox. Categorizing Land into 0.28–0.015 water, Built-up 0.015–0.14, Barren Land 0.14–0.18, Shrub and Grassland 0.18–0.27, Light Vegetation 0.27–0.36, Healthy Vegetation 0.36–0.74 [13]. NDVI has been calculated by the formula that is given below (Eq. 1). The methodology has been shown through a flowchart (Fig. 2).

$$\text{NDVI} = (\text{NIR} - \text{RED}) / (\text{NIR} + \text{RED}) \quad (1)$$

where RED indicates the reflectance of red range of spectrum and near-infrared spectrum is indicated by the NIR band. As per the reference of NASA, the scale ranges from −1.0 to + 1.0 in NDVI values. High vegetation (HV) area indicated by the NDVI value of + 1.0, vegetated (V) area represented by value 0 < & < + 1, No

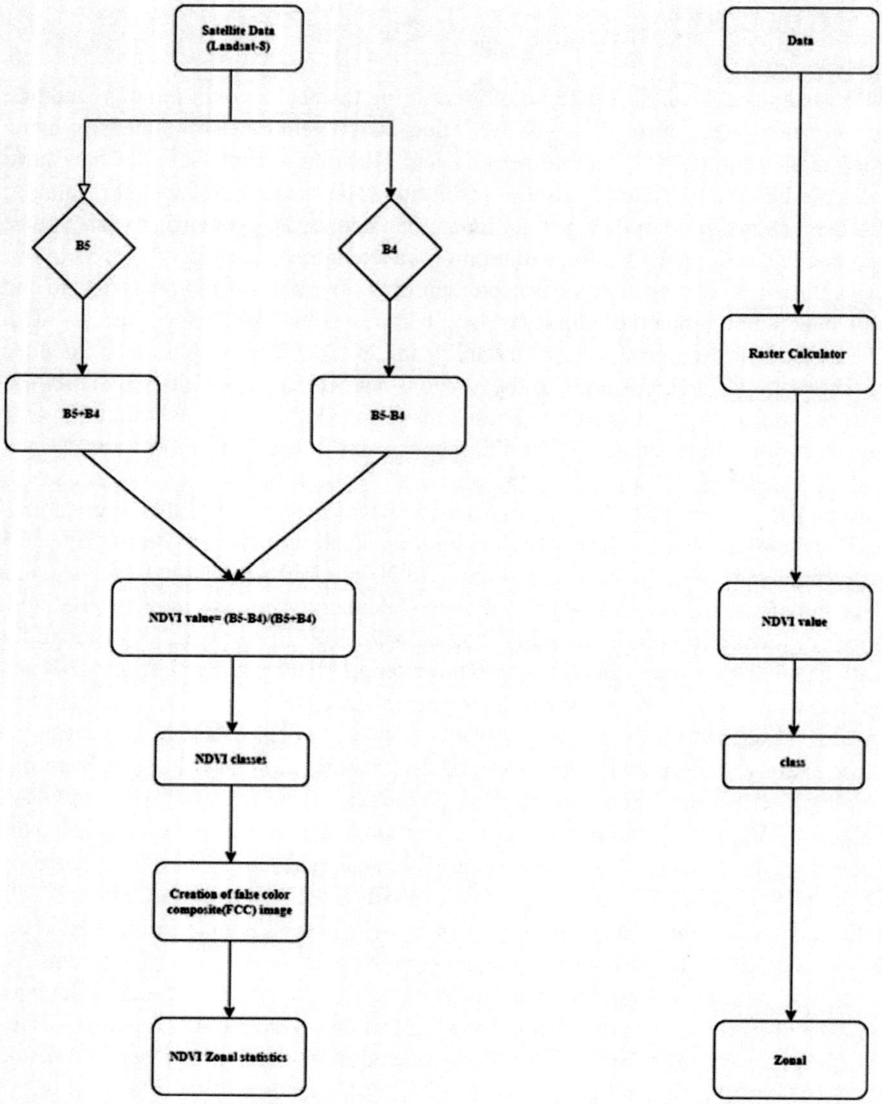

Fig. 2 Flowchart of methodology of the study area

Vegetation (NV) area indicated by value of 0 and water area represented by $-1 <=$ & < 0.

5 Result and Discussion of NDVI Analysis

NDVI analysis has been used for determine the vegetation cover and other land features in Indian Sundarbans. Six land cover features are detected in the NDVI analysis shown in Table 1 and Fig. 3 that have been identified from satellite data of Sundarbans. The outcome map of NDVI has been compared to Google Earth image for the accuracy of land features which improvised significantly. Six different features are Built-up/soil/sand, water, barren land, light vegetation shrub/grassland and healthy vegetation. The water area has been showing the NDVI range of 1 where the value determines as between −0.182 and −0.099. Built-up/sand/soil has been defined as range of 2 (value between −0.099 and 0.0244). Barren land considers the range of 3 in NDVI with scale range of 0.0244–0.162. Shrub/grassland considers with the NDVI range 4 (0.162–0.26). Light vegetation is indicated by the range of 4, scale value is 0.26–0.355. Finally, the highest NDVI range (6) is considered for healthy vegetation, value of 0.355–0.55. After the NDVI map analysis, the area has been calculated for each feature (Table 2 and Fig. 4). In the area analysis with the total area of 37,297 km^2, it has been observed that most of the region covered by water and built-up/soil/sand area, which are 11,506 and 10,713 km^2, respectively. Barren land occupied 5% of total area (1919 km^2). Almost the same portion of area has been covered by shrub/grassland and light vegetation, which is 15 and 16% (5698 and 5787 km^2), respectively. Healthy vegetation is occupied by the area of 4%, 1637 km^2 only.

6 Conclusion

The current study revealed the land cover variation and condition of different features in Indian Sundarbans through remote sensing (RS) and GIS technique applying satellite data. The research follows the Normalized Difference Vegetation Index (NDVI) object-based analytical method. This method has been employed with different threshold value for extracting various land cover features in Sundarbans mangrove region. The obtained result from NDVI has shown the percentage of vegetation cover. From the result, it has been concluded that most of the area cover with water and built-up region. Moderate part of light and healthy vegetation has been obtained where lowest percentage area is occupied by barren land and grassland finds out in major portion of study area. NDVI analysis gives a superior outcome for the vegetation density and scatter condition of vegetation using remote sensing techniques.

Table 1 NDVI range scale and value for each location

ID	Location	NDVI class	NDVI value
1	Basanti	6	0.390
2	Dayapur	3	0.155
3	Gosaba	3	0.087
4	Hamilton Island	4	0.251
5	Marichjhanpi	5	0.295
6	Hemnagar	5	0.313
7	Jogesganj	4	0.193
8	Madhabkati	4	0.245
9	Chalitabaria	3	0.049
10	Tegharia	2	−0.032
11	Mallik Kati	6	0.358
12	Paina	5	0.297
13	Hatimari	4	0.241
14	Dhuturdaha	4	0.242
15	Malancha	4	0.210
16	Tentultala	6	0.400
17	Basirhat	4	0.211
18	Berachampa	6	0.445
19	Gopalpur	3	0.127
20	Dakshin Chatra	5	0.344
21	Gopalpur	5	0.323
22	Gokulpur	3	0.082
23	Mallickpur	3	0.148
24	Bhomra	6	0.416
25	Madhobkati	5	0.344
26	Machkhola	4	0.223
27	Ghoshalati	4	0.231
28	Nazat	4	0.254
29	Bathuadanga	4	0.215
30	Gojuakati	5	0.269
31	Hingalganj	4	0.218
32	Assasuni	4	0.221
33	Debhata	6	0.378
34	Aksora	5	0.336
35	Gazipur	4	0.204
36	Kapilmuni	3	0.079
37	Katipara	5	0.302

(continued)

Table 1 (continued)

ID	Location	NDVI class	NDVI value
38	Kulla Benadanga	3	0.131
39	Jirbunia	2	−0.049
40	Paikgachha	3	0.087
41	Kinukity	5	0.320
42	Gajendrapur	4	0.212
43	Garuikhali	2	−0.054

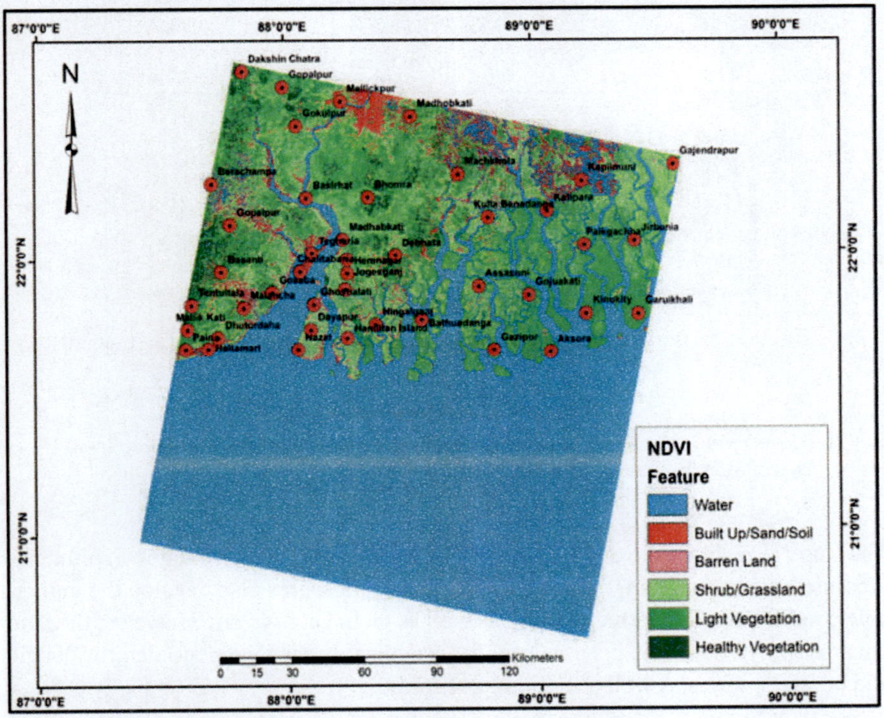

Fig. 3 Land cover map by NDVI

Table 2 Area statistics of each feature

NDVI class	Land cover feature class	Area (km^2)	Area (%)
1	Water	11,506	31
2	Built-up/sand/soil	10,713	29
3	Barren land	1919	5
4	Shrub or grassland	5698	15
5	Light vegetation	5787	16
6	Healthy vegetation	1673	4
		37,297	100

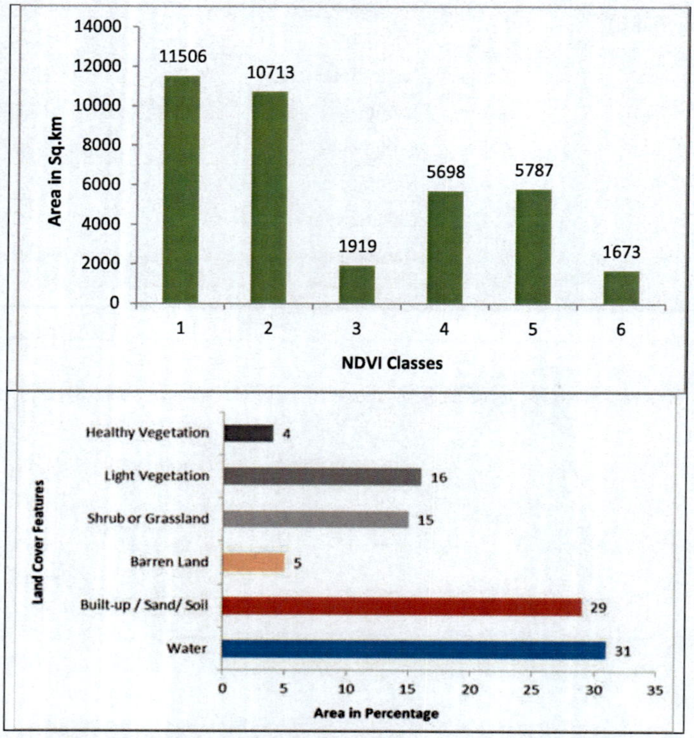

Fig. 4 Area of each feature in km^2 and percentage

The study also shows the expansion of urban land feature, which creates an immense pressure on existing mangrove vegetation. These pressures also develop the vulnerable condition on Sundarbans region as well as global ecosystem. However, the land cover map by NDVI has been helpful for establishing management plan for deltaic and coastal regions as well as agricultural study.

References

1. Mukhopadhyay A, Payo A, Chanda A, Ghosh T, Chowdhury SM, Hazra S (2018) Dynamics of the sundarbans mangroves in bangladesh under climate change. In: Ecosystem services for well-being in deltas. Palgrave Macmillan, Cham, pp 489–503
2. Akhter M, Iqbal Z, Chowdhury R (2008) Aster imagery of forest areas of sundarban damaged by cyclone sidr. ISME/GLOMIS Electron J 6(1):1–3
3. Thomas J, Arunachalam A, Jaiswal R, Diwakar P, Kiran B (2014) Dynamic land use and coastline changes in active estuarine regions-a study of sundarban delta. Int Arch Photogram Rem Sens Spat Inf Sci 40(8):133

4. Kumar D, Ghosh T (2021) Monitoring changes in land cover in the Indian sundarbans region, using geospatial sciences. In: Re-envisioning remote sensing applications. CRC Press, pp 137–150
5. Maurya K, Mahajan S, Chaube N (2021) Remote sensing techniques: mapping and monitoring of mangrove ecosystem—a review. Complex Intell Syst 7(6):2797–2818
6. J. W. Rouse Jr, R. H. Haas, J. Schell, and D. Deering, "Monitoring the vernal advancement and retrogradation (green wave effect) of natural vegetation," Tech. Rep., 1973.
7. Huete AR (1988) A soil-adjusted vegetation index (savi). Remote Sens Environ 25(3):295–309
8. Zha Y, Gao J, Ni S (2003) Use of normalized difference built-up index in automatically mapping urban areas from tm imagery. Int J Remote Sens 24(3):583–594
9. Xu H (2008) A new index for delineating built-up land features in satellite imagery. Int J Remote Sens 29(14):4269–4276
10. Maurya K, Mahajan S, Chaube N (2021) Remote sensing techniques: mapping and monitoring of mangrove ecosystem—a review. Complex Intell Syst 7:2797–2818. https://doi.org/10.1007/s40747-021-00457-z
11. Satyanarayana B et al (2018) Status of the undisturbed mangroves at Brunei Bay, East Malaysia: a preliminary assessment based on remote sensing and ground-truth observations. Peer J 6:1–22. https://doi.org/10.7717/peerj.4397
12. Po M, Kong H, Liu M (2018) Zonation and directional dynamics of mangroves forests derived from time-series satellite imagery in Mai Po, Hong Kong. Sustainability. https://doi.org/10.3390/su10061913
13. Debaine F (2019) Mapping the mangroves forest canopy using spectral unmixing of very high spatial resolution satellite images.Remote Sens. https://doi.org/10.3390/rs11030367
14. Pettorelli N et al (2017) Remote sensing in ecology and conservation: three years on. Rem Sens Ecol Conserv 3(2):53–56. https://doi.org/10.1002/rse2.53
15. Bhandari A, Kumar A, Singh G (2012) Feature extraction using normalized difference vegetation index (ndvi): a case study of Jabalpur city. Proc Technol 6:612–621
16. Shimu SA, Aktar M, Afjal MI, Nitu AM, Uddin MP, Al Mamun M (2019) Ndvi based change detection in Sundarban mangrove forest using remote sensing data. In: 2019 4th International conference on electrical information and communication technology (EICT). IEEE, pp 1–5
17. Sethi PS, Jain A, Kumar S, Tomar R (2022) Context-driven method for smarter and connected traffic lights using machine learning with the edge servers. In: Skala V, Singh TP, Choudhury T, Tomar R, Abul Bashar M (eds) Machine intelligence and data science applications. Lecture notes on data engineering and communications technologies, vol 132. Springer, Singapore. https://doi.org/10.1007/978-981-19-2347-0_43
18. Hayat A, Morgado-Dias F, Bhuyan BP, Tomar R (2022) Human activity recognition for elderly people using machine and deep learning approaches. Information 13:275. https://doi.org/10.3390/info13060275
19. Bhatia K, Arora S, Tomar R (2016) Diagnosis of diabetic retinopathy using machine learning classification algorithm. In: 2016 2nd International conference on next generation computing technologies (NGCT), Dehradun, India, 2016, pp 347–351. https://doi.org/10.1109/NGCT.2016.7877439
20. Akbar TA, Hassan QK, Ishaq S, Batool M, Butt HJ, Jabbar H (2019) Investigative spatial distribution and modelling of existing and future urban land changes and its impact on urbanization and economy. Rem Sens 11(2):105
21. Datta D, Deb S (2012) Analysis of coastal land use/land cover changes in the Indian Sunderbans using remotely sensed data. Geo-spat Inf Sci 15(4):241–250
22. Thakur S, Maity D, Mondal I, Basumatary G, Ghosh PB, Das P, De TK (2021) Assessment of changes in land use, land cover, and land surface temperature in the mangrove forest of Sundarbans, northeast coast of India. Environ Dev Sustain 23(2):1917–1943
23. Thakur S, Maity D, Mondal I et al (2021) Assessment of changes in land use, land cover, and land surface temperature in the mangrove forest of Sundarbans, northeast coast of India. Environ Dev Sustain 1917–1943 (2021). https://doi.org/10.1007/s10668-020-00656-7

24. Ricotta C, Avena G, Palma AD (1999) Mapping and monitoring net primary productivity with AVHRRNDVI time-series: statistical equivalence of cumulative vegetation indices. ISPRS J Photogramm Remote Sens 54:325–331
25. Zhang X, Hu Y, Zhuang D, Oi Y, Ma X (2009) NDVI spatial pattern and its differentiation on the Mongolian Plateau. J Geog Sci 19:403–415
26. Mukhopadhyay A, Mukherjee S, Hazra S, Mitra D (2011) Sea level rise and shoreline changes: a geoinformatic appraisal of Chandipur coast, Orissa. Int J Geol Earth Environ Sci 1(1):8–17
27. Joshi D, Patidar AK, Mishra A et al (2021) Prediction of sonic log and correlation of lithology by comparing geophysical well log data using machine learning principles. GeoJournal. https://doi.org/10.1007/s10708-021-10502-6
28. Choudhury T, Kandwal A (2021) Machine learning based analytical approach for geographical analysis and prediction of Boston City crime using geospatial dataset. GeoJournal. https://doi.org/10.1007/s10708-021-10485-4
29. Shukla A, Adwani N, Choudhury T et al (2021) Geospatial analysis for natural disaster estimation through Arduino and node MCU approach. GeoJournal. https://doi.org/10.1007/s10708-021-10496-1
30. Alex EC, Ramesh K, Sridevi H (2017) Quantification and under-standing the observed changes in land cover patterns in Bangalore. Int J Civ Eng Technol 8(4):597–603
31. Konda VGRK, Chejarla VR, Mandla VR, Voleti V, Chokkavarapu N (2018) Vegetation damage assessment due to Hudhud cyclone based on NDVI using landsat-8 satellite imagery. Arab J Geosci 11(2):1–11
32. Mohanty P, Shetty S, Mahendra R, Nayak R, Sharma L, Rama Rao EP (2021) Spatio-temporal changes of mangrove cover and its impact on bio-carbon flux along the west Bengal coast, northeast coast of India. Eur J Rem Sens 54(1):525–537

Author Index

A. Ramdane-Cherif et al. (eds.), *Machine Intelligence and Data Science Applications*,
Algorithms for Intelligent Systems, https://doi.org/10.1007/978-981-99-1620-7